Lecture Notes in Artificial Intelligence 12721

Subseries of Lecture Notes in Computer Science

More information about this subseries at http://www.springer.com/series/1244

Allan Tucker · Pedro Henriques Abreu ·
Jaime Cardoso · Pedro Pereira Rodrigues ·
David Riaño (Eds.)

Artificial Intelligence in Medicine

19th International Conference on Artificial Intelligence in Medicine, AIME 2021
Virtual Event, June 15–18, 2021
Proceedings

Editors
Allan Tucker
Brunel University London
Uxbridge, UK

Pedro Henriques Abreu
University of Coimbra
Coimbra, Portugal

Jaime Cardoso
University of Porto
Porto, Portugal

Pedro Pereira Rodrigues
University of Porto
Porto, Portugal

David Riaño
Universitat Rovira i Virgili
Tarragona, Spain

ISSN 0302-9743 ISSN 1611-3349 (electronic)
Lecture Notes in Artificial Intelligence
ISBN 978-3-030-77210-9 ISBN 978-3-030-77211-6 (eBook)
https://doi.org/10.1007/978-3-030-77211-6

LNCS Sublibrary: SL7 – Artificial Intelligence

This Springer imprint is published by the registered company Springer Nature Switzerland AG
The registered company address is: Gewerbestrasse 11, 6330 Cham, Switzerland

Preface

The European Society for Artificial Intelligence in Medicine (AIME) was established in 1986 following a very successful workshop held in Pavia, Italy, the year before. The principal aims of AIME are to foster fundamental and applied research in the application of artificial intelligence (AI) techniques to medical care and medical research, and to provide a forum at biennial conferences for discussing any progress made. The main activity of the society thus far has been the organization of a series of biennial conferences, held in Marseilles, France (1987), London, UK (1989), Maastricht, Netherlands (1991), Munich, Germany (1993), Pavia, Italy (1995), Grenoble, France (1997), Aalborg, Denmark (1999), Cascais, Portugal (2001), Protaras, Cyprus (2003), Aberdeen, UK (2005), Amsterdam, Netherlands (2007), Verona, Italy (2009), Bled, Slovenia (2011), Murcia, Spain (2013), Pavia, Italy (2015), Vienna, Austria (2017), Poznan, Poland (2019), and Minneapolis, USA (2020) - the latter hosted virtually due to the COVID-19 pandemic.

AIME 2021 was to be hosted in Portugal but, due to the ongoing pandemic, it was held virtually. This volume contains the proceedings of AIME 2021, the International Conference on Artificial Intelligence in Medicine, hosted virtually by the University of Coimbra, Portugal, during June 15–18, 2021.

The AIME 2021 goals were to present and consolidate the international state of the art of AI in biomedical research from the perspectives of theory, methodology, systems, and applications. The conference included two invited keynotes, full and short papers, tutorials, workshops, and a doctoral consortium. In the conference announcement, authors were invited to submit original contributions regarding the development of theory, methods, systems, and applications for solving problems in the biomedical field, including AI approaches in biomedical informatics, molecular medicine, and health-care organizational aspects. Authors of papers addressing theory were requested to describe the properties of novel AI models potentially useful for solving biomedical problems. Authors of papers addressing theory and methods were asked to describe the development or the extension of AI methods, to address the assumptions and limitations of the proposed techniques, and to discuss their novelty with respect to the state of the art. Authors of papers addressing systems and applications were asked to describe the development, implementation, or evaluation of new AI-inspired tools and systems in the biomedical field. They were asked to link their work to underlying theory, and either analyze the potential benefits to solve biomedical problems or present empirical evidence of benefits in clinical practice. All authors were asked to highlight the value their work, created for the patient, provider, or institution, through its clinical relevance.

AIME 2021 received 138 submissions across all categories. Submissions came from 34 countries, including submissions from Europe, North and South America, Asia, Australia, and Africa. All papers were carefully peer-reviewed by experts from the Program Committee, with the support of additional reviewers, and by members of the

Senior Program Committee (a new layer to the review process introduced in AIME 2020). Each submission was reviewed by at least three reviewers. The reviewers judged the overall quality of the submitted papers together with their relevance to the AIME conference, originality, impact, technical correctness, methodology, scholarship, and quality of presentation. In addition, the reviewers provided detailed written comments on each paper, and stated their confidence in the subject area. A Senior Program Committee member was assigned to each paper and they wrote a meta-review and provided a recommendation to the Organizing Committee.

A small committee consisting of the conference co-chairs, Allan Tucker, Pedro Henrique Abreu, and Jaime Cardoso, made the final decisions regarding the AIME 2021 scientific program.

This process began with virtual meetings starting in March 2021. As a result, 28 long papers (an acceptance rate of 23%) and 30 short papers were accepted. Each long paper was presented in a 20-minute oral presentation during the conference. Each regular short paper was presented in a 5-minute presentation and by a poster. The papers were organized according to their topics in the following main themes: (1) Deep Learning; (2) Natural Language Processing; (3) Predictive Modeling; (4) Image Analysis; (5) Unsupervised Learning; (6) Temporal Data Analysis; (7) Planning; and (8) Knowledge Representation.

AIME 2021 had the privilege of hosting two invited keynote speakers: Virginia Dignum, Wallenberg chair on Responsible Artificial Intelligence and Scientific Director of WASP-HS (Humanities and Society) based at Umeå University, Sweden, who gave the keynote exploring "The myth of complete AI-fairness" and Pearse Keane, Associate Professor at the UCL Institute of Ophthalmology and Consultant at Moorfields Eye Hospital, UK, who talked about "Transforming healthcare with artificial intelligence - lessons from ophthalmology".

AIME 2021 provided an opportunity for PhD students to present their research goals, proposed methods, and preliminary results at an associated doctorial consortium. A scientific panel consisting of experienced researchers in the field provided constructive feedback to the students in an informal atmosphere. The doctoral consortium was chaired by Dr. David Riaño.

Two workshops: (1) The 12th International Workshop on Knowledge Representation for Health Care - KR4HC 2021 (David Riaño, Mar Marcos, and Annette ten Teije); (2) Explainable Artificial Intelligence in Healthcare (Jose M. Juarez, Gregor Stiglic, Huang Zhengxing, and Katrien Verbert); and an interactive half-day tutorial: Evaluating Prediction Models (Ameen Abu Hanna), took place prior to the AIME 2021 main conference.

Prizes were awarded for best student paper, best bioinformatics paper, and a new rising star award for young researchers within the AIME community who are running labs with recent breakthrough papers in the area.

We would like to thank everyone who contributed to AIME 2021. First of all, we would like to thank the authors of the papers submitted and the members of the Program Committee together with the additional reviewers. Thank you to the Senior Program Committee for writing meta-reviews and to members of the Senior Advisory Committee for providing guidance during conference organization. Thanks are also due to the invited speakers, as well as to the organizers of the tutorials and doctoral consortium panel. Many thanks go to the Local Organizing Committee, who helped plan this conference and transition it to a virtual one. The free EasyChair conference system (http://www.easychair.org/) was an important tool supporting us in the management of submissions, reviews, and selection of accepted papers. We would like to thank Springer and the Artificial Intelligence Journal (AIJ) for sponsoring the conference. Finally, we thank the Springer team for helping us in the final preparation of this LNAI book.

June 2021

Allan Tucker
Pedro Henriques Abreu
Jaime Cardoso
Pedro Pereira Rodrigues
David Riaño

Organization

General Chairs

Allan Tucker	Brunel University London, UK
Pedro Henriques Abreu	University of Coimbra, Portugal
Jaime Cardoso	INESC TEC and University of Porto, Portugal

Local Organizing Chair

Pedro Pereira Rodrigues	University of Porto, Portugal

Doctoral Consortium Chair

David Riaño	Universitat Rovira i Virgili, Spain

Senior Program Committee

Ameen Abu-Hanna	University of Amsterdam, Netherlands
Riccardo Bellazzi	University of Pavia, Italy
Carlo Combi	University of Verona, Italy
Arianna Dagliati	University of Manchester, UK
Michel Dojat	INSERM, France
Adelo Grando	Arizona State University, USA
Milos Hauskrecht	University of Pittsburgh, USA
John Holmes	University of Pennsylvania, USA
Jose Juarez	University of Murcia, Spain
Elpida Keravnou-Papailiou	University of Cyprus
Xiaohui Liu	Brunel University London, UK
Martin Michalowski	University of Minnesota, USA
Mar Marcos	Universitat Jaume I, Spain
Stefania Montani	University of Piemonte Orientale, Italy
Robert Moskovitch	Ben Gurion University, Israel
Barbara Oliboni	University of Verona, Italy
Enea Parimbelli	University of Pavia, Italy
Nada Lavrač	Jozef Stefan Institute, Slovenia
Peter Lucas	Leiden University, Netherlands
Niels Peek	University of Manchester, UK
Silvana Quaglini	University of Pavia, Italy
Lucia Sacchi	University of Pavia, Italy
Yuval Shahar	Ben Gurion University, Israel
Stephen Swift	Brunel University London, UK
Annette ten Teije	Vrije Universiteit Amsterdam, Netherlands

Szymon Wilk	Poznan University of Technology, Poland
Blaz Zupan	University of Ljubljana, Slovenia

Program Committee

Syed Sibte Raza Abidi	Dalhousie University, Spain
Amparo Alonso-Betanzos	University of A Coruña, Spain
José Amorim	University of Coimbra, Portugal
Mahir Arzoky	Brunel University London, UK
Pedro Barahona	Universidade NOVA de Lisboa, Portugal
Isabelle Bichindaritz	State University of New York at Oswego, USA
Henrik Boström	KTH Royal Institute of Technology, Sweden
Alessio Bottrighi	Università del Piemonte Orientale, Italy
Ricardo Cardoso	University of Coimbra, Portugal
Kerstin Denecke	Bern University of Applied Sciences, Switzerland
Barbara Di Camillo	University of Padova, Italy
Georg Dorffner	Medical University Vienna, Austria
Inês Dutra	University of Porto, Portugal
Jan Egger	Graz University of Technology, Austria
Henrik Eriksson	Linköping University, Sweden
Ben Evans	Brunel University London, UK
Jesualdo Tomàs Fernàndez-Breis	Universidad de Murcia, Spain
Pedro Furtado	University of Coimbra/CISUC, Portugal
Josep Gomez	Pere Virgili Institute, Joan XXIII University Hospital, Spain
Zhe He	Florida State University, USA
Jaakko Hollmen	University of Stockholm, Sweden
Arjen Hommersom	Open University of the Netherlands, Netherlands
Zhengxing Huang	Zhejiang University, China
Nevo Itzhak	Ben Gurion University, Israel
Charles Kahn	University of Pennsylvania, USA
Eleni Kaldoudi	Democritus University of Thrace, Greece
Aida Kamisalic	University of Maribor, Slovenia
Haridimos Kondylakis	FORTH, Greece
Pedro Larranaga	University of Madrid, Spain
Giorgio Leonardi	Universita' del Piemonte Orientale, Italy
Michael Liebman	IPQ Analytics, USA
Beatriz López	University of Girona, Spain
Simone Marini	University of Florida, USA
Carolyn McGregor	Ontario Tech University, Canada
Alan McMillan	University of Wisconsin-Madison, USA
Paola Mello	University of Bologna, Italy
Alina Miron	Brunel University London, UK
Diego Molla	Macquarie University, Australia
Sara Montagna	University of Bologna, Italy

Sponsors

We would like to thank The AI Journal and Springer for their sponsorship of the conference.

Transforming Healthcare with Artificial Intelligence – Lessons from Ophthalmology (Abstract of Invited Talk)

Pearse A. Keane

UKRI Future Leaders Fellow
Consultant Ophthalmologist, Moorfields Eye Hospital NHS Foundation Trust
Associate Professor, Institute of Ophthalmology, University College London
pearse.keane1@nhs.net

Abstract. Ophthalmology is among the most technology-driven of the all the medical specialties, with treatments utilizing high-spec medical lasers and advanced microsurgical techniques, and diagnostics involving ultra-high resolution imaging. Ophthalmology is also at the forefront of many trailblazing research areas in healthcare, such as stem cell therapy, gene therapy, and - most recently - artificial intelligence. In July 2016, Moorfields announced a formal collaboration with the artificial intelligence company, DeepMind. This collaboration involves the sharing of >1,000,000 anonymised retinal scans with DeepMind to allow for the automated diagnosis of diseases such as age-related macular degeneration (AMD) and diabetic retinopathy (DR). In my presentation, I will describe the motivation - and urgent need - to apply deep learning to ophthalmology, the processes required to establish a research collaboration between the National Health Service (NHS) and a company like DeepMind, the initial results of our research, and finally, why I believe that ophthalmology could be first branch of medicine to be fundamentally reinvented through the application of artificial intelligence.

Contents

Temporal Data Analysis

Unsupervised Learning

Planning and Decision Support

Deep Learning

Natural Language Processing

Invited Talk

The Myth of Complete AI-Fairness

Virginia Dignum(✉)

Umeå University, Umeå, Sweden
virginia@cs.umu.se

Just recently, IBM invited me to participate in a panel titled *"Will AI ever be completely fair?"* My first reaction was that it surely would be a very short panel, as the only possible answer is 'no'. In this short paper, I wish to further motivate my position in that debate: "I will never be completely fair. Nothing ever is. The point is not complete fairness, but the need to establish metrics and thresholds for fairness that ensure trust in AI systems".

The idea of fairness and justice has long and deep roots in Western civilization, and is strongly linked to ethics. It is therefore not strange that it is core to the current discussion about the ethics of development and use of AI systems. Given that we often associate fairness with consistency and accuracy, the idea that our decisions and decisions affecting us can become fairer by replacing human judgement by automated, numerical, systems, is therefore appealing. However, as Laurie Anderson recently said[1] "If you think technology will solve your problems, you don't understand technology—and you don't understand your problems." AI is not magic, and its results are fundamentally constrained by the convictions and expectations of those that build, manage, deploy and use it. Which makes it crucial that we understand the mechanisms behind the systems and their decision capabilities.

The pursuit of fair AI is currently a lively one. One involving many researchers, meetings and conferences (of which FAccT[2] is the most known) and refers to the notion that an algorithm is fair, if its results are independent of given variables, especially those considered sensitive, such as the traits of individuals which should not correlate with the outcome (i.e. gender, ethnicity, sexual orientation, disability, etc.). However, nothing is ever 100% fair in 100% of the situations, and due to complex networked connection, to ensure fairness for one (group) may lead to unfairness for others. Moreover, what we consider fair often does depend on the traits of individuals. An obvious example are social services. Most people believe in the need for some form of social services, whether it is for children, for the elderly, for the sick or the poor. And many of us will benefit from social services at least once in our lives. Decision making in the attribution of social benefits is dependent on individual characteristics such as age, income, or chronic health problems. Algorithmic fairness approaches however overemphasize concepts such as equality and do not adequately address caring and concern for others.

[1] As quoted by Kate Crawford on Twitter https://twitter.com/katecrawford/status/1377551240146522115; 1 April 2021.

[2] https://facctconference.org/.

A. Tucker et al. (Eds.): AIME 2021, LNAI 12721, pp. 3–8, 2021.
https://doi.org/10.1007/978-3-030-77211-6_1

Many years ago, I participated in a project at my children basic school that was aimed at helping kids develop fairness standards, roughly modelled along Kohlberg's stages of moral development. It became clear quite quickly that children aged 6–12 easily understand that fairness comes in many 'flavours': if given cookies to divide between all kids of the class, the leading principle was equality, i.e. giving each kid the same amount of cookies. But they also understood and accepted the concept of equity: for instance in deciding that a schoolmate with dyslexia should be given more time to perform a school exam. Unfortunately, for the average algorithm, common sense and world knowledge is many light years away from that of a six year old, and switching between equity and equality depending on what is the best approach to fairness in a given situation, is rarely a feature of algorithmic decision making.

So, How Does Fairness Work in Algorithms and What is Being Done to Correct for Unfair Results?

Doctors deciding on a patient's treatment, or judges deciding on sentencing, must be certain that probability estimates for different conditions are correct for each specific subject, independent of age, race or gender. Increasingly these decisions are mediated by algorithms. Algorithmic fairness can be informally described as the probability of being classified in a certain category should be similar to for all that exhibit those characteristics, independently of other traits or properties. In order to ensure algorithmic fairness, given often very unbalanced datasets, data scientists use calibration (i.e. the comparison of the actual output and the expected output). Moreover, if we are concerned about fairness between two groups (e.g. male and female patients, or African-American defendants and Caucasian defendants) then this calibration condition should hold simultaneously for the set of people within each of these groups as well [4]. Calibration is a crucial condition for risk prediction tools in many settings. If a risk prediction tool for evaluating defendants is not calibrated towards race, for example, then a probability estimate could carry different meaning for African-American than for Caucasian defendants, and hence the tool would have the unintended and highly undesirable consequence of incentivizing judges to take race into account when interpreting the tool's predictions [7]. At the same time, ideally the incidence of false positives (being incorrectly classified as 'X') and false negatives (failing incorrectly to be classified as 'X') should be the same independently of other traits or properties. That is, fairness also means that, for instance, male and female candidates have the same chance of being offered a, for them irrelevant, service or product, or failing to receive for them relevant services or products.

Unfortunately, research shows that it is not possible to satisfy some of these expected properties of fairness simultaneously: calibration between groups, balance for false negatives, and balance for false positives. This means that if we calibrate data, we need to be prepared to accept higher levels of false positives and false negatives for some groups, and to deal with their human and societal impact [6]. Taking the diagnostic example, a false positive means that a patient is diagnosed with a disease they don't have. With a false negative that disease goes undiagnosed. The impact of either, both personal as well as societal, can be huge. In the same way, being wrongly classified as someone with a high risk to re-offend(false positive) has profound personal consequences, given that innocent people are held without bail, while incorrect classification as someone with a low risk to re-offend (false negative) has deep societal consequences, where people that pose a real criminal threat are let free[3].

Given these technical difficulties in achieving perfectly fair, data-driven, algorithms, it is high time to start a conversation about the societal and individual impact of false positives and false negatives, and, more importantly, about what should be the threshold for acceptation of algorithmic decisions, that, by their nature, will never be completely 'fair'.

Fairness is Not About Bias but About Prejudice

A commonly voiced explanation for algorithmic bias is the prevalence of human bias in the data. For example, when a job application filtering tool is trained on decisions made by humans, the machine learning algorithm may learn to discriminate against women or individuals with a certain ethnic background. Often this will happen even if ethnicity or gender are excluded from the data since the algorithm will be able to exploit the information in the applicant's name, address or even the use of certain words. For example, Amazon's recruiting AI system filtered out applications by women, because they lacked 'masculine' wording, commonly used in applications by men.

There are many reasons for bias in datasets, from choice of subjects, to the omission of certain characteristics or variables that properly capture the phenomenon we want to predict, to changes over time, place or situation, to the way training data is selected.

Much has been done already to categorize and address the many forms of machine bias [8]. Also, many tools are available to support to unbias AI systems, including IBM's AI Fairness 360[4] and Google's What If Tool[5]. Basically, these tools support the testing and mitigation of bias through libraries of methods and test environments. According to Google "[...] with the What If Tool you can test performance in hypothetical situations, analyse the importance of different data features, and visualize model behaviour across multiple models and subsets of input data, and for different ML fairness metrics." Note the focus on performance, a constant in much of the work on AI.

However, not all bias is bad, in fact, there are even biases in the way we approach bias. Bias in human data in not only impossible to fully eliminate, it is often there for a reason. Bias is part of our lives partly because, we do not

[3] This example is at the core of the well-known Propublica investigations of the COMPAS algorithms used by courts in the US to determine recidivism risk: www.propublica.org/article/how-we-analyzed-the-compasrecidivism-algorithm.

[4] https://github.com/Trusted-AI/AIF360.

[5] https://pair-code.github.io/what-if-tool/index.html#about.

have enough cognitive bandwidth to make every decision from ground zero and therefore need to use generalizations, or biases, as a starting point. Without bias, we would not been able to survive as a species, it helps us selecting from a myriad of options in our environment. But not all biases are bad. But that doesn't mean we shouldn't address them.

Bias is not the problem, prejudice and discrimination are. Whereas prejudice represents a preconceived judgment or attitude, discrimination is a behaviour. In society, discrimination is often enacted through institutional structures and policies, and embedded in cultural beliefs and representations, and is thus reflected in any data collected. The focus need be on using AI to support interventions aimed at reducing prejudice and discrimination, e.g. through education, facilitation of intergroup contact, targeting social norms promoting positive relations between groups, or supporting people identify their own bias and prejudices.

Facial analysis tools and recognition software have raised concerns about racial bias in the technology. Work by Joy Buolamwini and Timnit Gebru has shown how deep these biases go and how hard they are to eliminate [1]. In fact, debiasing AI often leads to other biases. Sometimes this is known and understood, such as the dataset they created as alternative to the datasets commonly used for training facial recognition algorithms: using what they called 'parliaments', Buolamwini and Gebru created a dataset of faces balanced in terms of race and gender, but notably unbalanced in terms of age, lighting or pose. As long as this is understood, this dataset is probably perfectly usable for training an algorithm to recognise faces of a certain age, displayed under the same lighting conditions and with the same pose. It will however not be usable if someone tries to train an algorithm to recognise children's faces. This illustrates that debiasing data is not without risks, in particular because it focus on those characteristics that we are aware of, which are 'coloured' by our own experience, time, place and culture.

AI bias is more than biased data. It starts with who is collecting the data, who is involved in selecting and/or designing the algorithms and who is training the algorithms and labelling the data. Moreover, decisions about which and whose data is collected and which data is being used to train the algorithms and how the algorithm is designed and trained also influence the fairness of the results. From labelling farms to ghost workers, the legion of poorly paid, badly treated and ignored human labourers, working behind the scenes of any working AI system, is huge and little is being done to acknowledge them and to improve their working conditions. Books such as 'Ghost work' [5] by Mary L. Gray and Siddharth Suri, or 'Atlas of AI' [2] by Kate Crawford, are raising the issue but, as often is the case, the 'folk is sleeping': it is easier to use the systems and profit from their results, than to question how these results are being achieved and at what cost. The question is thus: how fair is algorithm fairness for those that label, train and calibrate the data it needs to produce fair results and, more importantly, if we expect them to provide us with unbiased data, shouldn't we be treating them fairly?

Beyond Fairness

The fact that algorithms and humans, cannot ever be completely fair, does not mean that we should just accept it. Improving fairness and overcoming prejudice is partly a matter of understanding how the technology works. A matter of education. Moreover, using technology properly, fair treatment of those using and being affected by it, requires participation. Still many stakeholder are not invited to the table, not joining the conversation. David Sumpter describes the quest for algorithmic fairness as a game of 'whack-a-mole': when you try to solve bias in one place, it appears up somewhere else [9]. The elephant in the room is the huge blind spot we all have about our own blind spots. We correct bias for the bias we are aware of. An inclusive, participatory, approach to design and development of AI systems will facilitate a wider scope.

Lack of fairness in AI systems is often also linked to a lack of explanatory capabilities. If the results of the system cannot be easily understood or explained, it is difficult to assess its fairness. Many of the current tools that evaluate bias and fairness help identify where biases may occur, whether in the data or the algorithms or even in their testing and evaluation. Even if not all AI systems can be fully explainable, it is important to make sure that their decisions are reproducible and the conditions for their use are clear and open to auditing.

Current AI algorithms are built for accuracy and performance, or for efficiency. Improving the speed of the algorithm, minimizing its computational requirements and maximizing the accuracy of the results are the mantras that lead current computer science and engineering education. However, these are not the only optimization criteria. When humans and society are at stake, other criteria need be considered. How do you balance safety and privacy? Explainability and energy resources? Autonomy and accuracy? What do you do when you cannot have both? Such moral overload dilemmas are at the core of responsible development and use of AI [3].

Addressing them requires multidisciplinary development teams and involvement of the humanities and social sciences in software engineering education. It also requires a redefinition of incentives and metrics for what is a 'good' system. Doing the right thing, and doing it well means that we also need to define what is good and for whom.

Finally, it is important to keep continuous efforts to improve algorithms and data, define regulation and standardisation, and develop evaluation tools and corrective frameworks. But the same time, we cannot ignore that no technology is without risk, no action is without risk. It is high time to start the conversation on which AI-risks we find acceptable for individuals and for society as a whole, and how we distribute these risks, as well as the benefits of AI.

References

1. Buolamwini, J., Gebru, T.: Gender shades: intersectional accuracy disparities in commercial gender classification. In: Conference on Fairness, Accountability and Transparency, pp. 77–91. PMLR (2018)
2. Crawford, K.: The Atlas of AI. Yale University Press, London (2021)
3. Dignum, V.: Responsible Artificial Intelligence: How to Develop and Use AI in a Responsible Way. Springer, Cham (2019). https://doi.org/10.1007/978-3-030-30371-6
4. Flores, A.W., Bechtel, K., Lowenkamp, C.T.: False positives, false negatives, and false analyses: a rejoinder to machine bias: there's software used across the country to predict future criminals. And it's biased against blacks. Fed. Probation **80**, 38 (2016)
5. Gray, M.L., Suri, S.: Ghost Work: How to Stop Silicon Valley from Building a New Global Underclass. Eamon Dolan Books (2019)
6. Kleinberg, J., Mullainathan, S., Raghavan, M.: Inherent trade-offs in the fair determination of risk scores. arXiv preprint arXiv:1609.05807 (2016)
7. Pleiss, G., Raghavan, M., Wu, F., Kleinberg, J., Weinberger, K.Q.: On fairness and calibration. arXiv preprint arXiv:1709.02012 (2017)
8. Schnabel, T., Swaminathan, A., Singh, A., Chandak, N., Joachims, T.: Recommendations as treatments: debiasing learning and evaluation. In: International Conference on Machine Learning, pp. 1670–1679. PMLR (2016)
9. Sumpter, D.: Outnumbered: From Facebook and Google to Fake News and Filter-Bubbles-the Algorithms that Control Our Lives. Bloomsbury Publishing, London (2018)

Image Analysis

A Petri Dish for Histopathology Image Analysis

Jerry Wei[1(✉)], Arief Suriawinata[2], Bing Ren[2], Xiaoying Liu[2], Mikhail Lisovsky[2], Louis Vaickus[2], Charles Brown[2], Michael Baker[2], Naofumi Tomita[1], Lorenzo Torresani[1], Jason Wei[1], and Saeed Hassanpour[1]

[1] Dartmouth College, Hanover, NH, USA
saeed.hassanpour@dartmouth.edu
[2] Dartmouth-Hitchcock Medical Center, Lebanon, NH, USA

Abstract. With the rise of deep learning, there has been increased interest in using neural networks for histopathology image analysis, a field that investigates the properties of biopsy or resected specimens traditionally manually examined under a microscope by pathologists. However, challenges such as limited data, costly annotation, and processing high-resolution and variable-size images make it difficult to quickly iterate over model designs.

Throughout scientific history, many significant research directions have leveraged small-scale experimental setups as **petri dishes** to efficiently evaluate exploratory ideas. In this paper, we introduce a **m**inimalist **h**istopathology image analysis dataset (**MHIST**), an analogous petri dish for histopathology image analysis. MHIST is a binary classification dataset of 3,152 fixed-size images of colorectal polyps, each with a gold-standard label determined by the majority vote of seven board-certified gastrointestinal pathologists and annotator agreement level. MHIST occupies less than 400 MB of disk space, and a ResNet-18 baseline can be trained to convergence on MHIST in just 6 min using 3.5 GB of memory on a NVIDIA RTX 3090. As example use cases, we use MHIST to study natural questions such as how dataset size, network depth, transfer learning, and high-disagreement examples affect model performance.

By introducing MHIST, we hope to not only help facilitate the work of current histopathology imaging researchers, but also make the field more-accessible to the general community. Our dataset is available at https://bmirds.github.io/MHIST.

Keywords: Histopathology images · Deep learning · Medical image analysis

1 Introduction

Scientific research has aimed to study and build our understanding of the world, and although many problems initially seemed too ambitious, they were ultimately surmounted. In these quests, a winning approach has often been to break down large ideas into smaller components, learn about these components through experiments that can be iterated on quickly, and then validate or translate those ideas into large-scale applications. For example, in the Human Genome Project (which helped us understand much

A. Tucker et al. (Eds.): AIME 2021, LNAI 12721, pp. 11–24, 2021.
https://doi.org/10.1007/978-3-030-77211-6_2

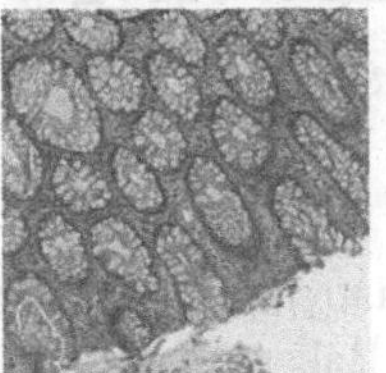
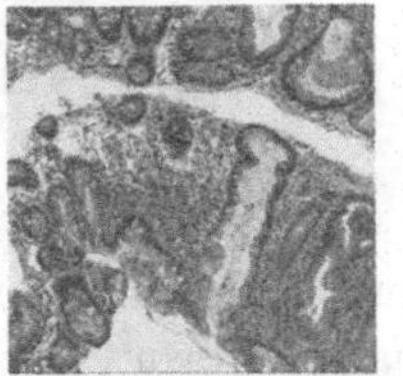
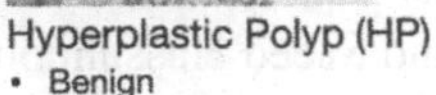

Fig. 1. Key features of our **m**inimalist **hist**opathology image analysis dataset (MHIST).

of what we know now about human genetics), many fundamental discoveries resulted from *petri dish* experiments—small setups that saved time, energy, and money—on simpler organisms. In particular, the Drosophila fruit fly, an organism that is inexpensive to culture, has short life cycles, produces large numbers of embryos, and can be easily genetically modified, has been used in biomedical research for over a century to study a broad range of phenomena [1].

In deep learning, we have our own set of petri dishes in the form of benchmark datasets, of which MNIST [2] is one of the most popular. Comprising the straightforward problem of classifying handwritten digits in 28 by 28 pixel images, MNIST is easily accessible, and training a strong classifier on it has become a simple task with today's tools. Because it is so easy to evaluate models on MNIST, it has served as the exploratory environment for many ideas that were then validated on large scale datasets or implemented in end-to-end applications. For example, many well-known concepts such as convolutional neural networks, generative adversarial networks [3], and the Adam optimization algorithm [4] were initially validated on MNIST.

In the field of histopathology image analysis, however, no such classic dataset currently exists due to many potential reasons. To start, most health institutions do not have the technology nor the capacity to scan histopathology slides at the scale needed to create a reasonably-sized dataset. Even for institutions that are able to collect data, a barrage of complex data processing and annotation decisions falls upon the aspiring researcher, as histopathology images are large and difficult to process, and data annotation requires the valuable time of trained pathologists. Finally, even after data is processed and annotated, it can be challenging to obtain institutional review board (IRB) approval for releasing such datasets, and some institutions may wish to keep such datasets private. As a result of the inaccessibility of data, histopathology image analysis has remained on the fringes of computer vision research, with many popular image datasets dealing with domains where data collection and annotation are more straightforward.

To address these challenges that have plagued deep learning for histopathology image analysis since the beginning of the area, in this paper, we introduce **MHIST**: a **m**inimalist **hist**opathology image classification dataset. MHIST is minimalist in that it comprises a straightforward binary classification task of fixed-size colorectal polyp images, a common and clinically-significant task in gastrointestinal pathology. MHIST contains 3,152 fixed-size images, each with a gold-standard label determined from the majority vote of seven board-certified gastrointestinal pathologists, that can be used to train a baseline model without additional data processing. By releasing this dataset publicly, we hope not only that current histopathology image researchers can build models faster, but also that general computer vision researchers looking to apply models to datasets other than classic benchmarks can easily explore the exciting area of histopathology image analysis. Our dataset is publicly available at https://bmirds.github.io/MHIST following completion of a simple dataset-use agreement form.

2 Background

Deep learning for medical image analysis has recently seen increased interest in analyzing histopathology images (large, high-resolution scans of histology slides that are typically examined under a microscope by pathologists) [5]. To date, deep neural networks have already achieved pathologist-level performance on classifying diseases such as prostate cancer, breast cancer, lung cancer, and melanoma [6–12], demonstrating their large potential. Despite these successes, histopathology image analysis has not seen the same level of popularity as analysis of other medical image types (e.g., radiology images or CT scans), likely because the nature of histopathology images creates a number of hurdles that make it challenging to directly apply mainstream computer vision methods. Below, we list some factors that can potentially impede the research workflow in histopathology image analysis:

- **High-resolution, variable-size images.** Because the disease patterns in histology slides can only occur in certain sections of the tissue and can only be detected at certain magnifications under the microscope, histopathology images are typically scanned at high resolution so that all potentially-relevant information is preserved. This means that while each sample contains lots of data, storing these large, high-resolution images is nontrivial. For instance, the slides from a single patient in the CAMELYON17 challenge [13] range from 2 GB to 18 GB in size, which is up to one-hundred times larger than the entire CIFAR-10 dataset. Moreover, the size and aspect ratios of the slides can differ based on the shape of the specimen in question—sometimes, multiple large specimens are included in one slide, and so some scanned slides may be up to an order of magnitude larger than others. As deep neural networks typically require fixed-dimension inputs, preprocessing decisions such as what magnification to analyze the slides at and how to deal with variable-size inputs can be difficult to make.
- **Cost of annotation.** Whereas annotating data in deep learning has been simplified by services such as Mechanical Turk, there is no well-established service for annotating histopathology images, a process which requires substantial time from

experienced pathologists who are often busy with clinical service. Moreover, access to one or two pathologists is often inadequate because inter-annotator agreement is low to moderate for most tasks, and so annotations can be easily biased towards the personal tendencies of annotators.

- **Unclear annotation guidelines.** It is also unclear what type of annotation is needed for high-resolution whole-slide images, as a slide may be given a certain diagnosis based on a small portion of diseased tissue, but the overall diagnosis would not apply to the normal portions of the tissue. Researchers often opt to have pathologists draw bounding boxes and annotate areas with their respective histological characteristics, but this comes with substantial costs, both in training pathologists to use annotation software and in increased annotation time and effort.
- **Lack of data.** Even once these challenges are addressed, it is often the case that, due to slides being discarded as a result of poor quality or to remove classes that are too rare to include in the classification task, training data is relatively limited and the test set is not sufficiently large. This makes it difficult to distinguish accurately between models, and models are therefore easily prone to overfitting.

To mitigate these challenges of data collection and annotation, in this paper we introduce a minimalist histopathology dataset that will allow researchers to quickly train a histopathology image classification model without dealing with an avalanche of complex data processing and annotation decisions. Our dataset focuses on the binary classification of colorectal polyps, a straightforward task that is common in a gastrointestinal pathologist's workflow. Instead of using whole-slide images, which are too large to directly train on for most academic researchers, our dataset consists only of 224 $\times$ 224 pixel image tiles of tissue; these images can be directly fed into standard computer vision models such as ResNet. Finally, for annotations, each patch in our dataset was directly classified by seven board-certified gastrointestinal pathologists and given a gold-standard label based on their majority vote.

Our dataset aims to serve as a petri dish for histopathology image analysis. That is, it represents a simple task that can be learned quickly, and it is easy to iterate over. Our dataset allows researchers to, without dealing with the confounding factors that arise from the nature of histopathology images, quickly test inductive biases that can later be implemented in large-scale applications. We hope that our dataset will allow researchers to more-easily explore histopathology image analysis and that this can facilitate further research in the field as a whole.

3 MHIST Dataset

In the context of the challenges mentioned in the above section, MHIST has several notable features that we view favorably in a minimalist dataset:

1. Straightforward binary classification task that is challenging and important.
2. Adequate yet tractable number of examples: 2,175 training and 977 testing images.
3. Fixed-size images of appropriate dimension for standard models.
4. Gold-standard labels from the majority vote of seven pathologists, along with annotator agreement levels that can be used for more-specific model tuning.

Table 1. Number of images in our dataset's training and testing sets for each class. HP: hyperplastic polyp (benign), SSA: sessile serrated adenoma (precancerous).

	Train	Test	Total
HP	1,545	617	2,162
SSA	630	360	990
Total	2,175	977	3,152

The rest of this section details the colorectal polyp classification task (Sect. 3.1), data collection (Sect. 3.2), and the data annotation process (Sect. 3.3).

3.1 Colorectal Polyp Classification Task

Colorectal cancer is the second leading cause of cancer death in the United States, with an estimated 53,200 deaths in 2020 [14]. As a result, colonoscopy is one of the most common cancer screening programs in the United States [15], and classification of colorectal polyps (growths inside the colon lining that can lead to colonic cancer if left untreated) is one of the highest-volume tasks in pathology. Our task focuses on the clinically-important binary distinction between hyperplastic polyps (HPs) and sessile serrated adenomas (SSAs), a challenging problem with considerable inter-pathologist variability [16–20]. HPs are typically benign, while SSAs are precancerous lesions that can turn into cancer if left untreated and require sooner follow-up examinations [21]. Pathologically, HPs have a superficial serrated architecture and elongated crypts, whereas SSAs are characterized by broad-based crypts, often with complex structure and heavy serration [22].

3.2 Data Collection

For our dataset, we scanned 328 Formalin Fixed Paraffin-Embedded (FFPE) whole-slide images of colorectal polyps, which were originally diagnosed on the whole-slide level as hyperplastic polyps (HPs) or sessile serrated adenomas (SSAs), from patients at the Dartmouth-Hitchcock Medical Center. These slides were scanned by an Aperio AT2 scanner at 40x resolution; to increase the field of view, we compress the slides with 8x magnification. From these 328 whole-slide images, we then extracted 3,152 image tiles (portions of size 224 × 224 pixels) representing diagnostically-relevant regions of interest for HPs or SSAs. These images were shuffled and anonymized by removing all metadata such that no sensitive patient information was retrievable from any images. All images contain mostly tissue by area (as opposed to white space) and were confirmed by our pathologists to be high-quality with few artifacts. The use and release of our dataset was approved by Dartmouth-Hitchcock Health IRB.

3.3 Data Annotation

For data annotation, we worked with seven board-certified gastrointestinal pathologists at the Dartmouth-Hitchcock Medical Center. Each pathologist individually and independently classified each image in our dataset as either HP or SSA based on the World

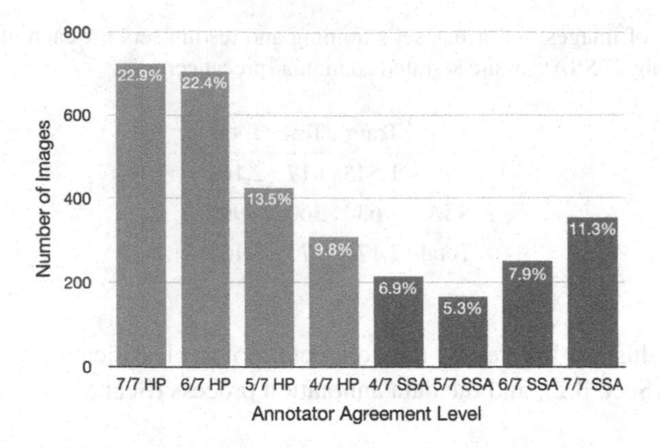

Fig. 2. Distribution of annotator agreement levels for images in our dataset.

Table 2. Model performance for five different ResNet depths. Adding more layers to the model does not improve performance. n indicates the number of images per class used for training. Means and standard deviations shown are for 10 random seeds.

	AUC (%) on test set by training set size			
ResNet	$n = 100$	$n = 200$	$n = 400$	Full
18	**67.4 ± 3.1**	73.6 ± 3.7	**79.3 ± 2.3**	84.5 ± 1.1
34	64.1 ± 2.2	**74.8 ± 3.1**	78.0 ± 2.4	**85.1 ± 0.7**
50	64.7 ± 3.1	72.2 ± 2.8	76.6 ± 1.9	83.0 ± 0.6
101	65.2 ± 5.6	73.2 ± 1.9	77.3 ± 0.9	83.2 ± 1.3
152	62.3 ± 3.3	73.3 ± 1.2	77.5 ± 1.9	83.5 ± 0.8

Health Organization criteria from 2019 [23]. After labels were collected for all images from all pathologists, the gold standard label for each image was assigned based on the majority vote of the seven individual labels, a common choice in literature [24–33]. The distribution of each class in our dataset based on the gold standard labels of each image is shown in Table 1.

In our dataset, the average percent agreement between each pair of annotators was 72.9%, and each pathologist agreed with the majority vote an average of 83.2% of the time. There is, notably, nontrivial disagreement between pathologists (approximately 16.7% of images have 4/7 agreement), which corresponds with the difficulty of our colorectal polyp classification task. The mean of the per-pathologist Cohen's κ was 0.450, in the moderate range of 0.41–0.60. Although not directly comparable with prior work, a similar evaluation found a Cohen's κ of 0.380 among four pathologists [20]. To facilitate research that might consider the annotator agreement of examples during training, for each image, we also provide the agreement level among our annotators (4/7, 5/7, 6/7, or 7/7). Figure 2 shows the distribution of agreement levels for our dataset.

4 Example Use Cases

In this section, we demonstrate example use cases of our dataset by investigating several natural questions that arise in histopathology image analysis. Namely, how does network depth affect model performance (Sect. 4.2)? How much does ImageNet pretraining help (Sect. 4.3)? Should examples with substantial annotator disagreement be included in training (Sect. 4.4)? Moreover, we vary the size of the training set to gain insight on how the amount of available data interacts with each of the above factors.

4.1 Experimental Setup

For our experiments, we follow the DeepSlide code repository [11] and use the ResNet architecture, a common choice for classifying histopathology images. Specifically, for our default baseline, we use ResNet-18 and train our model for 100 epochs (well past convergence) using stochastic data augmentation with the Adam optimizer [4], batch size of 32, initial learning rate of 1×10^{-3}, and learning rate decay factor of 0.91.

Table 3. Using weights pretrained on ImageNet significantly improves the performance of ResNet-18 on our dataset. n indicates the number of images per class used for training. Means and standard deviations shown are for 10 random seeds.

	AUC (%) on test set by training set size			
Pretraining?	$n = 100$	$n = 200$	$n = 400$	Full
No	67.4 ± 3.1	73.6 ± 3.7	79.3 ± 2.3	84.5 ± 1.1
Yes	**83.7 ± 1.7**	**89.3 ± 1.8**	**92.4 ± 0.7**	**92.7 ± 0.4**

For more-robust evaluation, for each model we consider the five highest AUCs on the test set, which are evaluated at every epoch. We report the mean and standard deviation of these values calculated over 10 different random seeds.

Furthermore, we train our models with four different training set sizes: $n = 100$, $n = 200$, $n = 400$, and Full, where n is the number of training images per class and Full is the entire training set. To obtain subsets of the training set, we randomly sample n random images for each class from the training set for each seed. We keep our testing set fixed to ensure that models are evaluated equally.

4.2 Network Depth

We first study whether adding more layers to our model improves performance on our dataset. Because deeper models take longer to train, identifying the smallest model that achieves the best performance allows for maximum accuracy with the least necessary training time.

We evaluate all five ResNet models proposed in [34]—ResNet-18, ResNet-34, ResNet-50, ResNet-101, and ResNet-152—on our dataset, and all hyperparameters (e.g., number of epochs, batch size) are kept constant; we only change the model depth.

As shown in Table 2, adding more layers does not significantly improve performance at any training dataset size. Furthermore, adding model depth past ResNet-34 actually decreases performance, as models that are too deep will begin to overfit, especially for small training set sizes. For example, when training with only 100 images per class, mean AUC decreases by 5.1% when using ResNet-152 compared to ResNet-18.

We posit that increasing network depth does not improve performance on our dataset because our dataset is relatively small, and so deeper networks are unnecessary for the amount of information in our dataset. Moreover, increasing network depth may increase overfitting on training data due to our dataset's small size. Our results are consistent with findings presented by Benkendorf and Hawkins [35]—deeper networks only perform better than shallow networks when trained with large sample sizes.

4.3 Transfer Learning

We also examine the usefulness of transfer learning for our dataset, as transfer learning can often be easily implemented into existing models, and so it is helpful to know whether or not it can improve performance.

Table 4. Removing high-disagreement images during training may slightly improve performance. Means and standard deviations shown are for 10 random seeds.

Training images used	AUC (%) on test set
Very easy images only	79.9 ± 0.8
Easy images only	83.1 ± 0.6
Very easy + Easy images	84.6 ± 0.8
Very easy + Easy + Hard images	**85.1 ± 0.8**
All images	84.5 ± 1.1

Because deeper models do not achieve better performance on our dataset (as shown in Sect. 4.2), we use ResNet-18 initialized with random weights as the baseline model for this experiment. We compare our baseline with an identical model (i.e., all hyperparameters are congruent) that has been initialized with weights pretrained on ImageNet.

Table 3 shows the results for our ResNet-18 model with and without pretraining. We find that ResNet-18 initialized with ImageNet pretrained weights significantly outperforms ResNet-18 initialized with random weights. For example, our pretrained model's performance when trained with only 100 images per class is comparable to our baseline model's performance when trained with the full training dataset. When both models are trained on the full training set, the pretrained model outperforms the baseline by 8.2%, as measured by mean AUC. These results indicate that, for our dataset, using pretrained weights can be extremely helpful for improving overall performance.

The large improvement from ImageNet pretraining is unlikely to result from our dataset having features expressed in ImageNet because ImageNet does not include histopathology images. Instead, the improvement is, perhaps, explained by our dataset's small size, as ImageNet pretraining may help prevent the model from overfitting. This would be consistent with Kornblith et al. [36], which found that performance improvements from ImageNet pretraining diminishes as dataset size increases.

4.4 High-Disagreement Training Examples

As many datasets already contain annotator agreement data [24–30,33,37–41], we also study whether there are certain ways of selecting examples based on their annotator agreement level that will maximize performance. Examples with high annotator disagreement are, by definition, harder to classify, so they may not always contain features that are beneficial for training models. For this reason, we focus primarily on whether training on only examples with higher annotator agreement will improve performance.

For our dataset, which was labeled by seven annotators, we partition our images into four discrete levels of difficulty: *very easy* (7/7 agreement among annotators), *easy* (6/7 agreement among annotators), *hard* (5/7 agreement among annotators), and *very hard* (4/7 agreement among annotators), following our prior work [42]. We then train ResNet-18 models using different combinations of images selected based on difficulty: very easy images only; easy images only; very easy and easy images; and very easy, easy, and hard images. For this experiment, we do not modify the dataset size like we did in Sects. 4.2 and 4.3, as selecting training images based on difficulty inherently changes the training set size.

As shown in Table 4, we find that excluding images with high annotator disagreement (i.e., hard and very hard images) during training achieves comparable performance to training with all images. Using only very easy images or only easy images, however, does not match or exceed performance when training with all images. We also find that training with all images except very hard images slightly outperforms training with all images. One explanation for this is that very hard images, which only have 4/7 annotator agreement, could be too challenging to analyze accurately (even for expert humans), so their features might not be beneficial for training machine learning models either.

5 Related Work

Due to the trend towards larger, more computationally-expensive models [43], as well as recent attention on the environmental considerations of training large models [44], the deep learning community has begun to question whether model development needs to occur at scale. In the machine learning field, two recent papers have brought attention to this idea. Rawal et al. [45] proposed a novel surrogate model for rapid architecture development, an artificial setting that predicts the ground-truth performance of architectural motifs. Greydanus [46] proposed MNIST-1D, a low-computational alternative resource to MNIST that differentiates more clearly between models. Our dataset falls within this direction and is heavily inspired by this work.

In the histopathology image analysis domain, several datasets are currently available. Perhaps the two best-known datasets are CAMELYON17 [13] and PCam [47]. CAMELYON17 focuses on breast cancer metastasis detection in whole-slide images (WSIs) and includes a training set of 1,000 WSI with labeled locations. CAMELYON17 is well-established, but because it contains WSIs (each taking up >1 GB), there is a large barrier to training an initial model, and it is unclear how to best pre-process the data to be compatible with current neural networks' desired input format.

PCam is another well-known dataset that contains 327,680 images of size 96 × 96 pixels extracted from CAMELYON17. While PCam is similar to our work in that it considers fixed-size images for binary classification, we note two key differences. First, the annotations in PCam are derived from bounding-box annotations which were drawn by a student and then checked by a pathologist. For challenging tasks with high annotator disagreement, however, using only a single annotator can cause the model to learn specific tendencies of a single pathologist. In our dataset, on the other hand, each image is directly classified by seven expert pathologists, and the gold standard is set as the majority vote of the seven labels, mitigating the potential biases that can arise from having only a single annotator. Second, whereas PCam takes up around 7 GB of disk space, our dataset aims to be minimalist and is therefore an order of magnitude smaller, making it faster for researchers to obtain results and iterate over models.

In Table 5, we compare our dataset with other previously-proposed histopathology image analysis datasets. Our dataset is much smaller than other datasets, yet it still has enough examples to serve as a petri dish in that it can test models and return results quickly. Additionally, our dataset has robust annotations in comparison to other histopathology datasets. Datasets frequently only have one or two annotators, but MHIST is annotated by seven pathologists, making it the least influenced by biases that any singular annotator may have.

Table 5. Comparison of well-known histopathology datasets. Our proposed dataset, MHIST, is advantageous due to its relatively small size (making it faster to obtain results) and its robust annotations. ROI: Regions of Interest.

Dataset	Images	Image type	Annotation type	Number of Annotators	Dataset size
MITOS (2012) [48]	50	High power fields	Pixel-level	2	~1 GB
TUPAC16 [49]	821	Whole-slide images	ROI	1	~850 GB
CAMELYON17 [13]	1,000	Whole-slide images	Contoured	1	~2.3 TB
PCam (2018) [47]	327,680	Fixed-sized images	Image-wise	1	~7 GB
BACH (2018) [50]	500	Microscopy images	Image-wise	2	>5 GB
LYON19 [51]	83	Whole-slide images	ROI	3	~13 GB
MHIST (Ours)	3,152	Fixed-sized images	Image-wise	7	~333 MB

6 Discussion

The inherent nature of histopathology image classification can create challenges for researchers looking to apply mainstream computer vision methods. Histopathology images themselves are difficult to handle because they have high resolutions and are variable-sized, and accurately and efficiently annotating histopathology images is a nontrivial task. Furthermore, being able to address these challenges does not guarantee a high-quality dataset, as histopathology images are difficult to acquire, and so data is often quite limited. Based on a thorough analysis of these challenges, we have presented MHIST, a histopathology image classification dataset with a straightforward yet challenging binary classification task. MHIST comprises a total of 3,152 fixed-size images that have already been preprocessed. In addition to providing these images, we also include each image's gold standard label and degree of annotator agreement.

Of possible limitations, our use of fixed-size images may not be the most-precise approach for histopathological image analysis, as using whole-slide images directly would likely improve performance since whole-slide images contain much more information than fixed-size images. Current computer vision models cannot train on whole-slide images, however, as a single whole-slide image can take up more than 10 GB of space. Thus, our dataset includes images of fixed-size, as this is appropriate for most standard computer vision models. Another limitation is that our dataset does not include any demographic information about patients nor any information regarding the size and location of the polyp (data that is often used in clinical classification). Our dataset contains purely image data and is limited in this fashion.

In this paper, we aim to have provided a dataset that can serve as a petri dish for histopathology image analysis. We hope that researchers are able to use MHIST to test models on a smaller scale before being implemented in large-scale applications, and that our dataset will facilitate further research into deep learning methodologies for histopathology image analysis.

References

1. Jennings, B.H.: Drosophila - a versatile model in biology & medicine. Mater. Today, **14**(5), 190–195 (2011). http://www.sciencedirect.com/science/article/pii/S1369702111701134
2. Lecun, Y., Bottou, L., Bengio, Y., Haffner, P.: Gradient-based learning applied to document recognition. Proc. IEEE, **86**(11), 2278–2324 (1998). https://ieeexplore.ieee.org/document/726791
3. Goodfellow, I.J., et al.: Generative adversarial networks (2014). https://arxiv.org/pdf/1406.2661.pdf
4. Kingma, D., Ba, J.: Adam: a method for stochastic optimization. In: International Conference on Learning Representations (2014). https://arxiv.org/pdf/1412.6980.pdf
5. Srinidhi, C.L., Ciga, O., Martel, A.L.: Deep neural network models for computational histopathology: a survey. Med. Image Anal. 101813 (2020). https://www.sciencedirect.com/science/article/pii/S1361841520301778
6. Arvaniti, E., et al.: Automated Gleason grading of prostate cancer tissue microarrays via deep learning. Nat. Sci. Rep. **8**(1), 1–11 (2018). https://www.nature.com/articles/s41598-018-30535-1
7. Bulten, W., et al.: Automated deep-learning system for Gleason grading of prostate cancer using biopsies: a diagnostic study. Lancet Oncol. **21**(2), 233–241 (2020). https://arxiv.org/pdf/1907.07980.pdf
8. Hekler, A., et al.: Pathologist-level classification of histopathological melanoma images with deep neural networks. Euro. J. Cancer, **115**, 79–83 (2019). http://www.sciencedirect.com/science/article/pii/S0959804919302758

9. Shah, M., Wang, D., Rubadue, C., Suster, D., Beck, A.: Deep learning assessment of tumor proliferation in breast cancer histological images. In: 2017 IEEE International Conference on Bioinformatics and Biomedicine (BIBM), pp. 600–603 (2017). https://ieeexplore.ieee.org/abstract/document/8217719
10. Ström, P., et al.: Pathologist-level grading of prostate biopsies with artificial intelligence. CoRR (2019). http://arxiv.org/pdf/1907.01368
11. Wei, J.W., Tafe, L.J., Linnik, Y.A., Vaickus, L.J., Tomita, N., Hassanpour, S.: Pathologist-level classification of histologic patterns on resected lung adenocarcinoma slides with deep neural networks. Sci. Rep. **9**(1), 1–8 (2019). https://www.nature.com/articles/s41598-019-40041-7
12. Zhang, Z., et al.: Pathologist-level interpretable whole-slide cancer diagnosis with deep learning. Nat. Mach. Intell. **1**, 236–245 (2019). https://www.nature.com/articles/s42256-019-0052-1
13. Bándi, P., et al.: From detection of individual metastases to classification of lymph node status at the patient level: the camelyon17 challenge. IEEE Trans. Med. Imaging, **38**(2), 550–560 (2019). https://ieeexplore.ieee.org/document/8447230
14. Colorectal cancer statistics. https://www.cancer.org/cancer/colon-rectal-cancer/about/key-statistics.html. Accessed 06 Jan 2021
15. Rex, D.K., et al.: Colorectal cancer screening: Recommendations for physicians and patients from the U.S. multi-society task force on colorectal cancer. Gastroenterology, **153**, 307–323 (2017). www.gastrojournal.org/article/S0016-5085(17)35599--3/fulltext
16. Abdeljawad, K., Vemulapalli, K.C., Kahi, C.J., Cummings, O.W., Snover, D.C., Rex, D.K.: Sessile serrated polyp prevalence determined by a colonoscopist with a high lesion detection rate and an experienced pathologist. Gastrointest. Endosc. **81**, 517–524 (2015). https://pubmed.ncbi.nlm.nih.gov/24998465/
17. Farris, A.B., et al.: Sessile serrated adenoma: challenging discrimination from other serrated colonic polyps. Am. J. Surg. Pathol. **32**, 30–35 (2008). https://pubmed.ncbi.nlm.nih.gov/18162767/
18. Glatz, K., Pritt, B., Glatz, D., HArtmann, A., O'Brien, M.J., Glaszyk, H.: A multinational, internet-based assessment of observer variability in the diagnosis of serrated colorectal polyps. Am. J. Clin. Pathol. **127**(6), 938–945 (2007). https://pubmed.ncbi.nlm.nih.gov/17509991/
19. Khalid, O., Radaideh, S., Cummings, O.W., O'brien, M.J., Goldblum, J.R., Rex, D.K.: Reinterpretation of histology of proximal colon polyps called hyperplastic in 2001. World J. Gastroenterol. **15**(30), 3767–3770 (2009). https://www.ncbi.nlm.nih.gov/pmc/articles/PMC2726454/
20. Wong, N.A.C.S., Hunt, L.P., Novelli, M.R., Shepherd, N.A., Warren, B.F.: Observer agreement in the diagnosis of serrated polyps of the large bowel. Histopathology, **55**(1), 63–66 (2009). https://pubmed.ncbi.nlm.nih.gov/19614768/
21. Understanding your pathology report: Colon polyps (sessile or traditional serrated adenomas). https://www.cancer.org/treatment/understanding-your-diagnosis/tests/understanding-your-pathology-report/colon-pathology/colon-polyps-sessile-or-traditional-serrated-adenomas.html. Accessed 06 Jan 2021
22. Gurudu, S.R., et al.: Sessile serrated adenomas: Demographic, endoscopic and pathological characteristics. World J. Gastroenterol. **16**(27), 3402–3405 (2010). https://www.ncbi.nlm.nih.gov/pmc/articles/PMC2904886/
23. Nagtegaal, I.D., et al.: The 2019 who classification of tumours of the digestive system. Histopathology, **76**(2), 182–188 (2020). https://onlinelibrary.wiley.com/doi/abs/10.1111/his.13975

24. Chilamkurthy, S., et al.: Deep learning algorithms for detection of critical findings in head CT scans: a retrospective study. Lancet, **392**(10162), 2388–2396 (2018). www.thelancet.com/journals/lancet/article/PIIS0140-6736(18)31645--3/fulltext
25. Gulshan, V., et al.: Development and validation of a deep learning algorithm for detection of diabetic retinopathy in retinal fundus photographs. JAMA, **316**(22), 2402–2410 (2016). https://jamanetwork.com/journals/jama/fullarticle/2588763
26. Irvin, J., et al.: Chexpert: a large chest radiograph dataset with uncertainty labels and expert comparison. In: Association for the Advancement of Artificial Intelligence (AAAI) (2019). http://arxiv.org/pdf/1901.07031
27. Kanavati, F., et al.: Weakly-supervised learning for lung carcinoma classification using deep learning. Nat. Sci. Rep. **10**(1), 1–11 (2020). https://www.ncbi.nlm.nih.gov/pmc/articles/PMC7283481/
28. Korbar, B., et al.: Deep learning for classification of colorectal polyps on whole-slide images. J. Pathol. Inform. **8** (2017). https://www.ncbi.nlm.nih.gov/pmc/articles/PMC5545773/
29. Sertel, O., Kong, J., Catalyurek, U.V., Lozanski, G., Saltz, J.H., Gurcan, M.N.: Histopathological image analysis using model-based intermediate representations and color texture: follicular lymphoma grading. J. Sig. Process. Syst. **55**(1), 169–183 (2009). https://link.springer.com/article/10.1007/s11265-008-0201-y
30. Wang, S., Xing, Y., Zhang, L., Gao, H., Zhang, H.: Deep convolutional neural network for ulcer recognition in wireless capsule endoscopy: experimental feasibility and optimization. Computat. Math. Meth. Med. (2019). https://www.ncbi.nlm.nih.gov/pmc/articles/PMC6766681/
31. Wei, J., Wei, J., Jackson, C., Ren, B., Suriawinata, A., Hassanpour, S.: Automated detection of celiac disease on duodenal biopsy slides: a deep learning approach. J. Pathol. Inform. **10**(1), 7 (2019). http://www.jpathinformatics.org/article.asp?issn=2153-3539;year=2019;volume=10;issue=1;spage=7;epage=7;aulast=Wei;t=6
32. Wei, J., et al.: Difficulty translation in histopathology images. In: Artificial Intelligence in Medicine (AIME) (2020). https://arxiv.org/pdf/2004.12535.pdf
33. Zhou, J., et al.: Weakly supervised 3D deep learning for breast cancer classification and localization of the lesions in MR images. J. Magn. Reson. Imaging, **50**(4), 1144–1151 (2019). https://onlinelibrary.wiley.com/doi/abs/10.1002/jmri.26721
34. He, K., Zhang, X., Ren, S., Sun, J.: Deep residual learning for image recognition. In: 2016 IEEE Conference on Computer Vision and Pattern Recognition (CVPR) (2015). http://arxiv.org/pdf/1512.03385
35. Benkendorf, D.J., Hawkins, C.P.: Effects of sample size and network depth on a deep learning approach to species distribution modeling. Ecol. Inform. **60**, 101137 (2020). http://www.sciencedirect.com/science/article/pii/S157495412030087X
36. Kornblith, S., Shlens, J., Le, Q.V.: Do better imagenet models transfer better? CoRR (2018). http://arxiv.org/pdf/1805.08974
37. Coudray, N., Moreira, A.L., Sakellaropoulos, T., Fenyö, D., Razavian, N., Tsirigos, A.: Classification and mutation prediction from non-small cell lung cancer histopathology images using deep learning. Nat. Med. **24**(10), 1559–1567 (2017). https://www.nature.com/articles/s41591-018-0177-5
38. Ehteshami Bejnordi, B., et al.: Diagnostic assessment of deep learning algorithms for detection of lymph node metastases in women with breast cancer. JAMA, **318**(22), 2199–2210 (2017). https://jamanetwork.com/journals/jama/fullarticle/2665774
39. Esteva, A., et al.: Dermatologist-level classification of skin cancer with deep neural networks. Nature, **542**(7639), 115–118 (2017). https://www.nature.com/articles/nature21056
40. Ghorbani, A., Natarajan, V., Coz, D., Liu, Y.: Dermgan: synthetic generation of clinical skin images with pathology (2019). https://arxiv.org/pdf/1911.08716.pdf

41. Wei, J.W., et al.: Evaluation of a deep neural network for automated classification of colorectal polyps on histopathologic slides. JAMA Netw. Open, **3**(4) (2020). https://jamanetwork.com/journals/jamanetworkopen/article-abstract/2764906
42. Wei, J., et al.: Learn like a pathologist: curriculum learning by annotator agreement for histopathology image classification. In: Winter Conference on Applications of Computer Vision (WACV) (2020)
43. Brown, T.B., et al.: Language models are few-shot learners. arXiv preprint arXiv:2005.14165 (2020). https://arxiv.org/pdf/2005.14165.pdf
44. Strubell, E., Ganesh, A., McCallum, A.: Energy and policy considerations for deep learning in nlp. arXiv preprint arXiv:1906.02243 (2019). https://arxiv.org/pdf/1906.02243.pdf
45. Rawal, A., Lehman, J., Such, F.P., Clune, J., Stanley, K.O.: Synthetic petri dish: a novel surrogate model for rapid architecture search. arXiv preprint arXiv:2005.13092 (2020). https://arxiv.org/pdf/2005.13092.pdf
46. Greydanus, S.: Scaling down deep learning. arXiv preprint arXiv:2011.14439 (2020). https://arxiv.org/pdf/2011.14439.pdf
47. Veeling, B.S., Linmans, J., Winkens, J., Cohen, T., Welling, M.: Rotation equivariant CNNs for digital pathology. In: Frangi, A.F., Schnabel, J.A., Davatzikos, C., Alberola-López, C., Fichtinger, G. (eds.) MICCAI 2018. LNCS, vol. 11071, pp. 210–218. Springer, Cham (2018). https://doi.org/10.1007/978-3-030-00934-2_24
48. Cireşan, D.C., Giusti, A., Gambardella, L.M., Schmidhuber, J.: Mitosis detection in breast cancer histology images with deep neural networks. In: Mori, K., Sakuma, I., Sato, Y., Barillot, C., Navab, N. (eds.) MICCAI 2013. LNCS, vol. 8150, pp. 411–418. Springer, Heidelberg (2013). https://doi.org/10.1007/978-3-642-40763-5_51
49. Veta, M., et al.: Predicting breast tumor proliferation from whole-slide images: the tupac16 challenge. Med. Image Anal. **54**, 111–121 (2019). https://doi.org/10.1016/j.media.2019.02.012, http://www.sciencedirect.com/science/article/pii/S1361841518305231
50. Aresta, G., et al.: Bach: Grand challenge on breast cancer histology images. Med. Image Anal.**56**, 122–139 (2019). https://doi.org/10.1016/j.media.2019.05.010, http://www.sciencedirect.com/science/article/pii/S1361841518307941
51. Swiderska-Chadaj, Z., et al.: Learning to detect lymphocytes in immunohistochemistry with deep learning. Med. Image Anal. **58**, (2019). https://doi.org/10.1016/j.media.2019.101547, http://www.sciencedirect.com/science/article/pii/S1361841519300829

fMRI Multiple Missing Values Imputation Regularized by a Recurrent Denoiser

David Calhas(✉) and Rui Henriques

Instituto Superior Técnico, Lisbon, Portugal
david.calhas@tecnico.ulisboa.pt

Abstract. Functional Magnetic Resonance Imaging (fMRI) is a neuroimaging technique with pivotal importance due to its scientific and clinical applications. As with any widely used imaging modality, there is a need to ensure the quality of the same, with missing values being highly frequent due to the presence of artifacts or sub-optimal imaging resolutions. Our work focus on missing values imputation on multivariate signal data. To do so, a new imputation method is proposed consisting on two major steps: spatial-dependent signal imputation and time-dependent regularization of the imputed signal. A novel layer, to be used in deep learning architectures, is proposed in this work, bringing back the concept of chained equations for multiple imputation [26]. Finally, a recurrent layer is applied to tune the signal, such that it captures its true patterns. Both operations yield an improved robustness against state-of-the-art alternatives. The code is made available on Github.

1 Introduction

The ability to learn from functional Magnetic Resonance Imaging (fMRI) data is generally hampered by the presence of artifact and limits on the available instrumentation and acquisition protocol, resulting in pronounced missingness. As MRI is collected in frequency space with the usual type of missing/corrupted values occurring at the frequency space. On the other hand, low-quality (voxel space) recordings prevents whole-brain analyzes in clinical settings and is specially pervasive among stimuli-inducing setups in research settings. Imputation of incomplete/noisy recordings is critical to classification [19], synthesis and enhancement tasks. For instance, given the unique spatial and temporal nature of each neuroimaging modality, synthesis between distinct modalities is a difficult task (particularly of multivariate time series nature), being imputation an important step of the process [16,22]. Finally, the integration of heterogeneous sources of fMRI recordings by multiple initiatives worldwide also drives the need to increase image resolutions and correct differences arriving from distinct setups.

This work reclaims the importance of a machine learning based model to perform imputation of multivariate signal data, as opposed to individual-specific imputation. In this context, we propose a novel layer that perform feature based imputation with the principle of chained equations [26]. Further, a recurrent

A. Tucker et al. (Eds.): AIME 2021, LNAI 12721, pp. 25–35, 2021.
https://doi.org/10.1007/978-3-030-77211-6_3

layer is proposed to serve as a denoiser to the spatially imputed signal. This two-step principled approach for imputation is illustrated in Fig. 1. Results on resting-state and stimuli-based fMRI recordings validate the robustness of the proposed approach against competitive alternatives.

2 Problem Setting

Multivariate Time Series (MTS) Missing Value Imputation is the focus of this work, specifically high-dimensional MTS data from fMRI recordings. The problem is divided into two parts: spatial imputation, in which missing values are sequentially predicted from the existing features; and time dimension regularization, where the imputed values are time tuned.

Consider an fMRI recording to be a multivariate time series $\mathbf{x} \in \mathbb{R}^{v \times t}$, being $v = (v_0, ..., v_{V-1})$ the voxel dimension with V voxels and $t = (t_0, ..., t_{T-1})$ the temporal dimension with T timesteps. An fMRI volume, at timestep t, is denoted as $\mathbf{x}^t = [\mathbf{x}_0^t, ..., \mathbf{x}_{V-1}^t]$ and a voxel time series, at voxel v, is denoted as $\mathbf{x}_v = [\mathbf{x}_v^0, ..., \mathbf{x}_v^{T-1}]$, each can be seen as the column and row of matrix $\mathbf{x}$, respectively.

In the problem of imputation, consider $\psi \in \mathbb{R}^{v \times t} \cup \{nan\}$, $\varphi \in \mathbb{R}^{v \times t}$ and $\mu \in \{0,1\}^{v \times t}$ as the variables involved in the learning phase. The *nan* symbol denotes a missing value in the fMRI instance ψ. μ is the mask, with 0 representing a complete value and 1 a missing value. φ is the complete fMRI instance. Illustrating[1], given $V = 5$ and $T = 7$,

$$\psi = \begin{bmatrix} 1 & 2 & 3 & 4 & 5 & 6 & 7 \\ nan & nan & nan & nan & nan & nan & nan \\ 15 & 16 & 17 & 18 & 19 & 20 & 21 \\ nan & nan & nan & nan & nan & nan & nan \\ nan & nan & nan & nan & nan & nan & nan \end{bmatrix}, \mu = \begin{bmatrix} 0\,0\,0\,0\,0\,0\,0 \\ 1\,1\,1\,1\,1\,1\,1 \\ 0\,0\,0\,0\,0\,0\,0 \\ 1\,1\,1\,1\,1\,1\,1 \\ 1\,1\,1\,1\,1\,1\,1 \end{bmatrix}, \varphi = \begin{bmatrix} 1 & 2 & 3 & 4 & 5 & 6 & 7 \\ 8 & 9 & 10 & 11 & 12 & 13 & 14 \\ 15 & 16 & 17 & 18 & 19 & 20 & 21 \\ 22 & 23 & 24 & 25 & 26 & 27 & 28 \\ 29 & 30 & 31 & 32 & 33 & 34 & 35 \end{bmatrix}$$

Consider the imputation of missing values is made by a model, I, and each imputed value is denoted as $\iota_v^t \in \mathbb{R}$. Continuing our example, an fMRI instance, ψ, after processing by I, is $\iota \in \mathbb{R}^{v \times t}$,

$$I(\psi) = \iota = \begin{bmatrix} 1 & 2 & 3 & 4 & 5 & 6 & 7 \\ \iota_1^0 & \iota_1^1 & \iota_1^2 & \iota_1^3 & \iota_1^4 & \iota_1^5 & \iota_1^6 \\ 15 & 16 & 17 & 18 & 19 & 20 & 21 \\ \iota_3^0 & \iota_3^1 & \iota_3^2 & \iota_3^3 & \iota_3^4 & \iota_3^5 & \iota_3^6 \\ \iota_4^0 & \iota_4^1 & \iota_4^2 & \iota_4^3 & \iota_4^4 & \iota_4^5 & \iota_4^6 \end{bmatrix}$$

Considering typical fMRI resolution, each voxel has a 3D euclidean point correspondence. As such, a spatial distance matrix, $d \in \mathbb{R}^{V \times V}$, is defined, where $d_{i,j}$ corresponds to the distance between voxels v_i and v_j, with $i, j \in \{0, ..., V-1\} \in \mathbb{N}$.

Missing voxels can occur at random or, in contrast, be spatially autocorrelated within a variable number of regions, resembling the characteristics of an artifact. Both modes are targeted in this work.

[1] ψ, μ, φ and ι values selected for simplicity sake, not necessarily resembling fMRI values.

3 Proposed Approach

In this section, two main steps are proposed to perform imputation of missing values from MTS data:

- **Spatial imputation**, where imputation is done by estimating missing values, $\mu_i^j = 1$, from complete features, $\mu_k^l = 0$, with $i, k \in \{0, ..., V-1\}, j, l \in \{0, ..., T-1\}$.
- **Time series regularization**, where the previously derived missing values are processed by a recurrent neural network.

This two-step approach is shown in Sect. 6 to outperform competitive baselines.

3.1 Spatial Imputation

We propose a novel neural network layer, Φ, that performs imputation inspired by the chained equations principle proposed by [26]. This imputation method consists of filling a missing value at a time, and using its estimate to guide the imputation of the remaining missing values.

The priority in which the values are filled is given by the pairwise voxel correlation matrix, C, computed from the training set data.

This layer, Φ, is characterized by a weight matrix W_Φ and bias B_Φ,

$$W_\Phi = \begin{bmatrix} 0 & w_0^1 & \dots & w_0^{V-1} \\ w_1^0 & 0 & \dots & w_1^{V-1} \\ \vdots & \vdots & \ddots & \vdots \\ w_{V-1}^0 & w_{V-1}^1 & \dots & 0 \end{bmatrix}, \; B_\Phi = \begin{bmatrix} b_0 \\ b_1 \\ \vdots \\ b_{V-1} \end{bmatrix}.$$

The activation function of Φ is linear, making the imputed values a linear combination of the already filled and complete values. The weight matrix has a zero-filled diagonal for each voxel to be described as a linear combination of the remaining voxels.

Since this neural function estimates a single value at a time, one only needs to compute the dot product of the missing value, v, with the corresponding column, W_Φ^v, and add the bias, B_Φ^v. The imputation operation of a missing value, v, is denoted as $\xrightarrow{\psi^0 W_\Phi{}^v + B_\Phi{}^v}$.

Let us consider the input presented in Sect. 2, with $\psi^0 = [1, nan, 15, nan, nan]$, and $\xrightarrow{\psi^0 W_\Phi{}^c + B_\Phi{}^c}$ as the operation made by layer Φ at each iteration to impute a missing value, c. Being $c = max(C_{\mu^0=1})$ the missing voxel that has the highest correlation with the complete and filled voxels. This scheme allows an imputation of missings under the chained equation principle, ϕ is the output of layer Φ, with ϕ_i corresponding to the v_i missing voxel imputation. A total of $\sum \mu^t$ iterations are necessary, corresponding to the number of missing values,

$$\Phi(\psi^0) = \begin{bmatrix} 1 \\ nan \\ 15 \\ nan \\ nan \end{bmatrix}^T \xrightarrow{\psi^0 W_\Phi{}^c + B_\Phi{}^c} \begin{bmatrix} 1 \\ \phi_1 \\ 15 \\ nan \\ nan \end{bmatrix}^T \xrightarrow{\psi^0 W_\Phi{}^c + B_\Phi{}^c} \begin{bmatrix} 1 \\ \phi_1 \\ 15 \\ nan \\ \phi_4 \end{bmatrix}^T \xrightarrow{\psi^0 W_\Phi{}^c + B_\Phi{}^c} \begin{bmatrix} 1 \\ \phi_1 \\ 15 \\ \phi_3 \\ \phi_4 \end{bmatrix}^T = \phi.$$

Algorithm 1 Φ chained imputation cycle

```
φ ← ψ^t
while Σ μ^t > 0 do
    c ← max(C_{μ^t=1})
    φ_c ← ψ^t W_Φ^c + B_Φ^c
    μ_c^t ← 0
end while
φ
```

Algorithm 1 presents the pseudocode for this imputation scheme. The Φ layer imputes all missing values per time frame, t, of an fMRI recording, ψ^t. Imputation is merely done accounting other features on the same time frame, therefore spatial. This layer contrasts with the traditional dropout layer for imputing missing values. In a dropout layer, each weights' column, W^v, shows intra-correlation, converging to a single target independently. However, there is no inter-column correlation/dependency. Φ layer forces the columns, W_Φ^v, to be inter-correlated, converging to the same target as a unit. Here, the estimates of a column (the imputed values) influence the estimates of the upcoming columns along the imputation process.

3.2 Time Dimension Regularization

Once spatial imputation is done for each voxel, v, of an fMRI recording, ϕ_v, the imputed values are fed to a recurrent layer, tweaking the signal in such a way that it emulates the target time series patterns. The recurrent layer removes the noise created by the spatial imputation method. We refer to this recurrent layer as the Denoiser, D, component of the imputation pipeline. An illustration of this noise removal is shown in Fig. 1.

D is a single layer Gated Recurrent Unit (GRU) [4]. This choice was motivated by results collected against its rival Long-Short Term Memory Layer on the target task, and further supported by studies showing that GRU performs well on datasets with limited observations [13], the common case when learning Neuroimaging datasets. [3,15] altered the internal function of a GRU layer to perform direct imputation on a multivariate time series. In contrast, we maintain the GRU layer as it was originally proposed [4] since our purpose is to remove noise and capture the desired temporal patterning properties of the (neurophysiological) signal. The imputation model we propose is denoted as $\Phi + D$, corresponding to the junction of the spatial

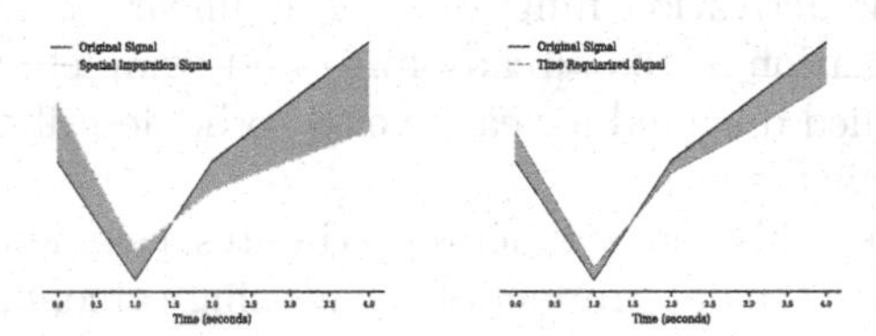

(a) Error between the true signal and spatially imputed signal (b) Error between the true signal and imputed signal after time regularization

Fig. 1. Time regularization (denoising) component.

imputation scheme described in Sect. 3.1 and the time regularization described in Sect. 3.2.

3.3 Validation and Hyperparameters

Bayesian Optimization (BO) [20] was used to find the hyperparameters that best fitted the pipeline. For each BO iteration a 6-Fold Cross Validation was ran and the Mean Absolute Error (MAE) was the metric targeted as alternative residue-based scores can overly focus on the minimization or large residues. Manual tweaking was performed before to check which optimizers should be used. In consensus, Adam optimizer [11] is used to optimize the D and Φ trainable parameters. Missing data generation (explained in Sect. 5.3) is made at every iteration of the BO algorithm, in order to avoid overfitting towards a certain missing values setting. Besides $\Phi + D$, BO was also ran for Φ, Dropout and Dropout+D. The hyperparameters were subjected to a total of 50 iterations. Their range spaces are: Φ learning rate, $L_{\Phi} \in [1e{-5}, 1e{-2}] \in \mathbb{R}$; D learning rate, $L_D \in [1e{-5}, 1e{-2}] \in \mathbb{R}$; Φ number of epochs, $E_{\Phi} \in \{2, 3, 4, 5\} \in \mathbb{N}$; D number of epochs, $E_D \in \{2, 3, 4, 5\} \in \mathbb{N}$; number of alternating epochs, $E \in \{2, 4, 8, 10\} \in \mathbb{N}$; D L1-norm regularization constant, $R_D \in [1e{-5}, 3] \in \mathbb{R}$; D Use of bias, $B_D \in \{0, 1\} \in \mathbb{N}$; D Dropout [21], $Dr_D \in [0, 3e{-1}] \in \mathbb{R}$; D Recurrent Dropout [21], $RDr_D \in [0, 3e{-1}] \in \mathbb{R}$.

4 Related Work

Smieja et al. [19] address the presence of missing values by training a Gaussian Mixture Model (GMM) of neural network activations. Missing values are imputed at the first hidden layer by computing the expected activation of neurons (instead of just calculating the expected input). Thus the imputation is not made by single values, instead it is modeled by a GMM. Although competitive, this approach performed worse than a Context Encoder (CE) [17] which, in contrast, is learnt from a loss function using the complete data. In this work, we perform chained imputation based on correlation ranking, instead of imputing values at one step by taking advantage of a GMM for each feature. Che et al. [3] proposed a variant of the Gated Recurrent Unit (GRU) to handle generic time series with missing values, claiming that, by placing masks and time intervals in accordance with the properties of missing patterns, their model, GRU-D, is able to take advantage of missing data to improve classification. Masking and time intervals in GRU-D [3] are represented using a decay term computed by a exponentiated negative rectifier function. Given a missing occurrence, the decay at that timestep is used over time to converge to the empirical mean. Cao et al. [2] perform missing imputation from two estimates produced from a spatial-based and a recurrent model. Results on air quality, health care and human activity datasets show superiority among baselines reaching 11.56, 0.278 and 0.219 MAE. The imputation task is mapped as a classification task to learn the target models. After imputation is made separately by these two models, a linear combination, defined by a parameter, is computed to produce the final estimate. In contrast, we take advantage of a recurrent model described in Sect. 3.2 to remove prediction noise from the spatial imputation model and strengthen the temporal consistency. Further, Cao et al. [2] further assume missing values occur sporadically in a feature time series of the multivariate time series, on the other hand our work focuses on the imputation of whole feature time series to resemble the characteristics missing neuroimaging data. Luo et al. [14] propose a disruptive model based on a Generative Adversarial Architecture [9]. The Discriminator and Generator components are both an internally tweaked version of GRU. The Discriminator classifies generated and real multivariate time series samples and the Generator performs imputation on samples with missing values. Results show classification superiority using the AUC metric. Fortuin et al. [8] use a Variational Autoencoder Architecture [12] to perform imputation. The Generative model is a Gaussian Process that generates samples from complete encoded feature representations. All the discussed works in this section perform multivariate time series imputation from incomplete data. For a more objective assessment, our work tests imputation methods over complete datasets with generated missing entries and regions according to the proposed validation scheme.

5 Experimental Setting

The imputation baselines, target fMRI datasets, missing generation procedures, and evaluation metrics are detailed in Sects. 5.1, 5.2, 5.3 and 5.4, respectively.

5.1 Baselines

For the sake of comparison, the following baselines were implemented to gather the results: kNN imputation [10]; Barycenter [18]; MICE [26]; Mean imputation; Context Encoder (CE) [17]; Dropout; Dropout with time regularization, denoted as Dropout+D; and Φ with no time regularization. kNN was used with a k=3 since there were no overall significant improvements for alternative k. Barycenter computes the average time series of a multivariate time series and imputation is made with the average time series under the Dynamic Time Wrapping (DTW) distance criterion. MICE, a.k.a. multiple imputation by chained equations, is a method similar to ours as it has its basis on the same rationale of imputing one missing value at a time. It is thus considered a baseline as well. Mean imputation method takes the mean of each feature from the training set and performs kNN imputation ($k = 3$) if there is no information about a voxel in the training set. Context Encoder (CE) is a simple Autoencoder with 2-Dimensional Convolution Layers. Dropout method drops the weights linked to the missing values. Dropout was also extended with the time regularization scheme presented in Sect. 3.2. Finally, we also considered comparing Φ alone against $\Phi + D$ to measure the impact of reshaping the imputed time series.

5.2 Datasets

EEG, fMRI and NODDI Dataset. This dataset [6,7] contains 16 individuals, with an average age 32.84 ± 8.13 years. Simultaneous EEG-fMRI recordings of resting state with eyes open (fixating a point) were acquired. The fMRI acquisition was done using a T2-weighted gradient-echo EPI sequence with: 300 volumes, TR of 2160 ms, TE of 30 ms, 30 slices with 3.0 millimeters (mm), voxel size of $3.3 \times 3.3 \times 4.0$ mm and a field of view of $210 \times 210 \times 120$ mm. For a more detailed description please see the dataset references [6,7]. The dataset is available for download in its original source at https://osf.io/94c5t/. Each individual recording was divided into 24 equally sized time series of fMRI volumes. Each time series is 28 s long and resampled to a 2 s period. The *training set* is composed of 12 individuals and the *test set* of 4 individuals per fold.

Auditory and Visual Oddball EEG-fMRI Dataset. This dataset [5,23, 24] contains 17 individuals. Simultaneous EEG-fMRI recordings were performed while the subjects laid down. Stimuli of auditory and visual nature were given to the subjects, which makes this a stimuli-based dataset. The fMRI imaging acquisition was made with a 3T Philips Achieva MR Scanner with: single channel send and receive head coil, EPI sequence, 170 TRs per run with a TR of 2000 ms and 25 ms TE, 170 TRs per run with a $3 \times 3 \times 4$ mm voxel size and 32 slices with no slice gap. For a more detailed description of the dataset please refer to [23]. The dataset is available at https://legacy.openfmri.org/dataset/ds000116/. Each individual recording is divided into 12 equally sized time series of fMRI volumes, each time series is 28 s long, sampled at 2 s period. The *training* set is composed of 12 individuals and the *test set* 5 individuals per fold.

Validation. In both datasets, the 6-Fold cross validation schema introduced in Sect. 3.3 is applied. One might argue that as datasets contain multiple individuals and not a single subject on the same scanner, it might be difficult to fit the correlation matrix, due to different line ups and brain sizes. To tackle it, we align and downsample the fMRI spatial resolution by a factor of 6, going from approximately 30K voxels to 100 voxels to represent the whole brain.

5.3 Random Value and Region Removal

This work performs missing data imputation using a supervised learning method. To guarantee an objective assessment, we operate on a complete dataset [6,7] and generate missing data using two distinct procedures: random value removal and random region removal, where the last captures the spatially correlated nature of artifacts. Random region removal can be further used to assess the applicability of supervised principles of imputation to facilitate the synthesis of images (e.g. EEG-to-fMRI). The occurrence of a missing on a voxel from a certain fMRI instance generally implies the absence of all values for that voxel along the time axis, $t = (0, ..., T-1)$. As such, We do not consider differentiated random removal across time frames (i.e. removing a different set of voxels for each time frame). The random value removal strategy generates missings from uniform space distribution. To remove random regions, a value is chosen at random and removed along with its adjacent values. The number of adjacent values is set by a removal rate, r, indicating the number of values to remove per region or, in alternative, by a maximum value of adjacent values. Adjacencies are identified from the 3-Dimensional fMRI voxel coordinates.

5.4 Evaluation Metrics

For comparing results, Mean Absolute Error (MAE) and Root Mean Squared Error (RMSE) are suggested, with RMSE as the preferred option for penalizing large imputation errors. Residuals are also subjected to statistical testing. Considering N individuals, S recordings per individual, and M missing voxels,

$$e = \frac{1}{N \times S} \sum_{i=0}^{N \times S} \text{RMSE}_i, \quad \text{with} \quad \text{RMSE}_i = \sqrt{\frac{\sum_{v=0}^{M} d(y_v, \hat{y}_v)^2}{M}},$$

where v is a missing voxel and d is the Manhattan distance (absolute differences).

6 Experiment Results

Table 2 presents the gathered results from assessing the imputation methods (Sect. 5.1) over the target datasets (Sect. 5.2) using the random region removal strategy (Sect. 5.3). The results outline the relevance of applying a recurrent layer D for the time-sensitive regularization of spatially imputed signals in comparison with Context Encoder and MICE alternatives. MICE does not scale with increases on missing rate, due to its inability to deal with features that have not been observed before. The chained imputation principle ($\Phi + D$) further shows slight improvements against weakly-correlated signals under a Dropout+D architecture, indicating the importance of identifying voxel priorities. Considering the general performance limits of kNN ($k = 3$) and DTW-based barycenter, the results further motivate the difficulty of the task at hands (Table 1).

Table 1. RMSE on time axis results on the NODDI EEG-fMRI Test Set.

Missing	10%	25%	50%	75%	90%
kNN	1.00 ± 0.79	0.99 ± 0.79	1.02 ± 0.84	1.01 ± 0.85	1.05 ± 0.93
Barycenter [18]	1.02 ± 0.81	1.02 ± 0.81	1.12 ± 0.92	1.10 ± 0.89	1.16 ± 0.94
Mean	1.01 ± 0.80	0.99 ± 0.79	1.03 ± 0.86	1.01 ± 0.85	1.02 ± 0.87
CE [17]	1.01 ± 0.78	1.01 ± 0.74	1.06 ± 0.80	1.04 ± 0.76	1.06 ± 0.76
MICE [26]	1.02 ± 0.80	1.02 ± 0.80	1.15 ± 0.93	4.63 ± 5.50	1.11 ± 0.93
Dropout	1.01 ± 0.80	1.02 ± 0.80	1.05 ± 0.84	1.00 ± 0.82	1.00 ± 0.84
Dropout+D	**0.81** ± 0.65	**0.85** ± 0.68	**0.94** ± 0.76	**0.95** ± 0.75	**1.00** ± 0.84
Φ	1.01 ± 0.80	1.01 ± 0.80	**1.04** ± 0.85	**1.00** ± 0.85	**1.02** ± 0.88
$\Phi + D$	**0.81** ± 0.65	**0.84** ± 0.67	**1.00** ± 0.83	**1.02** ± 0.87	**1.03** ± 0.89

Table 2. RMSE on time axis results on the Oddball EEG-fMRI Test Set.

Missing	10%	25%	50%	75%	90%
kNN	1.21 ± 0.86	1.21 ± 0.84	1.19 ± 0.81	1.11 ± 0.71	1.05 ± 0.64
Barycenter [18]	1.23 ± 0.88	1.25 ± 0.87	1.26 ± 0.85	1.23 ± 0.80	1.22 ± 0.77
Mean	1.21 ± 0.84	1.17 ± 0.80	1.14 ± 0.77	$\mathbf{1.05} \pm 0.66$	1.01 ± 0.61
CE [17]	$\mathbf{1.09} \pm 0.66$	$\mathbf{1.08} \pm 0.66$	$\mathbf{1.03} \pm 0.62$	$\mathbf{1.03} \pm 0.62$	$\mathbf{1.00} \pm 0.60$
MICE [26]	1.24 ± 0.86	1.27 ± 0.88	1.28 ± 0.86	3.33 ± 3.57	16.97 ± 21
Dropout	1.23 ± 0.86	1.22 ± 0.84	1.21 ± 0.82	1.11 ± 0.71	1.04 ± 0.64
Dropout+D	$\mathbf{0.72} \pm 0.44$	$\mathbf{0.81} \pm 0.50$	$\mathbf{0.90} \pm 0.57$	$\mathbf{0.93} \pm 0.58$	1.06 ± 0.65
Φ	$\mathbf{1.23} \pm 0.87$	$\mathbf{1.22} \pm 0.84$	$\mathbf{1.18} \pm 0.80$	$\mathbf{1.06} \pm 0.67$	$\mathbf{1.02} \pm 0.61$
$\Phi + D$	$\mathbf{0.70} \pm 0.43$	$\mathbf{0.80} \pm 0.49$	$\mathbf{0.87} \pm 0.55$	$\mathbf{0.98} \pm 0.58$	$\mathbf{0.99} \pm 0.58$

Results gathered using the alternative random value removal strategy (Sect. 5.3) are provided in *Appendix A*. Generally, these appended results yield similar ranks among the compared methods. Please also refer to *Appendix B* for results collected using alternative residue-based scores that offer complementary information on the spatial adequacy of the assessed methods, further supporting the relevance of the proposed $\Phi + D$ imputation approach.

Figure 2 provides a complementary view on the performance of imputation methods for the first dataset when considering a varying brain volume under analysis. The gathered results evidence the superiority of the proposed approach and suggest that performance is independent of the spatial extent.

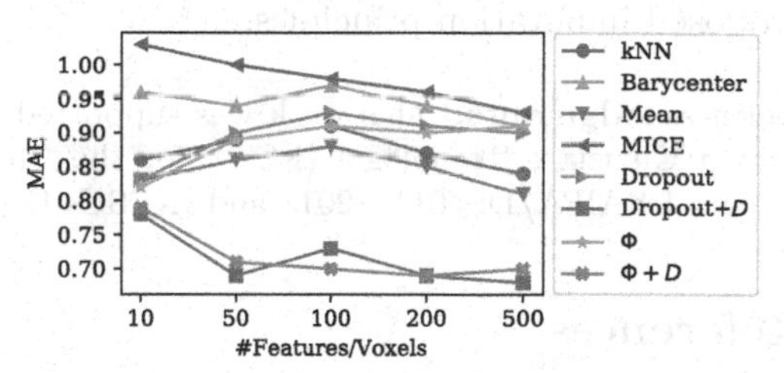

Fig. 2. MAE for varying brain volumes on the NODDI testing data under a 50% missing rate.

Figure 3 illustrates the denoising property of D on a randomly selected missing voxel – with coordinates [14, 29, 14] – from the first dataset. It compares, side by side, the error of a single time series imputed spatially, Φ, with the error of an imputed signal with time regularization, $\Phi + D$. This image, together with results (Table 2), show the importance of this recurrent layer to capture the neurophysiological temporal patterning of the signal.

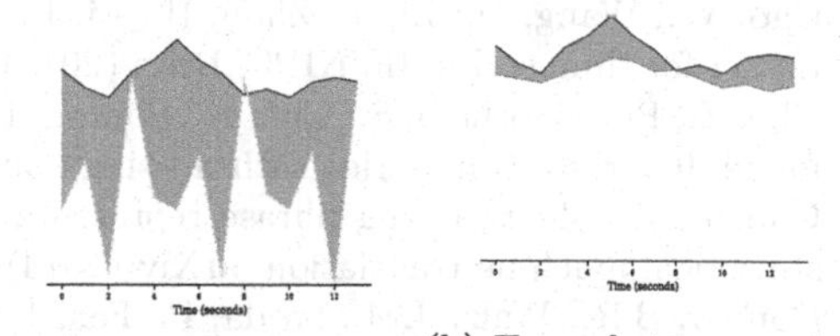

(a) Error between original signal and spatially imputed signal using Φ

(b) Error between original signal and time regularized signal using $\Phi + D$

Fig. 3. Impact of time regularization on the imputation of voxel with coordinates (14, 29, 14).

7 Concluding Remarks

The rich spatiotemporal nature of neuroimaging modalities such as fMRI, together with their high susceptibility to noise artifacts, create unique difficulties for missing data imputation for both resting-state and stimuli-induced settings [1]. This work highlights the role of combining time regularization, D, with expedite spatial imputation methods to achieve significant improvements on neuroimaging data with variable amount and structures of missing values. We further presented a chained imputation method applied in a neural network setting which achieves state-of-the-art results. In fMRI stimuli-induced settings, the importance of iterative imputation of missing time series in accordance with their pre-computed priority is highlighted by the stable performance of the Φ layer, even when the missing rate increases. Performing imputation all at once is illustrated by the Dropout baseline. Φ has the advantage of leveraging information from the already imputed information, showing robustness as the missing rate increases, aided by the removal of spatial prediction noise from the D layer. This stable performance is particularly interesting given the heightened differences between resting state and stimuli based fMRI [25]. As future work, we intend to extend the proposed imputation approach to aid modality transfer. The complexity of devising end-to-end approaches for image synthesis from modalities with different spatiotemporal resolutions can be guided under the proposed imputation principles.

Acknowledgments. This work was supported by national funds through Fundação para a Ciência e Tecnologia (FCT), for the Ph.D. Grant DFA/BD/5762/2020, ILU project DSAIPA/DS/0111/2018 and INESC-ID pluriannual UIDB/50021/2020.

References

1. Birn, R.M.: The role of physiological noise in resting-state functional connectivity. Neuroimage (2012)
2. Cao, W., Wang, D., Li, J., Zhou, H., Li, L., Li, Y.: Bidirectional recurrent imputation for time series. In: NIPS, Brits (2018)
3. Che, Z., Purushotham, S., Cho, K., Sontag, D., Liu, Y.: Recurrent neural networks for multivariate time series with missing values. Scientific reports (2018)
4. Cho, K., et al.: Learning phrase representations using RNN encoder-decoder for statistical machine translation. arXiv (2014)
5. Conroy, B.R., Walz, J.M., Sajda, P.: Fast bootstrapping and permutation testing for assessing reproducibility and interpretability of multivariate fMRI decoding models. PLoS ONE (2013)
6. Deligianni, F., Carmichael, D.W., Zhang, G.H., Clark, C.A., Clayden, J.D.: Noddi and tensor-based microstructural indices as predictors of functional connectivity. PLoS ONE (2016)
7. Deligianni, F., Centeno, M., Carmichael, D.W., Clayden, J.D.: Relating resting-state fMRI and EEG whole-brain connectomes across frequency bands. Front. Neurosci. (2014)

8. Fortuin, V., Baranchuk, D., Rätsch, G., Mandt, S.: GP-VAE: deep probabilistic time series imputation. arXiv (2019)
9. Goodfellow, I., et al.: Generative adversarial nets. In NIPS, Sherjil Ozair (2014)
10. Hastie, T., Tibshirani, R., Friedman, J.: The Elements of Statistical Learning. Springer, New York (2001)
11. Kingma, D.P., Ba, J.: Adam: a method for stochastic optimization. arXiv (2014)
12. Kingma, D.P., Welling, M.: Auto-encoding variational Bayes. arXiv (2013)
13. Lu, R., Duan, Z.: Bidirectional GRU for sound event detection. Detection and Classification of Acoustic Scenes and Events (2017)
14. Luo, Y., Cai, X., Zhang, Y., Xu, J., et al.: Multivariate time series imputation with generative adversarial networks. In: NIPS (2018)
15. Luo, Y., Cai, X., Zhang, Y., Xu, J., Xiaojie, Y.: Multivariate time series imputation with generative adversarial networks. In: NIPS (2018)
16. Pan, J.-Y., Yang, H.-J., Faloutsos, C., Duygulu, P.: Automatic multimedia cross-modal correlation discovery. In: ACM SIGKDD (2004)
17. Pathak, D., Krahenbuhl, P., Donahue, J., Darrell, T., Efros, A.A.: Context encoders: feature learning by inpainting. In: CVPR (2016)
18. Petitjean, F., Ketterlin, A., Gançarski, P.: A global averaging method for dynamic time warping, with applications to clustering. Pattern Recogn. (2011)
19. Śmieja, M., Struski, Ł., Tabor, J., Zieliński, B., Spurek, P.: Processing of missing data by neural networks. In: NIPS (2018)
20. Snoek, J., Larochelle, H., Adams, R.P.: Practical Bayesian optimization of machine learning algorithms. In: NIPS (2012)
21. Srivastava, N., Hinton, G., Krizhevsky, A., Sutskever, I., Salakhutdinov, R.: Dropout: a simple way to prevent neural networks from overfitting. JMLR (2014)
22. Tran, L., Liu, X., Zhou, J., Jin, R.: Missing modalities imputation via cascaded residual autoencoder. In: CVPR (2017)
23. Walz, J.M., Goldman, R.I., Carapezza, M., Muraskin, J., Brown, T.R., Sajda, P.: Simultaneous EEG-fMRI reveals temporal evolution of coupling between supramodal cortical attention networks and the brainstem. J. Neurosci. (2013)
24. Walz, J.M., Goldman, R.I., Carapezza,, M., Muraskin, J., Brown, T.R., Sajda, P.: Simultaneous eeg-fmri reveals a temporal cascade of task-related and default-mode activations during a simple target detection task. Neuroimage (2014)
25. Wehrl, H.F., et al.: Simultaneous pet-MRI reveals brain function in activated and resting state on metabolic, hemodynamic and multiple temporal scales. Nature Med. (2013)
26. White, I., Royston, P., Wood, A.: Multiple imputation using chained equations: issues and guidance for practice. Stat. Med. (2011)

Bayesian Deep Active Learning for Medical Image Analysis

Biraja Ghoshal(✉), Stephen Swift, and Allan Tucker

Brunel University London, Uxbridge UB8 3PH, UK
biraja.ghoshal@brunel.ac.uk
https://www.brunel.ac.uk/computer-science

Abstract. Deep Learning has achieved a state-of-the-art performance in medical imaging analysis but requires a large number of labelled images to obtain good adequate performance. However, such labelled images are costly to acquire in time, labour, and human expertise. We propose a novel practical Bayesian Active Learning approach using Dropweights and overall bias-corrected uncertainty measure to suggest which unlabelled image to annotate. Experiments were done on Brain Tumour MR images, Microscopic Cell Image classification, Fluoro-chromogenic cytokeratin-Ki-67 double staining cancer images and Retina fundus image segmentation tasks. We demonstrate that our active learning technique is equally successful or better than other existing active learning approaches in high dimensional data to reduce the image labelling effort significantly. We believe Bayesian deep active learning framework with very few annotated samples in a practical way will benefit clinicians to obtain fast and accurate image annotation with confidence.

Keywords: Bayesian Active Learning · Bias-corrected uncertainty · Dropweights · Image annotation · Semantic segmentation, classification

1 Motivation

Artificial Intelligence based medical diagnosis requires a large number of many labelled images to obtain a good performance. In the real-world, unlabelled medical images are available in abundance, however annotating the data with reliable class labels after careful inspection of numerous images can be very tedious, time-consuming, and expensive, as well as being subject to errors on the interpreter's part. Active learning is a mechanism that tries to minimise the amount of labelled data required to control the labelling process. Thus, developing active learning algorithms to learn from a small sample, high-dimensional labelled images, querying the highly informative unlabelled images, and minimising redundant examples with limited resources is of paramount practical importance.

There are many heuristic methods and numerous query strategies in active learning for medical image classification using traditional machine learning [1,6,9,12,13,16,17]. In deep active learning, uncertainty based acquisition function is heavily influenced by an average of the softmax probability values and

A. Tucker et al. (Eds.): AIME 2021, LNAI 12721, pp. 36–42, 2021.
https://doi.org/10.1007/978-3-030-77211-6_4

miscalibrated due to the diverse nature of the medical image samples, disease conditions and sampling bias.

We designed a novel Active Learning sample selection strategy for high dimensional image data to measure the confidence of the model uncertainty in classification and unbiased calibrated uncertainty weighted by the Euclidean Distance Transform (EDT) of the prediction for semantic image segmentation. A sample is selected based on the lowest uncertainty confidence score for labelling as highly informative and little redundancy. Using this metric can significantly reduce the number of labelled samples required compared to other selection strategies whilst achieving higher accuracy. In semantic segmentation, we have quantified aleatoric uncertainty and epistemic uncertainty by leveraging the functional relationship between the mean and variance of multinomial predictive probabilities from Bayesian neural networks (Fig. 1).

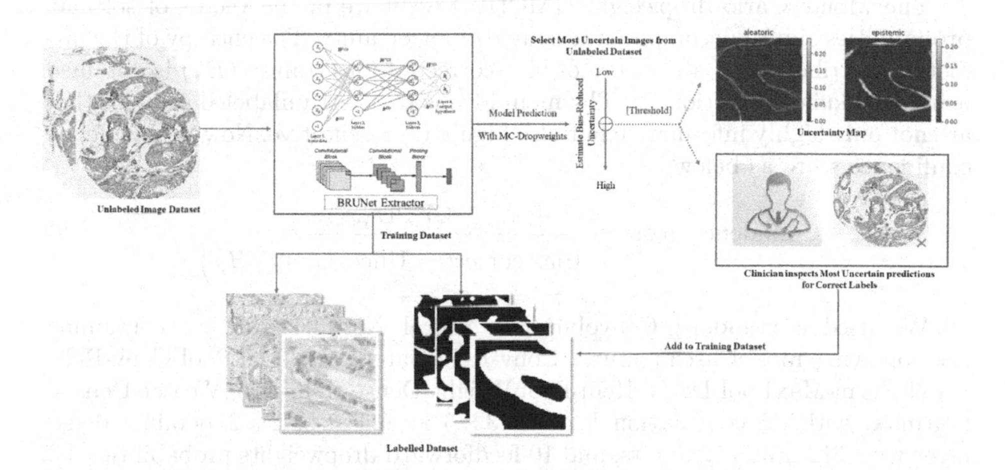

Fig. 1. Example - Active Learning Framework: At each round of active learning, the algorithm computes bias-corrected confidence score of uncertainty for all images in the unlabelled pool. Images with the least score value of uncertainty confidence is selected for the clinician to label, and then the corresponding images are added to the training set in the next round of the model training. Our method relies on the bias-corrected confidence score of uncertainty sampling, in which the algorithm selects the unlabelled images that it finds manually hardest to annotate.

2 Proposed Method

Active learning depends on the ability to select the right sample to be annotated to improve model performance and decrease model uncertainty. Therefore, defining the acquisition function is a real challenge. The most popular Dropout Bayesian Active Learning by Disagreemement (BALD) [5] maximises the mutual information between predictions and model posterior [9]. However it double counts Mutual Information (MI) between data points and overestimates the true MI. Estimation of entropy from the finite set of data suffers from a severe downward bias when the data is under-sampled.

2.1 Estimating Confidence in Image Classification

We employed the maximum Class Predictive Probability Distance (CPPD), which is the highest predictive probability value amongst the class probabilities as a measure of a representativeness heuristic. In a multi-class setting, given dataset $D = \{(x_n, y_n)\}_{n=1}^N$, the vector of softmax probabilities $\hat{y}_t = \text{Softmax}(f^{\hat{\theta}_t}(\hat{x}))$ obtained after the tth stochastic forward pass is denoted $p\left(\hat{y}_t|x^*, \hat{\theta}_t\right)$, where $\hat{\theta}_t$ denotes the sampled parameters resulting from DropWeights. Thus, the class probabilities of estimates are given by $\frac{1}{T}\sum_{t=1}^{T} p\left(\hat{y}_t|x^*, \hat{\theta}_t\right)$.

We obtained the Class Predictive Probability Distance (CPPD):

$$CPPD(x_i \in D_u) = p(y_{Best}|x_i) - p(y_{Second-Best}|x_i) \tag{1}$$

The Monte Carlo dropweights (MCDW) estimate of the vector of softmax probabilities aim to decompose the source of uncertainty. The entropy of the predictive distribution as a measure of bias-corrected uncertainty ($\hat{H}_J$) is obtained using Jackknife correction [7]. The main idea is to select unlabeled samples that are not only highly informative but also highly representative. Now we can define confidence score as below:

$$\text{Confidence Score} = \frac{\text{CPPD}\,(x_i)}{\text{Bias-corrected Uncertainty}\left(\hat{H}_J\right)} \tag{2}$$

We used a standard Convolutional Neural Networks (CNN) containing the following model architecture: Conv-Relu-BatchNorm-MaxPool-Conv-Relu-BatchNorm-MaxPool-Dense-Relu-DropWeight-Dense-Relu-DropWeight-Dense-Softmax, with 32 convolution kernels, 3×3 kernel size, 2×2 pooling, dense layer with 512 units, 128 units, and 10 feedforward dropweights probabilities 0.3. We optimised the model using Adam optimiser with the default learning rate of 0.001. We explored baseline acquisition functions (Random selection, BALD and Max Entropy) for image data classification on two datasets (Brain Tumor and Malaria).

2.2 Calibrated Uncertainty in Semantic Image Segmentation

The uncertainty obtained by Bayesian Neural Network [8] is prone to miscalibration, i.e. perfectly segmented image could have a higher uncertainty. The proposed method is presented threefold: First, uncertainty obtained by dropweights variational inference; second, pixel-wise estimated uncertainty is weighted by the Euclidean Distance Transform (EDT) [4,10] to standardise the importance of the pixel and so reduce overconfident prediction errors in dense pixels regions; and finally, calibrated uncertainty is compared with by random sample selection.

In semantic segmentation tasks, we demonstrated that the uncertainty estimates obtained from dropweights using the Bayesian residual U-Net (BRUNet) provide additional insight for clinicians with help from deep learners [8].

3 Application of Active Learning for Medical Image Analysis

3.1 Image Classification

A. Magnetic Resonance Imaging (MRI) Based Brain Tumor Classification: In order to validate the effectiveness of 'Confidence Score' in deep learning model accuracy and robustness of our proposed approach, we performed experiments on Brain MRI scan images of 3 brain tumour types (Astrocytoma, Glioblastoma, Oligodendroglioma) with an additional 2 categories (Healthy brain MRI and Unidentified tumour) obtained from three sources (REMBRANDT, MIRIAD and BRAINS) [2,3]. As shown in Fig. 2(a), our "Confidence Score" has performed better than Bayesian Max Entropy, and Random acquisitions function on the brain tumour dataset.

B. Microscopic Cell Image Classification: We performed experiments on Giemsa-stained microscopic cell image [11], using our Active learning framework. However, when we observed the results for the malaria parasite in thin blood smear images (Fig. 2(b)), which has a more realistic dataset with larger query batch sizes. We present here an advantage of our method for a smaller batch size over all other sampling methods.

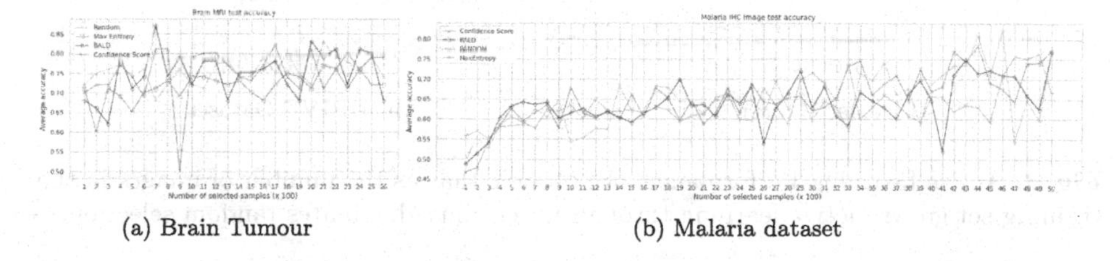

(a) Brain Tumour (b) Malaria dataset

Fig. 2. (a) Brain Tumour test accuracy as a function of the number of acquired images [Initial Training Size: 100; Pool Size: 5000; Batch Size: 10, AL Iterations: 50] (b) Malaria dataset test accuracy as a function of the number of acquired images [Pool Size: 5511; Initial Training Size: 20; Batch Size: 100, AL Iterations: 50]

3.2 Semantic Segmentation

Estimated aleatoric and epistemic uncertainty add additional insights to understanding the intrinsic uncertainty in deep learning. The value of each pixel represents the variance computed on MC samples. We used Dice loss function to train the network to highlight contrasting areas of the image and weighted uncertainty by distance transformation normalisation to address unreliable uncertainty mainly on class boundaries. Figure 3 and Fig. 4 illustrate the uncertainty map evolutions over active learning iterations for the Epithelial Cells in Breast Cancers and Retina Fundus images semantic segmentation.

A. Cytokeratin-Supervised Epithelial Cells in Breast Cancers Semantic Segmentation: Immunohistochemistry (IHC) staining's of oestrogen receptor (ER), progesterone receptor (PR), and proliferation antigen Ki-67 are routinely used for automated epithelial cell detection in breast cancer diagnostics. Manual annotation of complex IHC images to determine the proportion of non-malignant stromal or inflammatory cells in stained cells is extremely tedious, expensive and may lead to errors or inter-observer variability. Dataset included images from 152 patient samples stained with fluoro-chromogenic cytokeratin-Ki-67 double staining and sequential hematoxylin-IHC [15].

B. Digital Retinal Images for Blood Vessel Extraction (DRIVE) Semantic Segmentation: Diabetic retinopathy (DR) is one of the reasons for vision loss in diabetic patients due to the retinal blood vessels damage. Automatic detection and segmentation of retina fundus images are essential to prevent

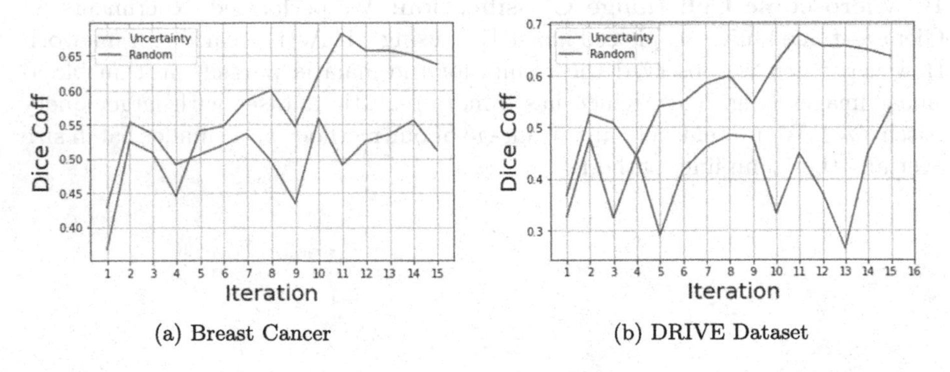

(a) Breast Cancer (b) DRIVE Dataset

Fig. 3. Active Learning Performance: We observe that as the number of images in the training set grows, active learning through uncertainty dominates random selection.

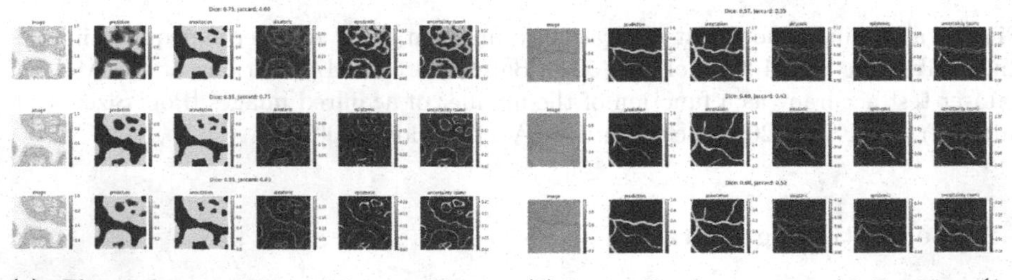

(a) Fluoro-chromogenic cytokeratin-Ki-67 Breast Cancers - AL iterations (3, 7 and 14) (b) Retina Fundus images - AL iterations (1, 5 and 10).

Fig. 4. Prediction with estimated aleatoric and epistemic uncertainty maps [8]. (a) We observe that the Dice coefficient increases as active learning iterations progress with more training images. (b) We observe that less variance is on thin boundary pixels and the model seems to be more confident where it can distinguish line shapes vs. round shapes.

vision loss in diabetic patients. The DRIVE database contains 40 images using 400 diabetic patients (seven of them have various pathological cases) [14]. All images also have corresponding manually segmented masks. The images have been divided into training and test set. Each part contains 20 images. We use testing images for performance evaluation. All images were resized to 96×96 pixels.

4 Conclusion and Future Research

We present a novel practical Bayesian deep active learning framework. It can significantly improve classification and semantic segmentation performances. The heatmaps of aleatoric and epistemic uncertainty in semantic segmentation along with prediction would help clinicians better understand spatial relations in images and where the model tends to fail the most. Future research includes an evaluation in which all samples are learned and when the model reaches the optimal learning performance level.

References

1. Aggarwal, C., Kong, X., Gu, Q., Han, J., Yu, P.: Active learning: a survey. In: Aggarwal, C.C. (ed.) Data Classification: Algorithms and Applications, pp. 571–606. CRC Press (2014)
2. Cheng, J., et al.: Retrieval of brain tumors by adaptive spatial pooling and fisher vector representation. PLoS ONE **11**(6), e0157112 (2016)
3. Cheng, J., et al.: Enhanced performance of brain tumor classification via tumor region augmentation and partition. PLoS ONE **10**(10), e0140381 (2015)
4. Di Scandalea, M.L., Perone, C.S., Boudreau, M., Cohen-Adad, J.: Deep active learning for axon-myelin segmentation on histology data. arXiv preprint arXiv:1907.05143 (2019)
5. Gal, Y., Ghahramani, Z.: Dropout as a Bayesian approximation: representing model uncertainty in deep learning. In: 33rd International Conference on Machine Learning, ICML 2016, vol. 3, pp. 1651–1660 (2016)
6. Gal, Y., Islam, R., Ghahramani, Z.: Deep Bayesian active learning with image data. arXiv preprint arXiv:1703.02910 (2017)
7. Ghoshal, B., Lindskog, C., Tucker, A.: Estimating uncertainty in deep learning for reporting confidence: an application on cell type prediction in testes based on proteomics. In: Berthold, M.R., Feelders, A., Krempl, G. (eds.) IDA 2020. LNCS, vol. 12080, pp. 223–234. Springer, Cham (2020). https://doi.org/10.1007/978-3-030-44584-3_18
8. Ghoshal, B., Tucker, A., Sanghera, B., Lup Wong, W.: Estimating uncertainty in deep learning for reporting confidence to clinicians in medical image segmentation and diseases detection. Comput. Intell. (2020)
9. Houlsby, N., Huszár, F., Ghahramani, Z., Lengyel, M.: Bayesian active learning for classification and preference learning. Stat **1050**, 24 (2011)
10. Ma, J., et al.: How distance transform maps boost segmentation CNNs: an empirical study. In: Arbel, T., Ayed, I.B., de Bruijne, M., Descoteaux, M., Lombaert, H., Pal, C. (eds.) Medical Imaging with Deep Learning. Proceedings of Machine Learning Research, vol. 121, pp. 479–492. PMLR (2020). http://proceedings.mlr.press/v121/ma20b.html

11. Rajaraman, S., et al.: Pre-trained convolutional neural networks as feature extractors toward improved malaria parasite detection in thin blood smear images. PeerJ **6**, e4568 https://doi.org/10.7717/peerj.4568 (2018)
12. Ren, P., Xiao, Y., Chang, X., Huang, P.Y., Li, Z., Chen, X., Wang, X.: A survey of deep active learning. arXiv preprint arXiv:2009.00236 (2020)
13. Settles, B.: Active learning literature survey. Computer Sciences Technical report 1648, University of Wisconsin, Department of Computer Science (2009)
14. Staal, J., Abramoff, M.D., Niemeijer, M., Viergever, M.A., Van Ginneken, B.: Ridge-based vessel segmentation in color images of the retina. IEEE Trans. Med. Imaging **23**(4), 501–509 (2004)
15. Valkonen, M., et al.: Cytokeratin-supervised deep learning for automatic recognition of epithelial cells in breast cancers stained for ER, PR, and Ki-67. IEEE Trans. Med. Imaging **39**(2), 534–542 (2019)
16. Wang, H., Yeung, D.Y.: Towards Bayesian deep learning: a survey. arXiv preprint arXiv:1604.01662 (2016)
17. Wang, K., Zhang, D., Li, Y., Zhang, R., Lin, L.: Cost-effective active learning for deep image classification. IEEE Trans. Circ. Syst. Video Technol. **27**(12), 2591–2600 (2016)

A Topological Data Analysis Mapper of the Ovarian Folliculogenesis Based on MALDI Mass Spectrometry Imaging Proteomics

Giulia Campi[1(✉)], Giovanna Nicora[1], Giulia Fiorentino[2], Andrew Smith[3], Fulvio Magni[3], Silvia Garagna[2], Maurizio Zuccotti[2], and Riccardo Bellazzi[1]

[1] Department of Electrical, Computer and Biomedical Engineering, University of Pavia, 27100 Pavia, Italy
giulia.campi01@universitadipavia.it

[2] Department of Biology and Biotechnology "Lazzaro Spallanzani", University of Pavia, 27100 Pavia, Italy

[3] Department of Medicine and Surgery, University of Milano-Bicocca, 20854 Vedano al Lambro, MB, Italy

Abstract. Matrix-Assisted Laser Desorption/Ionization Mass Spectrometry Imaging (MALDI-MSI), also referred to as molecular histology, is an emerging omics, which allows the simultaneous, label-free, detection of thousands of peptides in their tissue localization, and generates highly dimensional data. This technology requires the development of advanced computational methods to deepen our knowledge on relevant biological processes, such as those involved in reproductive biology.

The mammalian ovary cyclically undergoes morpho-functional changes. From puberty, at each ovarian cycle, a group of pre-antral follicles (type 4, T4) is recruited and grows to the pre-ovulatory (T8) stage, until ovulation of mature oocytes. The correct follicle growth and acquisition of oocyte developmental competence are strictly related to a continuous, but still poorly understood, molecular crosstalk between the gamete and the surrounding follicle cells.

Here, we tested the use of advanced clustering and visual analytics approaches on MALDI-MSI data for the *in-situ* identification of the protein signature of growing follicles, from the pre-antral T4 to the pre-ovulatory T8. Specifically, we first analyzed follicles MALDI-MSI data with PCA, tSNE and UMAP approaches, and then we developed a framework that employs Topological Data Analysis (TDA) Mapper to detect spatial and temporal related clusters and to pinpoint differentially expressed proteins. TDA Mapper is an unsupervised Machine Learning method suited to the analysis of high-dimensional data that are embedded into a graph model. Interestingly, the graph structure revealed protein patterns in clusters containing different follicle types, highlighting putative factors that drive follicle growth.

Keywords: Topological Data Analysis · Histology · MALDI-MSI

A. Tucker et al. (Eds.): AIME 2021, LNAI 12721, pp. 43–47, 2021.
https://doi.org/10.1007/978-3-030-77211-6_5

1 Introduction

Matrix-Assisted Laser Desorption/Ionization Mass Spectrometry Imaging (MALDI-MSI) is a powerful tool for mapping the native tissue localization of peptides and has great potential for biological applications. MALDI-MSI combines classical histology and molecular analysis, offering the flexibility of the simultaneous and label-free detection of a variety of molecules, including peptides, over a tissue section sampled in a grid of square regions (spots). In a histological section, mass spectra are recorded for all the spots, thus generating molecular images that map peptide distribution over the inspected regions. A single experiment generates highly dimensional data and can produce up to 50.000 spectra, each representing intensities measured at a large number of *m/z* bins and describing different peptides [1]. Raw spectra pre-processing (i.e., normalization, baseline correction, smoothing and peak picking) is essential for data analysis, although the intrinsic dimensionality of MSI data, containing both molecular and spatial information of each spot, still represents a challenge. For this reason, computational and statistical methods are essential for the identification of relevant protein signatures. Among unsupervised machine learning algorithms, factorization methods (including PCA), clustering approaches and manifold learning methods (such as tSNE and UMAP), as well as deep learning approaches, are widely applied for dimensionality reduction and MSI data analytics [2, 3]. Yet, most of these methods are based on the similarities of mass spectra alone, considering the spectra as independent, and do not take spatial information into direct account.

Up to date, a handful of studies attempted to investigate the gonads using MALDI-MSI in either physiological or pathological conditions [4], highlighting the potentialities of this approach to reveal molecular actors crucial for the correct function of the testis or ovary. The ovary is a dynamic organ, which houses folliculogenesis, the cyclic growth of the follicle together with the maturation of the enclosed oocyte. In the mouse, folliculogenesis progresses from primordial type 1 (T1) follicle, through the pre-antral (T4-T5), antral (T6-T7) and up to the fully-grown follicle (T8), ready to ovulate a mature oocyte [5]. Follicle growth and acquisition of oocyte developmental competence are strictly regulated by a complex, and still poorly understood, molecular crosstalk between the gamete and the surrounding follicle cells [6].

Here, we tested the use of advanced clustering and visual analytics approaches on MALDI-MSI data collected from mouse ovarian histological sections for the *in-situ* identification of the protein signature of growing follicles, from the pre-antral T4 to the pre-ovulatory T8 stage. We first applied dimensionality reduction methods, and then we developed a framework that uses Topological Data Analysis (TDA) Mapper. TDA Mapper is an emerging method based on algebraic topology and geometry, which allows pattern and shape recognition and that was successfully applied in the biomedical context [7]. The algorithm produces a graph by slicing the data-space in overlapping bins and performing clustering within each: the resulting clusters are then linked together to recover a simplified yet structured picture of the data topology [8]. Applied to folliculogenesis, TDA builds a continuous shape on top of the data, representing a "follicle-growth space", allowing for the analysis of follicle development as a continuum, in which data relative to different stages are connected. In our study, the TDA graph highlighted homogeneous clusters, across which enrichment on differentially expressed proteins was performed.

2 Exploratory Analysis and Dimensionality Reduction

We analyzed over 29.000 mass spectra of 550 different tissue sections acquired with MALDI-MSI from 59 follicles (from T4 to T8 stage). Pre-processed spectra are characterized by 55 different *m/z* values, corresponding to 94 putative proteins.

First, for visualization and dimensionality reduction purposes, we applied PCA, tSNE and UMAP to the mass spectra, each relative to a different measured spot, embedding data in three-dimensional reduced spaces. Multiple values for tSNE and UMAP parameters have been tested to achieve the optimal data embedding. The first three PCA components explained 70% of the total variance. Inspecting the loadings allowed for the identification of the most important proteins in calculating the components (Fig. 1D). However, none of these methods extracted clusters representative of single follicle types (Fig. 1A–C).

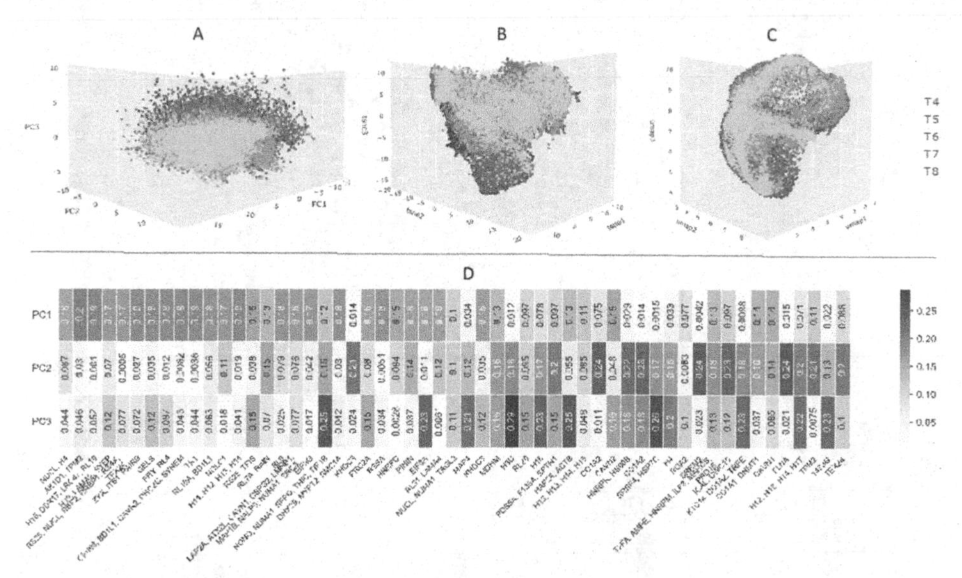

Fig. 1. Embedding obtained with (A) PCA, (B) tSNE with perplexity = 100, (C) UMAP with 250 neighbors and cosine distance metric. (D) Proteins importance in PCA in terms of the contribution of each *m/z* (corresponding to one or more proteins) to the three PCs.

3 TDA Mapper Approach

Then, the application of the TDA Mapper algorithm allowed to account for spots spatial dependency, by considering each tissue section as a k-dimensional image (being $k = 55$, the number of *m/z* bins). TDA Mapper requires the definition of a distance metric, one or more filter functions to project data into a lower dimensional space, and a set of resolution parameters, specifying the number of bins built on the projections and the overlapping percentage for the covering. We explored a wide combination of these parameters. Interesting results were obtained using the Kullback-Leibler divergence as distance metric between sections images, and two-dimensional UMAP as filter function. We identified several nodes that group sections pertaining to the same follicle type, highlighting their spectral similarity (Fig. 2, red circles). However, such nodes are distributed all over the graph and not always directly connected. Then, the graph was used as starting point for the detection of communities through a modularity optimization algorithm [9], identifying groups of nodes with strong internal connections. Based on their entropy, we selected three communities, each associated to T4, T6 or T8 follicles (Fig. 2, blue boxes).

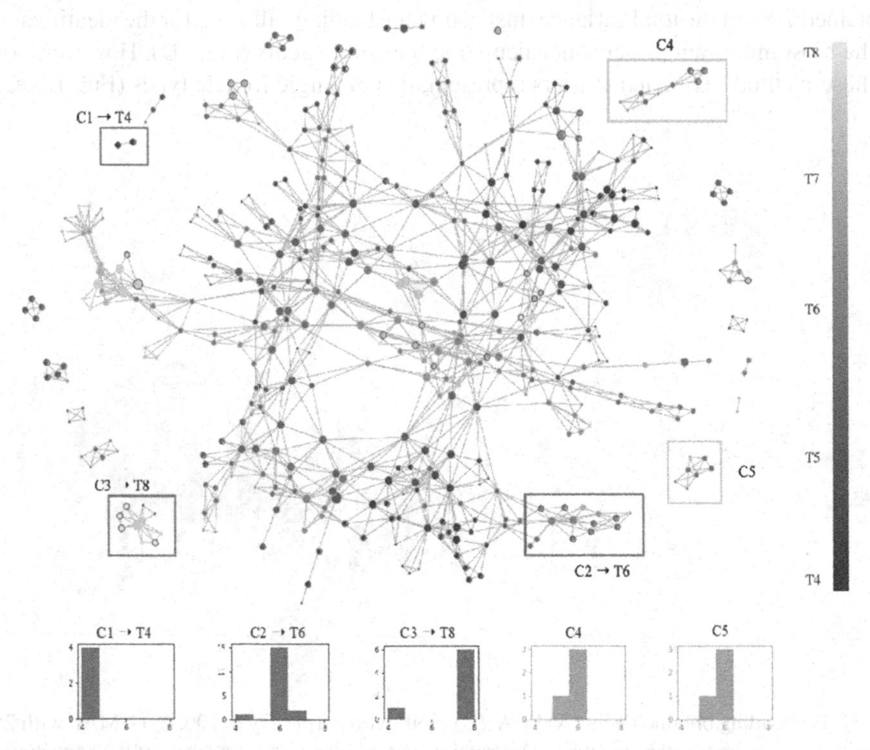

Fig. 2. TDA Mapper graph: each node is colored according to the clustered follicle sections. Red circles: nodes containing only sections from the same follicular types. Boxes: communities containing at least 4 distinct sections and with a maximum entropy of 0.68. Blue boxes: communities strongly associated with a specific follicular type (C1: T4, C2: T6, C3: T8). (Color figure online)

Enrichment analyses over the differentially expressed proteins (DEPs) across these three communities revealed matches with biological pathways (p-value ≤ 0.05). Specifically, DEPs between T4-T6 and T6-T8 communities overlapped with three pathways (*Establishment of Sister Chromatid Cohesion, Cohesin Loading onto Chromatin, Mitotic Telophase/Cytokinesis*), from the *Reactome_2016* set, associated with the process of cell division, crucial for follicle enlargement and growth. DEPs between T4-T8 communities matched the *Focal Adhesion* pathway from the *Kegg_2019_Mouse* set, particularly important for oocyte-follicle communication [10]. Also, T4-T8 comparison highlighted some proteins known to be involved in folliculogenesis: TPM2, contributing to follicle growth; NUCL, KHDC3, EIF3A, FLNA and ACTB, responsible for oocyte maturation and quality; ZYX, contributing to ovulation. Other interesting results are reported on GitHub (https://github.com/giuliacampi/MALDI_TDAMapper).

4 Conclusions

We analyzed MALDI-MSI data with advanced visual analytics and unsupervised machine learning techniques to identify proteins relevant during the ovarian folliculogenesis. Unsupervised techniques did not reveal clusters of data that could be linked to a follicular type. In comparison with standard techniques that treat each tissue spot as independent, our approach considered the spatial dependency of mass spectra by organizing data in k-dimensional images. Although the complexity of the underlying biological processes and the unavoidable data noise hampered the direct association of communities with single follicle types, the TDA Mapper resulted in a graph connecting clusters of sections, among which we identified interesting communities that reflect important stages of follicle growth. Enrichment analyses on communities DEPs confirmed the presence of proteins known to be important in folliculogenesis, highlighting the potentiality of this preliminary approach. Future work will involve the analysis of more data and the possibility to use other advanced techniques, such as deep learning.

References

1. Alexandrov, T.: MALDI imaging mass spectrometry: statistical data analysis and current computational challenges. BMC Bioinform. **13**, S11 (2012)
2. McCombie, G., et al.: Spatial and spectral correlations in MALDI mass spectrometry images by clustering and multivariate analysis. Anal. Chem. **77**, 6118–6124 (2005)
3. Behrmann, J., et al.: Deep learning for tumor classification in imaging mass spectrometry. Bioinforma. Oxf. Engl. **34**, 1215–1223 (2018)
4. Lagarrigue, M., et al.: Matrix-assisted laser desorption/ionization imaging mass spectrometry: a promising technique for reproductive research. Biol. Reprod. **86** (2012)
5. Pedersen, T., Peters, H.: Proposal for a classification of oocytes and follicles in the mouse ovary. J. Reprod. Fertil. **17**, 555–557 (1968)
6. Zuccotti, M., et al.: What does it take to make a developmentally competent mammalian egg? Hum. Reprod. Update **17**, 525–40 (2011)
7. Li, L., et al.: Identification of type 2 diabetes subgroups through topological analysis of patient similarity. Sci. Transl. Med. **7**, 311ra174 (2015)
8. Patania, A., et al.: Topological analysis of data. EPJ Data Sci. **6**, 1–6 (2017)
9. Clauset, A., et al.: Finding community structure in very large networks. Phys. Rev. E Stat. Nonlin. Soft Matter Phys. **70**, 066111 (2005)
10. McGinnis, L.K., Kinsey, W.H.: Role of focal adhesion kinase in oocyte-follicle communication. Mol. Reprod. Dev. **82**, 90–102 (2015)

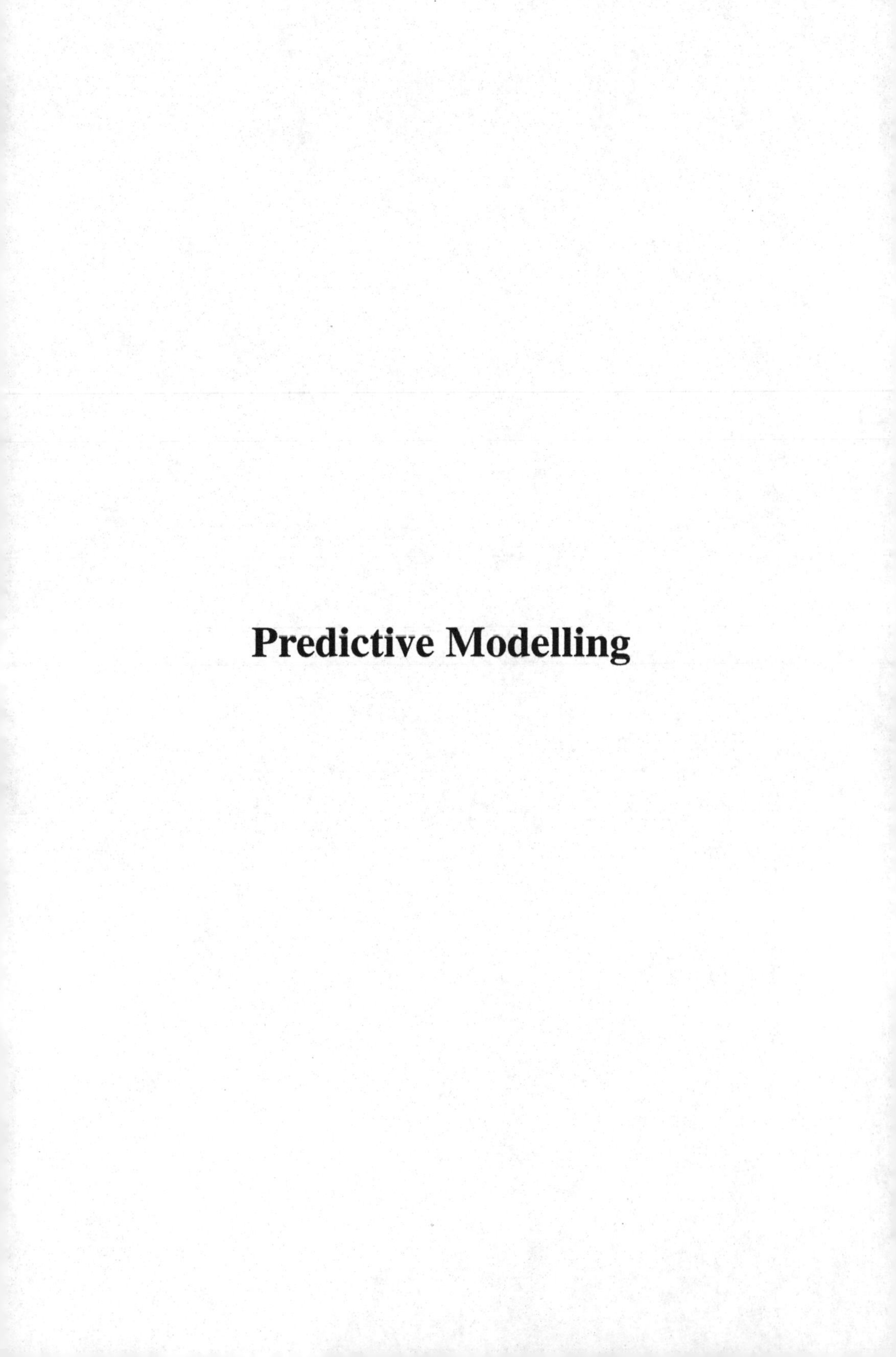

Predictive Modelling

Predicting Kidney Transplant Survival Using Multiple Feature Representations for HLAs

Mohammadreza Nemati[1], Haonan Zhang[1], Michael Sloma[1], Dulat Bekbolsynov[2], Hong Wang[3], Stanislaw Stepkowski[2], and Kevin S. Xu[1](✉)

[1] Department of Electrical Engineering and Computer Science, University of Toledo, 2801 W Bancroft St, Toledo, OH 43606, USA
kevin.xu@utoledo.edu

[2] Department of Medical Microbiology and Immunology, University of Toledo, Toledo, OH, USA

[3] Department of Engineering Technology, University of Toledo, Toledo, OH, USA

Abstract. Kidney transplantation can significantly enhance living standards for people suffering from end-stage renal disease. A significant factor that affects graft survival time (the time until the transplant fails and the patient requires another transplant) for kidney transplantation is the compatibility of the Human Leukocyte Antigens (HLAs) between the donor and recipient. In this paper, we propose new biologically-relevant feature representations for incorporating HLA information into machine learning-based survival analysis algorithms. We evaluate our proposed HLA feature representations on a database of over 100,000 transplants and find that they improve prediction accuracy by about 1%, modest at the patient level but potentially significant at a societal level. Accurate prediction of survival times can improve transplant survival outcomes, enabling better allocation of donors to recipients and reducing the number of re-transplants due to graft failure with poorly matched donors.

Keywords: Feature extraction · Human Leukocyte Antigens · Survival analysis · Graft survival

1 Introduction

Kidney transplantation is the therapy of choice for many people suffering from end-stage renal disease (ESRD). A successful kidney transplant can enhance a patient's living standards and diminish the patient's risk of dying. Although allograft (organ or tissue transplanted from one individual to another) and patient survival have improved because of new surgical technologies and effective immunosuppression, a transplant is not a lifetime treatment. Allografts, or simply grafts, will stop functioning over time, requiring re-transplantation for the patient after graft failure. There is a significant societal demand for kidney transplants, with over 90,000 people on the waiting list in the United States alone!

A. Tucker et al. (Eds.): AIME 2021, LNAI 12721, pp. 51–60, 2021.
https://doi.org/10.1007/978-3-030-77211-6_6

The time to graft failure or *graft survival time* is determined by a variety of factors, including the age, race, and overall health of the donor and recipient. The *compatibility* of the donor and recipient also plays a key role, particularly with respect to their Human Leukocyte Antigens (HLAs) [11].

In this paper, we aim to *predict* the graft survival time for a transplant given a variety of covariates on the donor and recipient. We focus on ways of incorporating HLA information into the predictor by examining multiple feature representations for HLAs. We incorporate these HLA features into several survival analysis models. By building a base model without HLA information and then comparing to models that contain more detailed representations of HLAs, we can identify whether the HLA information can improve prediction accuracy.

Our main contribution is 2 new feature representations for HLA types and pairs that account for *biological mechanisms* behind HLA compatibility and differences in categorization of HLAs. We find that incorporating HLA information can improve the accuracy of predicted graft survival time by about 1%, which is a modest improvement for an individual patient, but could translate to significant improvements at the societal level by increasing graft survival times, thus enabling more transplant *recipients* with the same number of donors, and potentially reducing the size of the waiting list.

2 Background and Motivation

Chronic kidney disease (CKD) is a public health issue and a general term for heterogeneous disorders affecting a kidney's function, which may lead to ESRD. According to 2019 reports of United States Renal Data System, CKD affects at least 10% of adults in the U.S., with nearly 750,000 Americans requiring kidney transplantation. In the absence of kidney donors, life support therapy for these patients is associated with exorbitant morbidity, mortality, and tremendous financial burden. Successful kidney transplantation may save about $55,000 per year in Medicare costs for every functioning transplant [15].

Human Leukocyte Antigens (HLAs): HLAs are a category of surface proteins encoded in a distinct gene cluster. These HLAs, which have multiple sequences, play a fundamental role in the body's immune system integrity. In organ transplantation, donor HLAs are also recognized as foreign to be attacked by the recipient's immune system [4]. Each human inherits 2 copies (1 maternal and 1 paternal) of each HLA gene. In the cluster, 3 specific loci, HLA-A, -B and -DR, are of utmost clinical significance for kidney transplantation. Thus, 6 HLAs (2 copies of each of HLA-A, -B, and -DR) are routinely typed in the clinic. An HLA is typically represented by the locus and a 2-digit number such as A1, which we call the HLA type. For example, a donor may have the following 6 HLA types: A1, A2, B1, B2, DR1, and DR2.

HLA antigen categorization has evolved, and some HLA types (broads) have actually been discovered to be multiple types (splits). For example, the splits and associated antigens of the broad antigen HLA-A9 are A23 and A24. Thus, some instances of the splits A23 and A24 may be coded as the broad A9.

The clinical importance of HLA stems from the sheer polymorphism [14], resulting in donor HLAs being in most instances different from recipient HLAs. Each HLA type present in the donor but not in the recipient leads to an HLA mismatch (MM). There may be 0 to 6 HLA-A/B/DR MM between a donor and recipient, with higher MM generally resulting in shorter graft survival times [11].

Survival Analysis: Survival analysis is a well-established technique in statistics used to predict time to an event of interest during a specific observed time interval. It is a form of regression where the objective is to predict the survival time, i.e. the time until an event of interest occurs. For many data points, however, the exact time of the event is unknown due to *censoring*, and thus, standard regression models are not well-suited to handle such time-to-event problems. Many survival analysis algorithms have been proposed to handle censored data—we refer readers to the survey [19]. We consider 3 machine learning-based survival analysis algorithms in this paper, which we describe in Sect. 4.2.

3 Data Description

This study uses data from the Scientific Registry of Transplant Recipients (SRTR) and includes data on all donors, wait-listed candidates, and transplant recipients in the U.S., submitted by the members of the Organ Procurement and Transplantation Network (OPTN).

Inclusion Criteria: We acquired 469,711 anonymous cases on all kidney transplants between 1987 and 2016 from the registry. We apply the following inclusion criteria to the data. We consider only transplants with deceased donors, recipients aged 18 years or older, and only candidates who are receiving their first transplant. We include only transplants between 2000 and 2016 due to the introduction of new therapy regimes and a new kidney allocation system [2] around the year 2000. Finally, we include only recipients with peak Panel Reactive Antibody (PRA) less than 80% since patients with high PRA levels experience increased acute rejection rate and graft failure [16]. After applying the inclusion criteria, subjects with missing values in any basic covariates except cold ischemia time (see Sect. 4.1) or HLAs are removed from the study. 106,372 transplants remain after the stages mentioned above to build the predictive models.

Target Variable: We use *death-censored graft loss* as the clinical endpoint (prediction target), which means that patients who died with a functioning graft are treated as censored since they did not exhibit the event of interest (graft loss). Graft loss is determined based on the record of either graft failure, return to maintenance dialysis, re-transplant, or listing for re-transplant. For censored instances, the censoring date is defined to be the last follow-up date.

4 Methods and Technical Solutions

Research Questions: We pose two main research questions in this study. First, does incorporating donor and recipient HLA information into a graft survival time predictor improve prediction accuracy? If so, what type of representation for the HLA information results in the highest prediction accuracy? We first describe the different HLA feature representations we propose in Sect. 4.1 and then discuss the survival analysis algorithms we use in Sect. 4.2. Our data processing pipeline is shown in Fig. 1.

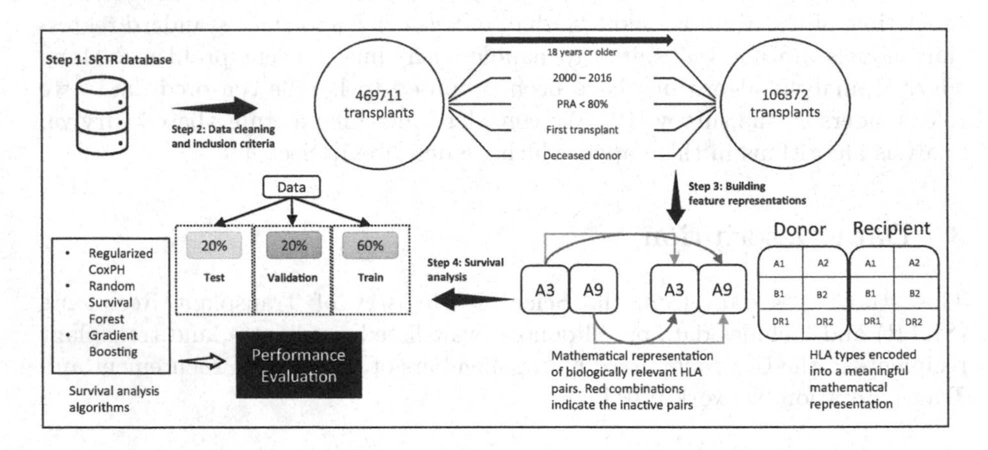

Fig. 1. Illustration of data processing and survival analysis pipeline.

4.1 Feature Representations

We consider 7 different feature sets ranging from 23 to 3,900 covariates. Some of the covariates are *pre-transplant covariates*, meaning that they are available prior to the transplant time, while others are post-transplant covariates, available only at the time of transplant or after a transplant has been performed and the patient has been discharged. We first consider prediction using only the pre-transplant covariates, as they can be used to predict graft survival prior to the transplant being performed and could potentially be used in the process of matching donors and recipients. We also consider prediction using both pre- and post-transplant covariates, which should be more accurate and can still be useful to a clinician.

Basic Features: Pre-transplant basic features consist of age, sex, race, and body mass index (BMI). Race is encoded using a one-hot representation. The post-transplant covariates employed are donor and recipient serum creatinine levels at the time of transplant, recipient serum creatinine at discharge time, whether the patient needs dialysis within the first week of the transplant, and the cold ischemia time (CIT), which denotes the amount of time the kidney was preserved after the blood supply has been cut off. Missing values for CIT were imputed using the mean over all other transplants. There are a total of 23 and 29 features for the pre- and post-transplant settings, respectively.

HLA Mismatches: We first consider the number of mismatches (MM) between donor and recipient, which has been found to be a significant factor in the time to graft failure. We consider two possible representations: the total number of MM (0 to 6), as well as the separate A-B-DR MM (0 to 2 each). These result in 1 and 3 features appended to the basic features, respectively.

HLA Types: We consider directly encoding the HLA types of the donor and recipient. The digits in an HLA type should be treated as categories and not numeric values, e.g. A2 and A1 differing by 1 does not imply that they are more similar than A2 and A23. One focus of our study is to address the methodological challenges arising from HLA broad and split antigens. We propose to encode HLA types using a one-hot-like encoding that also maps splits back to broads so that a split like A23 has a one in both the columns for A23 and A9. We encode donors and recipients separately so that each transplant has at least 12 ones (6 donor, 6 recipient), and possibly more due to splits. This encoding results in 229 features appended to the basic features.

HLA Pairs: We refer to the combination of a donor and a recipient HLA type as an HLA pair. For example, if a donor has HLA-A1 and a recipient has A2, then the HLA pair (A1, A2) is associated with the transplant. We can use a one-hot-like encoding for HLA pairs by placing a one in the column for each HLA pair associated with a transplant, similar to what we do for HLA types. However, this does not account for the biological mechanisms behind HLA compatibility.

The recipient's immune system may reject a transplant if the donor possesses HLA types not present in the recipient. However, if the recipient has HLA types that are not present in the donor, there is no issue, which creates an *asymmetry* in the roles of the donor and recipient HLA types. Thus, some HLA pairs are not biologically relevant. An example of inactive pairs is shown in Fig. 1 with 0 HLA-A mismatches resulting in only (A3, A3) and (A9, A9) as active pairs. We place a one in the columns of only active HLA pairs and consider broads and splits in the same manner as for HLA types. The HLA pair encoding results in 3,638 features appended to the basic features. Due to the large number of features, we also consider a smaller *frequent pairs* representation where we remove all HLA pairs observed in less than 1,000 transplants, which results in 180 features.

All: We consider also a combined feature set by concatenating all of the above feature representations.

4.2 Survival Analysis Algorithms

Coxnet: The Cox Proportional Hazards (Cox PH) model is one of the most widely used models for survival analysis. It models the hazard ratio using a weighted linear combination of covariates. The coefficient vector is estimated by maximizing the partial likelihood. We use a Cox PH model with combined ℓ_1 and ℓ_2 regularization, known as the elastic net, which leads to the Coxnet model [18]. The model has 2 hyperparameters: α, which controls the strength of regularization, and r, which denotes the ratio between the ℓ_1 and ℓ_2 penalties. We use a grid search with α uniformly distributed on a log scale between 10^{-4} and 10^{-2} and r uniformly distributed between 0.1 and 1.

Random Survival Forest: Random forest is a bootstrap aggregating (bagging) ensemble learning algorithm with decision trees as base learners. Ishavan et al. [5] proposed the random survival forest (RSF) algorithm that can handle right-censored data. We use an RSF with 500 trees and consider random selection of the square root of the number of features for each split. We perform a grid search on the maximum depth of each tree in the range $\{5, 10, 15, 20, 25, 30\}$.

Gradient Boosted Regression Trees: Gradient boosting (GB) is an ensemble learning technique that combines the predictions of many weak learners. Boosting algorithms using survival regression trees as their weak learners have been developed to be used in survival analysis problems [19]. We use stochastic gradient boosting with 200 trees using a 50% subsample to fit each tree. We perform a grid search on the maximum depth of each tree in the range $\{1, 2, 3\}$.

5 Empirical Evaluation

To evaluate the accuracy of our predictors, we randomly split the data into 3 sets: 60% training, 20% validation, and 20% testing. The validation set is used for hyperparameter tuning. For each algorithm, we choose the set of hyperparameters with the highest validation set C-index and then retrain it on the 80% set containing both the training and validation sets. We then finally evaluate each algorithm and feature set on the 20% test set, which was initially held out and not used at any point to prevent test set leakage. Our experiments are conducted using the scikit-survival Python package [12].

5.1 Evaluation Metric

We consider two metrics to evaluate the accuracy of our survival time predictions. First, we use Harrell's concordance index (C-index), which is perhaps the most widely used accuracy metric for survival prediction models [3]. The C-index is merely dependent on the *ordering* of predictions and is calculated by counting all possible pairs of samples and concordant pairs. A pair is a concordant pair if the risk $\eta_i < \eta_j$ and $T_i > T_j$, where T_i is the survival time for patient i.

We also consider the cumulative/dynamic area under receiver operating characteristic curve (AUC) metric that measures how accurately a model can predict the events that happen before and after a specific time t [7]. We consider the mean cumulative/dynamic AUC over 5 equally-spaced time points.

5.2 Effects of Feature Representations

The two main research questions are both centered around the effects of incorporating HLA information into graft survival time prediction. From the results in Table 1, notice that incorporating HLA information almost always results in an improvement in prediction accuracy in the pre-transplant prediction setting. The amount of improvement compared to the basic features varies for the differing feature representations and evaluation metrics.

Table 1. Survival prediction accuracy using only pre-transplant covariates. Best feature set for each predictor and each metric is listed in bold.

Feature set	Coxnet		Random surv. forest		Gradient boosting	
	C-index	Mean AUC	C-index	Mean AUC	C-index	Mean AUC
Basic	0.630	0.636	0.629	0.653	0.641	0.653
MM (total)	0.633	**0.643**	0.631	0.656	0.642	**0.657**
MM (A-B-DR)	0.633	0.643	0.631	0.655	**0.642**	0.657
Types	0.632	0.636	**0.636**	0.656	0.641	0.652
Pairs	0.633	0.642	0.627	0.656	0.641	0.653
Freq. pairs	0.633	0.642	0.635	**0.660**	0.642	0.655
All	**0.633**	0.641	0.624	0.652	0.642	0.656

HLA Mismatches: In all cases, adding HLA MM (either total or separate A-B-DR MM) improves the accuracy of the predictive models. We notice minimal differences in accuracy from including total MM and A-B-DR MM.

HLA Types: For Coxnet and GB, including HLA types resulted in worse accuracy than including HLA MM. Conversely, for RSF, including HLA types led to the highest C-index. Since the Coxnet is linear in the features, it cannot learn interactions between features, and thus, cannot learn compatibilities between different HLA types, so this result is not too surprising for Coxnet. On the other hand, the tree-based predictors should be able to learn donor-recipient HLA compatibilities, so it is somewhat surprising that GB also performs worse.

HLA Pairs: Unlike with HLA MM, the results with HLA pairs vary by model. The Coxnet benefits from the inclusion of all HLA pairs. Since it is linear in the features, it requires HLA pair features in order to learn compatibilities between donor and recipient HLAs. It is also robust to overfitting in high dimensions due to the elastic net penalty. On the other hand, the nonlinear predictors see no gain (GB) or even decreases (RSF) in accuracy from the inclusion of all HLA pairs. When restricting to just the most frequent HLA pairs, resulting in a much smaller number of HLA pair features (180 compared to over 3,600), the accuracy of RSF and GB now increase rather than decrease. This suggests that the high dimensionality may be causing a problem for the tree-based predictors. The high dimensionality results from the one-hot-like encoding mechanism we are using for HLA pairs, which can be disadvantageous for trees because it splits a single categorical variable into multiple variables, potentially requiring many splits for a single categorical variable with a large number of categories.

All: The accuracy when all features are included seems to be similar to that of including all HLA pairs, which contribute the highest number of features. The Coxnet model that achieves highest C-index includes all of the features, while RSF sees a decrease in C-index, and GB sees a slight increase.

Prediction with Post-transplant Covariates: For both models, the C-index and mean cumulative/dynamic AUC improve by about 0.3–0.4. The highest C-indices are 0.663, 0.676, and 0.675 for Coxnet, RSF and GB, respectively. The results indicate that integrating post-transplant covariates tremendously helps the survival prediction algorithms improve their accuracy, as one might expect.

6 Related Work

A broad group of studies has used data-driven statistical models to predict graft survival times or measure risk factors' impact on graft survival. Prior work includes multivariate analysis using Cox proportional hazards (Cox PH) models with a small number of covariates [1,13,20]. There has been more recent work on machine learning-based survival analysis applied to kidney transplantation, including an ensemble model that combines Cox PH models with random survival forests [10] and a deep learning-based approach [9].

Our results compare favorably to prior studies [1,9,13,20] using the same SRTR data we use in this study. Each study differs in inclusion criteria, time duration, and several other factors that prevent a direct comparison; however, we include their reported results here for reference. Two older studies [13] and [20] using Cox PH models without regularization achieved C-indices of 0.62 and 0.61, respectively. A more recent study also using a Cox PH model with only pre-transplant covariates [1] including HLA MM achieved a C-index of 0.64; however, their study included both living and deceased donors while ours considers only deceased donors. Transplant outcomes with living donors are much more favorable [1], which may result in easier prediction. Another recent study [9] used a deep learning approach applied to both pre- and post-transplant covariates to achieve a C-index of 0.655, less than the 0.676 we achieved.

Several other recent studies have focused on prediction of patient survival rather than graft survival, with [8] and [10] achieving C-indices of 0.70 and 0.724, respectively. Prediction of patient survival is much easier than prediction of graft survival, which we focus on in this paper. For example, [20] considered both patient and graft survival and achieved a C-index of 0.68 for patient survival compared to 0.61 for graft survival. We also argue that graft survival is the more relevant clinical endpoint, as a patient who survives a transplant but suffers a graft failure will require a re-transplant and returns to the waiting list.

7 Significance and Impact

Transplantation outcome prediction is instrumental for clinical decision-making, as well as allocation policy development. The kidney allocation policy by the OPTN was developed to encourage fairness (equal access to treatment) and effectiveness (the longest predicted survival) [6] in transplantation. Informed clinical decision making allows for avoidance of high-risk transplants and thus reduces number of graft losses. However, accurate prediction of transplant outcomes remains a daunting challenge due to the high complexity of human biology. By adding biologically-relevant representations of HLA, our models improved prediction accuracy, which may help to avoid transplants with poor survival times.

Despite the somewhat modest 1% improvement in prediction accuracy, the large number of patients and exorbitant cost of dialysis mean that a small improvement in kidney graft survival can result in a significant impact on the waiting list and financially. Illustrating this dependence, a kidney allocation simulation study showed the potential for saving $750 million if transplant rates would improve by 5.7% in a 4,000 patient pool [17]. Based on OPTN national data, 16,000–17,000 deceased donor transplants are performed in a single year. Thus, using sophisticated outcome prediction tools for extending median kidney graft survival by even a single year may save this many surgeries annually, with high economical and social impact. Our findings are thus useful for assisting clinical decision making aimed at improving long-term allograft survival.

Acknowledgements. The research reported in this publication was supported by the National Library of Medicine of the National Institutes of Health under Award Number R01LM013311 as part of the NSF/NLM Generalizable Data Science Methods for Biomedical Research Program. The content is solely the responsibility of the authors and does not necessarily represent the official views of the National Institutes of Health.

The data reported here have been supplied by the Hennepin Healthcare Research Institute (HHRI) as the contractor for the Scientific Registry of Transplant Recipients (SRTR). The interpretation and reporting of these data are the responsibility of the author(s) and in no way should be seen as an official policy of or interpretation by the SRTR or the U.S. Government. Notably, the principles of the Helsinki Declaration were followed.

References

1. Ashby, V.B., et al.: A kidney graft survival calculator that accounts for mismatches in age, sex, HLA, and body size. Clin. J. Am. Soc. Nephrol. **12**(7), 1148–1160 (2017)
2. Ashby, V., et al.: Transplanting kidneys without points for HLA-B matching: consequences of the policy change. Am. J. Transplant. **11**(8), 1712–1718 (2011)
3. Harrell, F.E., Califf, R.M., Pryor, D.B., Lee, K.L., Rosati, R.A.: Evaluating the yield of medical tests. JAMA **247**(18), 2543–2546 (1982)
4. Horton, R., et al.: Gene map of the extended human MHC. Nat. Rev. Genet. **5**(12), 889–899 (2004)
5. Ishwaran, H., Kogalur, U.B., Blackstone, E.H., Lauer, M.S.: Random survival forests. Ann. Appl. Stat. **2**(3), 841–860 (2008)

6. Israni, A.K., et al.: New national allocation policy for deceased donor kidneys in the united states and possible effect on patient outcomes. J. Am. Soc. Nephrol. **25**(8), 1842–1848 (2014)
7. Lambert, J., Chevret, S.: Summary measure of discrimination in survival models based on cumulative/dynamic time-dependent ROC curves. Stat. Methods Med. Res. **25**(5), 2088–2102 (2016)
8. Li, B., et al.: Predicting patient survival after deceased donor kidney transplantation using flexible parametric modelling. BMC Nephrol. **17**(1), 51 (2016)
9. Luck, M., Sylvain, T., Cardinal, H., Lodi, A., Bengio, Y.: Deep learning for patient-specific kidney graft survival analysis. arXiv preprint arXiv:1705.10245 (2017)
10. Mark, E., Goldsman, D., Gurbaxani, B., Keskinocak, P., Sokol, J.: Using machine learning and an ensemble of methods to predict kidney transplant survival. PLoS ONE **14**(1), e0209068 (2019)
11. Opelz, G., Wujciak, T., Döhler, B., Scherer, S., Mytilineos, J.: HLA compatibility and organ transplant survival. Collaborative transplant study. Rev. Immunogenet. **1**(3), 334–342 (1999)
12. Pölsterl, S., Navab, N., Katouzian, A.: Fast training of support vector machines for survival analysis. In: Appice, A., Rodrigues, P.P., Santos Costa, V., Gama, J., Jorge, A., Soares, C. (eds.) ECML PKDD 2015. LNCS (LNAI), vol. 9285, pp. 243–259. Springer, Cham (2015). https://doi.org/10.1007/978-3-319-23525-7_15
13. Rao, P.S., et al.: A comprehensive risk quantification score for deceased donor kidneys: the kidney donor risk index. Transplantation **88**(2), 231–236 (2009)
14. Robinson, J., Barker, D.J., Georgiou, X., Cooper, M.A., Flicek, P., Marsh, S.G.: IPD-IMGT/HLA database. Nucleic Acids Res. **48**(D1), D948–D955 (2020)
15. Salomon, D.R., Langnas, A.N., Reed, A.I., Bloom, R.D., Magee, J.C., Gaston, R.S., AST/ASTS Incentives Workshop Group (IWG) a: AST/ASTS workshop on increasing organ donation in the United States: creating an "arc of change" from removing disincentives to testing incentives. Am. J. Transplant. **15**(5), 1173–1179 (2015)
16. Schwaiger, E., et al.: Deceased donor kidney transplantation across donor-specific antibody barriers: predictors of antibody-mediated rejection. Nephrol. Dial. Transplant. **31**(8), 1342–1351 (2016)
17. Segev, D.L., Gentry, S.E., Warren, D.S., Reeb, B., Montgomery, R.A.: Kidney paired donation and optimizing the use of live donor organs. JAMA **293**(15), 1883–1890 (2005)
18. Simon, N., Friedman, J., Hastie, T., Tibshirani, R.: Regularization paths for Cox's proportional hazards model via coordinate descent. J. Stat. Softw. **39**(5), 1–13 (2011)
19. Wang, P., Li, Y., Reddy, C.K.: Machine learning for survival analysis: a survey. ACM Comput. Surv. **51**(6), 1–36 (2019)
20. Wolfe, R., McCullough, K., Leichtman, A.: Predictability of survival models for waiting list and transplant patients: calculating LYFT. Am. J. Transplant. **9**(7), 1523–1527 (2009)

Sum-Product Networks for Early Outbreak Detection of Emerging Diseases

Moritz Kulessa[1(✉)], Bennet Wittelsbach[1], Eneldo Loza Mencía[1], and Johannes Fürnkranz[2]

[1] Technische Universität Darmstadt, Darmstadt, Germany
{mkulessa,eneldo}@ke.tu-darmstadt.de,
bennet.wittelsbach@stud.tu-darmstadt.de
[2] Johannes Kepler Universität Linz, Linz, Austria
juffi@faw.jku.at

Abstract. Recent research in syndromic surveillance has focused primarily on monitoring specific, known diseases, concentrating on a certain clinical picture under surveillance. Outbreaks of emerging infectious diseases with different symptom patterns are likely to be missed by such a surveillance system. In contrast, monitoring all available data for anomalies allows to detect any kind of outbreaks, including infectious diseases with yet unknown syndromic clinical pictures. In this work, we propose to model the joint probability distribution of syndromic data with sum-product networks (SPN), which are able to capture correlations in the monitored data and even allow to consider environmental factors, such as the current influenza infection rate. Conversely to the conventional use of SPNs, we present a new approach to detect anomalies by evaluating p-values on the learned model. Our experiments on synthetic and real data with synthetic outbreaks show that SPNs are able to improve upon state-of-the-art techniques for detecting outbreaks of emerging diseases.

Keywords: Sum-product networks · Syndromic surveillance · Outbreak detection · Anomaly detection

1 Introduction

The spread of infectious disease outbreaks can be greatly diminished by early recognition which allows the application of suitable control measures. To that end, *syndromic surveillance* aims to identify illness clusters in syndromic data before diagnoses are confirmed and reported to public health agencies [4]. The conventional approach of syndromic surveillance is to define indicators for a particular infectious disease on the given data, also referred to as *syndromes*, which are monitored over time to detect unusually high number of cases. Rather than developing highly specialized algorithms which are based on monitoring specific syndromes, we argue that the task of outbreak detection should be viewed as a general anomaly detection problem where an outbreak alarm is triggered if the

A. Tucker et al. (Eds.): AIME 2021, LNAI 12721, pp. 61–71, 2021.
https://doi.org/10.1007/978-3-030-77211-6_7

distribution of the incoming syndromic data changes in an unforeseen and unexpected way. Therefore, we distinguish between *specific* syndromic surveillance, where factors related to a specific disease are monitored, and *non-specific* syndromic surveillance, where general, universal characteristics of the data stream are monitored for anomalies.

In previous work on non-specific syndromic surveillance [6], we have shown that statistical modeling techniques often outperform more elaborate algorithms. In this work, we transfer the idea of these modeling approaches to *sum-product networks* (SPNs) [8], a statistical and generative machine learning algorithm. Instead of fitting one particular distribution for each single syndrome, we use SPNs to model the joint probability distribution of syndromic data. We further introduce a technique that allows to detect anomalies by reasoning with the p-values of the SPN model. In addition, syndromic data can be enriched with information about environmental factors, such as the weather or the season, which can be used to condition the SPN on particular circumstances before p-values are computed. We experimentally compare our proposed approach to established algorithms for non-specific syndromic surveillance on a synthetic data set [2,14] and real data from a German emergency department in which we injected synthetic outbreaks. Our results demonstrate that SPNs can further improve upon the state-of-the-art statistical modeling techniques.

2 Non-specific Syndromic Surveillance

2.1 Problem Definition

Syndromic data can be seen as a constant stream of instances of a population $\mathcal{C}$ where each instance $\mathbf{c} \in \mathcal{C}$ is represented by a set of *response attributes* $\mathcal{A} = \{A_1, A_2, \ldots, A_m\}$ [14]. To be able to detect temporal changes, instances are grouped together according to pre-specified time slots, so that $\mathcal{C}(t) \subset \mathcal{C}$ denotes all patients arriving at the emergency department at day t. In addition, each group $\mathcal{C}(t)$ is associated with $\mathbf{e}(t) \in E_1 \times E_2 \times \ldots \times E_k$ where $\mathcal{E} = \{E_1, E_2, \ldots, E_k\}$ is a set of *environmental attributes*, which represent external factors that may influence the distribution of instances $\mathcal{C}(t)$. This allows, e.g., to model that flu-like symptoms are more frequent during the winter. We denote the history of the information available at time t as $\mathcal{H}(t) = ((\mathcal{C}(1), \mathbf{e}(1)), \ldots, (\mathcal{C}(t-1), \mathbf{e}(t-1)))$.

The goal of non-specific syndromic surveillance is to detect anomalies in the current time slot $\mathcal{C}(t)$ w.r.t. the history $\mathcal{H}(t)$ as potential indicators of an infectious disease outbreak. From the perspective of specific syndromic surveillance, the non-specific setting can be seen as monitoring all possible syndromes at the same time. The set of all possible syndromes can be defined as

$$\mathcal{S}_{all} = \left\{ \prod_{i \in \mathcal{I}} A_i \mid A_i \in \mathcal{A} \wedge \mathcal{I} \subseteq \{1, 2, \ldots, m\} \wedge |\mathcal{I}| \geq 1 \right\}$$

where $\prod_{i \in \mathcal{I}} A_i$ for $|\mathcal{I}| = 1$ is defined as $\{\{a\} \mid a \in A \wedge A \in \mathcal{A}\}$. In addition, we denote $\mathcal{S}_{\leq n} = \{s \mid s \in \mathcal{S}_{all} \wedge |s| \leq n\}$ as the set of all possible syndromes having a maximum of n conditions.

$\mathcal{C}(1) =$

gender	symptom
female	cough
male	fever
male	cough
...	...

$\mathbf{e}(1) = (summer)$

...

$\mathcal{C}(t-1) =$

gender	symptom
male	cough
female	cough
female	fever
...	...

$\mathbf{e}(t-1) = (winter)$

$\Longrightarrow$

time	gender		symptom		
slot	#male	#female	#fever	#cough	season
1	3	4	2	5	summer
2	2	2	1	3	summer
...	...	...	...	...	...
t-1	7	8	9	6	winter

Fig. 1. Example for the creation of a structured data set using syndromes $\mathcal{S}_{\leq 1}$.

2.2 Creation of Structured Data

We transform $\mathcal{H}(t)$ into a structured format, which facilitates the analysis with common machine learning algorithms. For a given set of syndromes $\mathcal{S} \subseteq \mathcal{S}_{all}$, we denote $f_{\mathcal{S}} : 2^{\mathcal{C}} \to \mathbb{N}^{|\mathcal{S}|}$ as the function that counts the number of occurrences $f_s(\mathcal{C}(i))$ for each syndrome $s \in \mathcal{S}$ in a given set of instances $\mathcal{C}(i)$ at time i. Based on the syndrome counts, we form a data set $\mathcal{D} = \{(f_{\mathcal{S}}(\mathcal{C}(i)), \mathbf{e}(i)) \mid (\mathcal{C}(i), \mathbf{e}(i)) \in \mathcal{H}(t)\}$ in which each instance represents a single time slot. Section 2.2 depicts an example of how the data set is created for syndromes $\mathcal{S}_{\leq 1}$. Note that in case of syndromes $\mathcal{S}_{\leq 2}$, the data set would additionally contain the columns #(male $\wedge$ cough), #(female $\wedge$ cough), #(male $\wedge$ fever), and #(female $\wedge$ fever).

2.3 Related Work

While specific syndromic surveillance is a well-studied research area, we found that only little research has been devoted to non-specific syndromic surveillance. Brossette et al. [1] adopts the idea of association rule mining to identify anomalous patterns in health data. Wong et al. [14] first learn a Bayesian network over historic health data and then compare a sample of historical cases to current cases $\mathcal{C}(t)$ to detect potential outbreaks. Fanaee-T and Gama [2] track changes in the data correlation structure using eigenspace techniques to identify anomalies. In particular, Wong et al. [14] and Fanaee-T and Gama [2] distinguish between indicator and environmental attributes to improve detection performance which is also known as contextual or conditional anomaly detection [10]. For more details on these methods, as well as an empirically comparison to common anomaly detectors and statistical modeling techniques, we refer to [6].

A particular result of Kulessa et al. [6] is that statistical techniques for a simultaneous and individual monitoring of syndromes $\mathcal{S}_{\leq 1}$ or $\mathcal{S}_{\leq 2}$ already achieve very competitive results and often outperform more elaborate algorithms. More precisely, for each syndrome s a distribution $P(X_s)$ is fitted on $\mathcal{H}(t)$ such that $f_s(C(t)) \sim P(X_s)$ where X_s denotes the random variable associated to syndrome s. The Poisson and the negative binomial distribution are natural choices but also the Gaussian distribution is used in practice [5]. However, this approach has two main limitations. Firstly, independence among the monitored syndromes is assumed and, secondly, environmental factors are not taken into account.

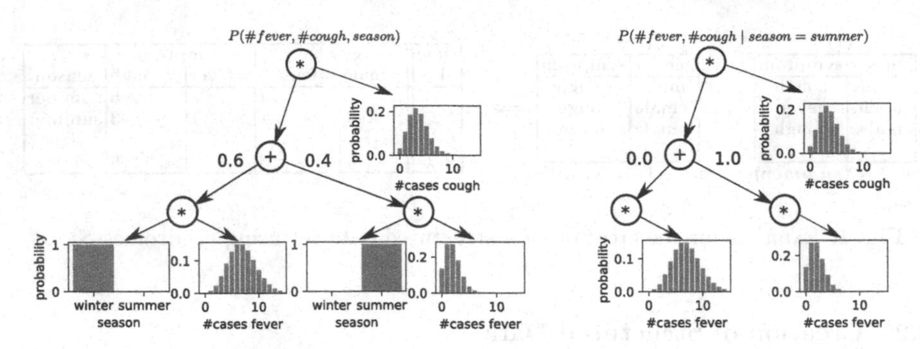

Fig. 2. The left SPN represents $P(\#fever, \#cough, season)$ while the right SPN represents $P(\#fever, \#cough \mid season = summer)$ derived from the left.

3 Sum-Product Networks for Syndromic Surveillance

Most statistical techniques, including those mentioned in the previous section, model the joint probability distribution as a product of individual syndrome distributions. Clearly, this is only valid if the syndromes are independent of each other. Sum-product networks [8] are an elegant way of extending this simple model by taking dependencies between the monitored syndromes and even dependencies to environmental factors into account.

A *sum-product network* (SPN) models the joint probability distribution $P(\mathcal{X})$ of a data set, where $\mathcal{X} = \{X_1, X_2, \ldots, X_m\}$ is a set of random variables, as a rooted directed acyclic graph of sum, product and leaf nodes. In this graph, the *scope* of a particular node is defined as the set of features appearing in the subgraph below that node. Formally, sum nodes provide a weighted mixture of distributions by combining nodes which share the same scope, whereas product nodes represent the factorization over independent distributions by combining nodes defined over disjunct scopes. Finally, each leaf node contains a univariate distribution $P(X)$ for a particular feature $X \in \mathcal{X}$. In this work, we use *mixed* SPNs [7] which allow to learn an SPN over both, continuous and discrete attributes. Figure 2 shows an exemplary SPN, representing the joint probability distribution $P(\#fever, \#cough, season)$. The top product node indicates that the distribution of *#cough* is independent of the other attributes. In contrast, the distribution of *#fever* depends on *season* and, therefore, it is split into two clusters by a sum node, one for the winter and one for the summer.

SPNs can be adapted to represent a conditional probability distribution $P(\mathcal{X} \setminus X_i \mid X_i = x_i)$ by evaluating one or more conditions $X_i = x_i$ in the leaves of X_i and propagating the resulting probabilities upwards. Whenever a sum node is passed, the weight for a child node is updated by multiplying it with the up-coming probability. In a final step, the weights for each sum node are normalized and the leaves for attribute X_i are removed from the SPN. Fig. 2 shows an SPN for the conditional probability distribution $P(\#fever, \#cough \mid season = summer)$. Note that the child node for the summer now has a weight of 100%.

3.1 Inference of p-values in Sum-Product Networks

The main advantage of SPNs over other probabilistic models is that inference for probabilistic queries is tractable and can be computed in linear time w.r.t. the size of the network [8]. For syndromic surveillance we are particularly interested in p-values, which express the chance of obtaining data at least as extreme under a given null hypothesis. In the following, we propose novel extensions to SPNs that allow them to properly reason with p-values.

To compute the p-value for a query $q \subseteq \{X_1 \geq x_1, \ldots, X_m \geq x_m\}$ for arbitrary x_i, the conditions of q are forwarded to the leaves of the SPN. In case q contains conditions only on a subset of attributes of $\mathcal{X}$, the SPN is marginalized beforehand by simply removing all leaves on attributes which are not contained in the query. In the remaining leaves, the p-value for the respective condition is evaluated and propagated upwards. At product nodes, we use either *Fisher's* or *Stouffer's* method for merging independent p-values [13]. At sum nodes, which encode a mixture of distributions over the same attributes, we need to merge dependent p-values. Vovk and Wang [12] recommend to use the *harmonic mean* in case of substantial dependence among the merging p-values and suggest to use the *geometric* or the *arithmetic mean* for stronger dependencies. We have implemented the weighted versions of these three merging functions in order to consider the weights of sum nodes during merging. As a result, the resulting value at the root node of the SPN can be seen as a composite p-value for query q.

3.2 Application to Non-specific Syndromic Surveillance

The key idea of our approach is to learn an SPN over a data set that is structured as described in Sect. 2.2. In particular, the SPN models the joint probability distribution $P(X_\mathcal{S}, X_\mathcal{E})$ where $X_\mathcal{S} = \{X_s \mid s \in \mathcal{S}\}$ and $X_\mathcal{E} = \{X_E \mid E \in \mathcal{E}\}$ are random variables associated with syndromes $\mathcal{S}$ and environmental attributes $\mathcal{E}$ respectively. For environmental attributes, categorical distributions are used in the leaves, whereas for the syndrome counts we either use Gaussian, Poisson or negative binomial distributions, which are commonly used for monitoring count data in syndromic surveillance.

To check for outbreaks in a given time slot t, we first condition the SPN on the current environmental setting to obtain $P(X_\mathcal{S} \mid X_{E_1} = e_1, \ldots, X_{E_k} = e_k)$ where e_i is the i-th element of $\mathbf{e}(t)$. The set of queries $\mathcal{Q}_1 = \{\{X_s \leq f_s(\mathcal{C}(t))\} \mid s \in \mathcal{S}\}$ is then evaluated on the conditioned SPN, which results in a p-value for each syndrome $s \in \mathcal{S}$. This sensitivity to changes for each individual syndrome is indeed important if the potential disease pattern for an outbreak is unknown beforehand. Moreover, the p-values can be used to generate a ranking of the most suspicious syndromes. Combined with the functionality of SPNs to compute expectations, a report can be provided to local health authorities in order to analyze and understand the found irregularities.

However, for our empirical study in Sect. 4 a single score for the evaluated time slot is required. Therefore, the p-values need to be aggregated under consideration of the multiple-testing problem. Following Roure et al. [9], we only

Table 1. Synthetic data.

$\mathcal{A}$	#values	$\mathcal{E}$	#values
Age	3	Weather	2
Gender	2	Flu level	4
Action	3	Day of week	3
Symptom	4	Season	4
Drug	4		
Location	9		

Table 2. Real data.

$\mathcal{A}$	#values	$\mathcal{E}$	#values
Age	3	Weather	2
Gender	2	Flu level	4
Symptom	28	Day of week	3
Fever	4	Season	4
Oxygen saturation	2		
Blood pressure	2		
Pulse	3		
Respiration	3		

report the minimum p-value for each time slot t since the Bonferroni correction can be regarded as a form of aggregation of p-values based on the minimum function. In particular, note that scale-free anomaly scores are sufficient for the purpose of identifying the most suspicious time slots. The complement of the selected p-value represents the anomaly score reported for time slot t.

3.3 Handling of Higher Order Syndromes

Note that an SPN modeled over frequency counts of syndromes of length 1 ($\mathcal{S}_{\leq 1}$) models the dependencies between the frequency counts of individual syndromes, but it does not model the frequency of their co-occurrence. For example, if both *cough* and *fever* occur with high frequency in the current window $\mathcal{C}(t)$, it does not imply that there are many patients that exhibit both symptoms at the same time. For modeling such interactions, we have two options: First, we can directly include syndromes of length two ($\mathcal{S}_{\leq 2}$) or even higher. The obvious disadvantage is that the number of possible syndromes grows exponentially with their length. Nonetheless, we can use the SPN for making a best guess. More specifically, if we only model syndromes of length 1 ($\mathcal{S}_{\leq 1}$), we can still form the query set $\mathcal{Q}_2 = \mathcal{Q}_1 \cup \{\{X_{s_1} \leq f_{s_1}(\mathcal{C}(t)), X_{s_2} \leq f_{s_2}(\mathcal{C}(t))\} \mid s_1 \neq s_2, s_1 \in \mathcal{S}, s_2 \in \mathcal{S}\}$, and use the resulting p-values as a heuristic best guess for the p-values of syndromes $\mathcal{S}_{\leq 2}$. We will evaluate both approaches in the experimental section.

4 Experiments and Results

The goal of our experimental evaluation is to demonstrate that modeling of syndromic data through an SPN can further improve state-of-the-art statistical modeling techniques. To that end, we conducted experiments on synthetic data [2,14] and on real data from a German emergency department. As the latter did not contain any information about real outbreaks, we injected synthetic outbreaks. This common practice allows the evaluation for arbitrary types of outbreak patterns in a controlled environment. The development of more realistic evaluation strategies —or alternatively the acquisition of complete and certain patient data— remains a major challenge for the research field.

4.1 Evaluation Setup

In both scenarios, we generate 100 data streams, where each data stream captures daily information $\mathcal{C}(t)$ over a time period of two years. The time slots of the first year are used for training, whereas the second year is reserved for testing only. Each test data stream contains exactly one simulated outbreak starting on a randomly chosen day. The *synthetic data* (Table 1) were generated as proposed by Wong et al. [14]. In each stream, an outbreak is simulated which lasts for 14 days, during which people have a higher chance of catching a particular disease. On average, 34 patients are reported per time slot.

The *real data* (Table 2) consist of fully anonymized patient data from a German emergency department. With the help of a physician, we have extracted a set of attributes and discretized them into meaningful categories. In addition, we enriched the syndromic data with environmental attributes matching the synthetic data. Information about the flu level has been obtained from *SurvStat*[1] and weather data from the *DWD*.[2] On average 165 patients are reported per day. To simulate an outbreak, we first uniformly sampled a syndrome from $\mathcal{S}_{\leq 2}$. In a second step, we sampled the size of the outbreak from a Poisson distribution with mean equal to the standard deviation of the daily patient visits. To avoid over-representing outbreaks on rare syndromes, we ensured that only 20 streams contain outbreaks with syndromes that have a lower frequency than one per day.

Preliminary experiments showed that statistical tests on low counts are often too sensitive to changes, causing many false alarms. As outbreaks are usually associated with a high number of infections, we reduced the sensitivity of statistical tests by setting the standard deviation σ (Gaussian) and the mean μ (Poisson and negative binomial) to 1. We compare our approach to the statistical benchmarks, WSARE and anomaly detection algorithms which performed best in previous work [6]. Parameters were tuned in a grid search using 1000 iterations of *bootstrap bias corrected cross-validation* [11] which integrates hyperparameter tuning and performance estimation into a single evaluation loop. The evaluated parameters combinations for all algorithms can be found in our repository.[3]

As a performance measure, we rely on the *activity monitor operating characteristic (AMOC)* [3], an adaptation of the *receiver operating characteristic* (ROC) in which the true positive rate is replaced by the *detection delay* measured in days. We report the partial area under AMOC-curve for a false alarm rate less than 5% (referred to as $AAUC_{5\%}$) because of the importance of very low false alarm rate in syndromic surveillance. Note that contrary to conventional AUC values in this case lower values represent better results. We report average $AAUC_{5\%}$ scores over all 100 data streams. Note that the worst possible result for $AAUC_{5\%}$ is 14 on the synthetic and 1 on the real data, respectively.

---

[1] Robert Koch-Institut: SurvStat@RKI 2.0, https://survstat.rki.de, 11.01.2021.
[2] Deutscher Wetterdienst: Open Data, https://www.dwd.de/opendata, 11.01.2021.
[3] Our code is publicly available at https://github.com/MoritzKulessa/NSS.

Table 3. Results $AAUC_{5\%}$.

Algorithm		Synthetic data $\mathcal{S}_{\leq 1}$	Synthetic data $\mathcal{S}_{\leq 2}$	Real data $\mathcal{S}_{\leq 1}$	Real data $\mathcal{S}_{\leq 2}$
Benchmark Gaussian		0.859	0.957	0.331	0.296
Benchmark Poisson		1.312	1.321	0.283	0.220
Benchmark neg. binomial		0.964	1.021	0.259	0.216
Without $\mathcal{E}$	Autoencoder	1.647	1.549	0.443	0.372
	One-class SVM	1.031	1.536	0.353	0.350
	Gaussian mix. models	1.128	3.601	0.332	0.449
	WSARE	0.907	1.066	0.333	0.281
	SPN($\cdot$, $\mathcal{Q}_1$)	0.913	1.082	0.271	0.200
	SPN($\mathcal{S}_{\leq 1}$, $\mathcal{Q}_2$)	1.102		0.250	
With $\mathcal{E}$	Autoencoder	2.523	1.629	0.452	0.365
	One-class SVM	1.519	1.427	0.392	0.347
	Gaussian mix. models	3.404	4.033	0.403	0.443
	WSARE	0.907	0.996	0.302	0.266
	SPN($\cdot$, $\mathcal{Q}_1$)	**0.647**	0.869	0.244	**0.190**
	SPN($\mathcal{S}_{\leq 1}$, $\mathcal{Q}_2$)	0.983		0.230	

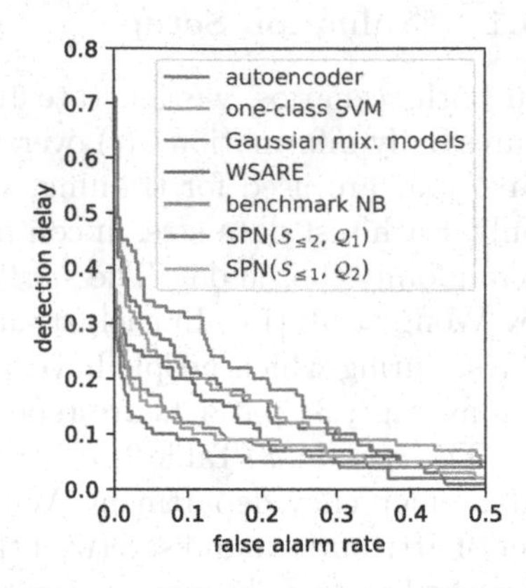

Fig. 3. AMOC-curve for monitoring $\mathcal{S}_{\leq 2}$ on the real data.

4.2 Results

Comparison of Algorithms. Table 3 shows the results of all algorithms for monitoring syndromes $\mathcal{S}_{\leq 1}$ and $\mathcal{S}_{\leq 2}$. As the consideration of environmental attributes is one of the main differences compared to the benchmarks, we performed additional evaluations without considering environmental attributes (c.f. without $\mathcal{E}$).

We can see that the SPNs with the conventional queries $\mathcal{Q}_1$ and taking environmental variables into account outperform all other competitors, on both syndrome sets $\mathcal{S}_{\leq 1}$, $\mathcal{S}_{\leq 2}$. The AMOC curve shown in Fig. 3 confirms this result. Clearly, SPN($\mathcal{Q}_1$) offers the best trade-off between detection delay and false alarm rate, also for the extended range from 0 to 0.5. Going into more detail, we can observe that the improvement over the benchmarks is more pronounced for the synthetic than for the real data. This is in line with previous findings that indicated a higher dependency on environmental factors in the synthetic data set [6]. Regarding the exclusion of environmental attributes, we expected an advantage of SPNs over the benchmarks due to the modeling of dependencies between syndromes. Conversely, in accompanying analyses we found that overfitting of the SPN can also result in less stable estimates. In fact, both effects seem to balance each other out in our comparison since the SPNs are on par with the benchmarks. Nonetheless, the results indicate that in our analyzed data SPNs can benefit from dependencies between the syndrome patterns if they are combined with environmental factors. Similar experiments have been conducted for the anomaly detectors. Contrary to the SPNs, we can observe that these approaches do not benefit from environmental information. An explanation could be the inability to condition on the given environmental attributes as all attributes are treated in the same manner by the anomaly detectors. For example, a rare environmental scenario can lead to an high anomaly score even though the observed syndromic situation might not be exceptional.

Table 4. Detailed SPN results

Setting	Distribution	Synthetic data			Real data		
		Average	Harmonic	Geometric	Average	Harmonic	Geometric
$\mathcal{S}_{\leq 1}$, $\mathcal{Q}_1$	Gaussian	0.711	0.862	0.894	0.267	0.332	0.332
	Poisson	0.660	1.317	1.130	0.276	0.244	0.243
	neg. binomial	**0.634**	0.962	1.112	**0.228**	0.235	0.234
$\mathcal{S}_{\leq 2}$, $\mathcal{Q}_1$	Gaussian	0.853	0.966	0.966	0.276	0.284	0.278
	Poisson	**0.805**	1.328	1.106	0.232	0.216	0.200
	neg. binomial	0.876	1.026	1.127	0.207	**0.178**	0.187
$\mathcal{S}_{\leq 1}$, $\mathcal{Q}_2$ Stouffer	Gaussian	1.045	1.152	1.152	0.276	0.314	0.263
	Poisson	**0.993**	2.322	1.549	0.261	0.250	0.252
	neg. binomial	1.091	1.500	1.683	0.263	0.256	**0.232**
$\mathcal{S}_{\leq 1}$, $\mathcal{Q}_2$ Fisher	Gaussian	0.943	1.110	1.112	0.265	0.295	0.268
	Poisson	**0.930**	2.099	1.414	0.262	0.254	0.253
	neg. binomial	0.968	1.372	1.538	0.253	0.256	**0.228**

Comparison Between $\mathcal{S}_{\leq 1}$ and $\mathcal{S}_{\leq 2}$. We can observe that outbreaks in the synthetic data are better detected when monitoring single condition syndromes $\mathcal{S}_{\leq 1}$ while monitoring $\mathcal{S}_{\leq 2}$ works better for the real data. As discussed in Sect. 3.3, we can approximate $\mathcal{S}_{\leq 2}$ results when using query $\mathcal{Q}_2$ on SPN($\mathcal{S}_{\leq 1}$) (last line of Table 3). We can see that in both cases, the approximation with SPN($\mathcal{S}_{\leq 1}, \mathcal{Q}_2$) does not reach the performance of directly modelling SPN($\mathcal{S}_{\leq 2}$, $\mathcal{Q}_1$) but in the case of real data it improves over SPN($\mathcal{S}_{\leq 1}$, $\mathcal{Q}_1$). Thus monitoring $\mathcal{Q}_2$ can be beneficial when the computational costs of direct modelling higher order syndromes are prohibitive.

Analysis of Parameters of the SPN. Table 4 shows the results of different methods for combining p-values with respect to the distributions used in the leaves of the SPN. The columns correspond to the method for merging p-values in sum nodes while rows represent the *setting* and the used *distribution*. Note that p-values in the product node are only merged if we evaluate $\mathcal{Q}_2$, in which case, we tested merging with Fisher's or Stouffer's method. Most notably, we observe that a simple weighted average of p-values works best on the synthetic data regardless of the other parameter settings. Following the theoretic results of Vovk andWang [12], we can only hypothesize that this is the case due to strong dependencies between the attributes. In contrast, the results are less clear on the real data set. For instance, regarding the negative binomial distribution, the arithmetic mean seems to be more preferable when using $\mathcal{S}_{\leq 1}$ whereas the harmonic mean achieves the highest score on $\mathcal{S}_{\leq 2}$. With respect to $\mathcal{Q}_2$ the results suggest a slight advantage of the Fisher's method over the Stouffer's method on both data sets. In summary, the results exhibit clear differences between the merging options, but these seem to be highly dependent on the data, distributions and architectures used. An approach that goes beyond the proposed parameter selection and makes these decisions at each inner node of the SPN in a data-driven way could be a way of further exploiting these gaps. We leave these extensions for future work.

5 Conclusion

In this work, we proposed the use of SPNs for modeling the joint probability distribution of syndromic data. The main technical contribution is a method for propagating p-values in SPNs in order to detect anomalies as potential indicators for an outbreak of an infectious disease. In addition, the SPN can consider environmental factors, such as the season, the weather, or the current level of influenza infections, which may increase or decrease the awareness of outbreaks with particular disease patterns. Our empirical study revealed that our proposed approach outperforms state-of-the-art algorithms in the field of non-specific syndromic surveillance, hence, on the task of detecting emerging diseases. In particular, by taking correlation between the monitored syndromes and environmental factors into account, the performance of our approach improved substantially.

Acknowledgments. We thank our project partners the *Health Protection Authority of Frankfurt*, the *Hesse State Health Office and Centre for Health Protection*, the *Hesse Ministry of Social Affairs and Integration*, the *Robert Koch-Institut*, the *Epias GmbH* and the *Sana Klinikum Offenbach GmbH* who provided insight and expertise that greatly assisted the research. This work was funded by the Innovation Committee of the Federal Joint Committee (G-BA) [ESEG project, grant number 01VSF17034].

References

1. Brossette, S., Sprague, A., Hardin, J., Waites, K., Jones, W., Moser, S.: Association rules and data mining in hospital infection control and public health surveillance. J. Am. Med. Inform. Assoc. **5**, 373–81 (1998)
2. Fanaee-T, H., Gama, J.: Eigenevent: an algorithm for event detection from complex data streams in syndromic surveillance. Intell. Data Anal. **19**(3), 597–616 (2015)
3. Fawcett, T., Provost, F.: Activity monitoring: noticing interesting changes in behavior. In: Proceedings of the 5th International Conference on Knowledge Discovery and Data Mining, pp. 53–62 (1999)
4. Henning, K.J.: What is syndromic surveillance? Morbidity and Mortality Weekly Report: Supplement, vol. 53, pp. 7–11 (2004)
5. Hutwagner, L., Thompson, W., Seeman, G., Treadwell, T.: The bioterrorism preparedness and response early aberration reporting system (EARS). J. Urban Health **80**(1), i89–i96 (2003)
6. Kulessa, M., Loza Mencía, E., Fürnkranz, J.: Revisiting non-specific syndromic surveillance. In: Proceedings of the 19th International Symposium on Intelligent Data Analysis (IDA) (2021)
7. Molina, A., Vergari, A., Mauro, N.D., Natarajan, S., Esposito, F., Kersting, K.: Mixed sum-product networks: a deep architecture for hybrid domains. In: Proceedings of 32nd AAAI Conference on Artificial Intelligence (AAAI), pp. 3828–3835 (2018)
8. Poon, H., Domingos, P.: Sum-product networks: a new deep architecture. In: Cozman, F.G., Pfeffer, A. (eds.) Proceedings of 27th Conference on Uncertain. AI, pp. 337–346 (2011)

9. Roure, J., Dubrawski, A., Schneider, J.: A study into detection of bio-events in multiple streams of surveillance data. In: Zeng, D., et al. (eds.) BioSurveillance 2007. LNCS, vol. 4506, pp. 124–133. Springer, Heidelberg (2007). https://doi.org/10.1007/978-3-540-72608-1_12
10. Song, X., Wu, M., Jermaine, C., Ranka, S.: Conditional anomaly detection. IEEE Trans. Knowl. Data Eng. **19**(5), 631–645 (2007)
11. Tsamardinos, I., Greasidou, E., Borboudakis, G.: Bootstrapping the out-of-sample predictions for efficient and accurate cross-validation. Mach. Learn. **107**(12), 1895–1922 (2018). https://doi.org/10.1007/s10994-018-5714-4
12. Vovk, V., Wang, R.: Combining p-values via averaging. Biometrika **107**(4), 791–808 (2020)
13. Whitlock, M.C.: Combining probability from independent tests: the weighted Z-method is superior to Fisher's approach. J. Evol. Biol. **18**(5), 1368–1373 (2005)
14. Wong, W., Moore, A., Cooper, G., Wagner, M.: What's strange about recent events (WSARE): an algorithm for the early detection of disease outbreaks. J. Mach. Learn. Res. **6**, 1961–1998 (2005)

Catching Patient's Attention at the Right Time to Help Them Undergo Behavioural Change: Stress Classification Experiment from Blood Volume Pulse

Aneta Lisowska[1](✉), Szymon Wilk[1], and Mor Peleg[2]

[1] Institute of Computing Science, Poznań University of Technology, Poznań, Poland
{aneta.lisowska,szymon.wilk}@put.poznan.pl
[2] Department of Information Systems, University of Haifa, Haifa, Israel
morpeleg@is.haifa.ac.il

Abstract. The CAPABLE project aims to improve the wellbeing of cancer patients managed at home via a coaching system recommending personalized evidence-based health behavioral change interventions and supporting patients compliance. Focusing on managing stress via deep breathing intervention, we hypothesise that the patients are more likely to perform suggested breathing exercises when they need calming down. To prompt them at the right time, we developed a machine-learning stress detector based on blood volume pulse that can be measured via consumer-grade smartwatches. We used a publicly available WESAD dataset to evaluate it. Simple 1D CNN achieves 0.837 average F1-score in binary *stress vs. non-stress* classification and 0.653 in *stress vs. amusement vs. neutral* classification reaching the state-of-art performance. Personalisation of the population model via fine-tuning on a small number of annotated patient-specific samples yields 12% improvement in *stress vs. amusement vs. neutral* classification. In future work we will include additional context information to further refine the timing of the prompt and adjust the exercise level.

Keywords: Blood volume pulse · Stress · Classification · Wearable · Fogg behavioral model

1 Introduction

Cancer patients frequently experience negative emotions such as stress, sadness and fear for the future, that hinder their emotional wellbeing [13], correlate with reduced treatment compliance [5], and increase risk of mortality [15]. Improving the emotional and physical wellbeing of patients is a goal of the Horizon 2020 CAncer PAtient Better Life Experience (CAPABLE) project. CAPABLE aims to develop and implement new persuasive computing methods that will provide cancer patients at home with continuous support for treatment adherence and the development of positive health habits that ultimately can improve

A. Tucker et al. (Eds.): AIME 2021, LNAI 12721, pp. 72–82, 2021.
https://doi.org/10.1007/978-3-030-77211-6_8

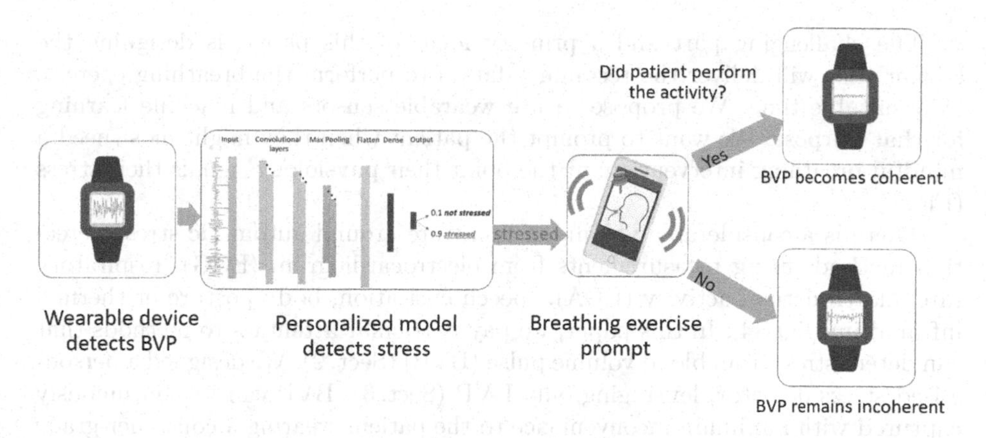

Fig. 1. Stress intervention system

their wellbeing. Patients are equipped with a smartwatch, which monitors their pulse, sleep, and physical activity, as well as a coaching system (a virtual coach), accessed via a mobile app. Our goal is to design an app-based mobile intervention facilitating patients with becoming engaged with their health and inspiring them to undergo behavioural change that builds emotional resilience. The virtual coach can suggest to patients evidence-based activities from the domain of mindfulness and positive psychology, known to have positive effects on one's wellbeing. Each (tiny) activity is meant to become a habit aligned with patients physical and psychological wellbeing goals.

Fogg's Tiny Habits Behaviour Model. [3] proposes that habits formation (i.e., performing a target activity) depends on three factors: motivation, ability to perform the task (which depends on the task's difficulty) and the presence of a trigger reminding the person to perform the target behaviour. For the purpose of this feasibility study, we assume that patients are motivated to achieve their wellbeing goals and that there are simple behaviours that match their abilities; therefore we concentrate on designing appropriately-timed prompts. To identify the simplest behaviour that all cancer patients have the ability to perform and that can impact their emotional wellbeing we draw inspiration from *Integrated Performance Model* [23]. According to Watkins, physiology impacts emotions, which in turn elicit feeling, thinking, and behaviour, leading to desired outcomes (results). Hence to achieve the best outcome, the starting point is changing one's physiology. Watkins points out that controlled rhythmic breathing can influence one's heart rate variability (HRV) and change one's physiological state, leading to a state of 'stable variability' called *coherence*[23]. This is in line with research that investigated the impact of breathing techniques on the autonomic and the central nervous systems [25]. Conscious slow breathing has been found to reduce negative emotions and stress [24] and increase emotional control [6]. Therefore, in CAPABLE, breathing exercise is the target behaviour that our cancer patients have the ability to perform and can benefit from [7].

The challenging part and a primary focus of this paper, is designing the trigger that will deliver the recommendation to perform the breathing exercise at a suitable time. We propose to use wearable sensors and machine learning for that purpose. We want to prompt the patient when they might most need a mindful breathing intervention; at the point their physiology signals their stress (Fig. 1).

There is a considerable volume of literature around automatic stress detection methods using measurements from electrocardiogram (ECG), respiratory rate, electrodermal activity (EDA), speech excitation, body posture or thermal infrared imaging [4]. In this paper, we pay particular attention to methods that can detect stress from blood volume pulse (BVP) (Sect. 2). We designed a personalised stress detector, leveraging only BVP (Sect. 3). BVP can be continuously captured with minimum inconvenience to the patient wearing a consumer-grade smartwatch. Additionally, BVP is altered by the breathing pattern [21], thus it can provide useful evaluation of effectiveness of the prompt and the breathing exercise intervention. In this paper we focus on experimental evaluation of the first part of the system, namely of our proposed stress detector (Sect. 4) that will dictate the appropriate timing for the breathing exercise prompt. Specifically, we intend to answer the following research questions:

- Is it possible to develop a stress detector using raw BVP data (without handcrafted features)?
- Does personalisation of a stress detector improve its performance?

2 Related Work

Stress is characterised by high physiological arousal [9]. Regardless of the person's age, tense arousal is associated with higher heart rate [16]. However, discriminating stress from other emotions, such as excitement or flow (an intrinsically rewarding experience of being immersed in a task [1]), is not straightforward, as both result in elevated physiological arousal and modulation of the heart period [14]. According to Russel, emotional states can be mapped into a 2D space [17], where arousal can be assigned to the first dimension and valence to the second one [10]. Valence reflects how positive or negative the emotion is. From the perspective of this study, it is important to discriminate between positive arousal states and stress, as both fall high in the arousal spectrum compared to relaxation but differ in the enjoy-ability of the experience (valence).

McCraty and Rees argue that positive emotions are characterised by the coherent pattern of the heartbeat rather than the heart rate [12]. Therefore, to distinguish between emotions it is important to acquire the full pulse signal and inspect fluctuation in the heartbeats rather than look only at the cumulative measure of the beats per minute. The heart beat-to-beat variability over time can be captured through measurement of the BVP using Photoplethysmography (PPG) sensor. The PGG sensors can be attached anywhere on the body, however, most commonly they are placed on a finger [22] or wrist [19] and importantly they are integrated into consumer-grade smartwatches [18].

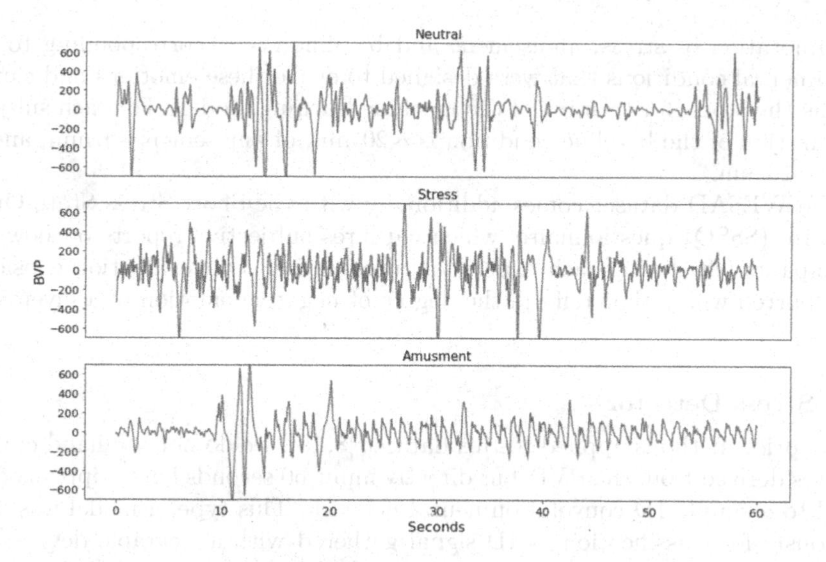

Fig. 2. An exemplar 1 min snippets of BVP measurements from one subject in the WESAD dataset captured during 3 different emotions: neutral, stress, amusement.

Schmit *et al.* utilise both wrist and chest-worn sensors to acquire a multimodal dataset for Wearable Stress and Affect Detection (WESAD)[19]. This data set is particularly interesting because it captures the positive arousal state of amusement alongside stress and neutral emotion from 15 subjects (Fig. 2). Constructively, Schmit et al. [19] provide baseline performance of multiple machine learning methods trained on features extracted from each of the sensors separately and in combination. When utilising only BVP, Linear Discriminant Analysis (LDA) achieved the best emotion classification performance in leave-one-out evaluation. The reported F1 score for the three-class problem is 0.547 from LDA, closely followed by Random Forest (0.538) and AdaBoost (0.533). In the reduced version of the problem limited to a binary classification of stress vs. not-stress, the F1 score from BVP reached 0.831. This is the state of the art performance of a generalised model on this dataset when leveraging only BVP. Nevertheless, given that people differ in their physiological response, personalised stress models might offer better performance [20]. Indikawati and Winiarti used all wrist sensor measurements from WESAD dataset to train personalised emotion classifier and obtained 88–99% classification accuracy using Random Forest [8]. However, they do not report the performance using BVP without the inclusion of EDA and temperature measurements.

3 Methods

3.1 Dataset

We use publicly available WESAD dataset [19] for development of our stress detector. The BVP measurements (64 Hz) from 15 subjects are accompanied by the annotation of stress, amusement and baseline state, corresponding to the experimental conditions that were designed to evoke these emotions and can be used as the labels for training of the machine learning models. For each subject, the duration of the baseline condition is ∼20 min, of amusement - 6 min, and of stress - 10 min.

The WESAD dataset comes additionally with the Short Stress State Questionnaire (SSSQ) questionnaire, which captures subjective reports on how the participants felt during each experimental condition. Our evaluation considers self-reported worry that reflects the degree of negative emotion of a given subject.

3.2 Stress Detector

Unlike prior methods applied to this dataset [8,19], we do not use hand crafted features derived from the BVP, but directly input 60 seconds long snippets of the signal to a simple 1D convolution neural network. This type of model was used previously for classification of 1D signal gathered with a wearable device [11]. The model is constructed from two convolutional layers with 16, 8 filters respectively and kernel size of 3, followed by max pooling layer and fully connected layer with 30 nodes and output layer, which size corresponded to the number of the classes (emotions). The number of filters and nodes was chosen empirical. Each convolutional layer has a ReLu activation function and output layer with softmax. The model is implemented in Keras with a TensorFlow backend.

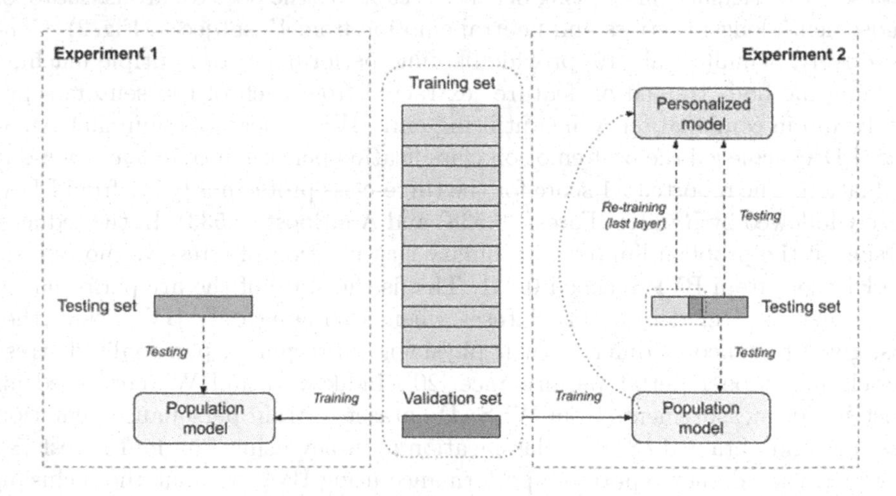

Fig. 3. Experimental set up.

3.3 Experiments

To answer our research questions we conduct two experiments (Fig. 3). The first experiment investigates if our stress detector trained on BVP signals reaches state-of-the-art classification performance. In the second experiment we examine if stress detector personalisation can provide further performance boost.

Experiment 1: When evaluating the generalised stress detector we follow Schmidt *et al.* [19] and adapt a leave-one-out (LOO) approach to directly benchmark against other previously reported approaches. The extracted examples from the test patient have a window size of 3840 samples and a step size of 1600 samples, yielding ~88 test examples (Fig. 4). We use weighted F1-score as a performance metric recommended for unbalanced classification tasks.

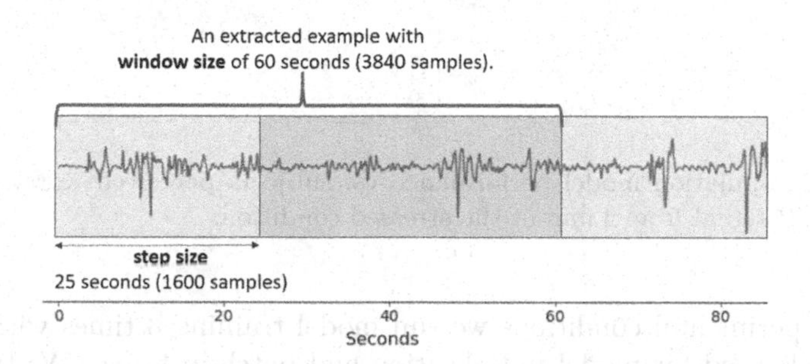

Fig. 4. An example of test example extraction from patient signal.

For population model training we extract training examples with window size of 3840 samples. In two-class problem the step size is 18 samples for non-stress condition and 12 samples for stress condition, which yields ~117000 training and 8600 validation examples. In three-class problem, we use a step size of 18, 10, 12 samples for baseline, stress and amusement conditions respectively, yielding ~130000 training and 9300 validation examples. The step size was varied during training to reduce the imbalance between classes.

Experiment 2: The number of annotated examples that can be obtained from the patient using the mobile app is very small, therefore training the model from scratch on the single patient data is not feasible. We suggest that the personalised model can be trained by fine-tuning of the population model on a small number of annotated samples from the patient of interest. All the layers of the population model except the last one are frozen and the first half of the patients' BVP signal from each emotional condition is used for the model retraining. Similarly as in population model training the step size for non-stress events was of a size 18 samples and 12 for stress, resulting in ~4000 retraining examples per patient. In three-class classification baseline examples were extracted with a step size 16, stress with 7 and amusement of 5 yielding ~6000 retraining examples. In both personalised fine tuning conditions 80% of retraining examples were used for training and 20% for validation. To be able to directly compare population and personalised models we applied both the second half of the patient's BVP signals. We used the same windows and step size (3840 and 1600 samples, respectively) as in experiment 1 and obtained ~43 test examples for each patient. Obtained performance is reported using micro F1.

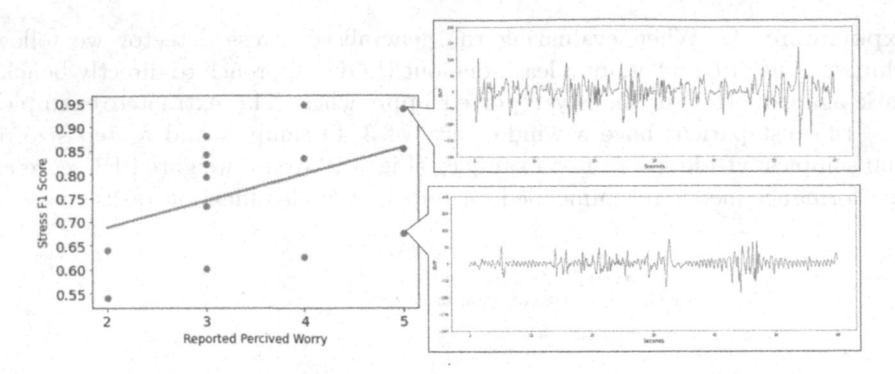

Fig. 5. Left: Population model performance vs. subjects perceived worry. Right: An exemplar BVP signal from 1 min of the stressed condition.

In all experimental conditions we run model training 3 times with different random seeds used for model initialisation and batch ordering. We trained the model with categorical cross-entropy using Adam optimiser and batch size 256. To avoid overfitting we use early stopping on validation data.

4 Results

Experiment 1: In discriminating *stress vs. non-stress* our model reached average across all subjects of 0.837 F1 score and in three-class condition 0.653. These results are slightly higher than the results obtained by LDA, the best generalised approach using BVP previously reported [19]. Nevertheless, the variation in performance between subjects is very high at 0.71–0.98 in a two-class problem. We hypothesised that the generalised model works well for subjects who are more stressed and is less accurate for moderately stressed subjects. To investigate whether the model performance reflects the degree of negative emotion of the subjects we plot the stress F1 score from the 3 class model against self-reported worry (Fig. 5). Note that the weak positive trend between model performance and subjects perceived worry is not significant ($r_s = 0.477, p = 0.07$). The stress detector performed better for the subjects, whose stress response manifested in very erratic BVP.

Experiment 2: Table 1 shows the F1 score for each patient using the small patient-specific testing set. In *stress vs. non-stress* classification, personalised model achieved better classification results for 10 subjects out of 15, there was no difference in performance for 1 subject and for 4 subjects personalised model yielded slightly worse results. The decrease in performance might be due to an insufficient number of samples retraining examples leading to overfitting of already well performing population models. On average personalised model achieved a higher score of 0.822 compared to the population model 0.813 with

Table 1. Mean F1 micro for each subject from 3 runs with different random seeds in personalised evaluation set up for both 2 class and 3 class classification problem. The better performing model is highlighted in grey.

subj.	***Stress vs. Non-stress***				***Baseline vs. Stress vs. Amusement***			
	Population Model Mean	[std]	Personalised Model Mean	[std]	Population Model Mean	[std]	Personalised Model Mean	[std]
2	0.894	[0.023]	0.813	[0.050]	0.715	[0.046]	0.732	[0.040]
3	0.821	[0.064]	0.829	[0.060]	0.520	[0.064]	0.585	[0.087]
4	0.870	[0.083]	0.911	[0.023]	0.870	[0.064]	0.870	[0.064]
5	0.937	[0.022]	0.937	[0.030]	0.468	[0.045]	0.754	[0.011]
6	0.762	[0.019]	0.778	[0.011]	0.651	[0.022]	0.683	[0.030]
7	0.889	[0.011]	0.944	[0.011]	0.548	[0.085]	0.635	[0.129]
8	0.849	[0.002]	0.794	[0.059]	0.595	[0.070]	0.500	[0.118]
9	0.675	[0.011]	0.714	[0.019]	0.373	[0.096]	0.619	[0.101]
10	0.742	[0.057]	0.735	[0.054]	0.462	[0.070]	0.636	[0.067]
11	0.891	[0.011]	0.915	[0.011]	0.535	[0.083]	0.814	[0.076]
13	0.845	[0.022]	0.791	[0.083]	0.636	[0.029]	0.659	[0.090]
14	0.752	[0.044]	0.760	[0.040]	0.535	[0.087]	0.783	[0.194]
15	0.620	[0.044]	0.674	[0.033]	0.194	[0.029]	0.543	[0.105]
16	0.814	[0.068]	0.891	[0.077]	0.845	[0.040]	0.868	[0.011]
17	0.833	[0.077]	0.841	[0.064]	0.818	[0.037]	0.894	[0.021]
Mean	0.813		**0.822**		0.584		**0.705**	
Std	0.084		0.081		0.176		0.119	

similar standard deviation (0.08). In a classification of *baseline (neutral) emotion vs. stress vs amusement*, the personalisation of the model leads to better performance for 13 subjects, for 1 there was no difference and for one the performance decreased. The average F1 score for the personalised model is 0.705 (0.119) and population model 0.585 (0.176). Note the variance of the population models in this condition is higher than of the personalised models. In three-class problem the advantage of the model personalisation is more apparent than in binary classification task.

5 Discussion

The CAPABLE project aims to support cancer patients with achieving their wellbeing goals. In this work, we take emotional health as a case study and propose a system for stress intervention. As a first step, we focus on empirical evaluation of the stress detector that we intend to utilize for prompting the patient to perform simple breathing exercise. Our generalized stress detector requires fewer preprocessing steps than previously used methods (as it is applied directly on the raw BVP signal) and achieves the state of the art performance on WESAD dataset. Nevertheless, there is a large variation in the appearance of the BVP signal in stress conditions between patients suggesting that personalization of the models could be beneficial. We find that personalisation of the generalised stress detector simply via fine-tuning on the small number of annotated samples shows an encouraging improvement in classification performance.

In practice, further improvements to the personalised models could be obtained with methods as active learning, where the patient is occasionally asked to report their emotion via annotation of their current emotional state. Similar techniques could drive improvements of the population model where annotated samples are gathered from multiple users and used to retrain the model in federated learning fashion. The best stress detectors might rely on a combination of the personalized models or assignment of the patient to the subpopulation model based on patients' similarity to that subpopulation.

However, from the perspective of the full stress intervention solution, the performance of the stress detector is not the only important factor; while others intended to simply detect stressful events, we try to determine when it is the best time to prompt a patient to perform a selected breathing activity. We hypothesize that prompting patients when they need the intervention the most (i.e., when they are stressed) will increase their compliance. We plan to evaluate this hypothesis by comparing the effectiveness against that of prompts sent randomly. Effectiveness could be concluded if coherent BVP would be measured soon after the patient clicked on the prompt reminding him to perform the breathing.

In future work, we will incorporate additional information such as GPS location, or time of the day, to try to predict the best timing of the prompt further. We also plan to follow *Fogg's 8 steps persuasive design process* [2], and figure out what is preventing users from performing target behaviour. Therefore, in case the activity was not performed, we will request the patient to specify whether the suggested exercise was adequate and whether the timing was good; the patient might have been stressed but the timing of the prompt could have been poor because the patient was performing some other activity (e.g. driving, shopping) preventing them from engaging in the suggested exercise.

Acknowledgments. The CAPABLE project has received funding from the European Union's Horizon 2020 research and innovation programme under grant agreement No 875052.

References

1. Csikszentmihalyi, M.: Flow: The Psychology of Optimal Experience, vol. 1990. Harper & Row, New York (1990)
2. Fogg, B.J.: Creating persuasive technologies: an eight-step design process. In: Proceedings of the 4th International Conference on Persuasive Technology, pp. 1–6 (2009)
3. Fogg, B.J.: Tiny Habits: The Small Changes That Change Everything. Houghton Mifflin Harcourt, Boston (2019)

4. Giannakakis, G., Grigoriadis, D., Giannakaki, K., Simantiraki, O., Roniotis, A., Tsiknakis, M.: Review on psychological stress detection using biosignals. IEEE Trans. Affect. Comput. (2019). https://doi.org/10.1109/TAFFC.2019.2927337
5. Greer, J.A., Pirl, W.F., Park, E.R., Lynch, T.J., Temel, J.S.: Behavioral and psychological predictors of chemotherapy adherence in patients with advanced non-small cell lung cancer. J. Psychosom. Res. **65**(6), 549–552 (2008)
6. Gross, M.J., Shearer, D.A., Bringer, J.D., Hall, R., Cook, C.J., Kilduff, L.P.: Abbreviated resonant frequency training to augment heart rate variability and enhance on-demand emotional regulation in elite sport support staff. Appl. Psychophysiol. Biofeedback **41**(3), 263–274 (2016). https://doi.org/10.1007/s10484-015-9330-9
7. Hayama, Y., Inoue, T.: The effects of deep breathing on 'tension-anxiety' and fatigue in cancer patients undergoing adjuvant chemotherapy. Complement. Ther. Clin. Pract. **18**(2), 94–98 (2012)
8. Indikawati, F.I., Winiarti, S.: Stress detection from multimodal wearable sensor data. In: IOP Conference Series: Materials Science and Engineering, vol. 771, p. 012028. IOP Publishing (2020)
9. Johnson, A.K., Anderson, E.A.: Stress and arousal. Principles of psychophysiology: Physical, social and inferential elements. Cambridge University Press (1990)
10. Lang, P.J.: The emotion probe: studies of motivation and attention. Am. Psychol. **50**(5), 372 (1995)
11. Lisowska, A., O'Neil, A., Poole, I.: Cross-cohort evaluation of machine learning approaches to fall detection from accelerometer data. In: HEALTHINF, pp. 77–82 (2018)
12. McCraty, R., Rees, R.A.: The central role of the heart in generating and sustaining positive emotions. In: Oxford Handbook of Positive Psychology, pp. 527–536 (2009)
13. Page, A.E., Adler, N.E., et al.: Cancer Care for the Whole Patient: Meeting Psychosocial Health Needs. National Academics Press, Washington, D.C. (2008)
14. Peifer, C., Schulz, A., Schächinger, H., Baumann, N., Antoni, C.H.: The relation of flow-experience and physiological arousal under stress–can u shape it? J. Exp. Soc. Psychol. **53**, 62–69 (2014)
15. Pinquart, M., Duberstein, P.: Depression and cancer mortality: a meta-analysis. Psychol. Med. **40**(11), 1797–1810 (2010)
16. Riediger, M., Wrzus, C., Klipker, K., Müller, V., Schmiedek, F., Wagner, G.G.: Outside of the laboratory: associations of working-memory performance with psychological and physiological arousal vary with age. Psychol. Aging **29**(1), 103 (2014)
17. Russell, J.A.: Affective space is bipolar. J. Pers. Soc. Psychol. **37**(3), 345–356 (1979)
18. Saganowski, S., et al.: Review of consumer wearables in emotion, stress, meditation, sleep, and activity detection and analysis. arXiv preprint arXiv:2005.00093 (2020)
19. Schmidt, P., Reiss, A., Duerichen, R., Marberger, C., Van Laerhoven, K.: Introducing WESAD, a multimodal dataset for wearable stress and affect detection. In: Proceedings of the 20th ACM International Conference on Multimodal Interaction, pp. 400–408 (2018)
20. Shi, Y., et al.: Personalized stress detection from physiological measurements. In: International Symposium on Quality of Life Technology, pp. 28–29 (2010)
21. Steffen, P.R., Austin, T., DeBarros, A., Brown, T.: The impact of resonance frequency breathing on measures of heart rate variability, blood pressure, and mood. Front. Public Health **5**, 222 (2017)

22. Udovičić, G., Đerek, J., Russo, M., Sikora, M.: Wearable emotion recognition system based on GSR and PPG signals. In: Proceedings of the 2ndInternational Workshop on Multimedia for Personal Health and Health Care, pp.53–59 (2017)
23. Watkins, A.: Coherence: The Secret Science of Brilliant Leadership. Kogan Page Publishers, London (2013)
24. Yu, X., et al.: Activation of the anterior prefrontal cortex and serotonergic system is associated with improvements in mood and EEG changes induced by Zen meditation practice in novices. Int. J. Psychophysiol. **80**(2), 103–111 (2011)
25. Zaccaro, A., et al.: How breath-control can change your life: a systematic review on psycho-physiological correlates of slow breathing. Front. Hum. Neurosci. **12**, 353 (2018)

Primary Care Datasets for Early Lung Cancer Detection: An AI Led Approach

Goce Ristanoski[1(✉)], Jon Emery[2,3], Javiera Martinez Gutierrez[2,4], Damien McCarthy[2], and Uwe Aickelin[1]

[1] School of Computing and Information Systems, The University of Melbourne, Melbourne, Australia
gri@unimelb.edu.au

[2] Department of General Practice and Centre for Cancer Research, Medicine, Dentistry and Health Sciences, The University of Melbourne, Melbourne, Australia

[3] Victorian Comprehensive Cancer Centre, Melbourne, Australia

[4] Department of Family Medicine, School of Medicine, Pontificia Universidad Católica de Chile, Santiago, Chile

Abstract. Cancer is one of the most common and serious medical conditions, with significant challenges in the detection of cancer originating from the non-specific nature of symptoms and very low prevalence. For general practitioners (GPs), this can be particularly important, as they are the primary contact for patients for most medical conditions. This places high significance on using the data available to a GP to design decision support tools that will aid GPs in detecting cancer as early as possible. With pathology data being one of the datasets available in the GP electronic medical record (EMR), our work targets this type of data in an attempt to incorporate an early cancer detection tool in existing GP practices. We focus on utilizing full blood count pathology results to design features that can be used in an early cancer detection model 3 to 6 months ahead of standard diagnosis. This research focuses initially on lung cancer but can be extended to other types of cancer. Additional challenges are present in this type of data due to the irregular and infrequent nature of doing pathology tests, which are also considered in designing the AI solution. Our findings demonstrate that hematological measures from pathology data are a suitable choice for a cancer detection tool that can deliver early cancer diagnosis up to 6 months ahead for up to 8 out of 10 patients, in a way that is easily incorporated in current GP practice.

Keywords: Early lung cancer detection · Primary care data · Explainable AI

1 Introduction

Through one's medical history we come in contact with our General Practitioners (GPs) far more often than we do with other medical staff, particularly specialists. The resources and technology available at the GP practices are, however, more limited to those in hospital specialist care. GPs play a key role in diagnosis of serious diseases, but this can be challenging due to the fact that symptoms alone are poorly predictive, especially

A. Tucker et al. (Eds.): AIME 2021, LNAI 12721, pp. 83–92, 2021.
https://doi.org/10.1007/978-3-030-77211-6_9

for uncommon conditions in primary care. Decision support tools could potentially contribute to flagging patients at increased risk of serious disease and prompting further referral or investigation.

One of the medical conditions that can have serious consequences on patient's lives depending on the time of diagnosis is cancer. In Australia, there are more than 144 000 cancer patients who were diagnosed in 2019 alone. Early detection of cancer by GPs is challenging if symptoms alone are used and patients existing history is underutilized. In the last 15 years, the research in the epidemiology of cancer symptoms in primary care data has grown, with many findings demonstrating how advanced analysis and combinations of different symptoms and tests from a patient's medical history can be used to assess cancer risk [1–3]. If patients have a regular GP they visit, having just 2-years' worth of patient data can be sufficient in some cases to combine several different tests into a risk prediction model that can provide the initial diagnosis around 3 months ahead of the current practice [3, 4]. This may not seem like too long a period at first, but with studies showing how every additional month of an undiagnosed cancer can increase the mortality rate for certain types of cancer [5], establishing early diagnosis at the GP's office is even more important.

Pathology results are one of the most common types of data that exist in the patient EMR that is readily available to a GP. This opens the opportunity to investigate if some of the blood tests can be associated with certain types of cancer. Recent research highlights raised platelet count (thrombocytosis) as a predictor of cancer risk [6, 7], but there have been no specific studies that focus on understanding how to introduce a more advanced AI component into cancer detection and how it can be adapted to current pathology data in the GP's EMR. Our work places a strong emphasis on this, allowing for both interpretability of our results and easy application and usage.

The full blood count test results we investigate as a potential input for a Machine Learning/AI model are: Platelet count, MCV (Mean Corpuscular Volume), MCH (Mean Corpuscular Hemoglobin - average mass of hemoglobin per red blood cell), MCHC (Mean Corpuscular Hemoglobin Concentration - concentration of hemoglobin in a given volume of packed red blood cell) and RDW (Red blood cell distribution width). Platelet count is already associated to lung cancer from other studies, and this set of features allows to develop an initial approach to use pathology results in cancer detection, providing opportunity to expand the list of pathology test metrics with more metrics in future work. We focused on lung cancer patients' pathology results as this is a common cancer and patients often have multiple pathology results; lung cancer has a high mortality rate and could benefit from an early cancer detection model.

The work we conducted places a heavy focus on delivering an initial cancer diagnosis early as possible, which is why we developed our models to make predictions 3-months and 6-months ahead of current practices. We attempted to design a model that could flag a diagnosis 6 months in advance specifically to aid in early detection of cancer for high risk patients - patients who did not survive the cancer, who potentially have most to gain from earlier detection.

We present our work with the following contributions:

- We discuss the ideas of using pathology results in an AI model for cancer detection and show reasoning behind this hypothesis.

- We demonstrate how the metrics listed above can be relevant to cancer patients.
- We address the type and structure of pathology results data available to a GP to design features that are easy to implement and use in a cancer detection model.
- We present AI models with performance that shows promising results in detecting cancer, especially for high risk patients.
- We list future opportunities that can improve this type of work even further with little modification and wide application.

2 Related Work

With more than 2.9 million deaths worldwide associated to lung cancer in 2018, it has become imperative to find additional ways to better detect the early symptoms of lung cancer and provide timely diagnosis [8]. The main challenge about the symptoms however is that they can vary from one patient to another and can take even up to 2 years for the symptoms to be visible enough to have them attributed to cancer [8]. Raised platelet count, or thrombocytosis, has been shown to be an indicator for cancer, with differences in the results for biological male vs female patients – male cancer patients were 50% more likely to have thrombocytosis than female patients [6, 10]. This resulted in the practice of referring patients with thrombocytosis to an x-ray scan in an attempt to detect the cancer patients promptly [11]. Anemia has also been shown to have association with lung cancer, with slightly higher presence in male patients as well [2, 10], which brings us to our hypothesis of investigating blood count results in combined scenario.

Designing risk prediction models with individual metrics have been investigated to a good extent [12–14], but without strong emphasis on combining several metrics into single model, or considering the application range of the prediction models in GP offices. Reviews on the use of primary care data for cancer prediction with other types of primary care datasets are also indicating that blood results are increasingly popular for the task [15, 16], and with indications that lung cancer patients tend to have blood tests more often [17], it provides fertile grounds for introducing AI models in the bigger picture.

3 Dataset Description

3.1 NPS MedicineInsight

The Australian Government Department of Health (DoH) established the NPS MedicineInsight initiative as a nationally representative primary care dataset that can be used by academic researchers in attempt to deliver new research findings that can improve medical practices. The NPS MedicineInsight contains patients records from more than 500 general practices and 5000 GP providers, which includes more than 8 million recorded diagnoses, 23 million prescriptions, 32 million encounters and 85 million pathology test results [18]. For our research work, we obtained the lung cancer patients cohort, as well as a non-cancerous patients' cohort as a control group.

With an extensive amount of records and results from pathology tests, we focused our work on the five blood test metrics listed earlier: Platelet count, MCHC, MCV, MCH and RDW. We looked at the out of range records for these metrics, with the standard range being: platelet count of 150–450 × 10^9/L, MCV of 80–98 fL, MCH of 28–32 pg/cell, MCHC of 330–370 g/L, RDW of 12.2%–16.1% F/11.8%–14.5% M. Our models used an out of range value in at least one of these metrics as a trigger for classification, meaning that lung cancer patients that have no out of range values unfortunately were not assessed for early detection. The analysis showed that around 20% of the patients per each metric had a record with an out of range value for that metric, so combining several metrics increased the total subset of lung cancer patients suitable for early detection. Subsequently, we only considered non-cancerous patients with out of range values as a control group.

3.2 Cancer Patient's Analysis

The available lung cancer patient cohort showed a very interesting pattern compared to other patients. One of the things we noticed initially was that not only did lung cancer patients had around 20% out of range value for a given blood test metric, they also had on average 3 times more tests taken in the two year period before cancer diagnosis than patients that had no out of range results, allowing both better quality in data and initial indication of use of pathology results for early diagnosis.

Another interesting aspect of our analysis showed that not only there were out of range tests for a good portion of cancer patients, but that the mortality rate for patients with out of range tests was much higher than for patients with no out of range results. Shown in Fig. 1 is an example for patients with out of range results for RDW compared to patients with no out of range results and the mortality figures per age group (group 2 = 20–29 y.o. etc.). We can observe higher mortality ratio for patients with out of range, showing that even if we can only include patients with out of range results in the final model, these patients are high risk patients and they may benefit from early cancer detection the most.

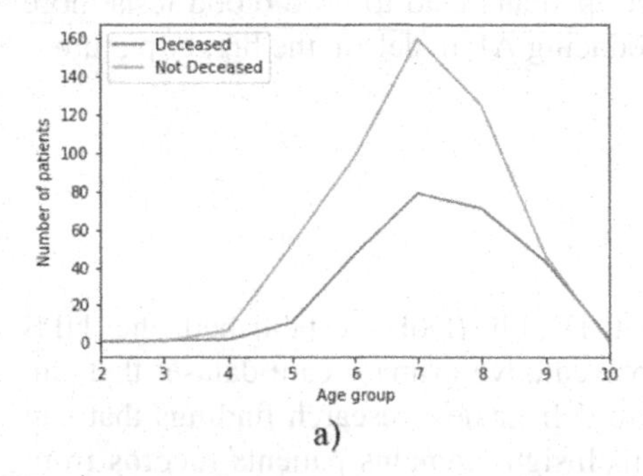

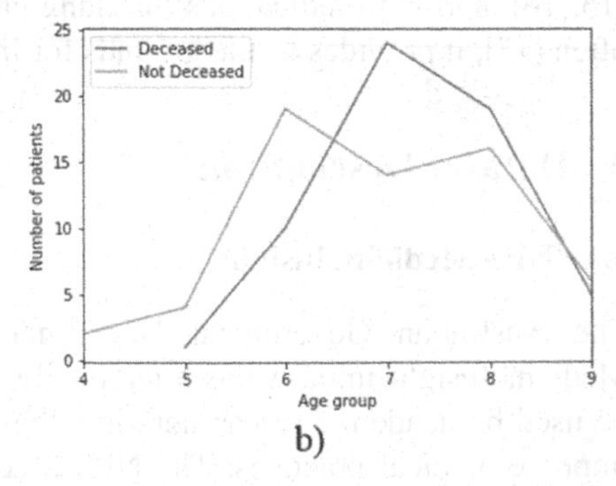

Fig. 1. Patients that survive vs. patients that did not survive cancer based on out of range results for RDW: a) Patients with no out of range results; b) Patients with out of range results 3 months prior diagnosis date

4 Features Design and Methods Selection

4.1 Uncertainty Based Features Design

Pathology tests are readily available to GP to order, but as we could see in the previous analysis, the number of tests per patient can vary – lung cancer patients that had out of range results had more tests than other patients which worked in favor of our work to some extent. The frequency and regularity of these tests is not a matter of standardized practices: GPs issue a request for test when patients visit them, and this is something that is irregular and driven by a range of factors. This poses some limitations on the quality of data as well as the use of this type of data for AI models.

Our approach was to handle this uncertainty by using time periods and occurrences of out of range tests results to incorporate some structure in the features and allow for use of pathology data without any special need for its format other than the current ones used in practice. For the lung cancer patients, from the initial diagnosis date recorded at the GP clinic, we took the pathology tests within the two-year period prior to that date. We then represented the occurrence of any out of range results for each of the five listed metrics in the periods of 24–18, 18–12, 12–6 and 6–3 months before the cancer diagnosis date. The occurrence of each metric per individual period formed one original feature, with 0 meaning no occurrence of out of range result for that metric for the given time period, and 1 meaning at least one occurrence of out of range result for that metric in that time period. This created data suitable for 3 months before diagnosis, and by removing the features with the 6–3 months period we could also perform a 6 months before diagnosis feature. For the control group of non-cancerous patients we selected the period of 2016–2017 and the same features were calculated for that period. We did not consider multiple occurrences within one time period as often pathology tests can be issued subsequently, and this would bring no new information to our models.

4.2 Soft Out of Range Results

The normal ranges for each of the hematological measures were listed earlier, and based on some of the test results, we included the results at the very end of the normal ranges as soft out of range. For example, platelet count is most commonly defined within normal range of 150–450 thousand platelets per microliter of blood, so patients with results of 150 will be within the range, but patients with results 149 are out of range. In order to allow patients with results of 150 or just above it to still be considered as out of range we defined soft range as being the 2.5% ends from within the lowest and highest values. Using the platelets example, 2.5% of the 450–150 = 300 is 7.5 units, so the soft ranges would be (157.5–442.5).

This means that we ended up with more test results being out of range and potentially more patients being suitable to be considered for early diagnosis. For the 3 months before diagnosis, using the standard range we had 592 patients within our dataset, and that number increased to 683 with the soft out of range definition.

4.3 Additional Features

Besides the original features for the five metrics for each time period listed above, we combined some more features based on those and other patient data to allow some more temporal and quantitative aspects to be included in the algorithm. These were:

- Summary of occurrences per blood test metric
- Summary of occurrences of any metrics over a 3- or 6-month pre-diagnosis period
- Separating the out of range values into two separate features for upper and lower threshold out of range
- Separating the previous features per biological sex
- Separating the previous features per age group

By using this feature set, we could get a clearer view of the importance of occurrences vs. frequency of out of range results, both total and per individual age group or biological sex.

5 Model Selection, Experiments and Results

5.1 Model Selection

The use of pathology data in the features listed above not only handled the uncertainty in the data that originated from the irregularity of pathology tests, but it also provided another crucial contribution: it allowed us to see if the individual original features or the combinatory ones had more useful information. In order to allow even more interpretability in both the final performance and the relevance of the features, we used decision tree style models: Decision Tree, AdaBoost, LightGBM and XGBoost. We also used an ensemble approach to check for additional performance evaluation: a stack model that uses the forecasts of the other classifiers as an input, as well a simple ensemble with the OR logic between all the classifiers.

We were interested to see how our models performed in correctly classifying the lung cancer patients. We wanted to achieve both high values for True Positive Rate (TPR) and True Negative Rate (TNR), and also from all the predicted positives we wanted the cancer patients to have the highest portion (Positive Predictive Value, PPV). We had a total of 592 patients for the 3 months ahead early diagnosis, and 683 patients for the same diagnosis with soft out of range features. For the 6 months ahead early diagnosis, we had 499 patients total for both standard range and soft out of range.

The use of different ratios of non-cancerous patients: cancer patients allowed us to see how the TPR and TNR changed and if we could avoid having lots of false positives. We used the ratio of 1:1, 1.5:1 and 2:1, and suggested not going higher than 4:1 in order to avoid issues due to an imbalanced dataset. The chi-squared statistic for ranking the top features was used, and we showed the average performance when using 41–54 features.

Our cancer patients were all 50+ years old, and we also added a subset the 50–79 years range to allow for better quality of data as patients aged 80+ had different frequencies of pathology tests and could have more health issues that made it difficult to differentiate between cancer based out of range pathology tests and other conditions out of range tests.

5.2 Experiments and Results

The results presented in Tables 1 and 2 show the average performance of the models with over 14 runs, with 41–54 features used per run. The standard deviation over each metric was rarely higher than 0.01, so the performance was quite consistent per each run. We investigated the impact of three ratios of non-cancerous patients to cancer patients (1:1, 1.5:1 and 2:1) and in all cases the ratio of 1:1 showed best results for TPR and only those figures are presented here. As we increased the ratio, the value of TPR dropped in all cases while TNR (True Negative Rate) increased to values of 0.9. The PPV value also shows good performance, and it is closely matched with the Negative Predictive Value (NPV).

Table 1. Performance metrics for data with regular range

All samples	3 months ahead				6 months ahead			
Classifier	TPR	TNR	PPV	NPV	TPR	TNR	PPV	NPV
AdaBoost	0.686	0.722	0.711	0.697	0.684	0.679	0.680	0.682
DecisionTree	0.613	0.747	0.708	0.659	0.586	0.692	0.656	0.626
Ensemble	0.807	0.619	0.679	0.763	0.784	0.565	0.643	0.723
LightGBM	0.705	0.710	0.708	0.707	0.684	0.686	0.685	0.684
Stack	0.705	0.710	0.708	0.707	0.684	0.686	0.685	0.684
XGB	0.722	0.748	0.742	0.730	0.671	0.715	0.702	0.685
Under80								
Classifier	TPR	TNR	PPV	NPV	TPR	TNR	PPV	NPV
AdaBoost	0.650	0.690	0.678	0.664	0.617	0.684	0.661	0.642
DecisionTree	0.594	0.730	0.688	0.643	0.541	0.670	0.621	0.594
Ensemble	0.770	0.573	0.643	0.714	0.755	0.569	0.637	0.701
LightGBM	0.664	0.717	0.701	0.681	0.650	0.653	0.652	0.651
Stack	0.665	0.718	0.702	0.682	0.649	0.653	0.652	0.651
XGB	0.655	0.762	0.733	0.689	0.652	0.717	0.697	0.704

The results confirm that not only we were able to provide 3 months ahead forecast with our pathology results with good accuracy (7–8 out of 10 correctly classified patients) but we could also provide similar accuracy with the 6 months forecast as well. The individual models when combined in an Ensemble with OR logic (if one model classifies a 1, the ensemble outputs 1) performed well for the 3 months ahead forecast, but not as good for the 6 months ahead. The Stack model did not seem to suffer this issue. Still, a 7 out of 10 forecast delivered 6 months ahead with only 5 metrics is very promising in the future use of the pathology results in primary care data for early cancer detection.

The performance of the 6 months ahead forecast was also satisfactory in the prediction of the high-risk patients task. We can observe from Fig. 2 that for both regular out of range and soft out of range, the percentage of deceased patients in the correctly classified cancer patients was higher than the percentage of deceased patients in the false negative forecasts: in some cases nearly 45% of the patients in the correct classifications were high risk patients that were deceased within 4 years of the cancer diagnosis, and this number was as low as 35% in the false negative forecasts. This shows that our models were able to detect the cancer patients that can benefit from an early diagnosis the most.

Table 2. Performance results for metrics with soft range

All samples	3 months ahead				6 months ahead			
Classifier	TPR	TNR	PPV	NPV	TPR	TNR	PPV	NPV
AdaBoost	0.659	0.760	0.733	0.690	0.665	0.712	0.698	0.680
DecisioTree	0.622	0.765	0.725	0.669	0.601	0.704	0.670	0.638
Ensemble	0.783	0.661	0.698	0.753	0.772	0.595	0.656	0.723
LightGBM	0.686	0.758	0.739	0.707	0.665	0.711	0.697	0.680
Stack	0.686	0.758	0.739	0.707	0.665	0.711	0.697	0.680
XGB	0.657	0.775	0.745	0.694	0.639	0.742	0.712	0.673
Under 80								
Classifier	TPR	TNR	PPV	NPV	TPR	TNR	PPV	NPV
AdaBoost	0.689	0.732	0.720	0.703	0.592	0.717	0.676	0.639
DecisioTree	0.591	0.689	0.656	0.628	0.537	0.646	0.603	0.583
Ensemble	0.781	0.613	0.668	0.737	0.738	0.585	0.640	0.692
LightGBM	0.675	0.680	0.679	0.677	0.620	0.649	0.639	0.631
Stack	0.675	0.680	0.679	0.677	0.620	0.649	0.638	0.631
XGB	0.670	0.741	0.721	0.692	0.628	0.704	0.680	0.655

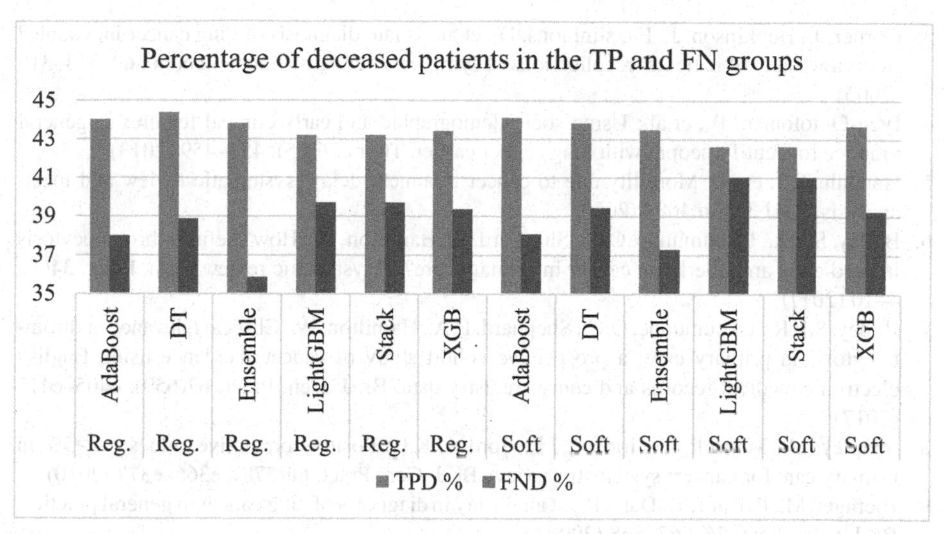

Fig. 2. Percentage of deceased patients in the True Positives (TP) and False Negatives (FN) groups per classification model and out of range type

6 Conclusion

The work presented in this paper demonstrates the opportunities to use currently underutilized set of data for early cancer detection: A primary care dataset containing pathology results. Not only do we justify the reasoning behind the use of full blood test metrics for early cancer detection, but we also handle the challenge of the data containing records at irregular and infrequent time periods. By using features that represent both temporal and quantitative values in the out of range results, we were able to predict lung cancer diagnosis up to 6 months ahead of time, with models that required no modification to current GP practices and would be relatively easy to implement in clinics for both lung cancer detection and other types of cancer as well.

This work opens opportunities for further research in areas such as more high-risk patient focused forecast, inclusion of other pathology tests, and potentially incorporating social and economic features in the AI models. Based on availability of additional data about the stage of cancer at the time of detection and hospital treatment, we may further deliver more insights by using pattern detection and visualization methods to determine the most descriptive features in the pathology tests per different type of cancer or patient cohort.

References

1. Hamilton, W.: The CAPER studies: five case-control studies aimed at identifying and quantifying the risk of cancer in symptomatic primary care patients. Br. J. Cancer **101**, S80–S86 (2009)
2. Hippisley-Cox, J., Coupland, C.: Identifying patients with suspected lung cancer in primary care: derivation and validation of an algorithm. Br. J. Gen. Pract. **61**(592), e715–e723 (2011)

3. Corner, J., Hopkinson, J., Fitzsimmons, D., et al.: Is late diagnosis of lung cancer inevitable? Interview study of patients' recollections of symptoms before diagnosis. Thorax **60**, 314–319 (2005)
4. Iyen-Omofoman, B., et al.: Using socio-demographic and early clinical features in general practice to identify people with lung cancer earlier. Thorax **68**(5), 451–459 (2013)
5. Hannah, T.P., et al.: Mortality due to cancer treatment delay: systematic review and meta-analysis. BMJ **371**, m4087 (2020)
6. Bailey, S.E.R., Ukoumunne, O.C., Shephard, E., Hamilton, W.: How useful is thrombocytosis in predicting an underlying cancer in primary care? A systematic review. Fam. Pract. **34**(1), 4–10 (2017)
7. Bailey, S.E.R., Ukoumunne, O.C., Shephard, E.A., Hamilton, W.: Clinical relevance of thrombocytosis in primary care: a prospective cohort study of cancer incidence using English electronic medical records and cancer registry data. Br. J. Gen. Pract. **67**(659), e405–e413 (2017)
8. Shapley, M., Mansell, G., Jordan, J.L., Jordan, K.P.: Positive predictive values of $\geq$5% in primary care for cancer: systematic review. Br. J. Gen. Pract. **60**(578), e366–e377 (2010)
9. Bjerager, M., Palshof, T., Dahl, R., et al.: Delay in diagnosis of lung cancer in general practice. Br. J. Gen. Pract. **56**, 863–868 (2006)
10. World Health Organization: Cancer Fact Sheets. https://www.who.int/news-room/fact-sheets/detail/cancer
11. Victoria Cancer Council: I-PACED (Implementing Pathways for Cancer Early Diagnosis). https://www.cancervic.org.au/downloads/resources/factsheets/AL1720_OCP_I-PACED_Lung_FINAL.pdf
12. ten Haaf, K., et al.: Risk prediction models for selection of lung cancer screening candidates: a retrospective validation study. PLoS Med. **14**(4), e1002277 (2017)
13. O'Dowd, E.L., et al.: What characteristics of primary care and patients are associated with early death in patients with lung cancer in the UK? Thorax **70**(2), 161–168 (2015)
14. Weller, D.P., Peake, M.D., Field, J.K.: Presentation of lung cancer in primary care. NPJ Prim. Care Respir. Med. **29**(1), 1–5 (2019)
15. Goldstein, B.A., et al.: Opportunities and challenges in developing risk prediction models with electronic health records data: a systematic review. J. Am. Med. Inform. Assoc. **24**(1), 198–208 (2017)
16. Schmidt-Hansen, M., et al.: Lung cancer in symptomatic patients presenting in primary care: a systematic review of risk prediction tools. Br. J. Gen. Pract. **67**(659), e396–e404 (2017)
17. Bradley, S.H., Martyn, P.T.K., Richard, D.N.: Recognizing lung cancer in primary care. Adv. Ther. **36**(1), 19–30 (2019)
18. NPS MedicineWise Annual Report 2019–20. https://www.nps.org.au/about-us/reports-evaluation

Addressing Extreme Imbalance for Detecting Medications Mentioned in Twitter User Timelines

Davy Weissenbacher[1(✉)], Siddharth Rawal[2], Arjun Magge[1], and Graciela Gonzalez-Hernandez[1]

[1] Department of Biostatistics, Epidemiology and Informatics (DBEI), University of Pennsylvania, Philadelphia, PA, USA
{dweissen,arjun.magge,gragon}@pennmedicine.upenn.edu

[2] School of Computing, Informatics, and Decision Systems Engineering (CIDSE), Arizona State University, Tempe, AZ, USA
sidrawal@asu.edu

Abstract. Tweets mentioning medications are valuable for efforts in digital epidemiology to supplement traditional methods of monitoring public health. A major obstacle, however, is to differentiate them from the large majority of tweets on other topics posted in a user's timeline: solving the infamous 'needle in a haystack' problem. While deep learning models have significantly improved classification, their performance and inference processing time remain low on extremely imbalanced corpora where the tweets of interest are less than 1% of all tweets. In this study, we empirically evaluate under-sampling, fine-tuning, and filtering heuristics to train such classifiers. Using a corpus of 212 Twitter timelines (181,607 tweets with only 0.2% tweets mentioning a medication), our results show that combining these heuristics is necessary to impact the classifier's performance. In our intrinsic evaluation, a classifier based on a lexicon and a BERT-base neural network achieved a 0.838 F1-score, a score similar to the score achieved by the best classifier on this task during the #SMM4H'20 competition, but it processed the corpus 28 times faster - a positive result, since processing speed is still a roadblock to deploying classifiers on large cohorts of Twitter users needed for pharmacovigilance. In our extrinsic evaluation, our classifier helped a labeler to extract the spans of medications more accurately and achieved a 0.76 Strict F1-score. To the best of our knowledge, this is the first evaluation of medications extraction from Twitter timelines and it establishes the first benchmark for future studies.

Keywords: Social Media · Medication detection · Text classification

1 Introduction

With more than 321 million monthly active users worldwide [11], Twitter is among the most influential Social Media platforms of the last decade. On Twitter, users discuss a great variety of subjects, including their health. These tweets

A. Tucker et al. (Eds.): AIME 2021, LNAI 12721, pp. 93–102, 2021.
https://doi.org/10.1007/978-3-030-77211-6_10

are now recognized as a valuable source of information in digital epidemiology to supplement traditional methods for population health surveillance [12]. A major obstacle, however, is to differentiate health-related tweets from the majority of tweets with other topics. Particularly relevant to a variety of health-related studies are those mentioning medications.

Previous studies on detecting medications on Twitter usually started by building their corpus. However, collection methods often biased such corpora by using a predefined list of medications [2,10], imposing co-occurrences of medications and diseases in tweets [8], or removing tweets with common terms ambiguous with medications [1]. This changed with the advent of the Social Media for Health Mining (#SMM4H) Shared Tasks, focused on deploying standard corpora to test and compare systems that extract health information in Social Media [9,16]. In 2018 and 2020, shared tasks included those to detect tweets mentioning medications and dietary supplements, with suitably annotated corpora made available. This task is often the first process applied in a pipeline for population health surveillance when mining Twitter data. In #SMM4H'18, the corpus was composed of randomly selected tweets and manually balanced. The task served as a proof of concept. In #SMM4H'20, the corpus consisted of 112 Twitter users' timelines and exhibited the real distribution of medication mentions in typical Twitter user timelines. This task was intended to measure the performance to expect in a "real world" application.

During the challenges, participants abandoned machine learning models that use hand-engineered features and largely adopted deep learning models, in particular, transformer models. While deep learning models significantly improved the accuracy of the systems, two limitations impede their use at a large scale in "real world" applications. First, their training is difficult on extremely imbalanced corpora. Corpora are imbalanced when negative examples outnumber positive examples, a very frequent occurrence in "real" data [7]. When classifiers are trained on these corpora, their optimizing algorithms tend to classify all examples as negative to reduce their losses. Imbalance was the main concern for the #SMM4H'20 participants who proposed various heuristics to improve their training: under-sampling and fine-tuning were the most popular. Ensemble learning, over-sampling, data augmentation, and cost-sensitive learning were proposed as well. Given the time constraints of the shared task, most participants focused on one heuristic at a time, and when they did combine heuristics, they usually did not evaluate methodically the individual contribution of each. Second, the speed of prediction of transformer models remains slow on very large corpora. When Twitter is used to identify cohorts with suitable statistical power, a large number of tweets has to be processed by the model to discriminate users of interest. It is not unusual to process billions of tweets to collect such cohorts. Despite hardware improvements, the inference times of transformer models are high [13]. This limits the size of the cohorts and the datasets that can be processed and studied. For example, it would take 1,750 h (72.9 days) for the BERT classifier used in our experiments to process the 1.5 billion tweets of our current collection of Twitter timelines where users are announcing a pregnancy [15].

In this study, our goal was to achieve F1-scores similar to those of the state-of-the-art deep learning classifiers for the task of finding tweets with medication mentions in user timelines but using a more agile system. To this end, we empirically evaluated three re-sampling heuristics and their combinations- under-sampling, fine-tuning, and filtering - to train a binary classifier. We experimented with two classifiers, a Convolutional Neural Network (CNN) and a BERT-based classifier. Among all possible heuristics to improve training on imbalanced datasets, we chose filtering because large lexicons of medications exist and the decisions made by the filter uniformly remove a large number of unlikely candidates, greatly speeding up the processing time of the classifier at run time while remaining reproducible. Under-sampling and fine-tuning are complementary heuristics to use with filtering since they allow for training a classifier on a balanced training set, removing the need for a corpus with the natural imbalance of the examples ratio, a corpus very expensive to annotate.

Our contributions are: 1) the release of 212 Twitter timelines annotated with medication spans as a benchmark for the extraction of medication mentions in Twitter, which will be used for a BioCreative shared-task [14]; 2) the design of a fast and efficient classifier to detect tweets mentioning medications; 3) an intrinsic evaluation of the classifier as well an extrinsic evaluation when it is used to help in the extraction of the spans of the medication mentions.

2 Methods

Data. We ran our experiments on three corpora. The first corpus was released during #SMM4H'18. It is composed of tweets mentioning medications or ambiguous terms that can be confused with medications. For example, 'Propel' is an English verb but it is also the brand name of a corticosteroid. The corpus was manually balanced with an equal number of positive and negative tweets (7,827 *vs* 7,178, respectively). The second corpus, released for #SMM4H'20, includes 98,959 tweets from 112 user timelines. This corpus has the natural distribution of tweets with very few tweets mentioning medications (258 positive *vs* 98,701 negative tweets). The inter-annotator agreements reported previously [15] for both corpora were high, with .892 Cohen's kappa for SMM4H'18 and .880 Cohen's kappa for SMM4H'20. With only 77 tweets mentioning medications in the test set, and common medications like Tylenol occurring multiple times, the SMM4H'20 corpus test set contains few positive examples. Therefore, we decided to create a larger corpus to evaluate our models more accurately. In the SMM4H'20+ corpus, we added 100 new timelines to the existing 112 timelines of the SMM4H'20 corpus (for a total of 442 positive and 181,165 negative tweets). With 2.5 h on average to annotate a timeline, the SMM4H'20+ corpus was very expensive to produce. We selected and annotated these 100 timelines by following the guidelines defined during the #SMM4H'20 shared-task. The SMM4H'20+ test set has 131 positive tweets and includes all examples of the SMM4H'20 test set.

Classifiers. We implemented three binary classifiers to detect tweets mentioning medications[1]. Each classifier receives raw tweets as input and returns 1 for tweets predicted to mention medications, 0 otherwise. As a baseline, we re-implemented the lexicon classifier proposed in [15]. This baseline classifier relies on an extensive list of 44,948 medications from RxNorm. We extended the lexicon with 231 generic mentions such as *statin, antibiotic and pain meds.* Since medications are often misspelled in Twitter, we automatically generated variants for our 44,948 medications by following the method described in [15] and manually removed variants that were too ambiguous with common English phrases, like *some*, a variant of *sone.* This classifier labels as positive all tweets with a phrase matching an entry in the lexicon. Besides, we implemented a CNN with word2vec embeddings trained on 400 million tweets [6]. Lastly, we implemented a classifier using BERT-base with no additional output layers and trained following the recommendations of the authors [5]. We chose BERT over more recent transformer-based classifiers because it is now a well-accepted milestone in text mining and it allows us to compare our performances with the ones of the best classifier of #SMM4H'20, which used an ensemble of 20 BERT classifiers [4]. Other participants did not perform better with more recent transformer-based classifiers such as RoBERTa or Electra. For simplicity, we will refer to these three classifiers as the lexicon, CNN, and BERT classifiers, respectively.

Training. We trained our CNN and BERT classifiers following different settings to evaluate the effects of our re-sampling heuristics: under-sampling, filtering, and fine-tuning. Our under-sampling heuristic is to train our classifiers on SMM4H'18. This corpus was under-sampled by manually keeping a balanced number of positive tweets, which were correctly predicted as mentioning medications by at least two of the four weak classifiers used in [15], and negative tweets, which were false positives of one weak classifier. The filtering heuristic also under-samples a corpus by removing all examples predicted as not mentioning a medication by the lexicon classifier. Our fine-tuning heuristic consists of training our classifiers for a few epochs on a corpus with a given ratio of negative/positive examples, and continuing their training on a corpus with a different ratio.

In all settings, the classifiers were trained on the official training set of the corpora mentioned and evaluated on the test set of SMM4H'20+. **Setting (a):** We trained our classifiers only on SMM4H'20+. This setting is the default training method for supervised classifiers: examples are randomly selected from users' timelines, and all examples sampled are annotated and used for training and evaluating the classifiers. Any improvements over the scores of our classifiers trained with setting (a) will show the benefits of the heuristics tested with other settings. **Setting (b):** We show the impact of under-sampling alone. We trained our classifiers on SMM4H'18, as it helps to learn the linguistic patterns to speak about medications and their homonyms. However, this training corpus is not representative of the test corpus, the SMM4H'20+, which exhibits the real distribution

[1] All classifiers are available at https://tinyurl.com/fo9u9xnn.

of tweets. **Setting (c):** We show the impact of filtering alone. We trained our classifiers on SMM4H'20+ and applied them only to the tweets of the test set of SMM4H'20+ matched by the lexicon classifier. For the tweets filtered out by the lexicon classifier, the label was set to 0. For the tweets filtered in, we applied the CNN or the BERT classifier to make the final decision on their label values. **Setting (d):** We combined the under-sampling and fine-tuning heuristics. We pre-trained our classifiers on SMM4H'18 and continued their training on SMM4H'20+. By continuing the training on SMM4H'20+, the classifiers learned the real distribution of positive examples in the test corpus. **Setting (e):** We combined the filtering and under-sampling heuristics. We trained our classifiers on SMM4H'18 and we applied them to the tweets of the test set of SMM4H'20+ filtered in by the lexicon classifier. **Setting (f):** We combined the three heuristics. We trained the classifiers on the SMM4H'20+ corpus providing more negative examples, fine-tuned on SMM4H'18, and applied the classifiers to the tweets of the test set of SMM4H'20+ filtered in by the lexicon classifier.

Evaluation. We intrinsically evaluated the performance of our classifiers with the Precision, Recall, and F1-score metrics. During this evaluation, all classifiers were evaluated on how well they individually perform the classification task and were ranked according to their F1-scores. We trained and evaluated our classifiers three times for each setting and reported the means of their scores in Table 1 to account for the variations due to the stochastic optimization of the neural networks. We kept in our experiments the results obtained on SMM4H'20 to compare our performance to that of the participants of the #SMM4H'20 shared-task.

We also extrinsically evaluated our best classifier. We measured the changes of performance of a baseline sequence labeler extracting medications with and without prefiltering the tweets with our best classifier. A medication labeler should return, for each medication occurring in a tweet, its span - i.e. the starting and ending positions. We evaluated the labeler with the Overlapping/Strict Precision, Recall, and F1-score metrics. In the overlapping evaluation, we rewarded the labeler if it found a span that overlaps with the span of a medication, and in the strict evaluation, only if it found the exact span of the medication.

To perform our extrinsic evaluation, we implemented three labelers. Our first baseline labeler applies our lexicon to a corpus and extracts the spans of every entry of the lexicon matched in the tweets without additional controls. For our second labeler, we converted into a sequence labeler the best classifier based on a lexicon and BERT in setting (f). This labeler extracts the spans of every entry of the lexicon matched in the tweets labeled as mentioning a medication by the BERT neural network. All other tweets do not contain medications for the labeler. Our third labeler is a standard labeler based on a bidirectional-LSTM using BERT embeddings and trained to recognized all tokens constituting a medication mention within a tweet using the IO annotation schema. We evaluated various settings to train the BERT labeler but we only report the two best settings due to space constraints. In setting (α), we trained the BERT labeler on the SMM4H'18 training set and fine-tuned it on the SMM4H'20+ training set. We evaluated the labeler on the full test set of SMM4H'20+, therefore performing the detection and extraction at the same time. In setting (β), we trained the BERT labeler on the SMM4H'18 training set but fine-tuned it only on the tweets of the SMM4H'20+ training set filtered in by our best classifier. During the evaluation, we applied the labeler only on the tweets of the SMM4H'20+ test set filtered in by our best classifier - making the decision that all tweets filtered out by the classifier did not contain any medications - thus, performing the detection independently from the extraction. Since the labelers in settings α and β only differ by filtering the tweets with the classifier during their training and evaluation steps, the differences between their performances measure the impact of the classifier.

3 Results and Discussion

3.1 Intrinsic Evaluation Results

An interesting aspect of Table 1 is the performance of our classifiers on SMM4H'18. With a 0.954 F1-score, our BERT classifier outperformed the ensemble of bi-LSTMs proposed in [15], which to the best of our knowledge, was the highest score achieved on this corpus with a 0.937 F1-score.

Another interesting finding is that under-sampling and filtering heuristics cannot be used alone, since settings (b) and (c) perform worse than the default setting (a) for both classifiers. When under-sampling is used alone, both classifiers became over-sensitive to words related to health but used in other contexts, such as *"been drinking"*, or tweets related to health but not mentioning any medications, such as *"I might as well show him since I'm at the OBGYN"*. Both classifiers have a good recall, higher than the recall of our lexicon classifier, but they have a very low precision, making their annotations unusable in upstream processes. When the filtering heuristic is used alone, the opposite phenomenon is observed. We expected our lexicon classifier to have a good recall but a low precision due to medication homonyms. Since our classifiers learned the contexts where medications occur, they should have detected false positives predicted by the lexicon classifier and corrected their labels. However, having seen too few positive examples in the training set, our classifiers were over-conservative and rejected valid patterns such as "prescribed me [medication]" or tweets mentioning unseen medications like *"femestra"*, resulting in low recalls of 0.49 and 0.46.

Our heuristics improved the training of our classifiers when they were combined. On SMM4H'20+, the best classifier was the CNN trained on the under-sampled corpus and fine-tuned with an 0.80 F1-score. The BERT classifier using under-sampling and filtering heuristics in setting (e) achieved a performance very close to that of the BERT classifier in setting (f), with 0.79 F1-score, showing a marginal help from the additional negative examples from the SMM4H'20+ corpus used in setting (f). However, the setting (e) has three main advantages over other settings. Our classifier does not require to be trained on SMM4H'20+, a corpus that is very expensive to produce, as annotation of each timeline takes

Table 1. Performance of binary classifiers on the SMM4H'18, SMM4H'20, and SMM4H'20+ test set corpus

Training setting	Classifier	SMM4H'18 P	SMM4H'18 R	SMM4H'18 F1			
	lexicon	67.96	92.60	78.39			
	[17]	93.3	90.4	91.8			
	[15]	95.1	92.5	93.7			
Training: SMM4H18	CNN	89.14	93.88	91.45			
	BERT	**95.47**	**95.30**	**95.38**			
		SMM4H'20			**SMM4H'20+**		
		P	R	F1	P	R	F1
	lexicon	29.49	83.12	43.54	25.32	75.57	37.93
	[4]	83.75	**87.01**	**85.35**	—	—	—
	[3]	77.11	83.12	80.00	—	—	—
(a) Training: SMM4H20+	CNN	78.87	61.04	68.41	75.44	58.78	65.81
	BERT	78.86	62.77	69.87	79.75	67.18	72.92
(b) training: SMM4H'18	CNN	3.40	81.82	6.56	3.22	83.97	6.20
	BERT	16.62	85.71	27.85	17.46	**90.08**	29.24
(c) training: SMM4H'20+ Filter: lexicon	CNN	**95.31**	52.38	67.60	**94.67**	49.11	64.65
	BERT	92.38	47.62	62.36	92.61	45.53	60.69
(d) Training: SMM4H'18, fine-tuning: SMM4H'20+	CNN	91.69	74.89	82.31	88.32	74.05	**80.43**
	BERT	81.39	64.50	71.96	83.16	68.96	75.36
(e) Training: SMM4H'18, Filter: lexicon	CNN	66.21	77.92	71.57	67.34	69.98	68.61
	BERT	88.85	78.79	83.50	87.50	72.77	79.45
(f) Training: SMM4H'20+, Fine-tuning: SMM4H'18, Filter: lexicon	CNN	79.44	78.35	78.88	79.11	70.99	74.82
	BERT	90.10	78.35	83.81	90.11	71.76	79.89

around 2–4 h. Our classifier was only trained on the SMM4H'18 corpus, a corpus composed of tweets that mention medication names rather than full timelines. This corpus was built semi-automatically and can be updated and annotated at a rate of 30–40 tweets per minute. With new medications released each year, it is preferable to train our classifiers only on the under-sampled corpus. With the setting (e), the classifier is fast. With parallel computing and indexing, our lexicon classifier can pre-filter millions of tweets in minutes; the BERT classifier is then only applied on a fraction of the initial tweets. Our classifier took 41 s on a MacBook Pro 2020 with a CPU to predict the labels of the 29,687 tweets (3.4 MB) of the SMM4H'20 test set. In comparison, the best classifier of #SMM4H'20, an ensemble of 20 BERT classifiers [4], took 19.4 min on Google Colab Pro with a GPU V100 to process this test set. Finally, since the lexicon classifier matches the spans of the medications in tweets, it is trivial to convert the classifier into a sequence labeler and perform the extraction of the medications, as we have done for the extrinsic evaluation.

On the SMM4H'20 test set, our best classifier was the BERT classifier trained with the three heuristics with an average of 0.838 F1-score, and 0.853 F1-score (0.924 Precision and 0.792 Recall) during its best iteration. Our classifier performs better than the BERT classifier proposed in [3] which ranked second during the shared task with 0.80 F1-score and achieved similar performance to the best system [4]. However, due to the small number of positive examples in the SMM4H'20 test set, we found the differences between these scores to be not significant using a McNemar test ($p = 0.05$).

3.2 Extrinsic Evaluation Results

The results reported in Table 2 show the good performance of our classifier by improving the scores of the extrinsic task. One may argue that developing an independent classifier is not needed since a labeler performs the medication detection and extraction tasks at the same time by discovering the spans of the medications; all efforts should rather focus on developing an efficient medication labeler. Our empirical results show that when processing very imbalanced data, it is better, to first detect the tweets of interest with a classifier, then apply the labeler only on the tweets detected to extract the concepts positions. A condition, however, is to fine-tune the labeler on a training set filtered by the classifier. Without the help of the classifier, the labeler extracts the medication spans on the full corpus and achieves 0.76 overlapping and 0.72 strict F1 scores. When the labeler is helped by the classifier and extracts the spans only in the tweets filtered in, its scores slightly improved to 0.77 overlapping and 0.76 strict F1 scores. However, the difference between the scores of the labeler achieved with, or without, the help of the classifier was not found to be significant using a McNemar test ($p = 0.05$). A possible explanation for this might be that it is easier to optimize two different loss functions: the classifier's, dedicated to representing the semantics of health-related tweets, and the labeler's, only focused on extracting the spans of medications.

Table 2. Performance of medications extraction on the SMM4H'20+ test set corpus

		SMM4H'20+					
		Strict			Overlapping		
Training setting	**Labeler**	P	R	F1	P	R	F1
	Lexicon	22.3	62.6	32.9	25.2	70.7	37.1
	Lexicon+BERT	77.4	60.5	67.9	87.0	68.0	76.3
(α) Training: SMM4H'18 Fine-tuning: SMM4H'20+	BERT	79.5	65.1	71.6	84.4	69.1	76.0
(β) Training: SMM4H'18 Fine-tuning: filtered SMM4H'20+ Filter: Lexicon+BERT	BERT	89.0	66.0	75.8	90.8	67.3	77.3

4 Conclusion

We propose an efficient classifier to detect tweets mentioning medications based on a lexicon and a BERT-base neural network. With a 0.838 F1-score, it also achieves performance comparable to those of the best classifiers on the SMM4H'20+ benchmark dataset, 212 timelines where only 0.2% of tweets mentioning medications. The intrinsic evaluation of the classifiers shows that training them on such imbalanced data is still a major challenge and underlines the need for dedicated training methods. We empirically evaluated three re-sampling heuristics - under-sampling, fine-tuning, and filtering - and showed that their combinations are required to be beneficial. Whereas under-sampling/filtering was not the best combination, it removes the need to train the classifier on a corpus exhibiting the real distribution of the data, a corpus very expensive to produce. It also improves the speed of the classifier by pre-filtering the tweets. Considering the difference in performances of our classifiers on balanced and imbalanced corpora, 10 F1-score points, there is still space for improvement. The extrinsic evaluation of our classifier on the medication extraction task shows that, when working with an imbalanced corpus, it is still preferable to perform the detection of the tweets mentioning medications independently from the extraction of their spans. By doing so, we achieved 0.773 overlapping F1-score and 0.758 strict F1-score on the medication extraction task. To the best of our knowledge, this is the first evaluation of this task on Twitter timelines and it establishes the first benchmark for future studies. A limitation of our study is that only three heuristics were tested. We plan to evaluate additional heuristics such as learning with generated data, cost-sensitive, transfer, few-shot, active, and distance learning. Another limitation is that few positive examples occur in our test sets, making the difference between the performances of the classifiers observed non-significant. We will combine the SMM4H'20+ test set with a larger collection of tweets manually imbalanced with the same 0.2% ratio to reassess the performances of our classifiers.

Funding. This work was supported by National Library of Medicine grant number R01LM011176 to GG-H. The content is solely the responsibility of the authors and does not necessarily represent the official view of National Library of Medicine.

References

1. Batbaatar, E., Ryu, K.H.: Ontology-based healthcare named entity recognition from twitter messages using a recurrent neural network approach. Int. J. Environ. Res. Public Health **16**(16:3628) (2019)
2. Carbonell, P., Mayer, M.A., Bravo, A.: Exploring brand-name drug mentions on twitter for pharmacovigilance. Stud. Health Technol. Inform. **210**, 55–59 (2015)
3. Casola, S., Lavelli, A.: FBK@SMM4H2020: RoBERTa for detecting medications on Twitter. In: Proceedings of the Fifth Social Media Mining for Health Applications (#SMM4H) Workshop & Shared Task (2020)

4. Dang, H.N., Lee, K., Henry, S., Uzuner, O.: Ensemble BERT for classifying medication-mentioning tweets. In: Proceedings of the Fifth Social Media Mining for Health Applications (#SMM4H) Workshop & Shared Task (2020)
5. Devlin, J., Chang, M.W., Lee, K., Toutanova, K.: BERT: pre-training of deep bidirectional transformers for language understanding. In: Proceedings of the 2019 Conference of the North American Chapter of the Association for Computational Linguistics: Human Language Technologies, Volume 1 (Long and Short Papers). Association for Computational Linguistics (2019)
6. Godin, F., Vandersmissen, B., De Neve, W., Van de Walle, R.: Multimedia lab @ ACL WNUT NER shared task: named entity recognition for Twitter microposts using distributed word representations. In: Proceedings of the Workshop on Noisy User-generated Text. Association for Computational Linguistics (2015)
7. Haixiang, G., Yijing, L., Shang, J., Mingyun, G., Yuanyue, H., Bing, G.: Learning from class-imbalanced data: review of methods and applications. Expert Syst. Appl. **73**, 220–239 (2017)
8. Jimeno-Yepes, A., MacKinlay, A., Han, B., Chen, Q.: Identifying diseases, drugs, and symptoms in twitter. Stud. Health Technol. Inform. **216**, 643–647 (2019)
9. Klein, A.Z., et al.: Overview of the fifth social media mining for health applications (#smm4h) workshop & shared task at coling 2020
10. Sarker, A., Gonzalez-Hernandez, G.: A corpus for mining drug-related knowledge from twitter chatter: language models and their utilities. Data Brief **10**, 122–131 (2017)
11. Shaban, H.: Twitter reveals its daily active user numbers for the first time (2019). https://www.ashingtonpost.com/technology/2019/02/07/twitter-reveals-its-daily-active-user-numbers-first-time/
12. Sinnenberg, L., Buttenheim, A.M., Padrez, K., Mancheno, C., Ungar, L., Merchant, R.M.: Twitter as a tool for health research: A systematic review. Am. J. Public Health **107**(1), e1–e8 (2017)
13. Turc, I., Chang, M.W., Lee, K., Toutanova, K.: Well-read students learn better: on the importance of pre-training compact models. arXiv preprint arXiv:1908.08962v2 (2019)
14. Weissenbacher, D.: Track 3 - automatic extraction of medication names in tweets (2020). https://biocreative.bioinformatics.udel.edu/tasks/biocreative-vii/track-3/
15. Weissenbacher, D., Sarker, A., Klein, A., O'Connor, K., Magge, A., Gonzalez-Hernandez, G.: Deep neural networks ensemble for detecting medication mentions in tweets. J. Am. Med. Inform. Assoc. **26**(12), 1618–1626 (2019)
16. Weissenbacher, D., Sarker, A., Paul, M.J., Gonzalez-Hernandez, G.: Overview of the third social media mining for health (SMM4H) shared tasks at EMNLP 2018. In: Proceedings of the 2018 EMNLP Workshop SMM4H: The 3rd Social Media Mining for Health Applications Workshop & Shared Task. Association for Computational Linguistics (2018)
17. Wu, C., Wu, F., Liu, J., Wu, S., Huang, Y., Xie, X.: Detecting tweets mentioning drug name and adverse drug reaction with hierarchical tweet representation and multi-head self-attention. In: Proceedings of the 2018 EMNLP Workshop SMM4H: The 3rd Social Media Mining for Health Applications Workshop & Shared Task. Association for Computational Linguistics (2018)

ICU Days-to-Discharge Analysis with Machine Learning Technology

David Cuadrado(✉) and David Riaño

Universitat Rovira i Virgili, Tarragona, Spain
david.cuadrado@estudiants.urv.cat, david.riano@urv.cat

Abstract. ICU management depends on the level of occupation and the length of stay of the patients. Daily prediction of the days to discharge (DTD) of ICU patients is essential to that management. Previous studies showed a low predictive capability of internists and ML-generated models. Therefore, more elaborated combinations of ML technologies are required. Here, we present four approaches to the analysis of the DTDs of ICU patients from different perspectives: heterogeneity quantification, biomarker identification, phenotype recognition, and prediction. Several ML-based methods are proposed for each approach, which were tested with the data of 3,973 patients of a Spanish ICU. Results confirm the complexity of analyzing DTDs with intelligent data analysis methods.

Keywords: ICU · Patient phenotyping · Days-to-discharge prediction · Feature selection

1 Introduction

According to the World Federation of Societies of Intensive and Critical Care Medicine, intensive care units (ICU) are "organized systems for the provision of care to critically ill patients that provides intensive and specialized medical and nursing care, an enhanced capacity for monitoring, and multiple modalities of physiologic organ support to sustain life during a period of acute organ system insufficiency" [1]. Patients attended in ICU use to have, or are at risk of having, some acute, severe, life-threatening organ dysfunction requiring critical care services which need to be provided in a continuous and specialized manner. Patients case-mix in ICUs uses to be heterogeneous [3] and it may include diverse patient types (e.g., medical or surgical), either scheduled or not, and resulting from conventional hospitalization, derived from other hospital areas (e.g., Emergency) or transferred from other hospitals.

It was estimated that in the US the mean ICU cost and length of stay were $31,574 ± 42,570 and 14.4 days ± 15.8 ($2,193 per day in average) for patients requiring mechanical ventilation, and $12,931 ± 20,569 and 8.5 days ± 10.5 ($1,521 per day in average) for those not requiring mechanical ventilation [4]. A multi-country study in Europe showed that ICU direct costs ranged, in average, from 1168 to 2025 per patient and day [8]. The European Hospital and Health

A. Tucker et al. (Eds.): AIME 2021, LNAI 12721, pp. 103–113, 2021.
https://doi.org/10.1007/978-3-030-77211-6_11

Care Federation reported that, in 2014, the average length of stay in acute care hospitals was 6.4 bed days in the EU-28 countries, ranging from 5.2 to 7.0 days. These important daily costs urge governments and ICU managers in a effort to reduce the length of stay of ICU patients without compromising healthcare quality.

In this context, it is desirable to count with reliable tools to predict the duration of patients in ICU for a better planning and also to optimize costs. There are two approaches to such predictive tools: static and dynamic. Static tools predict the length of stay (LOS) at the admission time (or 24–48 h after admission) whereas dynamic tools make predictions in a daily basis, as the patient evolves. Some studies found that physicians are not good at predicting ICU-LOS statically and they are poor at predicting stays longer than five days [9,10]. Moreover, a systematic review analyzed 31 ICU-LOS predictive models and concluded that they suffer from serious limitations [11]. Statistical and machine learning approaches (e.g., [12,13]) provide moderate predictions of short term LOS (1–5 days), but are unable to correctly predict long-term LOS (>5 days).

In order to confront the long-term prediction limitation of static methods, dynamic tools propose a day by day prediction of the days to discharge (DTD). So, a patient with total LOS $= x$, in day $y < x$ is expected to have a DTD prediction of $x - y$ days. This dynamic approach presents some conceptual benefits: (1) predictive errors in day y can be fixed in subsequent days and clinical decisions readjusted, (2) predictions are current and not based on the patient condition at the admission time days ago, (3) as patients approach discharge, their clinical parameters tend to be normal[1], and it is expected that the accuracy of dynamic predictions may improve with respect to the static predictions in the admission day.

However, not many works have been published on the prediction of DTD for ICU patients. An exception is [2], where Random Forest was applied to construct a DTD predictor for general ICU patients which achieved an average root mean square error of 1.73 days. Since then, our repeated attempts to produce a highly accurate predictive model that reduces the mean error to below one day have failed, but in the process we have carried out a series of studies with artificial intelligence technologies that may be useful for the analysis and understanding of DTD in ICU. These are: measuring DTD-based patient heterogeneity, identification of biomarkers for DTD prediction, DTD phenotype recognition, and construction of DTD predictive models.

Measuring DTD-Based Patient Heterogeneity: Heterogeneity is the quality of being diverse. In ICUs, patients are heterogeneous by nature. When dealing with a DTD patient group (i.e., all the ICU patients who are within the same days of being discharged), heterogeneity can be used to measure the variability within a group, or the similarity between different groups. DTD-based heterogeneity can explain, for example, the reason why two clinically similar patients have very different discharge days, and also why clinically different patients may have the

[1] This is only true for patients discharged alive for whom ICU discharge is due to stabilization of their vital signs.

same discharge day. In [3] we proposed four alternative methods to calculate DTD-based patient heterogeneity. They are summarized later.

Identification of Biomarkers for DTD Prediction: In ICUs, multiple clinical signs from patients are continuously recorded. Some of these values are aggregated to provide a daily clinical summary of the patient's condition, when combined with other data that are captured once a day. These could be the subject of a feature selection process to determine which clinical parameters are more relevant to determine the DTD of ICU patients, thus acting as DTD biomarkers.

DTD Phenotype Recognition: A consequence of DTD patient heterogeneity is that similar ICU patients may have very different DTDs, and the opposite. This defines DTD prediction in ICUs as a complicated task that could be simplified by the identification of patient phenotypes representing subgroups of patients, all of them with identical DTD. Technically, these phenotypes should have DTD precisions close to one, and DTD recalls as high as possible.

Construction of DTD Predictive Models: The benefits of having a predictive model of DTD for ICU patients are multiple in clinical, organizational, resource optimization, and analytical terms. For example, knowing when the current patients are leaving the ICU, provides useful information on when new patients could be admitted, and permits implementing better ICU strategies. The ultimate challenge, which we have not yet reached, is the construction of a model with a average error of the prediction of DTD of the ICU patient below one day.

In Sect. 2, we describe the methods that we followed for the analysis of these four previously described DTD issues and the results obtained. In Sect. 3, we discuss about the clinical implications and benefits these results can entail. The conclusions of the paper are exposed in Sect. 4.

2 Methods and Results

The analysis of the days to discharge (DTD) of patients in an ICU is a complex problem due to multiple factors: critical condition of the cases, possible unexpected complications, intervention of multiple practitioners, diversity of cases, etc. In this section, we address four DTD-related issues whose analysis can contribute to a better understanding and management of patients in ICUs. For each one of these issues, we describe the methods and artificial intelligence technologies that we have used to deal with them, and the results obtained.

All the studies were made on the same dataset containing data about all the patients admitted to the ICU of a Spanish tertiary hospital between 2014 and 2019. Survival was expected to be analyzed with specialized methods such as Kaplan-Meier, and our study focused exclusively on patients discharged alive. Patient data was treated individually.

2.1 The Dataset

Important information about patients admitted in the University Hospital Joan XXIII in Spain was captured in a database providing daily description of all the

ICU cases in terms of demographic, clinical, and treatment features which are essential in the professional decision of discharging patients from ICUs.

All the 3,973 patients discharged alive in 2014–19 were considered. The average LOS was 8.6 days, with 1 and 159 days the minimum and maximum LOS. Only information of patients in their seven last days in the ICU was considered. The average age was 59.9, with all patients between 18 and 99 years, and 1,424 (35.8%) being women. There were 2,878 (72.4%) medical cases and 1,095 (27.6%) surgical.

Forty-three parameters were considered: 8 categorical, 11 Boolean, and the rest numeric. Eleven were static in the sense that they remained constant during all the ICU stay of the patient. These were: "Age", "Gender", "DisYear", "PatType", "AdmType", "AdmWardGroup", "PrevHospDays", "APACHE_Adm_Group", "Pincipal_Diag_G", "APACHEII", and "CHE"[2]. Other parameters were taken at the rate of once a day, including signs (e.g., platelets, bilirubin, or creatinine), scales (e.g., NAS -nursing work load, EMINA -pain scale, STRATIFY -risk of falling, or SOFA scores), and treatment actions (most of them Boolean; e.g. arterial catheter, urinary catheter, central venous catheter, insulin treatment, vasoactive drugs, analgesics, antibiotics, mechanical ventilation -either invasive or not, etc.). The remaining five parameters whose frequency of observation was below one day were aggregated per day and represented by their mean, stdev, min, or max daily values (e.g., heart rate, temperature, or glucose).

Table 1 describes the dynamic numeric variables in the dataset.

2.2 Measuring Heterogeneity

The case-mix complexity of patients in ICUs is high. Patients can be medical or surgical, they may be scheduled or urgent, come from different clinical units (e.g., ER, surgery, conventional hospitalization, etc.), have different primary diagnosis (e.g., cardio-vascular, respiratory, digestive, infections, etc.), or they may show different levels of severity. This heterogeneity may affect the analysis of DTD of patients in ICUs, for example if two similar patients have different discharge days. In order to get some insight into the ICU patient heterogeneity, our previous work [4] proposed four methods to interpret heterogeneity and their corresponding measures to quantify the heterogeneity of intensive patients.

The first method (*clinical parameter analysis*) leveraged the mean and the standard deviation functions of each numeric parameter in the last days before discharge, grouped by DTD. The mean functions of clinical parameters were

[2] "Age": age of the patient at admission time, "Gender": female or male, "DisYear": year of ICU diagnosis, "PatType": patient type as medical or surgical, "AdmType": either scheduled or urgency, "AdmWardGroup": source among ER, surgery, other hospital areas, or other hospital, "PrevHospDays": number of days in hospital previous to ICU admission, "APACHE_Adm_Group": post-operations, heart failure, respiratory failure, trauma, etc., "Pincipal_Diag_G": principal diagnosis group (e.g., respiratory system, infection, external injuries, etc.), "APACHEII": APACHE II score value at ICU admission, and "CHE": Charlson comorbidity index.

Table 1. Statistical description of the dynamic numeric variables in the dataset: heart rate (HR); temperature (Tmp); glucose (minimal, maximal, and standard deviation within the day); nursing workload score (NAS); pain scale (EMINA); time since ICU admission (TSA); risk of falling (STRATIFY); SOFA scales (cardio, central nervous system, coagulation, liver, renal, respiratory, and total).

	HR	Tmp	Glu_min	Glu_max	Glu_std	NAS	EMINA	TSA	STRATIFY	SOFA_Cardio	SOFA_CNS	SOFA_Coag	SOFA_Liver	SOFA_Renal	SOFA_Resp	SOFA_Total
count	34026.00	34026.00	34026.00	34026.00	34026.00	34026.00	34026.00	34026.00	34026.00	34026.00	34026.00	34026.00	34026.00	34026.00	34026.00	34026.00
mean	114.61	37.01	108.74	153.68	18.91	53.52	8.94	13.22	2.78	0.88	1.29	0.33	0.16	0.28	0.67	3.61
std	23.04	0.77	26.40	47.93	16.09	17.57	2.86	17.37	1.12	1.27	1.36	0.76	0.55	0.73	0.93	2.99
min	45.00	32.40	20.00	63.00	0.00	0.00	0.00	1.00	0.00	0.00	0.00	0.00	0.00	0.00	0.00	0.00
25%	99.00	36.50	92.00	120.00	8.00	41.00	7.00	3.00	2.00	0.00	0.00	0.00	0.00	0.00	0.00	1.00
50%	112.00	36.90	105.00	142.00	14.57	49.00	9.00	6.00	3.00	0.00	1.00	0.00	0.00	0.00	0.00	3.00
75%	128.00	37.40	122.00	175.00	25.15	62.00	11.00	17.00	4.00	1.00	2.00	0.00	0.00	0.00	1.00	5.00
max	200.00	42.00	200.00	330.00	82.73	155.00	16.00	150.00	5.00	4.00	4.00	4.00	4.00	4.00	4.00	20.00

expected to converge to normality values and the standard deviation functions decrease as it gets closer to the discharge day. The results showed, however, that the variability of clinical parameters remained at similar mean values regardless of the DTD. Only SOFA-Cardio, SOFA-CNS and SOFA-Resp scales decreased as they approached to the patient's discharge day.

The second method (*severity scales analysis*) simplified the medical interpretation of ICU patients by merging multiple clinical parameters into one single parameter or score. EMINA, NAS, and SOFA-Total scores were studied. Their mean within each one of the last DTDs should trend to normality values (i.e., as the patient approaches discharge, her clinical parameters should progressively approach values that allow discharge). Their standard deviation should also decrease as patients approach discharge. Again, SOFA-Total and NAS were the scores that better adjusted to this expected behaviour of the mean and the standard deviation functions.

The third method (*confusion analysis*) used a similarity function between patients to calculate the number of ICU patients to be discharged in x days who are similar to other patients to be discharged in y days. The result was shown in a $n_\delta(x, y)$ confusion matrix for the last days of stay. It reflected a large proportion of patients (37% on average) who were similar to other patients discharged in the previous day and 26% of patients who were very similar to other patients discharged in later days. When we focused on the patients in their last ICU day, 63% of them were similar to other patients discharged in previous days.

The fourth method (*cluster analysis*) grouped all the patients with a same DTD value. Supposedly, patients in the same group should be more similar between them and less similar to patients in other groups. Cluster analysis metrics such as silhouette, Davies-Bouldin and Dunn scores [5,6] were computed to assess the quality of the clustering. The scores showed serious difficulties to differentiate between patients with different DTDs, either if we compare patients with close discharge days, or patients with distant discharge days.

2.3 Identification of DTD Biomarkers

The complexity of DTD analysis is not only related to the heterogeneity of ICU patients. A different approach could be taken, based on the clinical parameters used to describe the ICU patients. The relevance of these parameters in relation to the DTD was analyzed with several supervised feature selection methods. These methods were also used to identify low-relevant features.

Two approaches were followed: one which considered DTD as a n-ary variable and calculated the relevance of all parameters respect to this variable, and another one that binarized the DTD variable by applying dummy coding without comparison group. In this last one, n DTD binary new attributes DTD_x ($x = 1, ..., n$) were obtained, such that DTD_x was 1 for patients with DTD$= x$, and 0 for patients with DTD$\neq x$.

For the first method, filter-type feature selection was applied with five selection functions [7]: information gain, information gain ratio, correlation, Chi square, and Gini index. The results were min-max normalized to allow cross comparison. The ten best attributes were kept as the most significant in the study of DTD. These were SOFA-CNS, EMINA, sedative/analgesic, SOFA-Resp, arterial catheter, SOFA-Cardio, SOFA-Total, vasoactive drugs, central venous catheter, and NAS.

For the second method, we obtained the list of the most significant parameters of every DTD group individually. The normalized relevance of these features was used to select those which were among the 30% best ones to be considered for biomarkers. Table 3 shows the resulting features per DTD group (Table 2).

Table 2. Significant biomarkers for each of last 7 DTD with a 30% threshold.

DTD group	Most significant parameters for biomarkers
DTD = 1	EMINA, SOFA-Total, SOFA-CNS, SOFA-Resp
DTD = 2	EMINA, SOFA-Total, SOFA-CNS, SOFA-Resp, STRATIFY, Tmp
DTD = 3	EMINA, SOFA-Total, SOFA-CNS, SOFA-Resp, STRATIFY, Tmp, TSA
DTD = 4	EMINA, SOFA-CNS, SOFA-Total, TSA, Tmp, SOFA-Resp, STRATIFY
DTD = 5	Tmp, TSA, SOFA-CNS, EMINA, SOFA-Total, SOFA-Resp, STRATIFY
DTD = 6	TSA, Tmp, Glucose min
DTD = 7	EMINA, Glucose min, HR, TSA, NAS, APACHE-II, Sofa-Resp

The level of overlapping of values of the top three DTD possible biomarkers (i.e., EMINA, SOFA_Total, and SOFA_CNS) is represented with the boxplots in Fig. 1.

2.4 Phenotype Extraction

The diversity of ICU patients in every DTD group limits the possibility of finding general DTD prediction models, as we will see in the next section. In order to

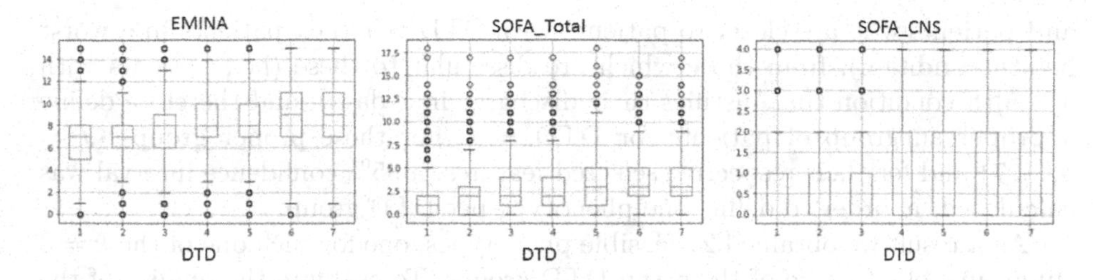

Fig. 1. Overlapping of EMINA, SOFA_Total, and SOFA_CNS values among DTD groups.

alleviate this problem of finding general descriptions for all patients in a DTD, we worked to identify patient phenotypes as descriptions of subgroups of patients with the same DTD who cannot be confused with patients with other DTDs.

To this end, we followed two methods. The first method applied subgroup discovery [14,15] in order to find interesting associations among different variables with respect to the DTD, our property of interest. The second method combined feature selection [16] and unsupervised clustering, focused on numeric parameters, in order to first reduce the dimension of the problem, and then, based on the selected features, find out interesting subgroups of patients (i.e., phenotypes), with the use of k-means.

The first method, which used all the parameters of the dataset, discovered a subgroup of patients for each DTD, corresponding to a feasible phenotype (see Table 3).

Table 3. Feasible phenotype for patients in their last seven DTD

DTD group	Feasible phenotype
DTD = 1	SOFA-CNS = 0, SOFA-Resp = 0
DTD = 2	SOFA-CNS = 0, SOFA-Resp = 0
DTD = 3	SOFA-Resp = 0
DTD = 4	SOFA-Resp = 0
DTD = 5	SOFA-Liver = 0, TSA < 2
DTD = 6	SOFA-Liver = 0, SOFA-Total = [2:4]
DTD = 7	SOFA-Total = [4:6]

The second method selected the most relevant numerical attributes for every DTD group (i.e., group of patients with the same DTD). These are the ones contained in Table 3. The k-means algorithm used these parameters to obtain three subclusters per DTD group. Our current selection of $k = 3$ for each group of patients with DTD $= x$ attends to our intention to separate those patients who are closer to patients with DTD $< x$ (i.e., patients in a better health condition), and patients who are closer to patients with DTD $> x$ (i.e., patients in a worst health condition), from those which are dissimilar to these (i.e., patients with a health condition that justifies their discharge in x days), and therefore define a proper subgroup of patients for DTD $= x$. For these proper groups ($x = 1, ..., 7$) and for their respective selected features, a 95% confidence interval was calculated, in order to define one phenotype per DTD group.

As a result we obtained 21 feasible phenotypes, one for each one of the $k = 3$ subgroups of each one of the seven DTD groups. To evaluate the quality of the phenotypes, the dataset was split in two subsets: patients admitted from 2014 to 2018 were used to obtain the phenotypes as previously explained, and patients admitted in 2019 were used to evaluate the phenotypes. Their low quality in terms of accuracy, sensitivity, and f1-value (75%, 23%, and 24%, respectively) confirmed the complexity of generating clinical decision models for the problem of DTDs in ICU.

2.5 Building DTD Predictive Models

Our initial approach to DTD prediction with standard supervised machine learning methods such as decision trees, logistic regression, naïve Bayes, and SVM did not obtain good predictive qualities. An exception was found in the Random Forest algorithm when trained with the patients in 2014–18, with a 10-fold-cross validation that reached a Mean Absolute Error (MAE) of 1.34, a Root Mean Square Error (RMSE) of 1.73, and a coefficient of determination $R2 = 0.61$ [2].

3 Discussion

Our experiences in the analysis of the days to discharge of patients in an ICU have shown that this is a complex area of work. We explored four different DTD-related issues with multiple methods and the help of a database on all the patients discharged alive in an interval of six years from an ICU.

The study of *patient heterogeneity* can confirm the complexity of the field, and having metrics to quantify this heterogeneity is not only good for benchmarking but also to gain insight on different interpretations of what heterogeneity in ICU-DTD means. When we gather all the patients in their x-th day before discharge in a DTD_x group, the longitudinal analysis of the progression of the means (and standard deviations) of their clinical parameters contributes to identify whether there are some parameters that we have to look at in order to determine if a patient is close to discharge or not. In [3], we found that SOFA-Cardio, SOFA-CNS, and SOFA-Resp are scores playing this role. Alternative scores such as SOFA-Total and NAS are also good at this purpose. An alternative interpretation of heterogeneity uses a distance function between ICU patients in order to calculate how confusing is to predict discharge for the patients in a DTD_x group by computing the quantity of patients in DTD_x who are similar to patients in other DTD groups. Our experiments showed that the degree of confusion is extremely high, which hinders the possibility of making good global DTD predictors, even for patients who are close to discharge.

The second issue analyzed concerns the identification of DTD *biomarkers* [17]. That is to say, measurable indicators concerning some biomedical condition of the ICU patients that could simplify the "diagnosis" of the patients' DTD group. Our application of several feature selection techniques identified SOFA-CNS, EMINA, and sedative/analgesic as best general biomarkers, but these may change if we focus on concrete DTD groups. For example, for patients in their 1, 2, 3, or 4 days to discharge EMINA, SOFA-Total, and SOFA-CNS seem to be the best biomarkers, but patient temperature (Tmp) becomes the most important for patients 5 or 6 days before discharge. According to [17], for an indicator to become a good biomarker, it must meet (1) analytical validity (i.e., be accurate and reproducible), (2) clinical validity (i.e., be medically meaningful and useful differentiating between groups), (3) clinical utility (i.e., improve health care), and (4) other validities (e.g., cost-effectiveness, psychological implications, or ethical implications). Currently, the indices out of our study fail to satisfy clinical validity due to the high level of overlapping between DTD groups (see Fig. 1).

The third issue addressed in this work is the extraction of *phenotypes* from the data that could identify subgroups of patients with a positive day to discharge. Our first approach with subgroup discovery techniques obtained phenotypes based on SOFA_CNS, SOFA_Resp, SOFA_Liver, TSA, and SOFA_Total. In the same way as for biomarkers, the number of patients with different DTDs in the same phenotype is high. Our attempt to overcome this problem with the generation of biomarkers for each DTD_x group by means of a binary analysis of all patients with $DTD = x$, versus all patients with $DTD \neq x$, obtained low-sensitive DTD phenotypes that require futher improvement before they can be of practical use.

Our final issue concerned the application of supervised machine learning to construct *predictive models* of DTD for ICU patients. Our first approach with regular algorithms such as decision trees, naïve Bayes, or logistic regression were soon discarded in favor of ensemble methods such as Random Forest, which so far, is the best approximation that we have obtained for the analysis and prediction of the days to discharge from the ICU [2].

4 Conclusion

Predicting DTDs of patients is essential to ICU management. Optimal prediction with models achieving an average error below one day is still far from being a reality. The high heterogeneity of ICU patients makes this a difficult objective. Here, we proposed four ways to analyze the DTD problem from different perspectives. For them, we suggested and implemented alternative ML-based methods that were tested with the data of the patients in an ICU. Results are commented and they confirm DTD analysis as a complex tasks for intelligent data analysis. In the future, a previous filtering might be convenient to reduce the diversity of patients before applying the methods proposed in this article. Time-series analysis methods will also be explored for a better DTD prediction.

Acknowledgements. Spanish Ministry of Science and Innovation (PID2019-1057 89RB-I00).

References

1. Marshall, J.C., Bosco, L., et al.: What is an intensive care unit? A report of the task force of the World Federation of Societies of Intensive and Critical Care Medicine. J. Critical Care **37**, 270–276 (2017). https://doi.org/10.1016/j.jcrc.2016.07.015
2. Cuadrado, D., et al.: Pursuing optimal prediction of discharge time in ICUS with machine learning methods. In: Riaño, D., Wilk, S., ten Teije, A. (eds.) AIME 2019. LNCS (LNAI), vol. 11526, pp. 150–154. Springer, Cham (2019). https://doi.org/10.1007/978-3-030-21642-9_20
3. Cuadrado, D., Riaño, D. Josep Gomez, J., Rodriguez, A., Bodi, M.: Methods and measures to quantify ICU patient heterogeneity. Submitted (2021)
4. Dasta, J., Mclaughlin, T., Mody, S., Piech, C.: Daily cost of an intensive care unit day: the contribution of mechanical ventilation*. Crit. Care Med. **33**, 1266–1271 (2005)
5. Lopez, C., Tucker, S., Salameh, T., Tucker, C.: An unsupervised machine learning method for discovering patient clusters based on genetic signatures. J. Biomed. Inform. **85**, 30–39 (2018). https://doi.org/10.1016/j.jbi.2018.07.004
6. Barak, S., Mokfi, T.: Evaluation and selection of clustering methods using a hybrid group MCDM. Expert Syst. Appl. **138**, 112817 (2019)
7. Jović, A., K. Brkić, K., Bogunović, N.: A review of feature selection methods with applications. In: 38th International Convention on Information and Communication Technology, Electronics and Microelectronics (MIPRO), Opatija, 2015, pp. 1200–1205. https://doi.org/10.1109/MIPRO.2015.7160458
8. Tan, S.S., Bakker, J., et al.: Direct cost analysis of intensive care unit stay in four European countries: applying a standardized costing methodology. Value Health **15**(1), 81–86 (2012). https://doi.org/10.1016/j.jval.2011.09.007
9. Nassar, A.P., Jr., Caruso, P.: ICU physicians are unable to accurately predict length of stay at admission: a prospective study. Int. J. Qual. Health Care **28**(1), 99–103 (2016). https://doi.org/10.1093/intqhc/mzv112
10. Gusmão Vicente, F., Polito Lomar, F., et al.: Can the experienced ICU physician predict ICU length of stay and outcome better than less experienced colleagues? Intensive Care Med. **30**(4), 655–659 (2004). https://doi.org/10.1007/s00134-003-2139-7
11. Verburg, I.W., et al.: Which models can i use to predict adult ICU length of stay? A Systematic Review. Crit. Care Med. **45**(2), e222–e231 (2017). https://doi.org/10.1097/CCM.0000000000002054
12. Kramer, A.A., Zimmerman, J.E.: A predictive model for the early identification of patients at risk for a prolonged intensive care unit length of stay. BMC Med. Inform. Decis. Mak. **10**, 27 (2010). https://doi.org/10.1186/1472-6947-10-27
13. Livieris, I.E., Kotsilieris, T., Dimopoulos, I., Pintelas, P.: Decision support software for forecasting patient's length of stay. Algorithms **11**, 199 (2018). https://doi.org/10.3390/a11120199
14. Herrera, F., Carmona, C.J., González, P., et al.: An overview on subgroup discovery: foundations and applications. Knowl. Inf. Syst. **29**, 495–525 (2011). https://doi.org/10.1007/s10115-010-0356-2

15. Helal, S.: Subgroup discovery algorithms: a survey and empirical evaluation. J. Comput. Sci. Technol. **31**(3), 561–576 (2016). https://doi.org/10.1007/s11390-016-1647-1
16. Chandrashekar, G., Sahin, F.: A survey on feature selection methods. Comput. Electr. Eng. **40**(1), 16–28 (2014)
17. Selleck, MJ, Senthil, M, Wall, NR.: Making meaningful clinical use of biomarkers. Biomark Insights **12** (2017). https://doi.org/10.1177/1177271917715236

Transformers for Multi-label Classification of Medical Text: An Empirical Comparison

Vithya Yogarajan(✉), Jacob Montiel, Tony Smith, and Bernhard Pfahringer

Department of Computer Science, University of Waikato, Hamilton, New Zealand
vy1@students.waikato.ac.nz

Abstract. Recent advancements in machine learning-based multi-label medical text classification techniques have been used to help enhance healthcare and aid better patient care. This research is motivated by transformers' success in natural language processing tasks, and the opportunity to further improve performance for medical-domain specific tasks by exploiting models pre-trained on health data. We consider transfer learning involving fine-tuning of pre-trained models for predicting medical codes, formulated as a multi-label problem. We find that domain-specific transformers outperform state-of-the-art results for multi-label problems with the number of labels ranging from 18 to 158, for a fixed sequence length. Additionally, we find that, for longer documents and/or number of labels greater than 300, traditional neural networks still have an edge over transformers. These findings are obtained by performing extensive experiments on the semi-structured eICU data and the free-form MIMIC III data, and applying various transformers including BERT, RoBERTa, and Longformer variations. The electronic health record data used in this research exhibits a high level of label imbalance. Considering individual label accuracy, we find that for eICU data medical-domain specific RoBERTa models achieve improvements for more frequent labels. For infrequent labels, in both datasets, traditional neural networks still perform better.

Keywords: Multi-label · Fine-tuning · Medical text · Transformers · Neural networks

1 Introduction

There has been a significant advance in natural language processing (NLP) in the last couple of years. Transformers such as BERT models (Bidirectional Encoder Representations from Transformers) have outperformed state-of-the-art (SOTA) results [4,6,7]. Such advancements are not restricted to general-domain tasks. Biomedical and health-related domains have also seen evidence of improvements in some medical domain-specific tasks such as question answering and

A. Tucker et al. (Eds.): AIME 2021, LNAI 12721, pp. 114–123, 2021.
https://doi.org/10.1007/978-3-030-77211-6_12

recognizing question entailment [2,3,9]. This research sets out to fill the gap in the use of transformers in multi-label medical domain-specific tasks for highly imbalanced datasets.

Multi-label problems predict multiple output variables for each instance. Consider a dataset $D = \{x^{(i)}, y^{(i)}\}_{i=1}^{N}$ with N samples, where $x^{(i)} = (x_1^{(i)}, ..., x_m^{(i)})$ and $y^{(i)} = (y_1^{(i)}, ..., y_l^{(i)})$. Each instance is associated with L labels, and each label is binary where $y_j^{(i)} \in 0, 1$. For example, given a patient admitted in a hospital with chest pain, any other medical condition that the patient has, such as cholesterol, blood pressure, or obesity, can be considered as labels.

This research focuses on electronic health records (EHR) from two distinctly different large publicly available medical databases: MIMIC-III contains huge documents in a free-form medical text; eICU has concise, compressed medical data presented in the semi-structured form. Automatically predicting medical codes is the down-stream task for this research where we fine-tune pre-trained transformer models, and we present results for multi-label medical code classifications with the number of labels being 18, 93, 158, 316, and 923.

The contributions of this work are: (i) we analyse the effectiveness of using transformers for the task of automatically predicting medical codes from EHRs for multiple document lengths and number of labels; (ii) we demonstrate that for documents with sequence length truncated at 512 tokens, medical domain-specific transformer models outperform SOTA methods for multi-label problems with 18, 93 and 158 labels for both datasets; (iii) it is shown that for longer documents, larger multi-label problems, and infrequent labels, transformer models' F1 scores are not as good as the traditional word-embeddings-based SOTA neural networks.

2 Related Work

This research is motivated by the recent advancements of transformer models which have shown substantial improvements in many NLP tasks, including BioNLP tasks. With minimum effort, transfer learning of pre-trained models by fine-tuning on down-stream supervised tasks achieves very good results [2,3]. For example, PubMedBERT [9] achieves SOTA performance on many biomedical natural language processing tasks such as named entity recognition, question answering and relation extraction and holds the top score on the Biomedical Language Understanding and Reasoning Benchmark (BLURB) [9].

Automatically predicting medical codes from EHRs has been studied over the years, where rule-based, machine learning-based and deep learning approaches have been proposed. Techniques including CNNs, RNNs and Hierarchical Attention Networks are some examples of deep learning approaches [2,16]. Mullenbach et al. (2018) [17] present Convolutional Attention for Multilabel classification (CAML) which uses the MIMIC III dataset for ICD-9 code predictions. As mentioned by the survey of deep learning methods for ICD coding of medical documents presented by Moons et al. (2020) [16] CAML is considered the SOTA method for automatically predicting medical codes from EHRs.

There is some evidence of the use of transformer models in automatically predicting medical codes such as submissions to CLEF eHealth 2019 ICD-10 predictions from German documents [2,19], and BERT and XLNet performance on most frequent ICD-9 codes from MIMIC III with a maximum number of tokens set at 512 [20]. However, it is unclear how well transformer models can perform with long clinical documents and in multi-label problems with a large number of labels [20]. Also, many studies [3,20] focus on high-frequency labels. Nonetheless, datasets such as MIMIC III and eICU consist of many infrequent labels where most codes only occur in a minimal number of clinical documents. This research presents results of multiple transformer methods and compares it with SOTA methods for various token lengths and number of labels.

For both word embeddings based networks and transformers, there is evidence to show domain-specific pre-trained models outperform general text pre-trained models [9,10,22]. This research uses word embeddings pre-trained on health-related text and transformers pre-trained on general and health-related data.

3 Data

Medical Information Mart for Intensive Care (MIMIC-III) [8,11] is a publicly available large database from the MIT with de-identified medical text data of more than 50,000 patients. We make use of free-form medical text from the discharge summaries. Figure 1 (top) presents a small sample of a discharge summary. MIMIC III discharge summary length varies between 50 to 8500 tokens with an average pre-processed text length of 1500 tokens. There are approximately 9000 unique ICD-9 codes associated with the hospital admissions in this database, with more than one code assigned to each patient.

Electronic Intensive Care Unit (eICU) is a database formed from the Philips eICU program [8,18], and contains de-identified data for more than 200,000 patients admitted to ICU. eICU data is found in tabular format with a drop-down

MIMIC III - Discharge Summary (sample text)
82 yo M with h/o CHF, COPD on 5 L oxygen at baseline, tracheobronchomalacia s/p stent, preseents with acute dyspnea over several days, and lethargy. This morning patient developed an acute worsening in dyspnea, and called EMS. EMS found patient tachypnic at saturating 90% on 5L. Patient was noted to be tripoding. He was given a nebulizer and brought to the ER.

eICU - Drop down menu (sample text)
Admission |Non-operative |Diagnosis |Cardiovascular |Sepsis, pulmonary |Non-operative Organ Systems |Was the patient admitted from the O.R. or went to the O.R. within 4 hours of admission? |No

Fig. 1. Sample data of MIMIC III (top) and eICU (bottom) obtained from the database. It includes acronyms and typos that are present in the data.

menu. Sample text data is presented in Fig. 1 (bottom). The length of medical text ranges from 10 to 1350, with an average of 130 tokens. eICU contains 883 unique ICD-9 codes.

The frequency of ICD-9 codes in both MIMIC III and eICU is unevenly spread with a large proportion of the codes occurring infrequently. For example, in MIMIC III and eICU only 0.02% and 0.2% of the codes are associated with at least 500 (1%) of the hospital admissions. One of the main reasons for the infrequent nature of medical codes in MIMIC III and eICU is because data are obtained from patients admitted in critical care. For this research, we consider each level of the ICD-9 hierarchy, as categorised by the World Health Organisation, as an individual flat multi-label problem. We remove all codes that occur in less than 10 unique hospital admissions. Consequently, our MIMIC III and eICU datasets contain 18 labels at level 1, 158 and 93 labels respectively at level 2, and 923 and 316 labels respectively at level 3.

4 Neural Network Algorithms

4.1 Transformers

Transformers [21] are one of the main recent developments in NLP which have achieved SOTA results in many language tasks [6,7,9]. Transformers are sequence-to-sequence models based on a self-attention mechanism. Given the linear projections Q, K, V, self-attention is computed as following [21]:

$$Attention(Q,K,V) = softmax\left(\frac{QK^T}{\sqrt{d_k}}\right)V \quad (1)$$

where the input queries and keys are of dimension d_k, and values of dimension d_v. See Vaswani et al. (2017) [21] for details of the transformer architecture.

BERT [7] is a deep neural network model that applies bidirectional training of the transformer encoder architecture [21] to language modelling. The BERT model relies on two pre-training tasks, masked language modelling and next sentence prediction. The 12-layer BERT-base model with a hidden size of 768, 12 self-attention heads, 110M parameter neural network architecture, was pre-trained on BookCorpus, a dataset consisting of 11,038 unpublished books and English Wikipedia.

ClinicalBERT model follows the same model architecture as the BERT-base model and was continually pre-trained on all notes from MIMIC III [1] from the BERT weights. PubMedBERT [9] uses the same architecture as the BERT-base model. However, unlike ClinicalBERT, PubMedBERT is domain-specifically pre-trained from scratch using abstracts from PubMed and full-text articles from PubMedCentral to enable better capturing of the biomedical language [9].

RoBERTa [14] is a robustly optimized BERT approach with improved training methodology and 160GB of general-domain training data in comparison to the 16GB data used in BERT. BioMed-RoBERTa-base [10] is based on the RoBERTa-base [14] architecture. RoBERTa-base was continuously pre-trained using 2.68 million scientific papers from the Semantic Scholar corpus starting with the RoBERTa-base weights.

Longformer [4] is a transformer model that is designed to handle longer sequences without the limitation on the maximum token size of 512 set by other transformers such as BERT. Longformer reduces the model complexity by reformulating the self-attention computation. This modified self-attention operation scales linearly with sequence length, instead of quadratically as in the original transformer models, making it possible to handle long documents. Longformer combines attention patterns such as sliding windows, dilated sliding windows and global attention (see Beltagy et al. (2020) [4] for more details). When compared to Eq. 1, Longformer uses two sets of projections, one to compute attention scores for a sliding window and another for global attention, providing the needed flexibility for the best performance of downstream tasks [4]. Longformers can be used for other NLP tasks in addition to language models. When compared to Transformer-XL [6], which can also handle long documents, Longformer is not restricted to the left-to-right approach of processing the documents.

After pre-training the models, the transformers are fine-tuned on task-specific data. All the parameters are fine-tuned end-to-end. Pre-trained transformer models learn good, context-dependent ways of representing text sequences which can be used on a specific downstream task. The models only need to fine-tune their representations to perform a particular task. Compared to the pre-training cost of transformers, the subsequent fine-tuning is relatively inexpensive.

4.2 Traditional Neural Networks

TextCNN [12] combines a single layer of one-dimensional convolutions with a max-over-time pooling layer and one fully connected layer. If $x_{i:i+j}$ is a concatenation of words from a sentence, each word, x_i, x_{i+1}, ... is mapped to its embeddings using the lookup table of word embeddings. The final prediction is made by computing a weighted combination of the pooled values and applying a sigmoid function. In our experiments, we use TextCNN with four different window sizes where each window takes 2, 3, 4 or 5 words with 100 feature maps each; the drop out rate is set to 0.2 and the learning rate to 0.003.

Gated Recurrent Units (GRU) [5] are a type of recurrent neural networks, with fewer parameters in comparison to long short-term memory (LSTM) networks. Bidirectional GRU (BiGRU) considers sequences from left to right, and right to left simultaneously. The learning rate used for our experiments is 0.003.

Mullenbach et al. (2018) [17] present CAML which achieves SOTA results for predicting ICD-9 codes from MIMIC III data [16]. CAML combines convolution networks with an attention mechanism. A secondary module is used to learn embeddings of the descriptions of ICD-9 codes to improve predictions of less frequent labels and are used as target regularization. For each word in a given document, word embeddings are concatenated into a matrix and a one dimensional convolution layer is used to combine these adjacent embeddings. The document is represented by matrix $\mathbf{H} \in R^{d_c \times N}$ where d_c is the size of convolutional filter and N is the length of the document. Then a per-label attention mechanism is applied, where $\mathbf{H}^T\mathbf{u}_l$ is computed for a given label l and a vector parameter $\mathbf{u}_l \in R^{d_c}$. The resulting vector is passed through a softmax operation with an output α_l. The vector representation for each label is calculated using $\mathbf{v}_l = \sum_{n=1}^{N} \alpha_{l,n} h_n$. The probability for l is calculated using a linear layer and a sigmoid transformation. A regularizing objective was added to the loss function of CAML with a trade-off hyperparameter. This variant is called Description Regularized-CAML (DR-CAML) [17]. The learning rate used for both CAML and DR-CAML in our experiments is 0.0001, and the regularization hyperparameter λ for DR-CAML is 0.01.

5 Experiments

We present results for multi-label medical code predictions for MIMIC III and eICU datasets. The number of labels being 18, 93, 158, 316, and 923. All experimental results presented are obtained from validations based on training-testing scheme, and are averaged over three runs. We explore a number of different transformer models and compare the performance to some traditional word embeddings based neural networks, including SOTA networks. The medical documents are truncated to a maximum number of tokens (512 and 4000). MIMIC III text was pre-processed by removing tokens that contain non alphabetic characters, including all special characters, and tokens that appear in fewer than three training documents. As eICU is already pre-processed extensively, no additional pre-processing was done for our research.

All neural network models presented in this research are implemented in PyTorch, and evaluations were done using sklearn metrics. All transformer implementations are based on the open-source PyTorch-transformer repository.[1] Transformer models are fine-tuned on all layers without freezing. As the optimizer we use Adam [13] with learning rates of 4e−6, or 4e−5. Training batch sizes were varied between 1 and 16, and the cut-off threshold was set to $t = 0.5$. Embeddings used for TextCNN, CAML, DR-CAML and BiGRU are health domain-specific fastText [15] pre-trained, skipgram word representation, 100-dimensional embeddings.

6 Results

Results for levels 1, 2 and 3 of the ICD-9 hierarchy, where each level is treated as an individual flat multi-label problem, for both eCIU and MIMIC III data are presented in Table 1. For eICU, we present results for 18, 93 and 316 labels. We find that using transformers for 18 and 93 labels, especially domain-specific models, result in performance improvements. We experimented with a maximum token length of 128, 512, and 1250 for eICU, and noticed a consistent improvement in performance between 128 and 512 tokens. However, there was no change between the micro and macro F1 scores for data truncated at 512 tokens and

[1] https://github.com/huggingface/transformers.

Table 1. Micro and macro F1 scores for multi-label problem with labels ranging from 18 to 923 are presented for eICU (left) and MIMIC III (right) datasets. Bold is used to indicate the highest scores within the grouping of networks, and underline to indicate the best score across all presented. Reported results are from validations based on training-testing scheme, averaged over three runs.

	eICU - 93 Labels		MIMIC III - 158 labels			
	512 tokens		512 tokens		4000 tokens	
	Micro-F1	Macro-F1	Micro-F1	Macro-F1	Micro-F1	Macro-F1
TextCNN	0.54	0.30	0.62	**0.32**	0.69	0.39
CAML	**0.57**	0.31	**0.64**	**0.32**	**0.72**	**0.42**
DR-CAML	**0.57**	**0.32**	**0.64**	**0.32**	**0.72**	**0.42**
BiGRU	0.56	**0.32**	0.60	0.31	0.70	**0.42**
Longformer	**0.60**	0.28	0.64	0.35	**0.70**	**0.38**
BERT-base	0.59	0.28	0.62	0.37	n/a	n/a
ClinicalBERT	0.59	0.28	0.64	0.36	n/a	n/a
BioMed-RoBERTa-base	**0.60**	**0.32**	0.64	0.40	n/a	n/a
PubMedBERT	0.58	0.24	**0.65**	**0.41**	n/a	n/a

	eICU - 512 tokens				MIMIC III - 512 tokens			
	18 labels		316 labels		18 labels		923 labels	
	Micro-F1	Macro-F1	Micro-F1	Macro-F1	Micro-F1	Macro-F1	Micro-F1	Macro-F1
TextCNN	0.63	0.48	0.43	0.17	0.79	**0.70**	0.50	0.18
CAML	**0.65**	**0.51**	0.50	**0.20**	0.79	0.69	**0.54**	**0.19**
DR-CAML	**0.65**	**0.51**	**0.51**	**0.20**	**0.80**	**0.70**	0.53	**0.19**
BioMed-RoBERTa-base	**0.68**	**0.52**	**0.50**	0.13	0.79	0.72	0.52	0.15
Pub-MedBERT	**0.68**	**0.52**	**0.50**	**0.14**	**0.81**	**0.74**	**0.53**	**0.16**

1250 tokens. Due to space limitations, we only present results for the maximum token length of 512. It is important to notice that only 0.2% of the eICU data contains medical text with a sequence length greater than 512. This might explain the small variation in neural network performances when the maximum sequence length is greater than 512 tokens. Compared to the word embeddings based methods, there is an improvement in micro-F1 when transformers are used. The overall best results are obtained using BioMed-RoBERTa-base for 93 labels, and Pub-MedBERT and BioMed-RoBERTa-base for 18 labels. However, for larger multi-label problem, such as the 316 labels, CAML and DR-CAML performs better with more significant differences in macro-F1 scores.

For MIMIC III, we present results for a maximum sequence length of 512 and 4000 tokens for 158 labels, and 512 tokens for 18 and 923 labels. As mentioned in Sect. 3, MIMIC III contains long documents and benefits from the increase in the length of maximum sequence size. Results using 4000 tokens are only presented for Longformer as the other transformer models are designed to handle a maximum of 512 tokens. Compared to the SOTA methods CAML and DR-CAML, most transformers show performance improvement for maximum sequence length of 512 tokens for 18 and 158 labels. For 158 labels macro-F1 of all transformers are considerably better than that of the SOTA methods, with

PubMedBERT setting a new SOTA results for ICD-9 code prediction. Similarly, with 18 labels, PubMedBERT results are better than that of word embeddings-based methods for 512 tokens. However, as observed for eICU with 923 labels, none of the transformers perform as well as the traditional neural networks, when the number of labels increases. However, we have only explored a subset of possible transformers. Future research might result in transformers that work well for multi-label problems with many infrequent labels.

Longformer is one of the very few transformers that can handle long documents. The model used in this research is pre-trained using general-domain data; however, like BERT and RoBERTa models, Longformer models trained on health domain-specific data may improve performance. To the best of our knowledge, there is no publicly available health domain-specific pre-trained Longformer, and it requires extensive resources to undertake such a task. Hence, we only present results for the general domain pre-trained publicly available model. It is essential to point out we also explored the option of using XLNet. However, a down-stream task for such large multi-label problem for text with tokens > 512 requires considerable computational power and time. Also, preliminary experiments with 18 labels for MIMIC data did not improve the performance of Longformer.

Figure 2 presents the winning F1 score and the differences between the two individual F1 scores for a given label for 93 labels for eICU and 158 labels for

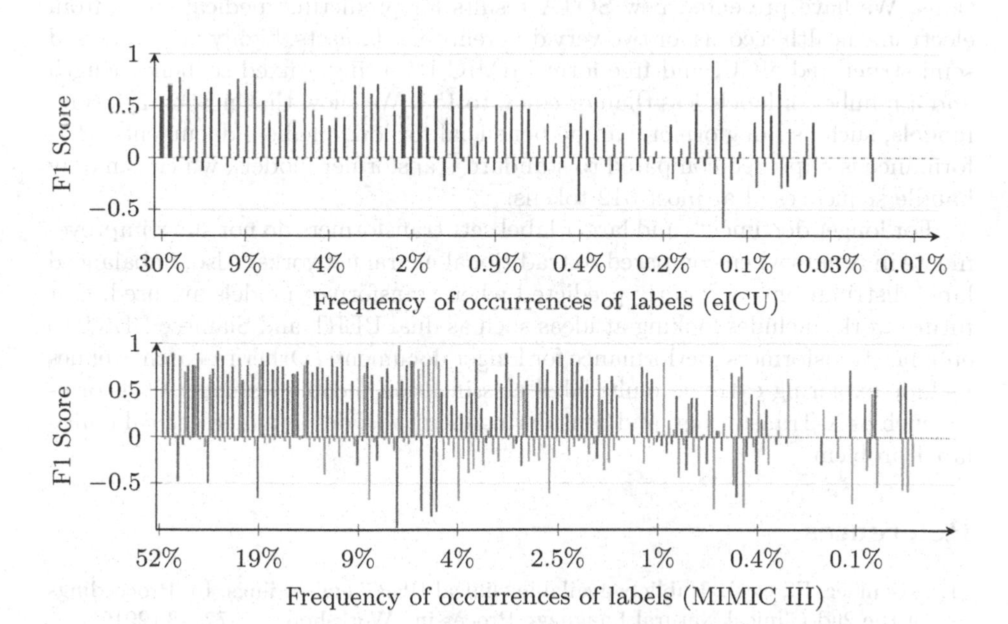

Fig. 2. The winning F1 score for each label, and the difference between the F1 scores from two networks are presented. Best F1 score is represented in the positive y-axis and the difference in the negative y-axis. F1 scores of 93 labels of eICU (top) where ▯▯ is BioMed-RoBERTa-base and ▯▯ is DR-CAML, and 158 labels of MIMIC III (bottom) where ▯▯ is CAML, and ▯▯ is Longformer.

MIMIC III data. The best performing (refer to Table 1) embeddings based neural network and transformers for each dataset is represented by different impulses in the Fig. 2. Positive F1 scores represent the best F1 score for each label of the two compared systems: Bio-Med-RoBERTa-base and DR-CAML for eICU, and Longformer and CAML for MIMIC III. The negative F1 scores represent the difference between the worst and the best compared F1 scores. Both data labels are ordered per frequency of occurrence. For eICU, for most labels with frequency $> 0.2\%$ F1 scores obtained using Bio-Med-RoBERTa-base are equal to or better than the DR-CAML ones. In some cases, for label frequencies between 0.7% to 0.2%, F1 scores obtained using DR-CAML are zero, while this is not the case for the transformer model. However, for infrequent labels, DR-CAML has a slight edge over transformer models. For MIMIC III data, for most labels F1 scores obtained using CAML model are better than the Longformer ones. Also, for rare labels the CAML model predicts some labels well, whereas Longformer's F1 scores are mostly zero.

7 Conclusions

This paper has shown that using transformers, especially domain-specific pre-trained models, can be highly beneficial in multi-label medical text classifications. We have presented new SOTA results for predicting medical codes from electronic health records for two very different text datasets, highly pre-processed semi-structured eICU, and free-form MIMIC III, using a fixed sequence length and a number of labels less than or equal to 158. We show that new transformer models, such as Longformer, can be beneficial for long medical documents. Performance is improved compared to standard transformer models, which can only handle sequences of at most 512 tokens.

For longer documents and larger label sets transformers do not show improvements in results when compared to traditional neural networks. Also, imbalanced label distributions are poorly predicted when transformer models are used. Our future works includes looking at ideas such as dual BERT and Siamese BERT to enhance transformers' performance for longer documents. Other research avenues include exploring extreme multi-label classification techniques using transformers such as X-Transformer, and considering medical codes as a hierarchical multi-label problem.

References

1. Alsentzer, E., et al.: Publicly available clinical BERT embeddings. In: Proceedings of the 2nd Clinical Natural Language Processing Workshop, pp. 72–78 (2019)
2. Amin, S., Neumann, G., Dunfield, K., Vechkaeva, A., Chapman, K.A., Wixted, M.K.: MLT-DFKI at CLEF eHealth 2019: multi-label classification of ICD-10 Codes with BERT. In: CLEF (Working Notes) (2019)
3. Amin-Nejad, A., Ive, J., Velupillai, S.: Exploring transformer text generation for medical dataset augmentation. In: Proceedings of The 12th Language Resources and Evaluation Conference, pp. 4699–4708 (2020)

4. Beltagy, I., Peters, M., Cohan, A.: Longformer: the long-document transformer. arXiv preprint arXiv:2004.05150 (2020)
5. Cho, K., van Merrienboer, B., Bahdanau, D., Bengio, Y.: On the properties of neural machine translation: Encoder-decoder approaches. In: Eighth Workshop on Syntax, Semantics and Structure in Statistical Translation (SSST-8), 2014 (2014)
6. Dai, Z., Yang, Z., Yang, Y., Carbonell, J., Le, Q.V., Salakhutdinov, R.: Transformer-XL: attentive language models beyond a fixed-length context. In: ACL (2019)
7. Devlin, J., Chang, M., Lee, K., Toutanova, K.: BERT: pre-training of deep bidirectional transformers for language understanding. In: NAACL-HLT (2019)
8. Goldberger, A.L., et al.: PhysioBank, PhysioToolkit, and PhysioNet: components of a new research resource for complex physiologic signals. Circulation **101**(23), e215–e220 (2000)
9. Gu, Y., et al.: Domain-specific language model pretraining for biomedical natural language processing. arXiv preprint arXiv:2007.15779 (2020)
10. Gururangan, S., et al.: Don't stop pretraining: adapt language models to domains and tasks. In: Proceedings of ACL (2020)
11. Johnson, A.E., et al.: MIMIC-III, a freely accessible critical care database. Sci. Data **3**, 160035 (2016)
12. Kim, Y.: Convolutional neural networks for sentence classification. In: Proceedings of the 2014 Conference on Empirical Methods in Natural Language Processing (EMNLP), pp. 1746–1751. Association for Computational Linguistics (2014)
13. Kingma, D.P., Ba, J.: Adam: a method for stochastic optimization. In: International Conference on Learning Representations (ICLR) (2015)
14. Liu, Y., et al.: RoBERTa: a robustly optimized BERT pretraining approach. arXiv preprint arXiv:1907.11692 (2019)
15. Mikolov, T., Grave, E., Bojanowski, P., Puhrsch, C., Joulin, A.: Advances in pre-training distributed word representations. In: Proceedings of the International Conference on Language Resources and Evaluation (LREC 2018) (2018)
16. Moons, E., Khanna, A., Akkasi, A., Moens, M.F.: A comparison of deep learning methods for ICD coding of clinical records. Appl. Sci. **10**(15), 5262 (2020)
17. Mullenbach, J., Wiegreffe, S., Duke, J., Sun, J., Eisenstein, J.: Explainable prediction of medical codes from clinical text. In: Proceedings of the 2018 Conference of the North American Chapter of the Association for Computational Linguistics: Human Language Technologies, Volume 1. ACL: New Orleans, LA, USA (2018)
18. Pollard, T.J., Johnson, A.E.W., Raffa, J.D., Celi, L.A., Mark, R.G., Badawi, O.: The eICU Collaborative Research Database, a freely available multi-center database for critical care research. Sci. Data **5**, 180178 (2018)
19. Sänger, M., Weber, L., Kittner, M., Leser, U.: Classifying german animal experiment summaries with multi-lingual BERT at CLEF eHealth 2019 Task 1. In: CLEF (Working Notes) (2019)
20. Schäfer, H., Friedrich, C.: Multilingual ICD-10 code assignment with transformer architectures using MIMIC-III discharge summaries. In: CLEF 2020 (2020)
21. Vaswani, A., et al.: Attention is all you need. In: Advances in Neural Information Processing Systems, vol. 30, pp. 5998–6008 (2017)
22. Yogarajan, V., Gouk, H., Smith, T., Mayo, M., Pfahringer, B.: Comparing high dimensional word embeddings trained on medical text to bag-of-words for predicting medical codes. In: Nguyen, N.T., Jearanaitanakij, K., Selamat, A., Trawiński, B., Chittayasothorn, S. (eds.) ACIIDS 2020. LNCS (LNAI), vol. 12033, pp. 97–108. Springer, Cham (2020). https://doi.org/10.1007/978-3-030-41964-6_9

Semantic Web Framework to Computerize Staged Reflex Testing Protocols to Mitigate Underutilization of Pathology Tests for Diagnosing Pituitary Disorders

William Van Woensel[1(✉)], Manal Elnenaei[2], Syed Ali Imran[3], and Syed Sibte Raza Abidi[1]

[1] NICHE Research Group, Faculty of Computer Science, Dalhousie University, Halifax, Canada
william.van.woensel@dal.ca

[2] Department of Pathology and Laboratory Medicine, Nova Scotia Health Authority, Halifax, Canada

[3] Division of Endocrinology, Department of Medicine, Dalhousie University, Halifax, Canada

Abstract. The complex and insidious presentation of certain health conditions, such as pituitary disorders, makes it challenging for primary care providers (PCP) to render a timely diagnosis—often delaying appropriate treatment for years. In contemporary clinical laboratories, laboratory interventions can appropriately add-on extra tests to help confirm or rule out complex disorders. For these protocols to be clinically valid and economically efficient, they require combining knowledge on abnormal test result patterns and patient health data to automatically "reflex" add-on tests and issue comments subsequent to their results. In this paper, we present a Semantic Web based framework for the computerization of reflex testing protocols. To avoid casting too wide a net in terms of add-on tests, a reflex (testing) protocol may include an arbitrary number of stages, where test result patterns in $stage_n$ can trigger add-on tests in $stage_{n+1}$. Our evaluation applies a computerized reflex protocol for pituitary dysfunction on 1-year retrospective data, and compares its accuracy and financial cost with a combined reflex/reflective approach that included manual laboratory clinician intervention.

Keywords: Semantic web · Reflex protocols · Pituitary disorders

1 Introduction

In recent decades, clinical biochemical laboratory services have grown to include laboratory interventions, such as interpreting initial test results, adding-on tests on patient samples, and providing comments to aid PCP in appropriate diagnoses [1, 2]. A common use case is pituitary disorder, as it is associated with clinical manifestations that are highly variable and insidious in onset. Endocrinologists are well aware of indicative abnormal patterns of basal pituitary hormones, but non-specialist PCPs are often challenged by test results that do not "flag" as abnormal, or when a stimulating hormone

A. Tucker et al. (Eds.): AIME 2021, LNAI 12721, pp. 124–134, 2021.
https://doi.org/10.1007/978-3-030-77211-6_13

level is inappropriately normal [3]. In prior work that applied laboratory interventions for pituitary disorders, we detected a higher incidence of pituitary dysfunction than the largest prospective study to date [4], which highlights the potential of laboratory interventions in early diagnosis of these conditions [5].

In general, there exist two types of protocols for laboratory interventions. Reflective testing protocols involve a Laboratory Clinician (LC) to manually "reflect" on initial test results and patient data, possibly issuing add-on laboratory tests on the patient sample and formulating reflective comments [1, 5, 6]. Reflex testing protocols are automated mechanisms, which "reflex" add-on laboratory tests by applying a set of fixed criteria on initial test results [2, 5, 6]. Reflex protocols have been found useful for early detection of conditions such as hypovitaminosis, haemochromatosis, and pituitary disorders [5, 6], but tend to be based exclusively on initial test results [6, 7]. In reflective protocols, LCs further consider patient data and prior test results from hospital/laboratory information systems [6, 7]. Compared to reflex protocols, reflective protocols issue fewer add-on tests, thus resulting in less marginal test cost and fewer false positives, and provide the same or higher levels of accuracy [6]. However, reflective protocols are also known to be very time-consuming: even in a combined reflex/reflective protocol [5, 6], i.e., where a reflex protocol is used to filter test results to be assessed for add-on by an LC, we estimated their time spent at 112 h over a year [5]. Given that financial factors will enter into organizational decision making, such high intervention costs call into question the feasibility of systematic, reflective laboratory interventions.

We present a generic Semantic Web framework for the computerization of reflex testing protocols as automated, economical and interpretable methods for adding-on laboratory tests, which can be specialized for different clinical domains. The computerization of the reflex testing protocols builds on our prior work of computerizing and executing clinical guideline protocols [8, 9]. To reduce the number of added-on tests compared to reflective protocols, while still casting a wide-enough net for clinical disorders, reflex protocols can include a series of stages: add-up tests in $stage_n$ are only issued when certain result patterns are encountered in $stage_{n-1}$. To offer a degree of interpretability by the requesting PCP, reflex protocols generate interpretive comments based on initial and add-on test results [10]. In part, we aim for these comments to act as a safety net against false positives caused by the relatively larger amount of add-on tests. Our Semantic Web framework is centered on the *ReflexOntology*, which includes a set of reflex rules that recommend follow-up tests in stages, based on initial and add-on test results, along with a rationale for the PCP. The *ReflexOntology* is freely available online [11] so others can re-use and build upon our work. In a retrospective evaluation, we compare the accuracy and cost of a computerized reflex protocol for pituitary disorders to a combined reflex/reflective protocol that includes manual LC intervention.

2 Reflex and Reflective Testing Protocols for Pituitary Disorder

Figure 1 illustrates a combined reflex/reflective pathway for the detection, intervention and follow-up of pituitary disorders, which was validated in a 12-month prospective study [5]. In this section, we highlight opportunities to refine this combined protocol, which paves the way for a fully reflexive protocol that does not require LC intervention.

In the reflex/reflective pathway, we setup reflex rules in the local Laboratory Information System (LIS) to reflex on possibly indicative test results (**1**), as exemplified in Table 1. Upon receiving a list of reflexed tests, the Laboratory Clinician (LC) excluded any test requested by a specialist (**2**), or where the Hospital Information System (HIS) yielded clear reasons for the abnormality (**3**). Then, the LC performed reflective add-on testing[1] (**4**), i.e., adding-on laboratory tests on the patient sample at their discretion.

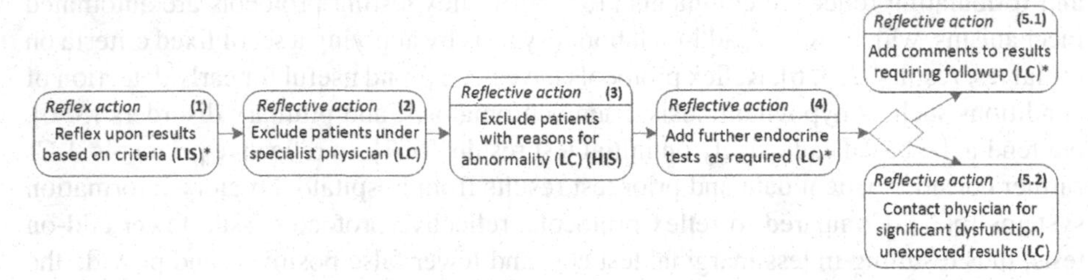

Fig. 1. Reflex/reflective pathway for detection, laboratory intervention and followup. Abbr. Laboratory Information System (LIS), Laboratory Clinician (LC), Hospital Information System (HIS) * As per Table 1.

Subsequently, the LC examined add-on test results and noted interpretive comments for those requiring further follow-up (**5.1**): listing possible non-pituitary causes for the result(s), and/or suggesting referral to Endocrinology. In cases of significant pituitary dysfunction, the LC directly contacted the requesting PCP (**5.2**).

Table 1 shows two reflex criteria together with all potential add-on tests[2]. These criteria are based on our local reference ranges and consensus between laboratory and endocrine physicians. We note that differences in testing platforms mean these values are not easily interchangeable; however, the underlying concepts are based on our own experience and review articles [3, 12–14].

These laboratory interventions, over a one year period, were successful in identifying 24 cases of pituitary dysfunction; these would have otherwise been overlooked, resulting in missed/delayed diagnosis. Subsequently, the overall incidence of pituitary disorders was found to be higher over that study period compared to the incidence in general populations, as reported by the largest prior prospective study to date [4]. However, even for a reflex/reflective pathway, i.e., where the LC is supported by a reflex algorithm, we found that the LC needs to spend considerable time to provide this service (estimated at 112 h [5]). However, a fully reflexive protocol, i.e., lacking LC reflection, would need to issue all suggested tests (Table 1, column 3) to capture all potential conditions: reducing cost-effectiveness and increasing likelihood of false positives. To pave the way for an efficient and cost-effective fully reflexive protocol, we refine Table 1 by distinguishing between first- and second-stage tests: certain second-stage tests only become relevant once first-stage test results are within certain bounds.

Table 2 illustrates this approach using the two reflex criteria from Table 1.

[1] Patient samples remained available for five days, as per our local laboratory policy.

[2] We refer to [5] for the full criteria and reflective tests (note these have been refined since).

Table 1. Criteria for Reflex capturing and Reflective testing protocol. Abbr.: FT4: free T4; TSH: thyroid stimulating hormone; FT3: free T3; FSH: follicle stimulating hormone; E2: estradiol; LH: luteinizing hormone; TST: testosterone; HRT: hormone replacement therapy

Criteria for reflexing tests	Exclusion criteria	Reflective tests	Results requiring clinical follow-up
FT4 < 9.5 & TSH < 4.3	- On thyroid medication - Non-thyroidal illness	- If female add FSH ± E2 - If male add LH, FSH, TST - If before noon add cortisol - Add prolactin	- Female > 55 yrs: FSH < 15 - Female < 55 yrs: FSH < 0.5 - Cortisol < 180 - Male: Low testosterone (<8), non-raised LH ± FSH - Raised prolactin
Female ≥ 55 yrs: FSH < 15	- On HRT	- Add E2 - Add TSH & FT4 - If before noon add cortisol - Add prolactin	- Low E2 -Non-raised TSH & low FT4 - before noon cortisol < 180 - Raised prolactin

Table 2. Criteria for Staged Reflex testing protocol. See Table 1 for Initial reflex and exclusion criteria and abbreviations.

Stage 1 Reflex tests	Criteria for Stage 2	Stage 2 Reflex tests	Results requiring clinical follow-up
- If female add FSH ± E2 - If male add LH, FSH,TST	- Female > 55 yrs: FSH < 15 and E2 < 100 - Female < 55 yrs: FSH < 0.5 and E2 < 100 - Male: Low TST & non-raised LH ± FSH	- If before noon add cortisol - Add prolactin	- Stage 1 results match Stage 2 criteria - Stage 2 results that match: Cortisol < 180 Raised prolactin
- Add E2	- Low E2	- Add TSH & FT4 - If before noon add cortisol - Add prolactin	- Stage 1 results match Stage 2 criteria - Stage 2 results that match: Non-raised TSH, low FT4 a.m. cortisol < 180 Raised prolactin

E.g., in the second row, 4 second-stage tests, i.e., TSH, FT4, a.m. cortisol and prolactin tests, will only be added in case E2 is low, else they will not be recommended.

Further, we formulated a-priori interpretative comments for each follow-up case (fourth column, Table 2), which delegate reflective elements to the PCP to aid them in acting on suggested referrals. Below, we show comments for the first row in Table 2:

Comment 1: Low TSH and FT4. If this patient is not on therapy for a thyroid condition, or is not unwell due to a systemic illness (non-thyroidal), suggest referral to the endocrine neuro-pituitary clinic to exclude secondary hypothyroidism.

Comment 2: Low TSH and FT4 which has triggered further endocrine testing. Results are suggestive of secondary hypothyroidism and other possible endocrine abnormalities.

If there is no clinical condition or therapy known in this patient to explain these results, referral to the endocrine neuro-pituitary clinic is warranted.

3 Semantic Framework for Computerizing Reflex Protocols

We present a knowledge-based approach for computerizing staged reflex protocols that encode clinical knowledge on test result patterns and health characteristics. We utilized Semantic Web technology, including OWL2 DL [15] and SWRL [16], to create a Reflex Ontology. The advantage of Semantic Web technology is that computerized reflex protocols have unambiguous formal semantics, are executable by any Semantic Web reasoner, and are interoperable with other biomedical ontologies.

Our Reflex Ontology can be extended to computerize reflex protocols for any clinical domain; for this project we extended the Reflex Ontology for pituitary disorders. Both the Reflex Ontology, and its pituitary disorder extension, can be found online [11]. In Fig. 2, we show the rule concepts from the Reflex OWL2 ontology:

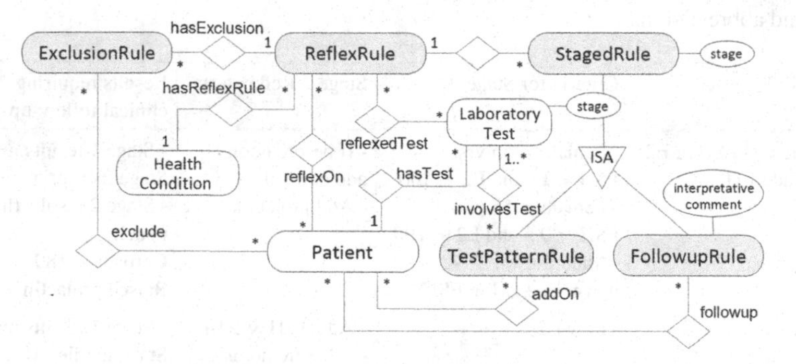

Fig. 2. Rules in the Reflex ontology.

A *ReflexRule* is triggered when initial test results lie within certain bounds and the patient matches certain criteria. A related *ExclusionRule* may exclude patients based on reasons for the abnormality (e.g., medications). Next, relevant *TestPatternRules* from $stage_n$ will recommend add-on tests. The followup rules will be tested in a sequential manner, whereby $stage_n$ *FollowupRules* will recommend followup when their results match certain criteria. In that case, *TestPatternRules* from $stage_{n+1}$ will be triggered, possibly followed by *FollowupRules* from $stage_{n+1}$, and so on. This process is implemented by domain-agnostic SWRL rules[3] described below (15 rules in total). Clinical domain criteria (e.g., Table 1) are implemented by domain-specific SWRL rules, and are exemplified later. A Semantic Web reasoner reasons over a *ReflexOntology* instance, including initial test results, to recommend add-on tests and follow-ups.

[3] For simplicity, the shown rules assume that the system reasons over one patient at a time.

In particular, a *HealthCondition* (e.g., pituitary disorder) is associated with (*hasReflexRule*) a set of *ReflexRules*. In case a reflex rule's domain-specific criteria are met (*meetsCriteria*), the rule will "reflex" on the patient (*reflexOn*):

$$\begin{aligned}&\mathit{HealthCondition}(?condition) \wedge \mathit{hasReflexRule}(?condition, ?reflexRule) \wedge \\ &\mathit{Patient}(?patient) \wedge \mathit{meetsCriteria}(?reflexRule, ?patient) \rightarrow \mathit{reflexOn}(?reflexRule, ?patient)\end{aligned} \tag{1}$$

If there are clear reasons for the abnormality, an *ExclusionRule* associated with the *ReflexRule* (*hasExclusion*) will be triggered (*exclude*):

$$\begin{aligned}&\mathit{reflexOn}(?reflexRule, ?patient) \wedge \mathit{hasExclusion}(?reflexRule, ?exclRule) \wedge \\ &\mathit{meetsCriteria}(?exclRule, ?patient) \rightarrow \mathit{exclude}(?exclRule, ?patient)\end{aligned} \tag{2}$$

At this point, the reflex protocol will recommend a series of (staged) add-on tests to confirm or rule out the health condition. After the *ReflexRule* was triggered[4], a $stage_1$*TestPatternRule* (case [**a**]), associated with the reflex rule (*hasTestPattern*), will recommend add-on tests when its domain criteria are met. (We discuss case [**b**] later.)

$$\begin{aligned}&\mathit{reflexOn}(?reflexRule, ?patient) \wedge \mathit{hasTestPattern}(?reflexRule, ?testRule) \wedge \\ &([\mathbf{a}]\ \mathit{hasStage}(?testRule, 1)[/a] \vee \\ &[\mathbf{b}]\ (\mathit{hasFollowupRule}(?reflexedRule, followupRule) \wedge \\ &\mathit{followup}(?followupRule, ?patient) \wedge \\ &\mathit{hasStage}(?testRule, ?stage) \wedge \mathit{hasStage}(?followupRule, ?stage - 1))[/\mathbf{b}]) \wedge \\ &\mathit{meetsCriteria}(?testRule, ?patient) \rightarrow \mathit{addOn}(?testRule, ?patient)\end{aligned} \tag{3}$$

After being triggered (*addOn*), the system uses the *involvesTest* predicate to identify recommended *LaboratoryTests* (Fig. 2); and follows the inverse *testPatternOf* to identify its *ReflexRule*. When an add-on laboratory test has been performed (*NewTest*), the system will represent it as follows (based on properties of the *TestPatternRule*):

$$\begin{aligned}&\mathit{hasTest}(Patient, NewTest) \wedge \mathit{stage}(NewTest, TestPatternStage) \wedge \\ &\mathit{reflexedTest}(ReflexRule, NewTest) \wedge \mathit{outcome}(NewTest, Outcome).\end{aligned} \tag{4}$$

In turn, a *FollowupRule* will recommend follow-up when added-on test results meet the domain criteria, together with an interpretative comment (e.g., *Comment 1*):

$$\begin{aligned}&\mathit{reflexOn}(?reflexRule, ?patient) \wedge \mathit{hasFollowup}(?reflexRule, ?followupRule) \wedge \\ &\mathit{meetsCriteria}(?followupRule, ?patient) \rightarrow \mathit{followup}(?followupRule, ?patient)\end{aligned} \tag{5}$$

At this point (see rule (**3**), case [**b**]), when a *FollowupRule* from the same reflex scenario (*hasFollowupRule*) recommended a follow-up (*followup*) at $stage_{n-1}$ (*hasStage*), a *TestPatternRule* at $stage_n$ within the same reflex scenario (*hasTestPattern*) will recommend further add-on laboratory tests when its domain criteria are met.

[4] SWRL (and OWL2) lack negation-as-failure, so it is left to the system to cope with exclusions.

Regarding laboratory test criteria (*meetsCriteria*), first, we formulate generic, domain-agnostic rules to ensure that criteria related to laboratory tests are properly checked. Reflex rule criteria only pertain to "stage-0" tests (i.e., initial tests): a reflex rule with two test-related criteria[5] and profile-related criteria is supported as follows:

$$\begin{aligned} &TwoTestReflex(?r) \wedge hasTest(?p, ?t_1) \wedge stage(?t_1, 0) \wedge hasTest(?p, ?t_2) \wedge stage(?t_2, 0) \wedge \\ &meetsProfile(?r, ?p) \wedge meetsTest_1(?r, ?t_1) \wedge meetsTest_2(?r, ?t_2) \rightarrow meetsCriteria(?r, ?p) \end{aligned} \tag{6}$$

Domain-specific SWRL rules (see below) will implement *meetsProfile* and *meetsTest*$_x$ predicates. Secondly, test-related criteria of follow-up rules must only be checked against *tests ordered in the context of the reflex rule.* In some cases, multiple reflex scenarios could lead to the same add-on tests. When a test is added-on within a particular reflex scenario, only follow-up rules within that scenario should be triggered:

$$\begin{aligned} &TwoTestFollowup(?f) \wedge reflexOn(?r, ?p) \wedge hasFollowup(?r, ?f) \wedge \\ &reflexedTest(?r, ?t_1) \wedge reflexedTest(?r, ?t_2) \wedge meetsTest_1(?f, ?t_1) \wedge meetsTest_2(?f, ?t_2) \wedge \\ &meetsProfile(?f, ?p) \rightarrow meetsCriteria(?f, ?p) \end{aligned} \tag{7}$$

I.e., a follow-up rule within a reflex scenario (*reflexOn*; *hasFollowup*) where two tests were reflexed (*reflexedTest*; see code (4)) will succeed if these tests meet the criteria. We similarly support *TestPatternRules* with test-related criteria (not shown).

To computerize a reflex protocol for a particular clinical domain, we first instantiate and link a set of reflex, exclusion, test-pattern and follow-up rule individuals (Fig. 2). E.g., for pituitary dysfunction, we created individuals $ReflexRule_1$, $ExclusionRule_{11}$, $TestPatternRule_{11}$ and $FollowupRule_{11}$ that partially represent the first row in Table 1 and Table 2. Secondly, we computerize their domain-specific criteria using SWRL:

$$\begin{aligned} &type(?t_1, FT4) \wedge outcome(?t_1, ?o) \wedge ?o < 9.5 \rightarrow meetsTest_1(ReflexRule_1, ?t_1) \\ &type(?t_2, TSH) \wedge outcome(?t_2, ?o) \wedge ?o < 4.3 \rightarrow meetsTest_2(ReflexRule_1, ?t_2) \end{aligned} \tag{8}$$

I.e., when the patient has two initial tests matching the test-related criteria from Table 1 (column 1), then the patient meets the criteria for triggering $ReflexRule_1$.

If the patient is on Thyroid medication, they meet the exclusion criterium:

$$medication(?p, ?m) \wedge type(?m, ThyroidMedication) \rightarrow meetsCriteria(ExclusionRule_{11}, ?p)$$

When the patient is female, they will meet profile-related criteria for $TestPatternRule_{11}$, which will recommend add-on tests of *FSH* and *E2* (rule not shown).

The criteria of $FollowupRule_{11}$ as per Table 2 (column 2) are represented as follows:

$$\begin{aligned} &type(?t_1, FSH) \wedge outcome(?t_1, ?o) \wedge ?o < 15 \rightarrow meetsTest_1(FollowupRule_{11}, ?t_1) \\ &type(?t_2, E2) \wedge outcome(?t_2, ?o) \wedge ?o < 100 \rightarrow meetsTest_2(FollowupRule_{11}, ?t_2) \\ &gender(?p, Female) \wedge hasAge(?p, ?a) \wedge ?a > 55 \rightarrow meetsProfile(FollowupRule_{11}, ?p) \end{aligned} \tag{9}$$

When a reflex scenario has completed, the system will indicate associated tests using *hadPriorTest* instead of *hasTest*, so they will no longer trigger any rules (e.g., rule (6)).

[5] This could be generalized using universal quantification, but this is not supported by SWRL.

4 Implementation of the Pituitary Disorder System

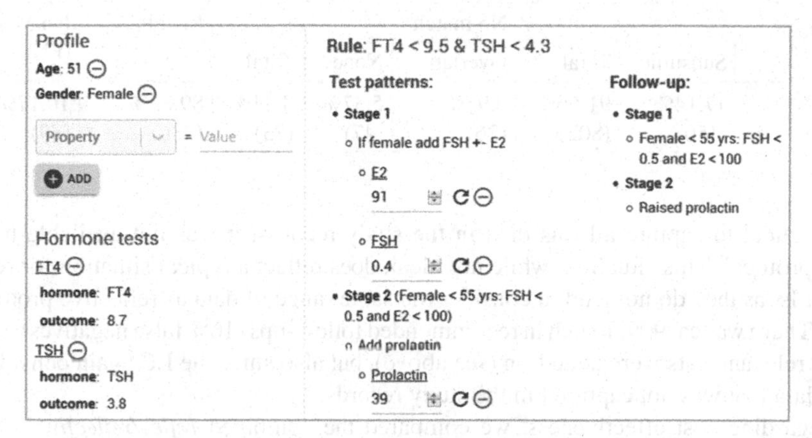

Fig. 3. Screenshot of the Pituitary Reflex system.

The prototype Reflex system was developed using Apache Jena[6] for storing, querying and reasoning over the Reflex ontology. In the left pane, the user enters profile details (e.g., age, gender) and initial tests results (FT4 = 8.7, TSH = 3.8). In the right pane, the system shows the fired reflex rule, its associated test patterns and any recommended follow-ups. In the prototype, results for add-on tests are directly entered (e.g., E2 = 91; FSH = 0.4). In this case, add-on test results met stage-1 follow-up criteria (see "follow-up" header), and a series of stage-2 test patterns were fired. In turn, results for these added-on tests were found to meet stage-2 follow-up criteria (Fig. 3).

5 Evaluation of Accuracy and Cost Effectiveness

We previously performed a 12-month prospective study of a combined reflex/reflective protocol for early detection of pituitary dysfunction [5]. From the clinical notes of the study LC, we extracted structured data on initial and added-on tests, follow-ups and profile data, which was utilized for a retrospective evaluation of our reflex protocol. Hence, a limitation of our evaluation is that the reflex protocol can only "reflex" on tests that were added-on in the prospective study, as it is based on that study's records.

Table 3 shows the matches between added-on laboratory tests and follow-ups by the reflex protocol vs. those from the combined reflex/reflective protocol.

In ca. (circa) 92% of cases, added-on tests from the reflex protocol directly matched or subsumed (i.e., included, but also extended) those issued by the reflex/reflective protocol. Hence, in ca. 8% of cases, test results that are possibly indicative of pituitary dysfunction may be missed. In the prospective study, the LC also considered other data, such as LIS hormone level trends and other HIS data (e.g., comorbidities); but since it was

6 https://jena.apache.org.

Table 3. Laboratory interventions of reflexive vs. combined reflex/reflective protocols.

Reflexed add-on tests						Reflexed follow-ups	
Match			No match			Match	No Match (FN)
Exact	Subsume	Total	Overlap	None	Total		
74.51% (652)	17.14% (150)	91.66% (802)	2.97% (26)	5.37% (47)	8.34% (73)	89.83% (786)	10.17% (89)

not practical to capture all this data in the study records, it was not available to the reflex protocol. This situation, while not ideal, does reflect a typical situation for reflex protocols, as they do not tend to consider the same range of data as reflective protocols [6, 7]. There was ca. 90% match in recommended follow-ups (10% false negatives); since not all relevant tests were added-on (see above), but also since the LC again considered extra data that was not captured in the study records.

Regarding cost effectiveness, we compared the combined *reflex/reflective*, *reflex: non-staged*, and *reflex: staged* protocols. Table 4 lists the total number of add-on tests, their total marginal[7] cost for the QE II hospital laboratory site (Nova Scotia Health Authority), and salary cost for the LC. For laboratories that lack a full track system, i.e., capable of automated add-on testing (including recall from storage), we also account for the salary of technical staff for manual handling of add-ons; we assume a total time of 7.5 min per set of add-on tests, with an avg. salary of 33 CAD/hour. Average LC salary for clinical/ medical biochemist was estimated at 120 CAD/hour.

Table 4. Financial impact of the evaluated testing protocols (costs are in CAD).

Protocol	# add-on tests	Marginal test cost	LC salary	Tech staff salary	Total cost
Reflex/reflective	244	463	13 440 (112 h)	512 (15.5 h)	14 415
Reflex: non-staged	1206	1721	0	883 (26.8 h)	2604
Reflex: staged	678	1123	0	1064 (32.3 h)	2187

The reflex protocol *significantly increases* the total number of add-on tests (factor ca. 4.95 for non-staged, ca. 2.8 for staged). Indeed, since the reflex protocol does not have access to the same comprehensive data as the LC, they need to cast a wider net to catch all relevant laboratory interventions. Nevertheless, the staged protocol only involves ca. 56% of tests required by the non-staged version. Due to the higher number of add-on tests, the reflex protocols yield higher marginal test and tech staff salary cost, but neither reflex protocol require LC involvement. Hence, the overall costs of reflex non-staged and staged protocols amount to only ca. 18% and ca. 15% of the combined reflex/reflective strategy. When an automated track system is available, total cost is only 12% (non-staged) and 8% (staged) of the reflex/reflective approach.

[7] I.e., extra cost of add-on tests on a patient sample (avg. CAD 1.15 per test).

At the same time, we observe that the significant increase in add-on tests will invariably result in an increase of false positives, due to their false positive rate. Tailored interpretive comments, i.e., generated on a case-by-case basis (Sect. 2, p. 4), are meant to act as a safety net against this, as they include an element for the PCP to reflect before acting on the recommended follow-up, but they are not a perfect solution.

6 Conclusions and Future Work

We presented a Semantic Web framework in the form of the Reflex Ontology for computerizing reflex protocols. These reflex protocols encode knowledge on abnormal test patterns and relevant patient data to automatically issue and follow-up on add-on laboratory tests for early detection of health conditions. These protocols (*a*) support "staged" reflex testing, where further stages of testing are only indicated for certain prior test results; and (*b*) generate interpretive comments for recommended follow-ups, delegating a degree of reflection to the PCP. We compared the accuracy and cost-effectiveness of a reflexive pathway to a combined reflex/reflective approach.

Improved reflex protocol accuracy is needed to avoid missed diagnoses, which themselves carry a long-term cost to the healthcare system. Given the evaluation results, we expect that a staged reflex protocol, with ties into HIS and LIS and outfitted with analysis tools for trending test results—i.e., with access to the same comprehensive data as the LC—will present an accurate and economical approach for early diagnosis of pituitary dysfunction or other health conditions. Utilizing a more expressive Semantic Web formalism, such as Notation3 [17], will allow us to improve the formal reflex semantics (e.g. Code (6), (7)) using (scoped) negation as failure. Finally, we aim to incorporate the Reflex system into clinical practice at the QE II laboratory site and connect it to the local LIS and HIS. A 12-month prospective study of the reflex protocol will compare pituitary diagnoses and added-on tests with the combined reflex/reflective protocol.

References

1. Verboeket-van de Venne, W.P.H.G., Aakre, K.M., et al.: Reflective testing: adding value to laboratory testing. Clin. Chem. Lab. Med. 50, 1249–52 (2012)
2. Fryer, A.A., Smellie, W.S.A.: Managing demand for laboratory tests: a laboratory toolkit. J. Clin. Pathol. **66**, 62–72 (2013)
3. Higham, C.E., et al.: Hypopituitarism. Lancet **388**, 2403–2415 (2016)
4. Regal, M., Páramo, C., et al.: Prevalence and incidence of hypopituitarism in an adult Caucasian population in northwestern Spain. Clin. Endocrinol. (Oxf) **55** (2001)
5. Elnenaei, M., Minney, D., Clarke, D.B., et al.: Reflex and reflective testing strategies for early detection of pituitary dysfunction. Clin. Biochem. **54**, 78–84 (2018)
6. Srivastava, R., Bartlett, W.A., et al.: Reflex and reflective testing: efficiency and effectiveness of adding on laboratory tests. Ann. Clin. Biochem. 47, 223–227 (2010).
7. Paterson, J.R., Paterson, R.: Reflective testing: how useful is the practice of adding on tests by laboratory clinicians? J. Clin. Pathol. **57**, 273–275 (2004)
8. Van Woensel, W., Abidi, S., Jafarpour, B., Abidi, S.S.R.: A CIG integration framework to provide decision support for comorbid conditions using transaction-based semantics and temporal planning. In: Michalowski, M., Moskovitch, R. (eds.) Artificial Intelligence in Medicine. AIME 2020. LNCS, vol. 12299, pp. 440–450. Springer, Cham (2020). https://doi.org/10.1007/978-3-030-59137-3_39

9. Jafarpour, B., Abidi, S.R., Van Woensel, W., Abidi, S.S.R.: Execution-time integration of clinical practice guidelines to provide decision support for comorbid conditions. Artif. Intell. Med. **94**, 117–137 (2019)
10. Barlow, I.M.: Are biochemistry interpretative comments helpful? Results of a general practitioner and nurse practitioner survey. Ann. Clin. Biochem. **45** (2008)
11. Van Woensel, W.: ReflexOntology. https://github.com/william-vw/ReflexOntology
12. Kim, S.Y.: Diagnosis and treatment of hypopituitarism. Endocrinol. Metab. **30**, 443–455 (2015)
13. Yip, C.-E., et al.: The role of morning basal serum cortisol in assessment of hypothalamic pituitary-adrenal axis. Clin. Invest. Med. **36**, E216–E222 (2013)
14. Essential Evidence Plus: Levels of Evidence. https://www.essentialevidenceplus.com/product/ebm_loe.cfm?show=grade
15. Hitzler, P., Krötzsch, M., Parsia, B., Patel-Schneider, P.F.: OWL 2 Web Ontology Language Primer, 2nd edn. https://www.w3.org/TR/owl2-primer/
16. Horrocks, I., Patel-Schneider, P.F., Boley, H., Tabet, S., Grosof, B., Dean, M.: SWRL: A Semantic Web Rule Language Combining OWL and RuleML. https://www.w3.org/Submission/SWRL/
17. Notation 3 (N3) Community Group. https://www.w3.org/community/n3-dev/

Using Distribution Divergence to Predict Changes in the Performance of Clinical Predictive Models

Mohammadamin Tajgardoon[1(✉)] and Shyam Visweswaran[1,2]

[1] Intelligent Systems Program, University of Pittsburgh, Pittsburgh, USA
{mot16,shv3}@pitt.edu
[2] Department of Biomedical Informatics, University of Pittsburgh, Pittsburgh, USA

Abstract. Clinical predictive models are vulnerable to degradation in performance due to changes in the distribution of the data (distribution divergence) at application time. Significant reductions in model performance can lead to suboptimal medical decisions and harm to patients. Distribution divergence in healthcare data can arise from changes in medical practice, patient demographics, equipment, and measurement standards. However, estimating model performance at application time is challenging when labels are not readily available, which is often the case in healthcare. One solution to this challenge is to develop unsupervised methods of measuring distribution divergence that are predictive of changes in performance of clinical models. In this article, we investigate the capability of divergence metrics that can be computed without labels in estimating model performance under conditions of distribution divergence. In particular, we examine two popular integral probability metrics, i.e., Wasserstein distance and maximum mean discrepancy, and measure their correlation with model performance in the context of predicting mortality and prolonged stay in the intensive care unit (ICU). When models were trained on data from one hospital's ICU and assessed on data from ICUs in other hospitals, model performance was significantly correlated with the degree of divergence across hospitals as measured by the distribution divergence metrics. Moreover, regression models could predict model performance from divergence metrics with small errors.

Keywords: Clinical predictive models · Electronic health records · Distribution divergence metrics · Concept drift · Dataset shift

1 Introduction

Prediction is central to many of the key activities in medicine. In medical practice, prediction is ubiquitous and includes evaluating the risk of developing disease in the future, diagnosis or establishing the presence of disease at the current time, prognosis or forecasting outcomes and complications related to ongoing disease, and estimating the response to therapeutics and other medical interventions [10,16,25]. Predictive models have immense potential to improve clinical

A. Tucker et al. (Eds.): AIME 2021, LNAI 12721, pp. 135–145, 2021.
https://doi.org/10.1007/978-3-030-77211-6_14

predictions and aid medical practitioners in clinical decision making. Typically, such models are derived from data using statistical and machine learning methods. Recent advances in the availability of big data, machine learning methods that can handle a wide range of data types, and abundant computing capability, have led to the development of a large number of predictive models for potential clinical use [16].

Many clinical predictive models are regression or classification models where the model predicts a target variable (or label) from several predictor variables. Such models are trained using supervised machine learning methods that make a fundamental assumption: the training data from which the models are derived, and the data on which the models will be applied, follow the same distribution. However, in medicine, as well as in other domains, properties of data change over time and across different geographical regions [17,26]. When the distribution changes, the performance of most predictive models is likely to change for the worse, and the models will likely need to be retrained using new data to regain the performance.

A straightforward mathematical solution to this generalization problem is to reconstruct the model from newer data or from data where the model will be used. However, practically, the collection of new data can be costly or even impossible. In clinical predictive models that are derived from observational data such as electronic health records (EHRs), obtaining labels may be expensive and time-consuming if clinical expertise is needed to determine them. Labels in EHRs may be unreliable or not available at all. For example, a code for pneumonia in an EHR may mean that the patient was screened for pneumonia rather than diagnosed with pneumonia [10]. As another example, whether a particular data element such as a hemoglobin laboratory test result is relevant to a clinical task cannot be readily obtained from EHRs [14]. Another solution is to use unsupervised domain adaptation [3] to improve predictive performance by including the new unlabeled data in the model training step. Nevertheless, evaluation at application time is not possible without obtaining labels. Therefore, it is necessary to develop approaches that evaluate the performance of predictive models that reduce or eliminate the need for collecting labels.

In machine learning, a significant difference between the distribution of training and test datasets is known as *distribution divergence* (or, *divergence*, for short), concept drift, or dataset shift [13,30]. As a result of divergence, the performance of a machine learning model may deteriorate on the test dataset if its distribution deviates significantly from the training dataset. Divergence between distributions is typically estimated empirically from samples drawn from those distributions. Popular metrics for distribution divergence are f-divergence metrics such as Kullback-Leibler (KL) divergence [15], and integral probability metrics (IPMs) including maximum mean discrepancy (MMD) [11] and Wasserstein distance (WD) [27]. IPMs are easier to implement in high-dimensional data and have better convergence rates compared to f-divergence estimators [24]. Note that these metrics are typically unsupervised, i.e., they do not use labels for computing the divergence, although supervised forms for some of these metrics have been developed [1].

In this article, we investigate the relation between unsupervised distribution divergence metrics and the performance of clinical predictive models that are derived using supervised learning. In particular, we investigate the effect of distribution divergence across intensive care units (ICUs) in 12 hospitals on the performance of predictive models that are derived from one hospital (training dataset) and evaluated on the other hospitals (test datasets). Our experiments include the prediction of mortality and long stays in the ICU. These labels are easily computable from the EHR data that we use and thus provide an abundant amount of labels for our experiments. However, in practice, we would use divergence metrics for clinical predictive models for which labels are not readily available and are expensive to obtain.

2 Related Work

2.1 Divergence Metrics to Predict Changes in Model Performance

Several investigators have explored the use of unsupervised divergence metrics to predict changes in model performance [5,8,12,22]. For example, Elsahar and Gallé [8] developed unsupervised methods to predict performance drop in models in the presence of distribution divergence. They used three classes of metrics: $\mathcal{H}$-divergence metrics, which are based on the performance of a domain classifier in distinguishing between training and test examples; confidence-based metrics, which measure the drop in the average predicted probabilities; and reverse model accuracy metrics, which use pseudo labels to derive a new model on the test dataset and compare its performance to that of the original model. As another example, Chuang et al. [5] used a set of domain-invariant features as a proxy to estimate the unknown labels in new data and estimate the model's performance in an unsupervised manner.

2.2 Distribution Divergence in Healthcare Data

A predictive model that is deployed in a clinical setting is not guaranteed to maintain its performance over time. Similarly, a model trained on data from one institution may perform poorly when deployed in another institution [10]. Model performance may suffer due to changes or differences in medical practice, patient demographics, equipment, and measurement standards [18]. Several studies have investigated distribution divergence in healthcare data and its effects on clinical predictive models [2,17,26]. For example, Davis et al. [6] studied the effects of changes in the patient population on the calibration of risk prediction models. The authors attributed calibration drift to changes in the patient population distribution rather than shifts in the marginal or conditional distribution of the outcome. Calibration drift was also investigated in the context of estimating the risk of hospital-acquired acute kidney injury [7]. The findings supported the need for continual model adaptations due to changes in the data distribution. In another study [29], distribution divergence due to differences in equipment between hospitals led to significant performance deterioration of models that predicted pneumonia from chest X-rays.

3 Divergence Metrics

3.1 Wasserstein Distance (WD)

Given $X \sim P$ and $Y \sim Q$ as i.i.d samples, WD between P and Q is the dual representation of Kantorovich metric on a given metric space (M, ρ), where $\rho(x, y)$ is a distance function for two instances $x \in X$ and $y \in Y$ in the set M:

$$\mathrm{WD}(P, Q) = \inf_{\mu \in \Gamma(P,Q)} \int_{M \times M} \rho(x, y)\, d\mu(x, y) \tag{1}$$

where $\Gamma(P, Q)$ is the set of all measures on $M \times M$ with marginals P and Q. In *optimal transport* theory, if $\mu(x, y)$ represents a randomized policy for transporting a unit quantity of some volume from location $x \in P$ to $y \in Q$ with the cost of $\rho(x, y)$, then $\mathrm{WD}(P, Q)$ is the minimum expected cost to transport the mass P to Q. Given $X = \{x_i\}_{i=1}^m$ and $Y = \{y_i\}_{i=1}^n$ as i.i.d samples, [24] proposed a non-parametric approach to compute WD based on an empirical estimate of its dual form:

$$\mathrm{WD}(P', Q') = \sup_{f \in \mathcal{F}} \left| \frac{1}{m} \sum_{i=1}^{m} f(x_i) - \frac{1}{n} \sum_{i=1}^{n} f(y_i) \right|, \tag{2}$$

where P' and Q' denote the empirical distributions of P and Q, respectively, and $\mathcal{F} = \{f : \|f\|_{\mathcal{L}} \leq 1\}$ where $\|f\|_L$ denotes the Lipschitz semi-norm of a real-values function f on M. Equation 2 can be solved by linear programming as shown in [24]. We used Python package POT [9] to implement the WD metric.

3.2 Maximum Mean Discrepancy (MMD)

Let $X \sim P$ and $Y \sim Q$ be two datasets in $\mathbb{R}^d$. The MMD metric measures the divergence between P and Q:

$$\mathrm{MMD}(P, Q) = \sup_{\|f\|_{\mathcal{H}} \leq 1} \left(\mathbb{E}_P\left[f(X)\right] - \mathbb{E}_Q\left[f(Y)\right] \right), \tag{3}$$

where function f is the unit ball in a reproducing kernel Hilbert space (RKHS) $\mathcal{H}$. Given $X = \{x_i\}_{i=1}^m$ and $Y = \{y_i\}_{i=1}^n$, an empirical unbiased estimate of MMD^2 is calculated as follows [11, Lemma6]:

$$\begin{aligned}\mathrm{MMD}^2(X, Y) = {} & \frac{1}{m(m-1)} \sum_{i=1}^{m} \sum_{j \neq i}^{m} k(x_i, x_j) + \frac{1}{n(n-1)} \sum_{i=1}^{n} \sum_{j \neq i}^{n} k(y_i, y_j) \\ & - \frac{2}{mn} \sum_{i=1}^{m} \sum_{j=1}^{n} k(x_i, y_j).\end{aligned} \tag{4}$$

A popular choice of k is Gaussian RBF kernel $k_\sigma(x, y) = \exp(-\frac{\|x-y\|^2}{2\sigma^2})$.

3.3 MMD p-value

A non-parametric test based on the $\text{MMD}^2(X, Y)$ estimator was proposed by [11] to distinguish between the null hypothesis $H_0 : P = Q$ and the alternative hypothesis $H_A : P \neq Q$. In our experiments, we used the p-value of the MMD test as an additional divergence metric. We implemented the MMD metrics using Python package hyppo [19].

4 Methods

4.1 Dataset

We used electronic health record (EHR) data from ICUs as a real-world EHR dataset. We obtained the data from the **High-Density Intensive Care** (HiDenIC) dataset that contains EHR data on 120,722 adult patients admitted to ICUs at 12 hospitals at the University of Pittsburgh Medical Center (UPMC) from February 2008 through December 2014. Table 1 shows the number of ICU stays per hospital. The dataset contains structured data including demographics, vital signs, and laboratory test values, and unstructured data that includes a variety of clinical text reports. For our experiments, we used only structured data.

Table 1. Number of ICU stays per hospital from February 2008 to December 2014.

	Hospital											
	H_1	H_2	H_3	H_4	H_5	H_6	H_7	H_8	H_9	H_{10}	H_{11}	H_{12}
No. ICU Stays	60,895	25,573	20,210	13,143	10,848	8,312	7,012	6,970	4,987	3,053	2,443	1,459

4.2 Experimental Setup

Distribution divergence is often found naturally in real-world datasets. For example, in a multi-hospital EHR dataset, the distribution of data usually changes from one hospital to another [2]. We take advantage of the multi-hospital nature of our dataset and partition the ICU stays by hospital. We trained models on each hospital's data in turn and evaluated the models on data from the remaining hospitals. The pseudo-code for our experiments is shown in Algorithm 1. First, we partitioned the data by hospital into 12 datasets $\{H_i\}_{i=1}^{12}$. Each H_i was used as the training set once and the remaining datasets were used as test sets. For more robust analysis, the experiment on each training dataset H_i was repeated for 10 iterations. At each iteration, a bootstrap sample of H_i, $H_i^{(k)}$, was obtained; a model f was trained on $H_i^{(k)}$ and evaluated on the remaining datasets $\{H_j\}_{j \neq i}^{12}$. The model performance results were stored in a matrix P. Pairwise divergence between the training set and test sets were stored in a matrix D. Matrices D and P are 10 by 11, with one row per bootstrap iteration and one column per test dataset H_j. We used D and P to fit a linear regression model that predicted model performance from distribution divergence.

Algorithm 1

Input: datasets $\{H_i\}_{i=1}^{12}$
Output: pairwise divergence results $\boldsymbol{D}$, model performance results $\boldsymbol{P}$
1: Let $\boldsymbol{D} = \{D_i\}_{i=1}^{12}$ and $\boldsymbol{P} = \{P_i\}_{i=1}^{12}$ be sets of empty matrices
2: **for** $i = 1$ **to** 12 **do** ▷ iterate over datasets $\{H_i\}_{i=1}^{12}$
3: **for** $k = 1$ **to** 10 **do** ▷ bootstrap iterations
4: $H_i^{(k)} = (X_i^{(k)}, y_i^{(k)}) \leftarrow bootstrap(H_i)$ ▷ a bootstrap sample of H_i
5: $f \leftarrow train(H_i^{(k)})$ ▷ train model f
6: **for** $j \neq i$ **to** 12 **do** ▷ iterate over test sets $\{H_j\}_{j \neq i}^{12}$
7: $(X_j, y_j) \leftarrow H_j$
8: $D_i \leftarrow D_i \cup divergence(X_i^{(k)}, X_j)$ ▷ unsupervised divergence metric
9: $P_i \leftarrow P_i \cup evaluate(f, H_j)$ ▷ evaluate f on test set H_j
10: return $\boldsymbol{D}$ and $\boldsymbol{P}$ ▷ return divergence and model performance results

We investigated two prediction tasks, two types of models, and two data representations, the details of which are described next.

Prediction Tasks. We used two prediction tasks including: (1) in-ICU **mortality** defined as death during the ICU stay, and (2) **long stay** defined as ICU stay that is longer than 3 days (following the definition used in [18]). For both tasks, the prediction was based on data from the first 24 h of ICU stay.

Models. We investigated two types of models: (1) **Random Forest (RF).** We implemented the RF model using Scikit-learn package [21]. We used 3-fold cross validation to select the optimal number of tree estimators from values $\{100, 500, 1000\}$. (2) **Multilayer perceptron (MLP).** We implemented MLP using PyTorch [20]. Grid search and a validation set were used to tune hyperparameters including model architecture and weight decay values $\{0.1, 0.01\}$ for the ADAM optimizer. The MLP model was selected from two architectures: Small MLP and Large MLP. Small MLP includes 3 linear layers (500, 100, 50 units), whereas Large MLP contains 4 linear layers (1000, 500, 200, 100 units). Hidden layers in both architectures were followed by ReLU activations and Dropout regularization. Softmax function was used at the output layer of both networks, and classification loss was measured by Binary Cross Entropy criterion. Univariate feature selection (ANOVA) was applied to both RF and MLP models to select 1,500 features during training.

Data Representation. Two data representations were used as input to divergence metrics: (1) **Raw representation.** This representation is high dimensional (1,500 vectorized features) and is the only representation that was used in experiments with the RF model. (2) **Embedding representation.** We used the output from the second layer of an MLP model as an embedding data representation (only used in experiments with MLP models). The embedding representation has been shown to contain meaningful and rich features that may result in more accurate divergence estimates [23].

4.3 Dataset Preparation

Cohort Selection. If a patient had multiple ICU stays, we included only the first ICU stay. Another inclusion criterion was that the stay should be a minimum of 36 h and a maximum of 10 days. Stays shorter than 36 h were excluded to enforce a minimum gap of 12 h between the prediction and outcome time. We required patients to be 15 years or older. We followed a MIMIC-III extraction process [28] to apply the selection criteria, which resulted in a dataset with 53,100 unique ICU stays.

Data Preprocessing. (1) Time-series discretization. Time-varying measurements such as laboratory test values and vital signs are unevenly spaced time-series observed during a stay at irregular time intervals. It is standard practice to discretize such time-series into equally-spaced hourly buckets [28]. We followed this approach and assigned the mean of values in each hour to the corresponding bucket. Each stay was censored to the first 24 hourly buckets. **(2) Normalization.** We normalized each numerical feature to have zero mean and unit standard deviation. In each experiment, the normalization was fit on the training set and applied to the test sets. **(3) Imputation.** We replaced missing values using the *Simple Imputation* method described in [4]. For each numerical feature, the method forward-fills the missing values and adds two additional features: a binary indicator denoting that the value was missing within an hourly bucket, and the time since the last measurement of the feature. **(4) Vectorization of time-series.** Standard machine learning algorithms such as RF and MLP expect a fixed-sized vector per stay. We vectorized each stay by transforming each time-varying feature into a vector of 24 values (each value corresponds to an hourly bucket). Demographic features were concatenated to the end of the vector. **(5) Features.** We used 70 features that included the 50 most frequent laboratory tests, 11 vital signs, and 9 demographic variables. Vital signs consisted of systolic and diastolic blood pressure, heart rate, respiratory rate, temperature, central venous pressure, intracranial pressure, and four Glasgow Coma Scale scores. Demographic features included age, height, weight, and one-hot encoded representations of gender and race.

4.4 Evaluation

Each experiment consisted of a training set from a specified hospital and test sets from the remaining hospitals, a specified prediction task, and a specified model type. For each experiment, we calculated three divergence metrics including WD, MMD, and MMD p-value, and we assessed the performance of the model on both the training set (using bootstrap samples) and the test sets with the area under the Receiver Operating Characteristic curve (AUC). We used linear regression to estimate the correlation between the values of a divergence metric and the AUC values. For example, for predicting mortality with RF using hospital H_1 as training data, we trained 10 RF models from 10 bootstrap samples of H_1 and applied them to data from the other 11 hospitals. This provided a total of $10 \times 11 = 110$ pairs of AUC and WD values that we used to fit a linear regression model (see Fig. 1).

For WD and MMD metrics, a negative value for the slope (β) of the regression line indicates that the higher the divergence of the test set from the training set, the lower is the model performance on the test set. The MMD p-value is inversely related to MMD and is expected to be positively correlated with the AUC when the MMD is negatively correlated.

To further assess the predictability of AUC values from divergence metrics, we used the AUC and divergence value pairs to train linear regression models with 10-fold cross-validation and reported the average mean absolute errors (MAEs). Lower MAE values indicate more accurate predictions (see Tables 2 and 3).

5 Results

Figure 1 shows example results where Hospital H_1 was used as the training hospital (i.e., H_i in Algorithm 1) for mortality prediction task with the RF model. As can be seen from the figure, WD and MMD are inversely correlated and MMD p-value is directly correlated with the AUC.

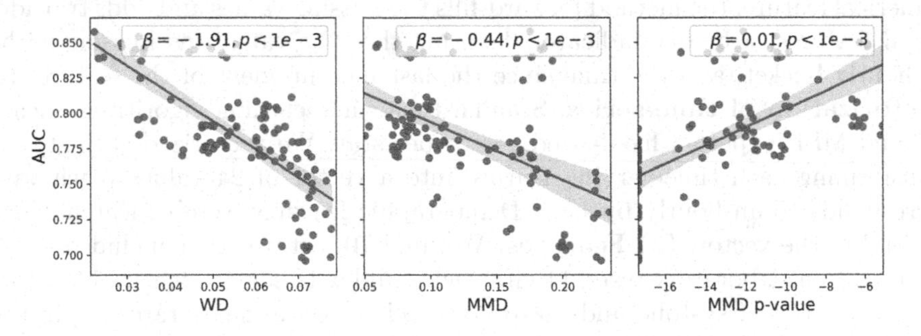

Fig. 1. Results from using Hospital H_1 as the training hospital for the mortality prediction task with the RF model. Each panel shows a linear regression line fitted to the values of a divergence metric and the corresponding AUC values. Note that MMD p-values are in log scale.

Tables 2 and 3 show the average slope (β) and average MAE of regression models for the three divergence metrics on the mortality and long stay prediction tasks. The WD metric more strongly correlated with the AUC compared to the other metrics, although the prediction performance (MAE) is similar across the three divergence metrics. Regression models could predict model AUC from divergence metrics with average MAEs of 0.02 to 0.05.

Table 2. Correlation between divergence and AUC for the mortality prediction task. Average slope (β) and average MAE of linear regression models to predict AUC from a divergence metric. The standard error of the mean is given for each mean value.

		WD		MMD		MMD p-value*	
Model	Representation	β	MAE	β	MAE	β	MAE
RF	Raw	-0.84 ± 0.25	0.03 ± 0.002	-0.18 ± 0.08	0.03 ± 0.002	0.004 ± 0.001	0.04 ± 0.003
MLP	Raw	-0.31 ± 0.10	0.05 ± 0.002	-0.06 ± 0.18	0.05 ± 0.002	0.001 ± 0.001	0.05 ± 0.002
	Embedding	-0.56 ± 0.26	0.05 ± 0.002	-0.01 ± 0.12	0.05 ± 0.002	0.003 ± 0.002	0.05 ± 0.002

* *MMD p-value was converted to log scale for regression analysis.*

Table 3. Correlation between divergence and AUC for the long stay prediction task. Average slope (β) and average MAE of linear regression models to predict AUC from a divergence metric. The standard error of the mean is given for each mean value.

		WD		MMD		MMD p-value*	
Model	Representation	β	MAE	β	MAE	β	MAE
RF	Raw	-0.60 ± 0.29	0.02 ± 0.002	-0.07 ± 0.06	0.02 ± 0.002	0.002 ± 0.001	0.05 ± 0.023
MLP	Raw	-0.57 ± 0.26	0.02 ± 0.001	-0.08 ± 0.07	0.02 ± 0.001	0.001 ± 0.001	0.02 ± 0.002
	Embedding	-0.41 ± 0.22	0.02 ± 0.001	-0.08 ± 0.06	0.02 ± 0.001	0.001 ± 0.001	0.02 ± 0.001

* *MMD p-value was used in log scale for regression analysis.*

6 Discussion and Conclusion

We investigated the correlation between unsupervised distribution divergence metrics and predictive model performance. Specifically, we analyzed the effects of divergence across ICU datasets from different hospitals on predictive modeling. The results show a significant inverse correlation between WD and MMD divergence metrics and model AUC values and a direct correlation between MMD p-values and model AUC values. In most of the experiments, the correlation was statistically significant (not included in the results). The direction of the correlation was more consistent, and the strength of the correlation was strongest for the WD metric.

There are several limitations to this study. First, we measured divergence only across the marginal distribution of predictor variables. These divergence metrics are likely to fail when changes occur in the distribution of labels without associated changes to the distribution of predictor variables. Second, we investigated only one scenario of the effect of divergence in EHR data which is the transfer of models across hospitals. Another scenario that we plan to examine in future work is the effect of divergence in the transfer of models from one period of time to another at the same hospital. Third, our study is limited to data from only one hospital system. In future studies, we plan to include EHR data from other hospital systems.

Acknowledgements. The research reported in this publication was supported by the National Library of Medicine of the National Institutes of Health under award number R01 LM012095, and a Provost Fellowship in Intelligent Systems at the University of Pittsburgh (awarded to M.T.). The content is solely the responsibility of the authors and does not necessarily represent the official views of the National Institutes of Health.

References

1. Alvarez-Melis, D., Fusi, N.: Geometric dataset distances via optimal transport. Adv. Neural. Inf. Process. Syst. **33**, 21428–21439 (2020)
2. Balachandar, N., Chang, K., Kalpathy-Cramer, J., Rubin, D.L.: Accounting for data variability in multi-institutional distributed deep learning for medical imaging. J. Am. Med. Inform. Assoc. **27**(5), 700–708 (2020)
3. Ben-David, S., Blitzer, J., Crammer, K., et al.: A theory of learning from different domains. Mach. Learn. **79**(1–2), 151–175 (2010)
4. Che, Z., Purushotham, S., Cho, K., Sontag, D., Liu, Y.: Recurrent neural networks for multivariate time series with missing values. Sci. Rep. **8**(1), 1–12 (2018)
5. Chuang, C.Y., Torralba, A., Jegelka, S.: Estimating generalization under distribution shifts via domain-invariant representations. In: International Conference on Machine Learning, pp. 1984–1994. PMLR (2020)
6. Davis, S.E., Lasko, T.A., Chen, G., Matheny, M.E.: Calibration drift among regression and machine learning models for hospital mortality. In: AMIA Annual Symposium Proceedings, pp. 625–634 (2017)
7. Davis, S.E., Lasko, T.A., Chen, G., Siew, E.D., Matheny, M.E.: Calibration drift in regression and machine learning models for acute kidney injury. J. Am. Med. Inform. Assoc. **24**(6), 1052–1061 (2017)
8. Elsahar, H., Gallé, M.: To annotate or not? Predicting performance drop under domain shift. In: Proceedings of EMNLP-IJCNLP, pp. 2163–2173 (2019)
9. Flamary, R., Courty, N.: POT python optimal transport library (2017)
10. Ghassemi, M., Naumann, T., Schulam, P., et al.: A review of challenges and opportunities in machine learning for health. AMIA Summits Transl. Sci. Proc. **2020**, 191–200 (2020)
11. Gretton, A., Borgwardt, K.M., Rasch, M.J., et al.: A kernel two-sample test. J. Mach. Learn. Res. **13**(1), 723–773 (2012)
12. Jaffe, A., Nadler, B., Kluger, Y.: Estimating the accuracies of multiple classifiers without labeled data. In: Artificial Intelligence and Statistics, pp. 407–415 (2015)
13. Kashyap, A.R., Hazarika, D., Kan, M.Y., Zimmermann, R.: Domain divergences: a survey and empirical analysis. arXiv preprint arXiv:2010.12198 (2020)
14. King, A.J., Cooper, G.F., Clermont, G., et al.: Using machine learning to selectively highlight patient information. J. Biomed. Informat. **100**, 103327 (2019)
15. Kullback, S.: Information theory and statistics. Courier Corporation (1997)
16. Miotto, R., Wang, F., Wang, S., et al.: Deep learning for healthcare: review, opportunities and challenges. Brief. Bioinform. **19**(6), 1236–1246 (2018)
17. Moons, K.G., Kengne, A.P., Grobbee, D.E., et al.: Risk prediction models: II. External validation, model updating, and impact assessment. Heart **98**(9), 691–698 (2012)
18. Nestor, B., McDermott, M.B., Boag, W., et al.: Feature robustness in non-stationary health records: caveats to deployable model performance in common clinical machine learning tasks. In: Machine Learning for Healthcare Conference, pp. 381–405. PMLR (2019)
19. S. Panda, S. Palaniappan, J. Xiong, et al. hyppo: A comprehensive multivariate hypothesis testing python package, 2020
20. Paszke, A., Gross, S., Massa, F., et al.: Pytorch: an imperative style, high-performance deep learning library. Adv. Neural. Inf. Process. Syst. **32**, 8026–8037 (2019)

21. Pedregosa, F., Varoquaux, G., Gramfort, A., et al.: Scikit-learn: machine learning in Python. J. Mach. Learn. Res. **12**, 2825–2830 (2011)
22. Platanios, E., Poon, H., Mitchell, T.M., Horvitz, E.J.: Estimating accuracy from unlabeled data: a probabilistic logic approach. Adv. Neural. Inf. Process. Syst. **30**, 4361–4370 (2017)
23. Rabanser, S., Günnemann, S., Lipton, Z.: Failing loudly: an empirical study of methods for detecting dataset shift. In: Advances in Neural Information Processing Systems, vol. 32, pp. 1396–1408 (2019)
24. Sriperumbudur, B.K., Fukumizu, K., Gretton, A., et al.: On integral probability metrics, φ-divergences and binary classification. arXiv:0901.2698 (2009)
25. Steyerberg, E.W., et al.: Clinical Prediction Models. Springer, Cham (2019). https://doi.org/10.1007/978-3-030-16399-010.1007/978-3-030-16399-0
26. Subbaswamy, A., Saria, S.: From development to deployment: Dataset shift, causality, and shift-stable models in health AI. Biostatistics **21**(2), 345–352 (2020)
27. Villani, C.: Optimal Transport: Old and Now, vol. 338. Springer, Heidelberg (2008). https://doi.org/10.1007/978-3-540-71050-9
28. Wang, S., McDermott, M.B., Chauhan, G., et al.: MIMIC-Extract: a data extraction, preprocessing, and representation pipeline for MIMIC-III. In: Proceedings of the ACM Conference on Health, Inference, and Learning, pp. 222–235 (2020)
29. Zech, J.R., Badgeley, M.A., Liu, M., et al.: Variable generalization performance of a deep learning model to detect pneumonia in chest radiographs: a cross-sectional study. PLoS Med. **15**(11):e1002683 (2018)
30. Žliobaitė, I., Pechenizkiy, M., Gama, J.: An overview of concept drift applications. In: Japkowicz, N., Stefanowski, J. (eds.) Big Data Analysis: New Algorithms for a New Society. SBD, vol. 16, pp. 91–114. Springer, Cham (2016). https://doi.org/10.1007/978-3-319-26989-4_4

Analysis of Health Screening Records Using Interpretations of Predictive Models

Yuki Oba[1], Taro Tezuka[2(✉)], Masaru Sanuki[3], and Yukiko Wagatsuma[3]

[1] Graduate School of Comprehensive Human Sciences, University of Tsukuba, Tsukuba, Japan
s2021662@s.tsukuba.ac.jp

[2] Faculty of Library, Information and Media Science, University of Tsukuba, Tsukuba, Japan
tezuka@slis.tsukuba.ac.jp

[3] Faculty of Medicine, University of Tsukuba, Tsukuba, Japan
{sanuki,ywagats}@md.tsukuba.ac.jp

Abstract. Health screening is conducted in many countries to track general health conditions and find asymptomatic patients. In recent years, large-scale data analyses on health screening records have been utilized to predict patients' future health conditions. While such predictions are significantly important, it is also of great interest for medical researchers to identify factors that could deteriorate patients' medical conditions in the future. For this purpose, we propose to use interpretations of trained predictive models. Specifically, we trained machine learning models to predict future diabetes stages, then applied permutation importance, SHapley Additive exPlanations (SHAP), and a sensitivity analysis to extract features that contribute to aggravation. Among the trained models, XGBoost performed best in terms of the Matthews correlation coefficient. Permutation importance and SHAP showed that the model makes good predictions using a number of attributes conventionally known to be related to diabetes, but also those not commonly used in the diagnosis of diabetes. A sensitivity analysis showed that the predictions' changes were mostly consistent with our intuition on how daily behavior affects type 2 diabetes's aggravation.

1 Introduction

Health screening, including annual health screening, are conducted in a number of countries due to their importance in finding early signs of diseases. Recently, there have been attempts to predict the onsets of disorders by training machine learning models by using electronic medical records and health screening records [5,6,9,15]. In many works, the goal was to increase the accuracy of prediction. However, it is also of interest for medical researchers to give an interpretation or explanation to trained predictive models [13]. Unfortunately, trained models of many machine learning methods are hard to interpret. They often fail to add new knowledge regarding the observed phenomena's underlining mechanism [2].

A. Tucker et al. (Eds.): AIME 2021, LNAI 12721, pp. 146–151, 2021.
https://doi.org/10.1007/978-3-030-77211-6_15

In this paper, we propose to use permutation importance, SHapley Additive exPlanations (SHAP), and a sensitivity analysis for health screening records to discover relationships between behavioral factors and stages of diseases.

2 Related Work

There have been a few studies that analyze annual health screening records by machine learning. Shimoda *et al.* predicted whether a patient who came to a health screening in a specific year would return the following year [12]. Ichikawa *et al.* predicted whether a patient should be given health guidance using health screening records [7]. Kim *et al.* used recurrent neural networks to predict missing values on the basis of previous records [6]. Tsunekawa *et al.* used a random forest to predict the onset of malignant neoplasms such as cancer [15]. Sisodia *et al.* trained models such as decision trees and SVMs to predict whether a patient would develop diabetes [14]. Zou *et al.* suggested that the use of fasting blood glucose as well as other attributes are necessary to improve prediction accuracy [16]. Garske *et al.* used deep neural networks to predict the onset of diabetes at different time scales [5]. Manini *et al.* used a Bayesian network to analyze a causal structure for clinical complications in type 1 diabetes [10].

3 Method

Dataset: We used a medical checkup dataset collected from Mito Kyodo General Hospital in Japan. The dataset consists of three annual medical checkup records. The number of samples (patients) for each year was 4,133, 4,261, and 4,269 for years 2016, 2017, and 2018, respectively. Since the goal was to predict a disease stage of the following year (output year) using the present year (input year), we only used data from patients who came consecutively for at least two years. The number of such patients were 2,396 for 2016–2017 and 2,530 for 2017–2018. We removed attributes that were missing in over 95% of the patients.

Classification Task: Our goal of training a prediction model is to predict whether a patient's stage of diabetes will aggravate. We formalized it as a two-class categorization problem. In our dataset, each patient is labeled with one of six possible stages of diabetes, namely *nothing particular*, *mild abnormality*, *follow-up*, *requires treatment*, *requires further testing*, and *under medical treatment*. These stages were defined by the Japan Society of Ningen Dock[1]. We only used data from patients whose stage in the input year is in *nothing particular*, *mild abnormality*, or *follow-up*. When the stage of the output year did not change or was alleviated, we labeled the patient as `stable`. When the stage of the output year was worse than that of the input year, we labeled the patient as `aggravated`. We calculated class weights using the compute_class_weight function in scikit-learn and obtained 0.6241 for `stable` and 2.5146 for `aggravated`.

[1] https://www.ningen-dock.jp/wp/wp-content/uploads/2018/06/Criteria-category.pdf.

Parameter Optimization for Prediction: We compared other prediction models used in related works, namely fully-connected neural networks, XGBoost [4], a random forest, logistic regression, and an SVM. We evaluated the models with and without batch normalization and dropout (rate: 0.2) for each fully-connected layer. We used ReLU as the activation function, the sigmoid as the output function, and the binary cross-entropy as the loss function. We employed the early-stopping strategy for the neural networks and XGBoost model, training them up till 200 and 1,000 epochs, respectively. We evaluated using multiple values for each method and selected the model that obtained the highest MCC.

Dataset Size and Preprocessing: For attributes taking continuous values, we replaced missing values with the average value. For attributes taking discrete values, a missing value is treated as an additional category. Attributes taking continuous values were standardized. Attributes taking discrete values were encoded as one-hot vectors. We split the whole dataset into training, validation, and testing datasets by a ratio of 8:1:1.

Evaluation Criteria: Permutation importance estimates how significant each attribute is to determine the prediction value [1]. The sensitivity analysis observes the amount of change in prediction results as the input is perturbed [11].

4 Evaluation

Prediction Model: The neural network having the highest MCC was a seven-layer fully-connected neural network whose structure is Dense(8) - Dense(8) - Dense(8) - Dense(8) - Dense(8) - Dense(4) - Dense(2) - Output(1), having a dropout layer between the fully-connected layers. Adam with a learning rate set to 0.0001 performed best. The best XGBoost model was max depth: 4, min child weight: 1, gamma: 0.4, colsample by tree: 0.6, and subsample: 0.9. The best random forest was criterion: entropy, max depth: 5, and min samples leaf: 1.

Comparison of Performance with Existing Methods: Table 1 summarizes the resulting prediction performance. Precision, recall, and F1 score were calculated for items in the `aggravated` category.

Table 1. Prediction performance of each method

Methods	Accuracy	Precision	Recall	F1 Score	MCC
Neural network	0.7431	0.3984	0.5698	0.4689	0.3149
XGBoost	**0.8148**	**0.5333**	0.5581	**0.5455**	**0.4294**
Random forest	0.7454	0.4063	0.6047	0.4860	0.3367
Logistic regression	0.6620	0.3295	0.6744	0.4427	0.2709
SVM	0.6389	0.3177	**0.7093**	0.4388	0.2657

Comparison of Permutation Importance and the SHAP Values: We calculated the accuracy-based permutation importance and the SHAP values for each attribute using the testing data. To obtain the latter, we used TreeExplainer [8]. Table 2 shows that the status of diabetes mellitus, hemoglobin A1c, FBS, BMI, age, and weights were ranked high for both permutation importance and SHAP values. The results match our intuition that these attributes are likely to contribute to improved prediction performance. Attributes showing a high degree of importance include creatinine, hematocrit, and total cholesterol. Creatinine is an indicator to diagnose the state of renal function. Hematocrit is the volume of red blood cells and is an indicator when diagnosing the state of blood. Total cholesterol is an indicator to diagnose the state of blood lipids. These indicators are usually not used to diagnose diabetes directly; however, it might be of clinical interest that our result suggests their contribution to enable better predictions.

Table 2. Top 10 items in permutation importance and SHAP

Rank	Permutation importance	SHAP
1	Hemoglobin A1c	Hemoglobin A1c
2	FBS	Creatinine
3	Age	C-reactive protein
4	Status of Diabetes mellitus (Follow-up)*	Status of Diabetes mellitus (Follow-up)*
5	Creatinine	FBS
6	C-reactive protein	Status of Diabetes mellitus (Nothing particular)*
7	Uric acid	White blood cell count
8	Height	BMI
9	Cholinesterase (ChE)	Neutral fat
10	γ-Glutamyl Transpeptidase	Age

* One of the category variables

Results of Each Question in the Sensitivity Analysis: Figs. 1, 2 and 3 show the results of the sensitivity analysis. Figures 2 and 3 show that maintaining weight and having enough sleep can have a positive effect. Weight gain, or obesity, is well-known to be a high-risk factor for diabetes. Moreover, it has been pointed out that skipping breakfast relates to the onset of type 2 diabetes [3].

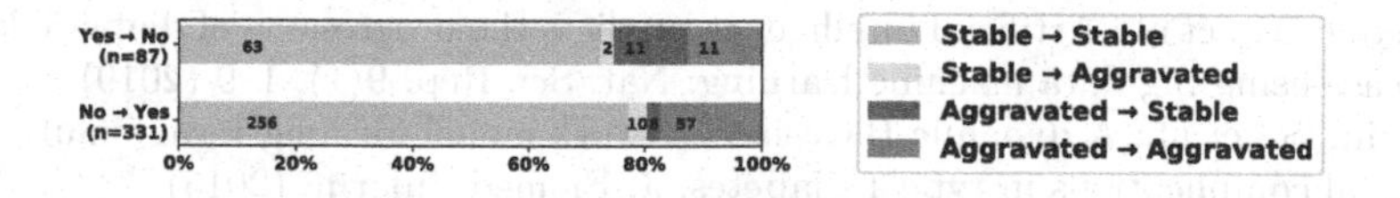

Fig. 1. Changes in Q8: Have you smoked in the last month?

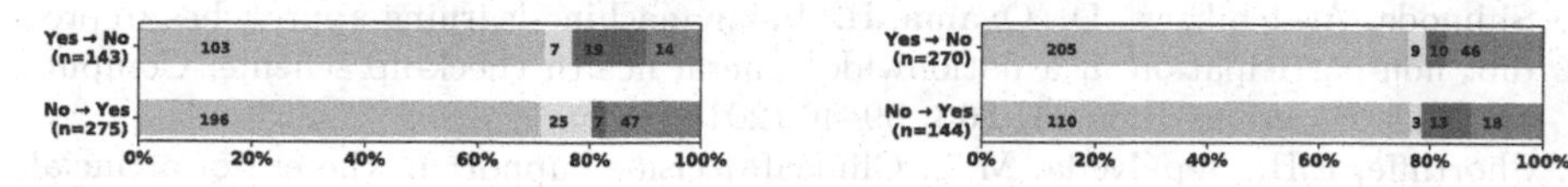

Fig. 2. Changes in Q9: Have you put on weight by 10 kg since your 20s?

Fig. 3. Changes in Q20: Do you sleep enough?

5 Conclusion

We developed a method of analyzing relationships between test items and the stages of diabetes using trained predictive models. We obtained interpretations of the predictive models. The result of permutation importance and SHAP showed that the predictor was primarily affected by attributes already known to be related to diabetes, but also by attributes that are usually not used for monitoring stages of diabetes.

Acknowledgements. This work was supported by JST COI Grant Number JPMJCE1301 and JSPS KAKENHI Grant Number JP16K00228, JP16H02904.

References

1. Altmann, A., Toloşi, L., Sander, O., Lengauer, T.: Permutation importance: a corrected feature importance measure. Bioinformatics **26**(10), 1340–1347 (2010)
2. Arrieta, A.B., et al.: Explainable artificial intelligence (XAI): concepts, taxonomies, opportunities and challenges toward responsible AI. Inf. Fusion **58**, 82–115 (2020)
3. Bi, H., Gan, Y., Yang, C., Chen, Y., Tong, X., Lu, Z.: Breakfast skipping and the risk of type 2 diabetes: a meta-analysis of observational studies. Public Health Nutr. **18**(16), 3013–3019 (2015)
4. Chen, T., Guestrin, C.: XGBoost: a scalable tree boosting system. In: Proceeding ACM International Conference on Knowledge Discovery and Data Mining, pp. 785–794 (2016)
5. Garske, T.: Using Deep Learning on EHR Data to Predict Diabetes. Ph.D. thesis, University of St. Thomas (2018)
6. Kim, H.-G., Jang, G.-J., Choi, H.-J., Kim, M., Kim, Y.-W., Choi, J.: Recurrent neural networks with missing information imputation for medical examination data prediction. In: International Conference on Big Data and Smart Computing (2017)
7. Ichikawa, D., Saito, T., Oyama, H.: Impact of predicting health-guidance candidates using massive health check-up data: a data-driven analysis. Int. J. Medical Informatics **106**, 32–36 (2017)
8. Lundberg, S.M., et al.: From local explanations to global understanding with explainable AI for trees. Nature Machine Intell. **2**(1), 56–67 (2020)
9. Makino, M., et al.: Artificial intelligence predicts the progression of diabetic kidney disease using big data machine learning. Nat. Sci. Rep. **9**(1), 1–9 (2019)
10. Marini, S., et al.: A dynamic Bayesian Network model for long-term simulation of clinical complications in type 1 diabetes. J. Biomed. Inform. (2015)
11. Mussone, L., Bassani, M., Masci, P.: Analysis of factors affecting the severity of crashes in urban road intersections. Accident Analysis & Prevention **103** (2017)
12. Shimoda, A., Ichikawa, D., Oyama, H.: Using machine-learning approaches to predict non-participation in a nationwide general health check-up scheme. Comput. Methods Programs Biomed. **163**, 39–46 (2018)
13. Shortliffe, E.H., Sepúlveda, M.J.: Clinical decision support in the era of artificial intelligence. J. Am. Med. Assoc. **320**(21), 2199–2200 (2018)
14. Sisodia, D., Sisodia, D.S.: Prediction of diabetes using classification algorithms. Procedia Comput. Sci. **132**, 1578–1585 (2018)

15. Tsunekawa, M., Oka, N., Araki, M., Shintani, M., Yoshikawa, M., Tanigawa, T.: Prediction of the onset of lifestyle-related diseases using regular health checkup data. In: Proceedings of the Annual Conference of the Japan Social for Artificial Intelligence (2019)
16. Zou, Q., Qu, K., Luo, Y., Yin, D., Ju, Y., Tang, H.: Predicting diabetes mellitus with machine learning techniques. Front. Genet. **9**, 515 (2018)

Seasonality in Infection Predictions Using Interpretable Models for High Dimensional Imbalanced Datasets

Bernardo Cánovas-Segura(✉), Antonio Morales, Jose M. Juárez, and Manuel Campos

Computer Science Faculty, University of Murcia, Murcia, Spain
{bernardocs,morales,jmjuarez,manuelcampos}@um.es

Abstract. Seasonality plays a significant role in the prevalence of infectious diseases. We evaluate the performance of different approaches used to deal with seasonality in clinical prediction models, including a new proposal based on sliding windows. Class imbalance, high dimensionality and interpretable models are also considered since they are common traits of clinical datasets.

We tested these approaches with four datasets: two created synthetically and two extracted from the MIMIC-III database. Our results corroborate that clinical prediction models for infections can be improved by considering the effect of seasonality. However, the techniques employed to obtain the best results are highly dependent on the dataset.

Keywords: Seasonality · Concept drift · Clinical prediction models

1 Introduction

It is widely accepted that seasonal variations are a common trait for many of infectious diseases [3]. Seasonality is usually considered in epidemiological studies (i.e. to predict the spread of a disease over a population), although it is rarely examined in clinical prediction models (i.e. to predict whether a particular patient will suffer a disease) [2,5].

In this paper, we explore the most promising approaches as regards dealing with the problem of seasonality in prediction models for infectious diseases. We focus on classification problems and use the approach of simply ignoring season in the models as a gold standard. We consider common approaches for dealing with seasonality and propose a new algorithm based on sliding windows. We also consider their combination with interpretable models and techniques focused on solving high dimensionality and class imbalance, since these issues can reduce the usability of models [1]. The effects of these approaches were studied in synthetic

This work was partially funded by the SITSUS project (Ref: RTI2018-094832-B-I00), given by MCIU/AEI/FEDER, UE.

A. Tucker et al. (Eds.): AIME 2021, LNAI 12721, pp. 152–156, 2021.
https://doi.org/10.1007/978-3-030-77211-6_16

datasets and in data related to infectious diseases extracted from the MIMIC-III database.

The contributions of this research are the following: 1) a new approach to deal with seasonality based on sliding windows and 2) a study of the best combinations of techniques for the creation of interpretable models in the presence of seasonality, high dimensionality and imbalanced datasets.

2 Methods

In this work we focus on interpretable models since they are the most used in clinical practice. As consequence, we use logistic regression with LASSO as the basic technique for creating prediction models.

High dimensionality and class imbalance are also considered since they are common problems in clinical datasets. We study in this work the performance of P-value filtering, Fast correlation-based filter (FCBF) [10], random undersampling and random oversampling with different class ratios when dealing with these problems. These approaches have in common their negligible impact on model interpretability.

The problem of seasonality in data is commonly addressed by including the season as an additional feature to explore [9] or by building different models, one per season [5].

Additionally, we propose a new algorithm to deal with the problem of seasonality in those datasets in which observations do not follow a strict temporal order (e.g. owing to de-identification processes) yet the month in which the observation was made is available.

Our algorithm is based on the sliding window approach. Let assume that we need to predict an observation belonging the month m. We first classify our training dataset according to the month of the year in which the data point was obtained. We then select a window of L-months in length, which includes training data points for the month m, along with the previous and subsequent $\frac{L-1}{2}$ months, assuming a circular disposition and where L is an odd number. The model created with only these data is used to predict the observation. Figure 1 shows a graphical representation of the process explained.

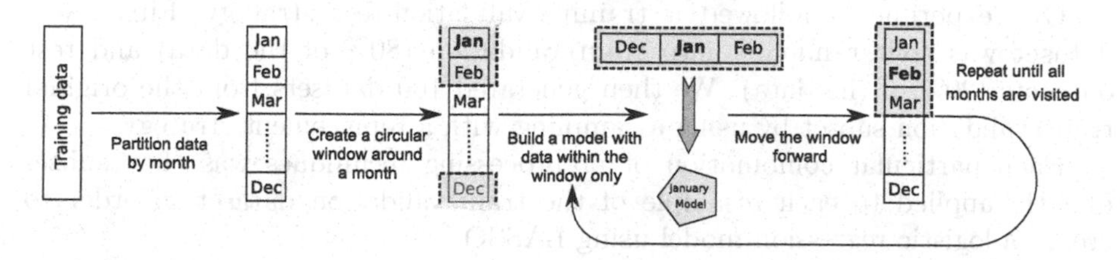

Fig. 1. Sliding window approach for seasonality with window size of 3 months.

3 Experiment

We tested the aforementioned methods and combinations in two synthetic and another two clinical datasets extracted from the MIMIC-III database [7].

In the synthetic datasets we assumed the model defined by the equation $k_1x_1 + k_2x_2 = y$, where x_1 and x_2 are two random variables in the interval $[-10, 10]$. The values of k_1 and k_2 are the unknown factors of the model, with values in $[0, 1]$, and being $k_2 = 1 - k_1$. Since we focus on classification models, a binary outcome *class* is generated with two values: *non-negative* if $y \geq 0$ and *negative* otherwise. We varied k_1 according to a *timestamp* attribute in order to simulate seasonality, following a Gaussian curve whose maximum ($k_1 = 1$) is centred in the middle of the winter season. Furthermore, we included 10 random variables and 20 variables correlated with x_1 or x_2 to simulate high dimensionality. The problem of class imbalance is also simulated by randomly removing samples until obtaining 10 *negative* samples for each *non-negative* sample.

Two different synthetics datasets were created: In the *condensed dataset* k_1 reaches 0 in the boundaries of the winter season, whereas in the *sinusoidal dataset* k_1 decreases slightly until reaching its minimum in the middle of the summer season, and therefore the seasonal effects are present throughout the year.

In the clinical datasets we gathered the first positive microbiology test, for each microorganism and type of sample, available in the patient's stays of the MIMIC-III database. We extracted as features: date and time (already de-identified in MIMIC-III) when the test was performed, specimen tested, type of admission, location of the admission, patient's insurance, marital status, ethnicity, patient's gender, patient's age group, minimum, mean and maximum values of white blood cells in $[-24, +24]$ hours around test time, minimum, mean and maximum values of lactate in $[-24, +24]$ hours around test time, time since the patient was admitted to the Intensive Care Unit when the test was ordered (discretised) and hospital service in which the test was ordered.

Two different clinical datasets were created with the extracted data, each of them focused on clinically relevant bacteria species: *Acinetobacter* species and *Streptococcus pneumoniae*. Therefore, in the *Acinetobacter dataset* we assumed as *positive* those microbiology tests whose microorganism found belongs to the *Acinetobacter* species and *negative* the rest of them, and the *S. pneumoniae dataset* was created with the same strategy.

Our experiments followed a training-validation-test strategy. First, each dataset was split randomly into train/validation (80% of the data) and test datasets (20% of the data). We then generated 100 datasets from the original train/validation subset by using a sampling with a replacement strategy.

Each particular combination of preprocessing techniques was then subsequently applied to each resample of the train/validation dataset in order to create a logistic regression model using LASSO.

The resulting model was then used to predict the class of observations in the test dataset. Once all the predictions were available, they were collected and used to calculate the AUC of the experiment and the mean number of predictors included in these models.

All the experiments were performed on the R platform, version 4.0.2 and RStudio, version 1.3.1093. The LASSO models were fitted using the *glmnet* R package [4,6], with parameters $\alpha = 1.0 - 10^{-5}$ and $\lambda = \lambda_{1se}$ as recommended in [6]. The *Biocomb* package [8] was used for the experiments with FCBF with *threshold* parameter set to 0 as suggested in [10].

4 Results

Table 1 shows the 3 best combinations of techniques, ranked by AUC and including the mean number of features per model as an approximate measure of model complexity.

Table 1. Best results (and 95% confidence intervals) obtained in each dataset.

	Filter	Balancing	Seasonality	Mean AUC	Mean features per model
Condensed dataset	FCBF	Oversample 2:1	Model per season	0.995 (0.994, 0.995)	1.08 (0.99, 1.17)
	FCBF	Undersample 2:1	Model per season	0.994 (0.994, 0.995)	0.62 (0.58, 0.66)
	FCBF	Oversample 2:1	3-month window	0.993 (0.992, 0.994)	1.33 (1.28, 1.39)
	...	...	...	...	...
	None	None	None	0.959 (0.959, 0.959)	7.79 (7.41, 8.17)
Sinusoidal dataset	None	None	3-month window	0.992 (0.991, 0.992)	7.91 (7.67, 8.15)
	P value	None	3-month window	0.991 (0.990, 0.993)	6.11 (5.93, 6.29)
	P value	Undersample 2:1	5-month window	0.989 (0.988, 0.989)	8.74 (8.57, 8.91)
	...	...	...	...	...
	None	None	None	0.919 (0.918, 0.919)	9.60 (9.11, 10.07)
Acinetobacter dataset	None	None	7-month window	0.671 (0.664, 0.678)	10.53 (10.03, 11.03)
	None	None	5-month window	0.670 (0.664, 0.677)	10.79 (10.36, 11.23)
	P value	None	7-month window	0.667 (0.660, 0.675)	9.56 (9.28, 9.84)
	...	...	...	...	...
	None	None	None	0.660 (0.654, 0.666)	11.05 (10.19, 11.91)
S. pneumoniae dataset	None	Oversample 2:1	Season as a feature	0.762 (0.759, 0.765)	35.31 (34.88, 35.77)
	P value	Oversample 2:1	Season as a feature	0.762 (0.759, 0.764)	33.79 (33.35, 34.23)
	None	None	Season as a feature	0.760 (0.758, 0.763)	21.31 (20.52, 22.10)
	...	...	...	...	...
	None	None	None	0.755 (0.753, 0.757)	18.07 (17.31, 18.83)

5 Discussion and Conclusions

Despite the fact that all the experiments used LASSO, a feature selection technique before modelling reduced model complexity even more in most cases. FCBF drastically reduced the number of features, as will be noted in the results obtained for the *condensed* dataset, yet it was not among the best techniques in the remaining experiments. On the contrary, the *P*-value filter attained a smaller reduction of complexity yet did not drastically affect model performance. When reducing the complexity of the model is a critical requisite, extra filtering techniques would, therefore, be advisable even in the presence of seasonality.

Our approach clearly obtained the best results in two datasets, the *sinusoidal* and the *Acinetobacter* datasets. This might seem odd in the latter dataset, bearing in mind that the months in the data obtained from the MIMIC-III database had been randomised to ensure patient confidentiality. However, since the season was preserved, the month of the shifted date should be close to the real month, and this may explain the good performance of these methods. Moreover, the drift in data might not occur precisely within an astronomical season, but may be delayed respect to its boundaries. Our proposed monthly window may, therefore, be a good option in these cases.

In all datasets, the use of a combination of techniques improved the resulting models as regards both AUC and model simplicity when compared to the application of only LASSO and logistic regression. However, this improvement clearly depends on the dataset and the best combination of techniques also varied, which led us to the conclusion that the best approach for dealing with seasonality is highly dependent on the dataset and that several of them should, therefore, be tested in future studies in order to obtain better clinical prediction models.

As future work, we intend to explore ensemble methods and extend our experiments to include other interpretable AI modelling techniques.

References

1. Cánovas-Segura, B., et al.: Exploring antimicrobial resistance prediction using post-hoc interpretable methods. In: Marcos, M., Juarez, J.M., Lenz, R., Nalepa, G.J., Nowaczyk, S., Peleg, M., Stefanowski, J., Stiglic, G. (eds.) KR4HC/TEAAM -2019. LNCS (LNAI), vol. 11979, pp. 93–107. Springer, Cham (2019). https://doi.org/10.1007/978-3-030-37446-4_8
2. Christiansen, C., Pedersen, L., Sørensen, H., Rothman, K.: Methods to assess seasonal effects in epidemiological studies of infectious diseases-exemplified by application to the occurrence of meningococcal disease. Clin. Microbiol. Infect. **18**(10), 963–969 (2012)
3. Dowell, S.F.: Seasonal variation in host susceptibility and cycles of certain infectious diseases. Emerg. Infect. Dis. **7**(3), 369–374 (2001)
4. Friedman, J., Hastie, T., Tibshirani, R.: Regularization paths for generalized linear models via coordinate descent. J. Stat. Softw. **33**(1) 1 (2010)
5. Godahewa, R., Yann, T., Bergmeir, C., Petitjean, F.: Seasonal averaged one-dependence estimators: a novel algorithm to address seasonal concept drift in high-dimensional stream classification. In: Proceedings of the International Joint Conference on Neural Networks (2020)
6. Hastie, T., Qian, J.: Glmnet ignette. Technical report. https://web.stanford.edu/~hastie/Papers/Glmnet_Vignette.pdf. Accessed 24 Jan 2021
7. Johnson, A.E., et al.: MIMIC-III, a freely accessible critical care database. Sci. Data **3**(1) (2016)
8. Novoselova, N., Wang, J., Pessler, F., Klawonn, F.: Biocomb: Feature Selection and Classification with the Embedded Validation Procedures for Biomedical Data Analysis. https://cran.r-project.org/web/packages/Biocomb/index.html. Accessed 24 Jan 2021
9. Williams, D.J., et al.: Predicting severe pneumonia outcomes in children. Pediatrics **138**(4), e20161019 (2016)
10. Yu, L., Liu, H.: Efficient feature selection via analysis of relevance and redundancy. J. Mach. Learn. Res. **5**, 1205–1224 (2004)

Monitoring Quality of Life Indicators at Home from Sparse, and Low-Cost Sensor Data

Dympna O'Sullivan[1(✉)], Rilwan Basaru[2], Simone Stumpf[2], and Neil Maiden[2]

[1] ASCNet Reseach Group, School of Computer Science, TU Dublin, Dublin, Ireland
dympna.osullivan@tudublin.ie
[2] City, University of London, London, UK
{remilekun.basaru.1,simone.stumpf.1,neil.maiden.1}@city.ac.uk

Abstract. Supporting older people, many of whom live with chronic conditions, cognitive and physical impairments to live independently at home is of increasing importance due to ageing demographicssss. To aid independent living at home, much effort is being directed at reliably detecting activities from sensor data to monitor people's quality of life or to enhance self-management of their own health. Current efforts typically leverage large numbers of sensors to overcome challenges in the accurate detection of activities. In this work, we report on the results of machine learning models based on data collected with a small number of low-cost, off-the-shelf passive sensors that were retrofitted in real homes, some with more than a single occupant. Models were developed from sensor data to recognize activities of daily living, such as eating and dressing as well as meaningful activities, such as reading a book and socializing. We found that a Recurrent Neural Network was most accurate in recognizing activities. However, many activities remain difficult to detect, in particular meaningful activities, which are characterized by high levels of individual personalization.

Keywords: Activity recognition · Sensors · Machine learning · Independent living

1 Introduction

An understanding of a person's activities and the extent to which activities are being achieved or not can be used to improve self-monitoring and self-care at home, including their quality of life [1]. There are two main challenges to implementing activity recognition at home. First, there is the challenge of retrofitting residences with sensors. Typically, smart home solutions have hundreds of sensors with the aim of collecting data to recognize a range of different activities. The cost and complexity of such installations often prevents their take-up in real-world applications. Second, even with large amounts of sensor data, there are challenges to developing machine learning models for activity recognition. These include noisy sensor data, large numbers of false positives, difficulty training activity recognition algorithms on data collected in homes with a different layout and the multiple-occupancy problem. Research has typically focused on detecting

A. Tucker et al. (Eds.): AIME 2021, LNAI 12721, pp. 157–162, 2021.
https://doi.org/10.1007/978-3-030-77211-6_17

activities of daily living (ADLs), which are tasks that people undertake routinely in their everyday lives, for example, eating, sleeping and grooming [2]. There is less research on monitoring meaningful activities, i.e. physical, social, and leisure activities that provide the patient with "emotional, creative, intellectual, and spiritual stimulation" [3], as an important indicator of quality of life.

To address these challenges, we developed and investigated a toolkit composed of a small number of low-cost off-the-shelf passive sensors, typically up to 10, which were retrofitted into real, sometimes multiple-occupancy homes to detect both ADLs and meaningful activities. To measure meaningful activities we employed beacons sensors, small devices that broadcast packets of data over Bluetooth, and are placed on objects in the home that residents interact with frequently. This allows capturing more minute details on a resident's activities. For example, in [4] the authors demonstrated how accelerometer data captured from beacon sensors could detect not only the presence of residents interacting with the objects, but also the way the objects were moved (e.g. placing a knife on the table vs. using the knife to cut food in its preparation). Niu et al. [5] propose a similar approach using BLE (Bluetooth Low Energy) beacons to measure movement and achieved an accuracy of 70% averages across seven activities. There are challenges with the use of such small sensors affixed to objects. In addition to the size constraints of the sensors themselves, energy consumption can be a problem, as analyzing accelerometer data requires a high transmission rate in order to capture the movements effectively with machine learning techniques. However, accuracy of detection using beacons is relatively high and they are suited to multiple occupancy environments as they can provide specific location accuracy allowing to identify who is interacting with a device. We collected data from five users in five different homes, each over a period of one week. Three of the homes were multi occupancy homes. We used this data to train five machine learning algorithms and evaluated their accuracy in recognizing ADLs and meaningful activities. In this paper we present the methods employed in our study, including how we collected data and ground truth labels from human participants, and how we trained and evaluated the machine learning models. We present our results, focusing on the overall accuracy of the machine learning models as well as the accuracy in recognizing individual activities. We conclude by discussing the potential implications of our work, as well as directions for future research.

2 Methods

2.1 Data Collection

We recruited 5 participants (3 males, 2 females), all aged 18 and above, without any cognitive or physical impairments to take part in a pilot study. We received ethics approval prior to commencing the study and obtained informed consent from all participants. Participants were able to choose a set of activities (Table 2), agreed between the researcher and each participant, with a mixture of ADLs and meaningful activities. Participants carried out a set of activities over the course of one week in their own homes. To detect interaction with objects around the home six main sensor types were used - motion, door, power, ambient (temperature and humidity), pressure and beacon sensors. The motion, door, and pressure sensors are binary sensors that can detect motion in an environment, for example, opening of a door and the application of a pressure on a surface such as a bed respectively. The temperature, humidity, and power sensors are continuous sensors that detect changes in temperature, humidity and power surges. Finally, the beacon sensor is a binary sensor that detects the disturbance of any object or surface it is attached to. For example, they were attached to bookmarks and the remote control for the TV. Based on the selected set of activities, the appropriate set of sensors was provided and installed by the researcher, who noted down the location on a rough sketch of the floor plan of the participant's home. During the study, data collected from the sensors was stored in a database on a Raspberry Pi. Because of the time-dependent nature of the data being stored, we used InfluxDB [6], an open-source time series database framework, optimized for fast, storage and retrieval of time series data. The motion, door and ambient sensors were from the same manufacturer, Xiaomi. The pressure and power sensors interfaced with the Raspberry Pi, using a z-wave communication protocol. We used the Home Assistant open-source framework [7] as a service for asynchronously listening for sensor readings and updating the InfluxDB database. A typical kit was composed of 25 sensors and cost on average £412 including the hub components.

Data collection took place over February and March 2019. Participants recorded a log of activities using a journaling app called ATracker [8] on an Android tablet to record the start and end time of activities as they were completed. These logs were used as ground truth labels for the sensor data. We collected data for 14 activities. There was high variation in the frequency and the duration of completing each task. Sleeping, was recorded the most frequently (11 times) and recorded the most (95.34 h), followed by Going Out (10 times, 30.44 h). Food preparation was the most frequently recorded activity (24 times) but on average took much less time (0.29 h). On the other end of activity frequency and duration were Vacuuming (3 times), Nail Care (2 times), Grooming (2 times), Laundry (3 times) and Playing Board Games (1 times); these activities only happened infrequently and also recorded the least amount of time overall. To reduce bias in the prediction models (such that models developed would not be biased towards classes with higher frequency or duration), we removed infrequent activities where there are not enough training and testing data (playing board games) and we applied a class weight to "boost" activities with lower frequencies.

2.2 Model Development and Evaluation

We used the ScikitLearn Python machine learning library to implement SVM, Naïve Bayes, Logistic Regression, and Perceptron models. The Naïve Bayes was multinomial and trained with an adaptive smoothing parameter (alpha) of 0.01. The SVM model was trained with 5 maximum epochs. The Perceptron model was trained with a stopping criterion of 1e-3. The RNN was implemented with the TensorFlow framework and trained with a learning rate of 0.001, weight decay of 0.005 and under 2 epochs. Data was split into training and validation sets by a 75:25 ratio. Model performance was measured by comparing predicted activities with ground truth gathered via the Atracker app. We calculated accuracy for each algorithm as a ratio of all correctly labelled data point to all test data points. To take into consideration the imbalanced nature of the data, we also computed micro and macro averages for precision, recall, and F1-score.

3 Results

3.1 Model Accuracy

RNN achieved the highest average accuracy, correctly recognizing 65.59% of the activities from the dataset followed closely by Perceptron on 65.09%. The other models performed as follows - SVM (59.3), Logistic Regression (58%) and Naïve Bayes (53.95%). The micro-average, macro-average and F1 scores for each classifier and shown in Table 1. The RNN yields the highest scores, with macro-average precision and recall scores of 0.88 and 0.41 respectively, and an F1-score of 0.46. It significantly outperforms the other classifiers at correctly recognizing a range of activities, achieving a macro average F1-score that is 228.5% higher than the Perceptron. A McNemar test with alpha = 0.05.indicated that the prediction performance of the RNN was statistically significant compared with the other four models. We hypothesize that the superiority of RNN is owed to its inherent feedback architecture, which allows it to hold latent information about the previous state of the model in memory. For example, the Sleeping activity is typically completed in 8 h, hence a suitable time window for tracking the Sleeping activity will be too large for tracking Laundry (which typically takes 40 min). The RNN model is better able to adjust its weight (during the training step) to adaptively retain information.

Table 1. Micro-averaged and macro-averaged precision, recall and F1-scores

Algorithms		Precision	Recall	F1-Score
SVM	Micro average	0.59	0.59	0.59
	Macro average	0.18	0.10	0.09
Naïve Bayes	Micro average	0.54	0.54	0.54
	Macro average	0.04	0.07	0.05
Logistic regression	Micro average	0.58	0.58	0.58
	Macro average	0.11	0.10	0.09
Perceptron	Micro average	0.65	0.65	0.65
	Macro average	0.16	0.14	0.14
RNN	Micro average	0.56	0.56	0.56
	Macro average	**0.88**	**0.41**	**0.46**

3.2 Activity Accuracy

We explored the performance of the models across the different activities (Table 2). The Perceptron had high overall accuracy and a high micro-average accuracy, however, the macro-average accuracy showed that it is not very good at recognizing a variety of activities. In comparison, RNN can recognize a much wider range of activities reliably than the Perceptron. Seven activities had low F1-scores across all algorithms - Washing Dishes, Mealtime, Food Prep, Watching TV, Sleeping, Reading and Grooming.

Table 2. F1-scores per activity, decreasing order of RNN's F1 score

	SVM	Naïve Bayes	Logistic regression	Perceptron	RNN
Nailcare	0.00	0.00	0.00	0.00	1.00
Laundry	0.00	0.00	0.00	0.00	0.98
Housekeeping	0.00	0.00	0.00	0.00	0.85
Bathing	0.00	0.00	0.00	0.00	0.82
Mealtime	0.00	0.00	0.00	0.00	0.22
Dressing	0.00	0.00	0.00	0.00	0.75
No activity	0.72	0.70	0.72	0.76	0.71
Wash dishes	0.00	0.00	0.00	0.00	0.46
Food prep	0.00	0.00	0.00	0.00	0.14
Watching TV	0.00	0.00	0.00	0.31	0.11
Sleeping	0.16	0.00	0.00	0.00	0.04
Going out	0.40	0.00	0.51	0.88	0.04
Reading	0.00	0.00	0.00	0.00	0.00
Grooming	0.00	0.00	0.00	0.00	0.00

4 Discussion and Conclusions

Our results demonstrate that an RNN model shows promise given a limited number of cheap off-the-shelf sensors and a low number of training examples. Activities that involved a number of distinct subtasks were difficult to detect, e.g. meal times may involve laying a table with cutlery or plates and sitting at a table. Furthermore, real users may have different routines, for example, breakfast may be a faster event and involve fewer tasks than eating dinner. This suggests careful consideration needs to be given to the set and combination of sensors to capture activities. Furthermore, high levels of personalization are likely to be necessary for detecting meaningful activities, which can be learned from datasets collected over longer periods to analyze user habits. In future work we are interested in addressing our limitations in using BLE sensors. Rather we propose the use of conventional Bluetooth. Although this will consume more energy, they can be detected by sensors on mobile devices more consistently. This approach can detect location and therefore activity recognition in multi-occupancy scenarios and may also help recognizing more personalized meaningful activities.

References

1. Majumder, S., et al.: Smart homes for elderly healthcare—recent advances and research challenges. Sensors **17**, 2496 (2017)
2. Wallace, M., Mary, S.: Katz index of independence in activities of daily living (ADL). Urologic Nursing J. **27**(1), 93–94 (2007)
3. NICE. Mental wellbeing of older people in care homes (2013) [Online]

4. Civitarese, G., Belfiore, S., Bettini, C.: Let the objects tell what you are doing. In: The ACM International Joint Conference on Pervasive and Ubiquitous Computing, Heidelberg, Germany (2016)
5. Niu, L., Saiki, S., Bousquet, L.D., Nakamura, M.: Recognizing ADLs based on non-intrusive environmental sensing and BLE beacons. In: 8th International Conference on Indoor Positioning and Indoor Navigation, Sapporo, Japan (2017)
6. InfluxDB. InfluxDB: Purpose-Built Open Source Time Series Database. https://www.influxdata.com/. Accessed 12 Apr 2020
7. Home-Assistant. Home Assistant. https://home-assistant.io. Accessed 21 Sep 2019
8. Wonderapps. ATracker - Daily Task and Time Tracking. https://www.wonderapps.se/atracker/home.html. Accessed 19 Mar 2020

Detection of Parkinson's Disease Early Progressors Using Routine Clinical Predictors

Marco Cotogni[1,2](✉), Lucia Sacchi[1], Dejan Georgiev[2,3], and Aleksander Sadikov[2]

[1] Department of Electrical, Computer and Biomedical Engineering, University of Pavia, Pavia, Italy
marco.cotogni01@universitadipavia.it, lucia.sacchi@unipv.it

[2] Faculty of Computer and Information Science, University of Ljubljana, Ljubljana, Slovenia
dejan.georgiev@kclj.si, aleksander.sadikov@fri.uni-lj.si

[3] Department of Neurology, University Medical Centre Ljubljana, Ljubljana, Slovenia

Abstract. Parkinson's disease (PD) is a progressive, neurodegenerative disease characterised by the presence of motor and non-motor symptoms and signs. The symptoms of PD tend to begin very gradually and then become progressively more severe. The rate of PD progression is hard to predict and is different from one person to another. Namely, while in some patients the disease develops fast in just a few years from the diagnosis, in some the disease takes a more idle course and progresses slowly. We aimed to identify patients that develop severe motor symptoms within four years from PD diagnosis (early progressors) and separate them from those in whom severe symptoms develop beyond this point. We used data from the Parkinson's Progression Markers Initiative (PPMI) dataset to calculate motor progression of the disease by the use of motor scores as assessed by MDS-UPDRS III. The predictors were defined as baseline scores of selected clinical variables and the difference between motor scores at 1-year after enrolment in the study and the same scores at baseline. The rationale for predictor selection was that they should be readily available in routine clinical practice. We tested four different classifiers: logistic regression, decision tree, random forest, and gradient boosting. The best performing classifier was the logistic regression with an area under the ROC curve of 81%. We believe this can be the basis for a reliable and explainable classifier, using only standard clinical variables, for identifying early progressors with high recall (80%) three years in advance.

Keywords: Parkinson's disease · Early progression · Clinical predictors · Artificial intelligence · Machine learning

1 Motivation

Parkinson's disease (PD) is a chronic, progressive, neurodegenerative disease characterised by the presence of motor and non-motor symptoms and signs. The symptoms of PD tend to begin very gradually and then become progressively more severe. The rate of PD progression is hard to predict and is different from one person to another. Namely, while in some patients the disease develops fast in just a few years from the

A. Tucker et al. (Eds.): AIME 2021, LNAI 12721, pp. 163–167, 2021.
https://doi.org/10.1007/978-3-030-77211-6_18

diagnosis, in some the disease takes a more idle course and progresses slowly. Understanding progression of PD is of outmost importance to improve clinical management of the patients and to offer patients accurate information regarding progression of the disease in the early stages of PD. In addition, accurate baseline/early disease predictors can also improve clinical research by optimizing trial efficiency (e.g. trial duration, selection criteria, sample size estimates). Different clinical predictors of fast motor progression in PD have been identified so far, such as higher age at onset, symmetrical disease at onset, postural instability and gait disorder type of the disease, cognitive decline, bradykinesia score, and female gender [1–3]. However, it still not known how reliable these predictors are, alone or combined with other clinical and paraclinical investigations to predict the rate of progression of PD.

Different approaches have been tried to explore the prognostic factors of PD progression. In addition to the use of the classical statistical approaches [4] recent studies have employed machine learning approaches on clinical [5], or a combination of both clinical and paraclinical data [6]. However, so far studies have identified too many different clinical and paraclinical predictors to be of practical use in routine clinical practice. In addition, in practice, it is very difficult to predict the progression of the disease based on the initial neurological examination only. Therefore, the aim of this study was to identify patients that develop severe motor symptoms within four years from PD diagnosis (early progressors) and separate them from those in whom severe symptoms develop beyond this point by the use of a more rational approach with only a limited number of routine clinical predictors.

2 Patients

We used data from the Parkinson's Progression Markers Initiative (PPMI) [7]. PPMI is a longitudinal observational clinical study including various cohorts of PD patients. In this paper we considered the De Novo PD patients cohort. These are patients with a diagnosis of PD for at most two years and without yet starting medication treatments and ideally fit the purpose of this research.

We divided the patients into two groups based on their results of the Movement Disorder Society-Unified Parkinson's Disease Rating Scale (MDS-UPDRS) part III. During this examination, composed of several tests, the patient is asked to perform several physical tasks (e.g. finger tapping test, hand movement test, etc.). To avoid any kind of bias we excluded patients that performed the MDS-UPDRS part III examination on PD medication. For each patient, we computed the MDS-UPDRS part III overall score (sum of scores of all tasks) both at baseline and at the fourth year. There were 310 patients that have reached the fourth year of the study and have completed the MDS-UPDRS part III without medication. Next, we computed the difference between the fourth-year overall score and the baseline overall score.

Finally, based on the input of a neurologist, specialised in PD, we defined two groups of patients: early progressors (difference>20) and non-early progressors (difference <15). Patients with the difference between 15 and 20 were excluded from further study since they represent borderline cases that could always be interpreted as correct even if misclassified. We also decided to exclude all the patients with at least one missing value in order to avoid imputing. The final dataset was thus composed of 209 patients (162 non-early progressors and 47 early progressors).

3 Predictors

Our choice of possible predictors of early progression was guided by the final utility of the predictive model in routine clinical practice. As such we decided on a selection of predictors commonly used in routine clinical practice. We also decided not to consider each score from each test but to compute some overall scores that are usually used in clinical practice. Furthermore, in the opinion of our neurology expert, it is improbable to be able to estimate the progression rate of the disease from only a single visit at baseline. Thus, we decided to consider both baseline data and 1-year data. However, instead of including 1-year predictors in the models directly, we computed the difference between each predictor at 1-year and at baseline in order to simulate a follow-up of 1-year (called Delta (Δ) predictors). The final selection of predictors, approved by a neurologist, is given in Table 1.

Table 1. List of routine clinical predictors used for training the classifiers. The total number of predictors considered was 62: 36 from baseline data and 26 delta predictors.

Category	Routine clinical predictors
Vital parameters	Body Mass Index*, Systolic supine blood pressure*, Heart Rate Supine*
Demographic data	Age at baseline, Gender, Education years, Age on set symptoms, Age at diagnosis, Dominant side
MDS-UPDRS part I	Cognitive*, Sleep*, Autonomic Nervous System*,
MDS-UPDRS part II	Bulbar*, Common daily activities*, Bed*, Gait*,
MDS-UPDRS part III	Axial I*, Axial II, Limb rigidity*, Limb bradykinesia, Tremor*, Resting Tremor, Appendicular, overall score**
Activities of daily living	Modified Schwab & England activities of daily life: overall score*
Neuropsychological	Benton Judgment of line orientation: derived score*, Hopkins verbal learning test: derived total*, Letter number sequencing PD: derived score*, MoCA: overall score*, Symbol digit Modalities: derived score*
Neurobehavioral	QUIP: overall score*, Geriatric depression scale short: overall score*, STAI overall score*
Autonomic	SCOPA-AUT: overall score*
Sleep disorders	Epworth Sleepiness Scale: overall score*, REM Sleep disorder questionnaire: overall score*
Smell disorders	University of Pennsylvania smell id test: overall score

*For this predictor also the equivalent delta version was computed.
**For this predictor only the delta version was computed.

4 Methods

Spearman's correlation coefficient (SCC) was used to further reduce the number of considered predictors. We deleted one predictor from pairs with SCC higher than 0.8; keeping the more meaningful one from a clinical point of view. Furthermore, we performed a filtered feature selection based on univariate feature ranking method using the Mann-Whitney U test (for continuous variables) and $\chi 2$ test (for categorical variables). The nested 10-fold cross validation was used for feature selection and for estimating the performance of the trained models. At each iteration of the inner loop, feature selection was performed on the training set: we selected the best n[1] features on the training set. At the end of the inner loop, we obtained 10 sets of n features. Among these sets, we counted the occurrences of each feature and took the n most often selected ones.

Since the ratio between both classes was one-to-four we decided to balance the dataset before training the models. This can improve learning the minority class (early predictors) – the class we are particularly interested in. We used Synthetic Minority Oversampling Technique (SMOTE) for balancing the training set. We followed the recommendation by Chawla et al. [8] to first undersample the majority class in order reduce the ratio to 1:2 and then create samples from the minority class increasing the ratio to 1:1.

5 Results and Discussion

We compared four classifiers: decision tree (DT), logistic regression (LR), gradient boosting (GB) and random forest (RF). The results are reported in Table 2. The best-performing classifier is the logistic regression that achieves the best results for all the metrics considered. The relatively high recall (0.80) means the majority of early progressors will be detected. On the other hand, comparatively low precision (0.50) does not modify the risks-benefit trade-off of the model. In the clinical practice the model would be used as a warning for the clinician who could pay extra attention to potential early progressors. In addition, these patients and their family would be warned in advance and could prepare from a psychological point of view. They could also prepare their living environment in order to increase the safety of the patient.

The final logistic regression model was built with the six most selected variables during the feature selection procedure: Δ MDS-UPDRS III, Δ Axial I, Δ Tremor, Δ Limb rigidity, Δ Common daily activity, Δ Bulbar. As we can see these are only Δpredictors, i.e., differences between one-year and baseline. The results obtained are reasonable from a medical point of view since all the selected variables come from MDS-UPDRS scores and PD is characterized principally by motor signs that are among the first to worsen during the years.

[1] Since we had a validation set in the inner cross validation loop, we decided to use it to tune this parameter. We tried to increase the number of features n until the mean performances of the chosen algorithms on the inner loop continued to increase. When, rising n, they started to decrease we found the optimal number of features $n = 6$.

Table 2. Results comparison (10-fold CV). The simple accuracy was avoided due to the unbalanced situation of the test sets as SMOTE/undersampling were applied only on training folds.

	Balanced Accuracy	AUC_ROC	Recall	Precision
DT	0.67	0.60	0.63	0.40
LR	**0.76**	**0.81**	**0.80**	**0.50**
RF	0.71	0.72	0.66	0.44
GB	0.69	0.56	0.39	0.27

Finally, the fact that the most selected predictors are routinely computed sub-scores allows for the inexpensive and clinically useful tool able to help the clinicians in recognising early-progressing patients three years in advance.

Acknowledgements. Data used in the preparation of this article were obtained from the Parkinson's Progression Markers Initiative (PPMI) database (www.ppmi-info.org/data). For up-to-date information on the study, visit www.ppmi-info.org. PPMI–a public-private partnership – is funded by the Michael J. Fox Foundation for Parkinson's Research and funding partners (The list with full names of all of the PPMI funding partners found at www.ppmi-info.org/fundingpartners). The research was supported by the Slovenian Research Agency (ARRS) under the Artificial Intelligence and Intelligent Systems Programme (ARRS No. P2–0209).

References

1. Post, B., Merkus, M.P., de Haan, R.J., Speelman, J.D., CARPA Study Group: Prognostic factors for the progression of Parkinson's disease: a systematic review. Mov. Disord. Off. J. Mov. Disord. Soc. 22(13), 1839–1988 (2007)
2. Reinoso, G., Allen, J.C., Jr., Au, W.L., Seah, S.H., Tay, K.Y., Tan, L.C.: Clinical evolution of Parkinson's disease and prognostic factors affecting motor progression: 9-year follow-up study. Eur. J. Neurol. **22**(3), 457–463 (2015)
3. Venuto, C.S., Potter, N.B., Dorsey, E.R., Kieburtz, K.: A review of disease progression models of Parkinson's disease and applications in clinical trials. Mov. Disod. Off. J. Mov. Disord. Soc. **31**(7), 947–956 (2016)
4. Iddi, S., et al.: Estimating the evolution of disease in the Parkinson's progression markers initiative. Neuro-Degenerative Dis. **18**(4), 173–190 (2018)
5. Tsiouris, K.M., Konitsiotis, S., Koutsouris, D.D., Fotiadis, D.I.: Prognostic factors of rapid symptoms progression in patients with newly diagnosed Parkinson's disease. Artif. Intell. Med. **103**, 101807 (2020)
6. Chahine, L.M., et al.: Predicting progression in Parkinson's disease using baseline and 1-year change measures. J. Parkinson's Dis. **9**(4), 665–679 (2019)
7. PPMI Site, https://www.ppmi-info.org, last accessed 2021/01/22
8. Chawla, N.V., Bowyer, K.W., Hall, L.O., Kegelmeyer, W.P.: SMOTE: Synthetic minority over-sampling technique. J. Artif. Intell. Res. **16**, 321–357 (2002)

Detecting Mild Cognitive Impairment Using Smooth Pursuit and a Modified Corsi Task

Alessia Gerbasi[1,2(✉)], Vida Groznik[2,3,4], Dejan Georgiev[2,5], Lucia Sacchi[1], and Aleksander Sadikov[2,3]

[1] Department of Electrical, Computer and Biomedical Engineering, University of Pavia, Pavia, Italy
alessia.gerbasi01@universitadipavia.it, lucia.sacchi@unipv.it
[2] Faculty of Computer and Information Science, University of Ljubljana, Ljubljana, Slovenia
{vida.groznik,aleksander.sadikov}@fri.uni-lj.si, dejan.georgiev@kclj.si
[3] NEUS Diagnostics, d.o.o., Ljubljana, Slovenia
[4] Faculty of Mathematics, University of Primorska, Natural Sciences and Information Technologies, Koper, Slovenia
[5] Department of Neurology, University Medical Centre Ljubljana, Ljubljana, Slovenia

Abstract. Over 50 million people today live with some form of dementia as it is the most common neurodegenerative disease in the world. Mild cognitive impairment (MCI) is a stage before dementia symptoms overtly manifest. An estimated 10–15% of patients diagnosed with MCI annually convert to Alzheimer's dementia. Early detection of MCI is imperative as disease-modifying therapies in development could have the potential to significantly delay disease progression before dementia symptoms develop. There is evidence that observing oculomotor movements during different neuropsychological tasks can serve as a biomarker for MCI. A clinical study with 105 participants was performed at several centres in Ljubljana, Slovenia. All the participants underwent an extensive neurological and psychological evaluation and were, on the basis of this evaluation, divided into two groups: cognitively impaired and healthy controls. At the same time the participants performed several short tasks on the computer screen, including smooth pursuit dot tracking and a modified version of the Corsi block-tapping test. During the tasks, performed using their gaze alone, their eye movements were recorded with an eye-tracker. The eye-tracking data was analysed and a number of features describing the gaze behaviour was proposed. These features were used to construct several machine learning models to predict whether a person exhibits signs of cognitive impairment or not. A model based on random forest classifier achieved the best performance with 80% classification accuracy and an area under the ROC curve of 85%.

Keywords: Mild cognitive impairment (MCI) · Early detection · Eye-tracking · Smooth pursuit · Corsi block-tapping test · Machine learning

A. Tucker et al. (Eds.): AIME 2021, LNAI 12721, pp. 168–172, 2021.
https://doi.org/10.1007/978-3-030-77211-6_19

1 Introduction

Over 50 million people today worldwide live with some form of dementia as it is the most common neurodegenerative disease in the world. The concept of mild cognitive impairment (MCI) represents a state of cognitive function between that seen in normal aging and dementia in which, despite the cognitive impairment, the activities of daily living and the quality of life are not substantially affected. An estimated 10–15% of patients diagnosed with MCI annually convert to Alzheimer's dementia (AD) [1]. Early detection of MCI is imperative as disease-modifying therapies in development could have the potential to significantly delay or even stop disease progression before dementia symptoms develop.

Different approaches, based on clinical, biochemical and imaging grounds, have been used for early detection of MCI [2]. However, these methods are in general time consuming, expensive and require highly specialised staff to execute them. This renders them unsuitable for a wider use to detect cognitive impairment on a population level. Therefore, the quest for sensitive measures to assess subtle cognitive decline in MCI is still ongoing.

There is evidence that observing eye movements during different neuropsychological tasks can serve as a biomarker for MCI and AD [eg. 3]. In addition to saccadic, impairments of smooth pursuit eye movements have been shown in AD [4]. However, the rate and quality of smooth pursuit abnormalities in MCI is not known, just as it is not known whether they can be useful biomarkers for early detection of MCI.

The aim of this study was to explore the usefulness of smooth pursuit in combination with a modified eye-tracking compatible version of the Corsi block-tapping memory test [5] to detect MCI.

2 Patients and Methods

The data of 105 participants from an ongoing clinical study in Slovenia was analysed. The participants underwent an extensive neurological and psychological evaluation on the basis of which they were assigned into two groups: cognitively impaired (denoted as MCI + and consisting of borderline cases, MCI, and dementia) and healthy controls (HC). As the study is ongoing, we initially had data from 67 participants (37 MCI + and 30 HC) that we used as a training set. The remaining 38 participants (15 MCI + and 23 HC) were used as a test set. Furthermore, all the participants' eye movements were recorded while they performed a digitalised neuropsychological test battery composed of several short tasks. A combination of two tasks, smooth pursuit and Corsi block-tapping, was analysed in this paper to understand whether it is possible to detect MCI based on the features derived from these tasks.

2.1 Smooth Pursuit Test

The smooth pursuit test consists of following a moving dot on the screen. The movement is sinusoidal in time and it can be either vertical or horizontal, while the speed can be low (period of 4800 ms), medium (2400 ms) or high (1600 ms). All combinations of speed and direction are presented to the user.

The eye-tracker records the spatial coordinates (x, y) on the screen of both eyes at each timestamp (90 Hz). The raw data was pre-processed: (1) tasks with less than 90% of valid data points were discarded; (2) initial 10% of the recording was removed to analyse only the most stable part of the data; (3) a median filter of window size 5 was applied to reduce the noisy artefacts; and (4) only the best of the two eyes in terms of each considered feature was used for machine learning. A number of features were constructed from the pre-processed recordings: mean squared error (MSE), range, speed difference, #predictions, #fixations, and latency error.

MSE measures how well the user follows the dot. It is computed as the mean of the squared differences between the position of the dot and the gaze in the direction of the given task. The range feature compares the amplitude of the dot oscillation with the gaze amplitude in order to understand if the users are following the dot until the very end of the oscillation or if they are overshooting/undershooting. The difference in speed was computed as the sum of the absolute speed differences between the eye and the dot speed point by point in the direction of the dot movement. It is known from the literature that the sinusoidal smooth pursuit is one of the best stimulus waves to observe the user's predictive behaviour [6]. When the user stops following the dot with a smooth pursuit movement and just anticipates its next extreme end position, we can see a jump in the space curve that corresponds to a speed peak. The feature #predictions measures the number of speed peaks in each gaze curve. The #fixations feature is the number of detected eye fixations during the task; these were defined as consecutive samples with an inter-sample distance of less than 30 pixels and a duration of more than 200 ms. Finally, the latency error is computed as the MSE for the first 700 ms for each task (for this feature the first 10% of the recording was not removed). In humans there is normally a latency of ~100–130 ms before smooth pursuit movement starts [6] and this initial delay can be bigger in cognitively impaired subjects even if sinusoidal target trajectory is not the best one to observe the typical smooth pursuit latency [7].

2.2 Modified Corsi Block-Tapping Test

In our eye-tracking version of the Corsi block-tapping test the user sees 9 circles arranged in a 3 × 3 grid. There are two different versions of the test; one forward and one reverse. In the forward test, when the task starts, the user sees two of the 9 circles lighting up in a random sequence, one at a time, and after that, the user is asked to repeat the sequence by looking at the circles in the same order. For the reverse test the user is asked to repeat the displayed sequence in the *reverse* order. There are 8 different levels depending on the sequence length. The first level consists of a sequence of two circles. If the level is cleared, the sequence length increases by one – up to a maximum sequence length of 9 circles. Each level is repeated twice (with a different sequence of the given length) and one correct answer is enough to proceed to the next level. The first level is an exception; it is repeated three times.

Measures related to how the users move their eyes in the case of the Corsi task are likely not a direct and simple indicator of memory functioning. Therefore, as Corsi features we simply used the highest levels (forward, reverse) reached; considering a level "cleared" if at least one of the sequences of that level was correctly repeated.

Most of the cognitively impaired participant (78% for the forward task and 76% for the reverse task) did not reach the second level. Therefore, we used a threshold of level two to binarize both features, forward and reverse maximum level reached, to maximize the differences between the classes.

3 Results

A deep learning approach for the automatic extraction of the features was avoided in order to identify few clinical and explainable predictors based on the knowledge-domain. Therefore, we used the statistically significant features from smooth pursuit and both Corsi features (15 features in total) when training the machine learning models. We also decided against imputing missing data, therefore we deleted five subjects with one or more missing values. The final dataset was thus composed of 100 participants (50 MCI + and 50 HC): 65 participants (36 MCI + and 29 HC) in the training set and 35 (14 MCI + and 21 HC) in the test set. Continuous variables were standardised before training and one of the two features (the one with a higher p-value in the initial analysis) for the couples with an absolute Spearman's correlation coefficient ≥ 0.7, was discarded, bringing the final number of features to 12.

We used 10-fold cross validation on the training data to select the best features, performing a ranking at each round of the cross validation and choosing as final features only the most often selected ones. The ranking was made on the basis of the p-value returned by a statistical test (Mann-Whitney U test for continuous features and Fisher's exact test for binary features). The seven most frequently selected features were: forward Corsi level (10/10), reverse Corsi level (10/10), MSE 4800 y (10/10), speed 2400 x (10/10), MSE 4800 x (9/10), speed 4800 y (8/10), MSE 1600 x (7/10).

Table 1. The performance on the previously unseen test set (n = 35).

Model	ROC AUC	Accuracy (CI 95%)	Precision	Recall
Decision tree	0.65	0.63 ± 0.16	0.52	0.79
Logistic regression	0.79	0.71 ± 0.15	0.64	0.64
Random forest	**0.85**	**0.80 ± 0.13**	**0.77**	**0.71**
Gradient boosting	0.82	0.69 ± 0.15	0.57	0.86

We compared four machine learning algorithms: Decision Tree, Logistic Regression, Random Forest, and Gradient Boosting. For the last two we tuned the parameters using a randomised search cross validation followed by an exhaustive grid search cross validation. The evaluation results on the previously unseen test set are presented in Table 1. The best classifier is Random Forest with an AUC of 85% and a classification accuracy of 80%. It is also the best model in terms of combined precision and recall with both above 70%.

4 Discussion and Conclusions

Due to aging population a fast and non-invasive screening test would be of great benefit to health care system, patients, and care takers. The usefulness of saccadic eye-movements in diagnosing deviations from normal cognitive functioning is well established, but there is very little information on the use of smooth pursuit and Corsi-computer based test in the context of eye-tracking based detection of MCI. The results show good potential of using eye-tracking technology in combination with digitalised neuropsychological tests for detecting MCI. If used in the clinical practice, subject classified as MCI + would be referred for neurological examination so that any misclassification would be corrected. In addition, we believe that if preformed periodically in people after a certain critical age (e.g. 50), it could even detect MCI more accurately. Another way to improve the accuracy of the system would be to integrate additional neuropsychological tests that cover other cognitive domains.

Further research on the topic is of outmost importance because of the potential the proposed technology holds. Since it requires only a technician to carry out the test it could be used as a screening test at a large-scale, populational level to detect MCI as early as possible. This will specially become important in near future when neuroprotective drugs become available.

Acknowledgements. The research has received funding under project NEUS from the European Institute of Innovation and Technology (EIT) Health KIC. This body of the European Union receives support from the European Union's Horizon 2020 research and innovation programme.

References

1. Farias, S.T., Mungas, D., Reed, B.R., Harvey, D., DeCarli, C.: Progression of mild cognitive impairment to dementia in clinic-vs community-based cohorts. Arch. Neurol. **66**(9), 1151–1157 (2009)
2. Chehrehnegar, N., et al.: Early detection of cognitive disturbances in mild cognitive impairment: a systematic review of observational studies. Psychogeriatr. Off. J. Japan. Psychogeriatr. Soc. **20**(2), 212–228 (2020)
3. Seligman, S., Giovannetti, T.: The potential utility of eye movements in the detection and characterization of everyday functional difficulties in mild cognitive impairment. Neuropsychol. Rev. **25**(2), 199–215 (2015). https://doi.org/10.1007/s11065-015-9283-z
4. Molitor, R.J., Ko, P.C., Ally, B.A.: Eye movements in Alzheimer's disease. J. Alzheimer's Dis. JAD **44**(1), 1–12 (2015)
5. Guariglia, C.C.: Spatial working memory in Alzheimer's disease: a study using the Corsi block-tapping test. Dement. Neuropsychol. **1**(4), 392–395 (2007)
6. Barnes, G.R.: Cognitive processes involved in smooth pursuit eye movements. Brain Cogn. **68**(3), 309–326 (2008)
7. Bahill, A.T., McDonald, J.D.: Smooth pursuit eye movements in response to predictable target motions. Vis. Res. **23**(12), 1573–1583 (1983)

Temporal Data Analysis

Neural Clinical Event Sequence Prediction Through Personalized Online Adaptive Learning

Jeong Min Lee(✉) and Milos Hauskrecht

Department of Computer Science, University of Pittsburgh, Pittsburgh, PA, USA
jlee@cs.pitt.edu, milos@cs.pitt.edu

Abstract. Clinical event sequences consist of thousands of clinical events that represent records of patient care in time. Developing accurate prediction models for such sequences is of a great importance for defining representations of a patient state and for improving patient care. One important challenge of learning a good predictive model of clinical sequences is patient-specific variability. Based on underlying clinical complications, each patient's sequence may consist of different sets of clinical events. However, population-based models learned from such sequences may not accurately predict patient-specific dynamics of event sequences. To address the problem, we develop a new adaptive event sequence prediction framework that learns to adjust its prediction for individual patients through an online model update.

1 Introduction

Clinical event sequence data based on Electronic Health Records (EHRs) consist of thousands of clinical events representing records of patient condition and its management, such as administration of medications, records of lab tests and their results, and various physiological signals. Developing accurate temporal prediction models for such sequences is extremely important for understanding the dynamics of the disease and patient condition under different interventions and detection of unusual patient-management actions, and it may ultimately lead to improved patient care [6]. One important challenge of learning good predictive models for clinical sequences is patient-specific variability. Depending on the underlying clinical condition specific to a patient combined with multiple different management options one can choose and apply in patient care, the event patterns may vary from patient to patient. Unfortunately, many modern event prediction models and assumptions incorporated into training of such models may prevent one from accurately representing such a variability. The main challenge, which is also the main topic of this paper, is how to recover at least some of the patient-specific behavior of such models.

We study this critical challenge in context of neural autoregressive models. Briefly, neural temporal models based on RNN, LSTM, and attention mechanism have been widely used to build models for predicting clinical event time-series [3,10–13,15]. However, when built from complex multivariate clinical event

A. Tucker et al. (Eds.): AIME 2021, LNAI 12721, pp. 175–186, 2021.
https://doi.org/10.1007/978-3-030-77211-6_20

sequences, aforementioned neural models may fail to accurately model patient-specific variability due to their limited ability to represent distributions of dynamic event trajectories. Briefly, the parameters of neural temporal models are learned from many patients data through Stochastic Gradient Descent (SGD) and are shared across all types of patient sequences. Hence, the population-based models tend to average out patient-specific patterns and trajectories in the training sequences. Consequently, they are unable to predict all aspects of patient-specific dynamics of event sequences and their patterns accurately.

To address the above problem, we propose, develop, and study two novel event time-series prediction solutions that attempt to adjust the predictions for individual patients through an online model update. First, starting from the population model trained on a broad population of patients, we adapt (personalize) the model to individual patients to better fit patient-specific relations and predictions based on the current history of observations made for that patient. We refer to this model as the patient-specific model. However, one concern with the patient-specific model and adaptation is that it may lose some flexibility by being fit too tightly to the specific patient and its recent condition. To address this, we also investigate a model switching approach that learns how to adaptively switch among multiple prediction models that may consist of both population and patient-specific models. These solutions extend RNN based multivariate sequence prediction to support personalized clinical event sequence prediction. We demonstrate the effectiveness of both solutions on clinical event sequences derived from real-world EHRs data from MIMIC-3 Database [9].

2 Related Work

Patient-Specific Models. The problem of fitting patient-related outcomes and decisions as close as possible to the target individual has been an important topic of biomedical research and personalized medicine. One classic approach identifies a small set of traits or features that help to define a subpopulation the patient belongs to and applies a model built specifically for that subpopulation [7,8]. More flexible patient-specific models [4,22,26] identify the subpopulation of patients relevant to the target patient by using a patient similarity measure, and then build and apply the model online when the prediction is needed.

Online Adaptation Methods. However, in many sequential prediction scenarios, the models that are applied to the same patient more than once create an opportunity to adapt and improve the prediction from its past experiences and predictions. This online adaptation lets one to improve the patient-specific models and their prediction in time gradually. The standard statistical approach can implement the adaptation process using the Bayesian framework where population-based parameter priors combined with the history of observations and outcomes for the target patient are used to define parameter posteriors [1]. Alternative approaches for online adaptation developed in literature use simpler residual models [16] that learn the difference (residuals) between the past predictions made by population models and observed outcomes on the current patient. Liu and Hauskrecht [16] learn these patient-specific residual models for continuous-valued clinical time-series and achieve better forecasting performance.

Online Switching Methods. The online switching (selection) method is a complementary approach that has been used to increase prediction performance of online personalization models by allowing multiple (candidate) models to be used together [14,24]. At each time in a sequential process, a switching decision is made based on recent prediction performance of each candidate model. For example, for continuous-valued clinical time-series prediction, Liu and Hauskrecht [17] have a pool of population and patient-specific time-series models and at any point of time the switching method selects the best performing model.

Neural Clinical Event Sequence Prediction. EHR-derived clinical event sequence data consists of thousands of sparse and infrequently occurring clinical events. In recent years, neural-based models have become the most popular and also the most successful models for representing and predicting EHR-derived clinical sequence data. The advantages of such models are their flexibility in modeling latent structures, feature representation, and their learning capability. Specifically, word embedding methods [20] are effectively used to learn low-dimensional compact representation (embedding) of clinical concepts [2] and predictive patient state representations [25]. For autoregressive event prediction task, hidden state-space models (e.g., RNN, GRU) and attention mechanism are applied to learn latent dynamics of patient states progression and predict clinical variables such as diagnosis codes [18,19], ICU mortality risk [27], heart failure onset [3], and multivariate future clinical event occurrences [10–13,15]. For neural-based personalized clinical event prediction, most works focus on using patient-specific feature embedding obtained from patient demographics features [5,28]. A limitation of the approach is that complex transitions of patient states in time cannot be modeled in a personalized way through static feature embeddings. In this work, we develop and investigate methods for adapting modern autoregressive models based on RNN that have been successfully applied to various complex clinical patient states and prediction models.

3 Methodology

3.1 Neural Autoregressive Event Sequence Prediction

Our goal is to predict occurrences of multiple target events in clinical event sequences. We aim to build an autoregressive model ϕ that can predict, at any time t, the next step (target) event vector $\boldsymbol{y}'_{t+1}$ from a history of past (input) event vectors $\boldsymbol{H}_t = \{\boldsymbol{y}_1, \ldots, \boldsymbol{y}_t\}$, that is, $\hat{\boldsymbol{y}}'_{t+1} = \phi(\boldsymbol{H}_t)$. The event vectors are binary $\{0, 1\}$ vectors, one dimension per an event type. The input vectors are of dimension $|E|$ where E are different event types in clinical sequences. The target vector is of dimension $|E'|$, where $E' \subset E$ are events we are interested in predicting.

One way to build a neural autoregressive prediction model ϕ is to use Recurrent Neural Network (RNN) with input embedding matrix $\boldsymbol{W}_{emb}$, output linear projection matrix $\boldsymbol{W}_o$, bias vector $\boldsymbol{b}_o$, and sigmoid (logit) activation function σ. At each time step t, the RNN-based autoregressive model ϕ reads new input $\boldsymbol{y}_t$, updates hidden state $\boldsymbol{h}_t$, and generates prediction of the target vector $\hat{\boldsymbol{y}}'_{t+1}$:

$$\boldsymbol{v}_t = \boldsymbol{W}_{emb} \cdot \boldsymbol{y}_t \qquad \boldsymbol{h}_t = \mathrm{RNN}(\boldsymbol{h}_{t-1}, \boldsymbol{v}_t) \qquad \hat{\boldsymbol{y}}'_{t+1} = \sigma(\boldsymbol{W}_o \cdot \boldsymbol{h}_t + \boldsymbol{b}_o)$$

$\boldsymbol{W}_{emb}, \boldsymbol{W}_o, \boldsymbol{b}_o$, and RNN's parameters are learned through SGD with loss function $\mathcal{L}$ defined by the binary cross entropy (BCE):

$$\mathcal{L} = \sum_{s \in \mathcal{D}} \sum_{t=1}^{T(s)-1} e(\boldsymbol{y}'_{t+1}, \hat{\boldsymbol{y}}'_{t+1}) \tag{1}$$

$$e(\boldsymbol{y}'_t, \hat{\boldsymbol{y}}'_t) = -[\boldsymbol{y}'_t \cdot \log \hat{\boldsymbol{y}}'_t + (\mathbf{1} - \boldsymbol{y}'_t) \cdot \log(\mathbf{1} - \hat{\boldsymbol{y}}'_t)] \tag{2}$$

where $\mathcal{D}$ is training set and $T(s)$ is length of a sequence s. This neural autoregressive approach has several benefits when modeling complex high-dimensional clinical sequences: First, low-dimensional embedding with $\boldsymbol{W}_{emb}$ helps us to obtain a compact representation of high-dimensional input vector $\boldsymbol{y}$. Second, complex dynamics of observed patient state sequences are modeled through RNN which is capable of modeling non-linearities of the sequences. Furthermore, latent variables of neural models typically do not assume a specific probability form. Instead, the complex input-output association is learned through SGD based end-to-end learning framework which allows more flexibility in modeling complex latent dynamics of observed sequence.

However, the neural autoregressive approach cannot address one important characteristic of the clinical sequence: the variability in the dynamics of sequences across different patients. Typically, EHR-derived clinical sequences consist of medical history of several tens of thousands of patients. The dynamics of one patient's sequence could be significantly different from the sequences of other patients. For typical neural autoregressive models, parameters of the trained model are used to process and predict sequences of *all* patients which consist of individual patients who can have different types of clinical complications, medication regimes, or observed sequence dynamics.

3.2 Online Adaptation of Model Parameters

To address the patient variability issue, we propose a novel learning framework that adapts the parameters of the neural autoregressive model to the current patient sequence via SGD. For simplicity, we denote population model ϕ^P as a model trained on all training set patient data and patient (instance)-specific model ϕ^I that adapted to the current patient sequence at the prediction (test) stage. As described in Algorithm 1, the online model adaptation procedure at time t for the current patient starts by creating a patient-specific model ϕ^I from

Algorithm 1: Online Model Adaptation

Input : Population model ϕ^P, Current patient's history of **observed** input sequence $\boldsymbol{H}_t = \{\boldsymbol{y}_1, \ldots, \boldsymbol{y}_t\}$ and target sequence $(\boldsymbol{y}'_1, \ldots, \boldsymbol{y}'_t)$
Initialize patient-specific model ϕ^I from ϕ^P; $\tau = 0$; $\mathcal{L}^*_t(0) = \infty$;
repeat
  $\tau = \tau + 1$;
  $\mathcal{L}^*_t(\tau) = \sum_{i=1}^{t-1} e(\boldsymbol{y}'_{i+1}, \hat{\boldsymbol{y}}'_{i+1}) \cdot K(t,i)$ where $\hat{\boldsymbol{y}}'_{i+1} = \phi^I(\boldsymbol{H}_i)$;
  Update parameters of ϕ^I with $\mathcal{L}^*_t(\tau)$ via SGD;
until $\mathcal{L}^*_t(\tau - 1) - \mathcal{L}^*_t(\tau) < \epsilon$;
Output: Patient-specific model ϕ^I

the population model ϕ^P. They have identical model architecture and values of parameters in ϕ^I are initialized from ϕ^P. Then, we compute an online error $\mathcal{L}^*_t = \sum_{i=1}^{t-1} e(\boldsymbol{y}'_{i+1}, \hat{\boldsymbol{y}}'_{i+1})K(t,i)$ that reflects how much the prediction of ϕ^I deviates from the already observed target sequence for the current patient. With $\mathcal{L}^*_t$, we iteratively update parameters of ϕ^I via SGD. Stopping criterion for the iterative update is: $\mathcal{L}^*_t(\tau-1) - \mathcal{L}^*_t(\tau) < \epsilon$ where τ denotes the epoch of adaptation update and ϵ is a positive threshold.

Discounting. Please note that our adaptation-based loss $\mathcal{L}^*_t$ combines prediction errors for all time steps of the sequence. However, in order to better fit it to the most recent patient-specific behavior, it also biases the loss more towards recent clinical events. This is done by weighting prediction error for each step $i < t$ with $K(t,i)$ that is based on its time difference from the current time t. More specifically, $K(t,i)$ defines an exponential decay function:

$$K(t,i) = \exp\Big(-\frac{|t-i|}{\gamma}\Big) \tag{3}$$

where γ denotes the bandwidth (slope) of exponential decay; if γ is close to $+\infty$, errors at all time steps have the same weight.

Online Adaptation of Model Components. The RNN model may have too many parameters, and it may not help to adapt to all of them at the same time. One solution is to relax and permit to adapt only a subset of parameters. We experiment with and compare the adaptation of output layer parameters $(\boldsymbol{W}_o, \boldsymbol{b}_o)$ and transition model (RNN) parameters.

3.3 Adaptation by Model Switching

One limitation of online patient-specific adaptation is that it tries to modify the dynamics to fit more closely the specifics of the patient. However, when the patient state changes suddenly due to recent events (e.g., a sudden clinical complication such as sepsis), the parameters of the patient-specific model ϕ^I may not be able to adapt quickly enough to these changes. In such a case, switching back to a more general population model could be more desirable.

Algorithm 2: Online Model Switching

Input : ϕ^P, ϕ^I, $\boldsymbol{H}_t = \{\boldsymbol{y}_1, \ldots, \boldsymbol{y}_t\}$,$(\boldsymbol{y}'_1, \ldots, \boldsymbol{y}'_t)$
$\mathcal{L}^I = \sum_{i=1}^{t} e(\boldsymbol{y'}_{i+1}, \hat{\boldsymbol{y}}'^{I}_{i+1}) \cdot K(t, i)$ where $\hat{\boldsymbol{y}}'^{I}_{i+1} = \phi^I(\boldsymbol{H}_i)$;
$\mathcal{L}^P = \sum_{i=1}^{t} e(\boldsymbol{y}'_{i+1}, \hat{\boldsymbol{y}}'^{P}_{i+1}) \cdot K(t, i)$ where $\hat{\boldsymbol{y}}'^{P}_{i+1} = \phi^P(\boldsymbol{H}_i)$;
if $\mathcal{L}^P \geq \mathcal{L}^I$ **then**
  $\hat{\boldsymbol{y}}'_{t+1} = \hat{\boldsymbol{y}}'^{I}_{t+1}$
else
  $\hat{\boldsymbol{y}}'_{t+1} = \hat{\boldsymbol{y}}'^{P}_{t+1}$
end
Output: Prediction at time step $t + 1$: $\hat{\boldsymbol{y}}'_{t+1}$

Model switching framework [17,24] can resolve this issue by dynamically switching among a patient-specific model and the population model. Driven by the recent performance of models, it can switch to the best performing model at each time step. Algorithm 2 implements the model switching idea. Given a trained population model ϕ^P, a patient-specific model ϕ^I trained via online adaptation, and the current patient's observed sequence, we can compute discounted losses $\mathcal{L}^P, \mathcal{L}^I$ for both models on the past data. By comparing the two losses, we select the model that gives the best error and use it for predicting the next step.

4 Experimental Evaluation

4.1 Experiment Setup

Clinical Sequence Generation. We extract 5137 patients from publicly available MIMIC-3 database [9] using the following criteria: (1) age is between 18 and 99, (2) length of admission is between 48 and 480 h, and (3) clinical records are stored in Meta Vision system, one of the systems used to create MIMIC-3. We generate train and test sets using 80/20% split ratio. From the extracted records, we generate multivariate event sequences with a sliding-window method. We segment all sequences with a time window $W = 24$ h. All events that occurred in a time-window are aggregated into a binary vector $\boldsymbol{y}_i \in \{0, 1\}^{|E|}$ where i denotes a time-step of the window and E is a set of event types. At any point of time t, a sequence of vectors created from previous time-windows defines an (input) sequence. A vector representing events in the next time window defines the prediction target.

Feature Extraction. We use medication administration, lab results, procedures, and physiological results to define events. For the first three categories, we remove events that were observed in less than 500 different patients. For physiological events, we select 16 important event types with the help of a critical care physician. Lab test results and physiological measurements with continuous values are discretized to high, normal, and low values based on normal ranges compiled by clinical experts. In terms of prediction targets, we only consider and represent events corresponding to occurrences of such events, and we do not predict their normal or abnormal values. This process results in 65 medications, 44 procedures, 155 lab tests, and 84 physiological events as prediction targets, for the total target vector size of 348. The input vectors are of size 449.

Baseline Models. We compare proposed models to the following baselines:

- **GRU-based POPulation model (GRU-POP)**: For RNN-based time-series modeling described in Sect. 3.1, we use GRU. (λ = 1e−05) The patient (INstance)-specific model (**GRU-IN**) has the same architecture.
- **REverse-Time AttenTioN (RETAIN)**: RETAIN is a representative work on using attention mechanism to summarize clinical event sequences, proposed by Choi et al. [3]. It uses two attention mechanisms to comprehend the history of GRU-based hidden states in reverse-time order. For multi-label output, we use a sigmoid function at the output layer. (λ = 1e−05)
- **Logistic regression based on Convolutional Neural Network (CNN)**: This model uses CNN to build predictive features summarizing the event history of patients. Following Nguyen et al. [21], we implement this CNN-based model with a 1-dimensional convolution kernel followed by ReLU activation and max-pooling operation. To give more flexibility to the convolution operation, we use multiple kernels with different sizes (2, 4, 8) and features from these kernels are merged at a fully-connected (FC) layer. (λ = 1e−05)

Model Parameters. We use embedding dimension 64, hidden state dimension 512, for all neural models. The population model, RETAIN, and CNN use learning rate 0.005 and patient-specific models use 0.005. To prevent over-fitting, we use L2 weight decay regularization during the training of GRU-POP, RETAIN, and CNN, and the weight λ is determined by the internal cross-validation set (range: 1e−04, 1e−05, 1e−06, 1e−07). For the SGD optimizer, we use Adam. For the early stopping criteria parameter, we set ϵ = 1e−04. For γ, we use fixed value 3.0.

Evaluation Metric. We use the area under the precision-recall curve (AUPRC) as the main evaluation metric. AUPRC is known for presenting a more accurate assessment of performance of models for a highly imbalanced dataset [23].

4.2 Results on Online Adaptation vs. Population Model

We first compare the prediction performance of the population model (GRU-POP) and the proposed method on a patient-specific online adaption model that adapted all parameters (GRU-IN) as described in Algorithm 1. As shown in Fig. 1, patient-specific model clearly outperforms population-based model across all time-steps. Especially in earlier days of admissions (day = 1–3), the performance gap is smaller, but as time progresses on, the gap is increasing. It shows patient-specific online adaptation models can learn to more accurately predict patient-specific dynamics of event sequences compared to the population-based model.

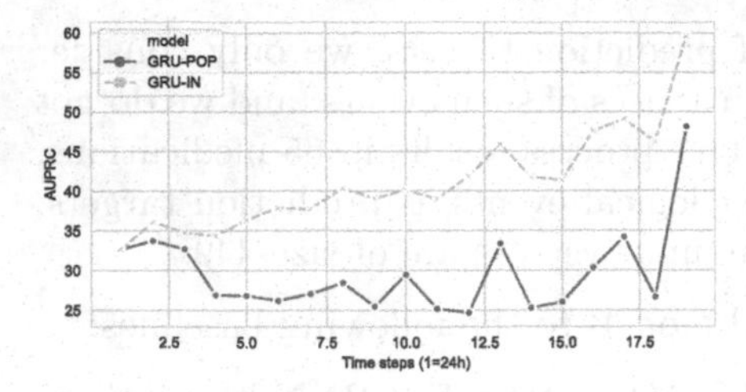

Fig. 1. Prediction performance (AUPRC) of online adaptation method (GRU-IN) and population-based model (GRU-POP).

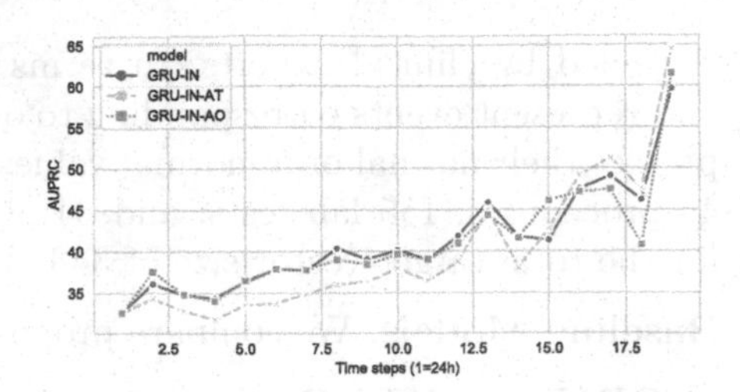

Fig. 2. Performance of online adaptation methods on all parameters (GRU-IN) and two subsets of parameters (GRU-IN-AT, GRU-IN-AO).

4.3 Results on Adaptation on Partial Components

Next, we relax the online adaptation procedure to update only subsets of parameters: GRU-IN-AO is only adapting the output weight layer ($\boldsymbol{W}_o, \boldsymbol{b}_o$) and GRU-IN-AT is only adapting parameters of the transition layer (GRU) only. As shown in Fig. 2, the overall performance of GRU-IN-AO is close to GRU-IN which adapts all parameters. Since GRU-IN-AO is more efficient, it offers the best overall approach for patient-specific model adaptation.

Table 1. Prediction results of all models averaged across all time steps

	CNN	RETAIN	GRU-POP	GRU-IN	GRU-IN-SW	GRU-IN-AO-SW
AUPRC	30.81	29.67	29.61	41.13	42.14	42.62

4.4 Results for Online Switching-Based Adaptation

We also experiment with online switching-based adaptation approach. It chooses the best predictive model from among a pool of available prediction models. We run the method to choose between a population-based model and a patient-specific adaptation model. We try the switching model in combination with the population model and two patient-specific models GRU-IN and GRU-IN-AO. The switching model results use post-fix '-SW'. As shown in Fig. 3, models that rely on multiple models and online switching outperform baseline models of GRU-POP, CNN, and RETAIN. When the prediction performance is averaged across all time steps, we can observe that GRU-IN-AO-SW outperforms all models as shown in Table 1. Particularly, GRU-IN-AO-SW's AUPRC is +43% higher than GRU-POP and RETAIN models. Compared to GRU-IN-AO (averaged AUPRC: 40.89), the online switching adaptation method increases AUPRC by +4.2% and this reveals the benefit added by the online switching method.

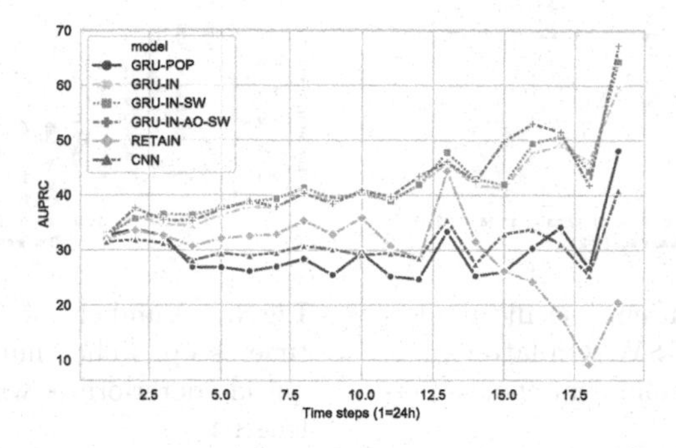

Fig. 3. Performance of online switching methods (-SW) with population and patient-specific adaptation models. Online switching methods clearly outperform baseline models (GRU-POP, RETAIN, CNN)

When the Model Switches? To have a better understanding of the behavior of online switching-based adaptation, we investigate when the model switches to a patient-specific model and to the population model. First, we analyze how many times the online switching mechanism selects a patient-specific model (instead of a population model) over time and report the ratio of it. As shown in Fig. 4, in the early time steps, the online switching mechanism chooses the population model. However, at later time steps, the switching mechanism selects patient-specific models. This can be explained by the fact that patient-specific models need enough observations to adapt the patient-specific variability which is not possible with shorter sequences. To properly interpret the results, Fig. 5 shows the number of patients in each time step. This number can also be interpreted as the length of patient sequences and their volume. We can clearly see that the number of patients with longer sequences is very small, as the majority of sequences are very short. For example, patients with sequences longer than 13 days of admission are only about 12% of all patients in test set. From this, we can conclude that the population model is often biased towards the dynamics and characteristics of shorter patient sequences. Meanwhile, patient-specific models can effectively learn and adapt better to the dynamics of longer sequences.

Predicting Repetitive and Non-repetitive Events. To perform this analysis, we divide event occurrences into two groups based on whether the same type of event has or has not occurred before. We compute AUPRC for each group as shown in Table 2. The results show that for non-repetitive events, the performance of the patient-specific model is the lowest among all models. This is expected because with no previous occurrence of a target event, a patient-specific model could have difficulty making an accurate prediction for the new target event. In this case, we can also see the benefit of the online switching mechanism: the prediction of the population model is more accurate than the patient-specific model, and the online switching mechanism correctly chooses the population model. More specifically, GRU-IN-SW recovers most of the

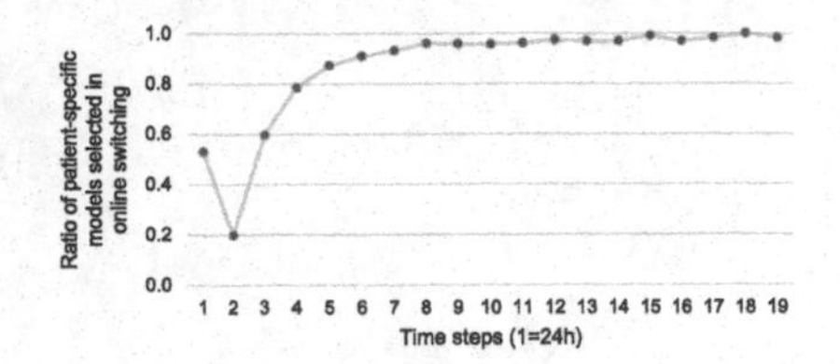

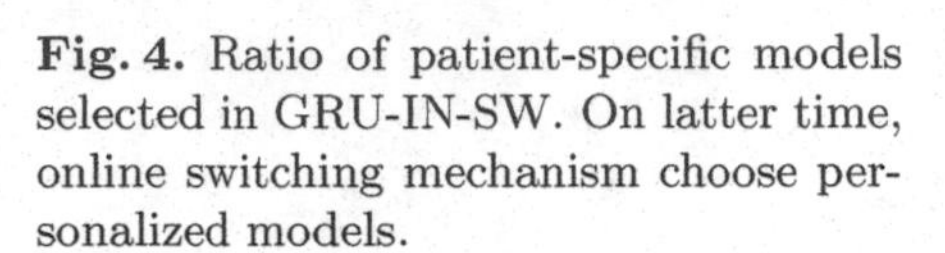

Fig. 4. Ratio of patient-specific models selected in GRU-IN-SW. On latter time, online switching mechanism choose personalized models.

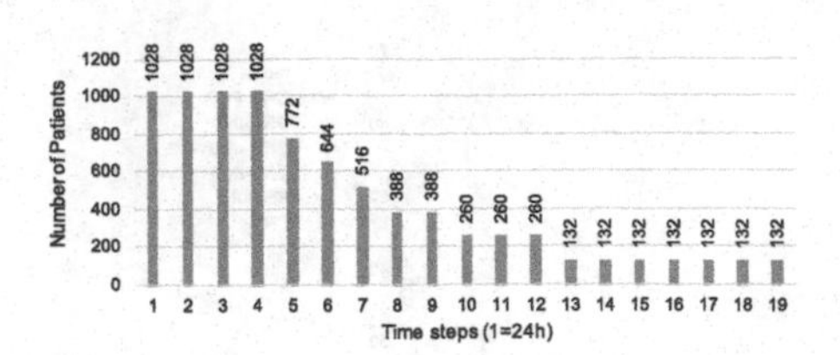

Fig. 5. Number of patients in each time step. The number of patients quickly deteriorates with longer sequence lengths.

Table 2. Prediction result on non-repetitive and repetitive event groups. For non-repetitive events, the performance of patient-specific models (GRU-IN) is the lowest. The online switching approaches (GRU-IN-SW, GRU-IN-AO-SW) recover the predictability by switching to the population model.

	CNN	RETAIN	GRU-POP	GRU-IN	GRU-IN-SW	GRU-IN-AO-SW
Non-repetitive	14.13	15.54	<u>15.85</u>	11.11	15.55	14.37
Repetitive	45.16	50.30	52.04	52.83	53.73	<u>53.91</u>

predictability of GRU-POP for non-repetitive event prediction. For repetitive event prediction, we can see that the patient-specific model (GRU-IN) outperforms the population-based models. However, the online switching approaches (GRU-IN-SW, GRU-IN-AO-SW) are the best and outperform all other approaches.

5 Conclusion

In this work, we have developed methods for patient-specific adaptation of predictive models of clinical event sequences. These models are of a great importance for defining representations of a patient state and for improving care. We demonstrate the improved performance of our models through experiments on MIMIC-3, a publicly available dataset of electronic health records for ICU patients.

Acknowledgement. The work presented was supported by NIH grant R01GM 088224. The content of this paper is solely the responsibility of the authors and does not necessarily represent the official views of NIH.

References

1. Berzuini, C., et al.: Bayesian networks for patient monitoring. Artif. Intelli. Med. **4**, 243–260 (1992)
2. Choi, E., et al.: Multi-layer representation learning for medical concepts. In: The 22nd ACM SIGKDD (2016)
3. Choi, E., et al.: RETAIN: an interpretable predictive model for healthcare using reverse time attention mechanism. In: Advances in NeurIPS (2016)
4. Fojo, A.T., et al.: A precision medicine approach for psychiatric disease based on repeated symptom scores. J. Psychiatr. Res. **95**, 147–155 (2017)
5. Gao, J., et al.: CAMP: co-attention memory networks for diagnosis prediction in healthcare. In: ICDM (2019)
6. Hauskrecht, M., et al.: Outlier-based detection of unusual patient-management actions: an ICU study. J. Biomed. Inform. **64**, 211–221 (2016)
7. Huang, Z., et al.: Medical inpatient journey modeling and clustering: a Bayesian hidden Markov model based approach. In: AMIA, vol. 2015 (2015)
8. Huang, Z., et al.: Similarity measure between patient traces for clinical pathway analysis: problem, method, and applications. IEEE J-BHI **18**, 4–14 (2013)
9. Johnson, A.E.W., et al.: MIMIC-III, a freely accessible critical care database. Sci. Data **3**, 160035 (2016)
10. Lee, J.M., Hauskrecht, M.: Recent context-aware LSTM-based clinical time-series prediction. In: International Conference on AI in Medicine (AIME) (2019)
11. Lee, J.M., Hauskrecht, M.: Clinical event time-series modeling with periodic events. In: The 33rd International FLAIRS Conference (2020)
12. Lee, J.M., Hauskrecht, M.: Multi-scale temporal memory for clinical event time-series prediction. In: Michalowski, M., Moskovitch, R. (eds.) AIME 2020. LNCS (LNAI), vol. 12299, pp. 313–324. Springer, Cham (2020). https://doi.org/10.1007/978-3-030-59137-3_28
13. Lee, J.M., Hauskrecht, M.: Modeling multivariate clinical event time-series with recurrent temporal mechanisms. Artif. Intell. Med. **112**, 102021 (2021)
14. Littlestone, N., et al.: The weighted majority algorithm. Inf. Comput. **108**(2), 212–261 (1994)
15. Liu, S., Hauskrecht, M.: Nonparametric regressive point processes based on conditional Gaussian processes. In: Advances in NeurIPS (2019)
16. Liu, Z., Hauskrecht, M.: Learning adaptive forecasting models from irregularly sampled multivariate clinical data. In: The 30th AAAI Conference (2016)
17. Liu, Z., Hauskrecht, M.: A personalized predictive framework for multivariate clinical time series via adaptive model selection. In: ACM CIKM (2017)
18. Malakouti, S., Hauskrecht, M.: Hierarchical adaptive multi-task learning framework for patient diagnoses and diagnostic category classification. In: IEEE BIBM (2019)
19. Malakouti, S., Hauskrecht, M.: Predicting patient's diagnoses and diagnostic categories from clinical-events in EHR data. In: Riaño, D., Wilk, S., ten Teije, A. (eds.) AIME 2019. LNCS (LNAI), vol. 11526, pp. 125–130. Springer, Cham (2019). https://doi.org/10.1007/978-3-030-21642-9_17
20. Mikolov, T., et al.: Distributed representations of words and phrases and their compositionality. In: Advances in NeurIPS, pp. 3111–3119 (2013)
21. Nguyen, P., et al.: Deepr: a convolutional net for medical records. IEEE J. Biomed. Health Inform. **21**(1), 22–30 (2016)

22. Rizopoulos, D.: Dynamic predictions and prospective accuracy in joint models for longitudinal and time-to-event data. Biometrics **67**, 819–829 (2011)
23. Saito, T., Rehmsmeier, M.: The precision-recall plot is more informative than ROC plot when evaluating binary classifiers on imbalanced datasets. PLoS ONE **10**, e0118432 (2015)
24. Shalev-Shwartz, S., et al.: Online learning and online convex optimization. Found. Trends Mach. Learn. **4**, 107–194 (2011)
25. Tran, T., et al.: Learning vector representation of medical objects via EMR-driven nonnegative restricted Boltzmann machines. JBI **54**, 96–105 (2015)
26. Visweswaran, S., Cooper, G.F.: Instance-specific Bayesian model averaging for classification. In: Advances in NeurIPS (2005)
27. Yu, K., et al.: Monitoring ICU mortality risk with a long short-term memory recurrent neural network. In: Pacific Symposium on Biocomputing. World Scientific (2020)
28. Zhang, J., et al.: Patient2Vec: a personalized interpretable deep representation of the longitudinal electronic health record. IEEE Access **6**, 65333–65346 (2018)

Using Event-Based Web-Scraping Methods and Bidirectional Transformers to Characterize COVID-19 Outbreaks in Food Production and Retail Settings

Joseph Miano[1,2], Charity Hilton[1(✉)], Vasu Gangrade[3], Mary Pomeroy[3], Jacqueline Siven[3], Michael Flynn[3], and Frances Tilashalski[3]

[1] Georgia Tech Research Institute, Atlanta, GA 30318, USA
Charity.Hilton@gtri.gatech.edu
[2] Georgia Institute of Technology, Atlanta, GA 30332, USA
[3] Centers for Disease Control and Prevention, Atlanta, GA 30333, USA

Abstract. Current surveillance methods may not capture the full extent of COVID-19 spread in high-risk settings like food establishments. Thus, we propose a new method for surveillance that identifies COVID-19 cases among food establishment workers from news reports via web-scraping and natural language processing (NLP). First, we used web-scraping to identify a broader set of articles (n = 67,078) related to COVID-19 based on keyword mentions. In this dataset, we used an open-source NLP platform (ClarityNLP) to extract location, industry, case, and death counts automatically. These articles were vetted and validated by CDC subject matter experts (SMEs) to identify those containing COVID-19 outbreaks in food establishments. CDC and Georgia Tech Research Institute SMEs provided a human-labeled test dataset containing 388 articles to validate our algorithms. Then, to improve quality, we fine-tuned a pre-trained RoBERTa instance, a bidirectional transformer language model, to classify articles containing ≥1 positive COVID-19 cases in food establishments. The application of RoBERTa decreased the number of articles from 67,078 to 1,112 and classified (≥1 positive COVID-19 cases in food establishments) articles with 88% accuracy in the human-labeled test dataset. Therefore, by automating the pipeline of web-scraping and COVID-19 case prediction using RoBERTa, we enable an efficient human in-the-loop process by which COVID-19 data could be manually collected from articles flagged by our model, thus reducing the human labor requirements. Furthermore, our approach could be used to predict and monitor locations of COVID-19 development by geography and could also be extended to other industries and news article datasets of interest.

Keywords: COVID-19 · Public health · Web-scraping · Natural language processing

A. Tucker et al. (Eds.): AIME 2021, LNAI 12721, pp. 187–198, 2021.
https://doi.org/10.1007/978-3-030-77211-6_21

1 Introduction

Though several studies have captured outbreaks of COVID-19 among employees in congregate settings [18,22,31], to our knowledge few reports [14,28,32] comprehensively characterize outbreaks among essential workers in food production and retail settings. Local news reports can be an important source in identifying outbreaks. This project applies machine learning-based web-scraping and a RoBERTa (A Robustly Optimized Bidirectional Encoder Representations from Transformers Pretraining Approach) [21] language model to locate, quantify, and characterize outbreaks among food system workers via local news reports. Our system highlights the utility of media reports in characterizing COVID-19 outbreaks and can be extended to other industries and congregate settings.

2 Background

2.1 COVID-19 and Food Establishments

Understanding the scope of outbreaks of COVID-19 identified in non-healthcare settings, such as food settings, is crucial to both informing public health decision-making and prevention messaging tailored to reducing transmission among worker populations and identifying potential health equity issues in workplaces. Though outbreaks in food settings have been reported through various state surveillance mechanisms [1,2,7], in web-based portals[1] and in scientific reports [3,14,20,28,32], which specifically highlight incidence in meat processing facilities, little to no systematic and ongoing data collection exists for all food system sectors, including restaurants or grocery settings. Additionally, public, and occupational health data collection systems often do not collect information on social determinants of health (such as race, ethnicity, nativity, language spoken, etc.) in workplaces [26]. Workers in food settings are predominantly critical infrastructure workers, who contribute to the security and well-being of our food supply. Collecting these data may aid in the early identification of workplace transmission of SARS-CoV-2, the virus that causes COVID-19. This creates potential opportunities for successful mitigation efforts and improving overall worker safety and health.

2.2 Bidirectional Transformers

Transformer language models have seen widespread development and use [13,21,24,30] over the past several years and achieve state-of-the-art results across various natural language processing tasks, like text generation and classification [12,29,34]. Introduced in 2017 by Vaswani et al. [30], the Transformer is a purely attention-based model (i.e., with no convolutional layers or recurrence) that is faster to train and achieves better results than previous techniques on a

[1] https://public.tableau.com/profile/leah.douglas#!/vizhome/CumulativeCovid-19casesbysector/Dashboard1.

language translation task [30], and its architectural details can be found in the original 2017 paper [30].

Several variations on the original Transformer architecture and training process have been developed since 2017 [12,13,21,30], including Bidirectional Transformer models like BERT (Bidirectional Encoder Representations from Transformers) [13] and RoBERTa [21]. BERT extended previous transformer-based models [24,30] by incorporating bidirectional pre-training; i.e., BERT is pre-trained using the masked language model (MLM) objective, by which it learns to predict a masked word given the context words that come before and after [13]. Devlin et al. [13] also leveraged next sentence prediction (NSP) to pre-train BERT, which is an objective by which the model attempts to predict the next sentence given a current sentence input [13]. Together, these pre-training objectives build a robust language model that can be fine-tuned for a target task like text classification [29].

RoBERTa [21] shares the same model architecture as BERT but modifies the BERT pre-training process to omit the NSP task and leverages additional data during pre-training (e.g., CC-News, which consists of 63 million English news articles) not used by BERT. The result is a model that outperforms BERT across several tasks [21] and that may be especially suited to the task of news article classification, due to the news article data used during RoBERTa pre-training.

3 Methods

Our proposed method (Fig. 1) leverages NLP-Based Web-Scraping and a RobERTa model [21] to classify news articles as containing mention of ≥ 1 cases of COVID-19 in food establishments. Once a model is trained and evaluated, a daily list of relevant news articles could be output for human review.

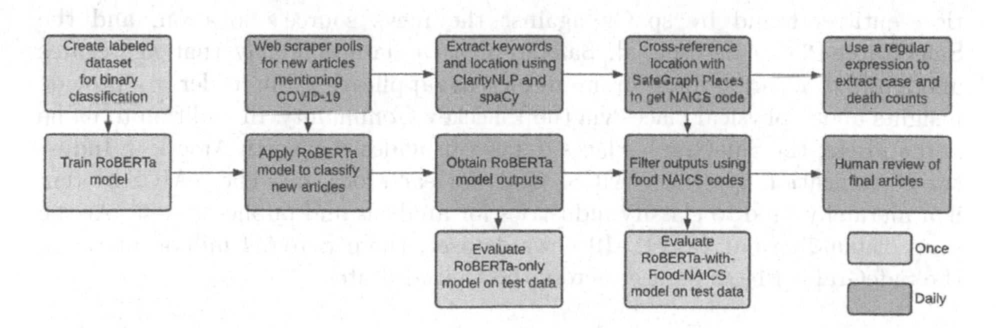

Fig. 1. A flow diagram of what our method would look like if it were deployed. In our experiments, we ran each of these steps on our overall dataset once, but we did not deploy our approach to produce real-time daily outputs.

3.1 NLP-Based Web-Scraping

Since April 2020, we tracked keyword mentions of COVID-19 in the news media from articles reported from March 15, 2020, to September 30, 2020 to develop a COVID-19 news dataset. From aggregating lists of news media sources from web sites, such as Wikipedia, we created a list of over 1,700 news sources [8–10]. Each news source was visited daily by an automated Python script, using a web-scraping library to discover publicly available news articles mentioning COVID-19 [25]. Next, we ran each article through two NLP pipelines, ClarityNLP and spaCy [15–17].

NLP Pipelines. ClarityNLP is an open-source project written in a modern, Python-based stack, and it leverages a user-friendly query language, NLPQL [16,17]. ClarityNLP allows for building custom NLP pipelines with a focus on biomedical text. To build our COVID-19 news dataset, we utilized ClarityNLP's NLPQL query language to find non-negated mentions of keywords relevant to our dataset. These keywords were deemed relevant to potential COVID-19 outbreaks in congregate settings, such as food processing facilities, long-term care facilities, and correctional facilities. Each keyword was reviewed by SMEs from the CDC. Additionally, we developed a regular expression algorithm to extract case counts and death counts from each news article.

We utilized the advanced NLP library, spaCy to identify the location of the news article [15]. SpaCy provides a named-entity recognition feature to extract categories of entities from sections of text. We used the location (LOC), organization (ORG), geographic (GPE), and facility (FAC) categories to identify the location of the potential COVID-19 outbreak.

Location Identification and NAICS Codes. We cross-referenced the location entities found by spaCy against the news source's location, and the SafeGraph Places dataset [4]. SafeGraph is a data company that aggregates anonymized location data from numerous applications in order to provide insights about physical places via the Placekey Community. In addition to name and address, the SafeGraph Places dataset provides the North American Industry Classification System (NAICS) Code for each location. The NAICS system is a hierarchy used to classify industries for analysis and publication [5]. At the time of building our COVID-19 news dataset, there were 5.4 million places in the SafeGraph Places dataset across the United States.

Persistence. Finally, we saved the metadata, case/death count, NAICS code, and location variables for each news article in a MySQL database. Currently, there are over 93,000 keyword mentions in our news dataset. These articles were the basis for training the RoBERTa model[2].

[2] At the time we trained the RoBERTa model, we evaluated just over 67,000 articles.

3.2 Data Validation

We validated the extracted variables and removed duplicate facilities mentioned in multiple articles by manual review. We specified "potential" outbreaks if there were $\geq$2 positive cases identified as employees of a single food setting. However, we could not always confirm that the infection was attributable to workplace exposure as the sources for the data were only news articles. We also could not verify the reporting mechanism of the case; whether a case was identified as a case by confirmatory laboratory testing or whether it was self-reported. News articles may not always include how the establishment was made aware of employee cases–whether through laboratory confirmation of a diagnosis or through self-reporting. Self-reporting without lab confirmation might have led to introduction of biases which could lead to misclassification (e.g., an employee reports COVID-19-like symptoms to workplace but has not confirmed infection [11]). While validating, we also extracted more qualitative information; for example, if these establishments opted to close their doors for 1–2 weeks after an employee or employees were identified as having COVID-19.

3.3 Applying RoBERTa to Article Classification

Implementation. We implemented our RoBERTa model training and evaluation using Python with the PyTorch [23] and HuggingFace [34] libraries between October 2020 and January 2021.

Training. We trained a RoBERTa model to automatically assign a positive tag to an article, based on its contents, if it contains mention of $\geq$1 cases of COVID-19 in food establishments, which we also combined with a NAICS code filter for food-related industries. To train our RoBERTa model, we first randomly divided our overall human-labeled dataset (n = 2061) into a training dataset (n = 1189), validation dataset (n = 471), and test dataset (n = 388). The training dataset is what the RoBERTa model sees at every training iteration in order to update its weights. In contrast, the validation dataset determines which model weights should be kept after completing all the training epochs (i.e., we keep the model weights with the lowest loss on the validation dataset). The test dataset is used for model evaluation, which we discuss in the next section. We used cross-entropy loss, with class weights set to be balanced using scikit-learn's compute_class_weight function as our loss function and trained for 300 epochs at a batch size of 32 using the Adam optimizer [19]; here, an epoch is defined as a single complete pass through the training dataset (i.e., the model seeing every training example once constitutes one epoch) and the batch size refers to the number of training examples used to update the model's weights at each iteration of training.

Evaluation. We evaluated our models by computing various metrics on our human-labeled test dataset, including accuracy, precision, recall, and F1-score (terms are defined below). We also counted the number of positive examples flagged by our models in the overall 67,078 articles; for this evaluation, an article is considered a ground truth positive example if it contains mention of ≥1 positive cases of COVID-19 in food establishment employees and a ground truth negative example otherwise. Accuracy was computed as the number of correctly classified examples divided by the total number of examples classified (n = 388 in the test dataset). Precision is computed as: $\left(\frac{\text{true positives}}{\text{true positives+false positives}}\right)$, where a true positive refers to a ground truth positive example, as defined above, that the model correctly classifies as positive, a true negative refers to a ground truth negative example that the model correctly classifies as negative, a false positive refers to a ground truth negative example that the model falsely classifies as positive, and a false negative refers to a ground truth positive example that the model falsely classifies as negative. Next, recall is computed as: $\left(\frac{\text{true positives}}{\text{true positives+false negatives}}\right)$. Finally, the F1-score is computed as: $\left(\frac{2}{\frac{1}{\text{precision}}+\frac{1}{\text{recall}}}\right)$.

4 Results

4.1 Web-Scraping, Manual Data Validation, and Visualization

This paper summarizes results from a web-scraping analysis of news articles and media reports, where one or more employees in specific food settings were identified as having COVID-19. Through an automated web-scraping process followed by manual cleaning and data validation, we identified 276 facilities with 30,734 employees who reportedly tested positive for COVID-19, according to the news articles (see Table 1)[3].

Ninety percent of these cases were identified among the animal slaughtering and processing facility workers, with some plants reporting as many as 1,000 positive cases at a single location. Although most articles were about restaurants among the consumer-facing establishments, news articles reported significantly more employees per facility with COVID-19 in grocery store settings. News reports mentioned only five food distribution centers, but they had 151 employees with COVID-19. Of note, articles reported that most restaurants opted to close their doors for 1–2 weeks after an employee or employees were identified as having COVID-19. We also categorized the outbreaks by geographic region

[3] The US map, which can be accessed at: https://www.google.com/maps/d/u/0/viewer?mid=1ymY4bzI70AOCeFzRYvfe4HPWVvgPBoJh&ll=45.80359787060013%2C-114.35715944999998&z=4 shows the food-setting locations found using manual validation of news articles. The legend on the left within the map shows the different types of food settings based on NAICS codes. The user may access more details about each facility including the title, a link to the article, and descriptors including case and death counts by clicking on a chosen location on the map.

Table 1. Below, we show a summary of the reported facility types organized by NAICS code descriptors. This includes facility type, number of facilities (with ≥1 case reported), number of potential outbreaks (≥2 cases) of COVID-19, and number of cases linked to the type of facility.

NAICS code	Industry classification/type of facility	No. of facilities	No. of potential outbreaks	No. of cases
3116	Animal slaughtering and processing	132	130	27704
7225	Restaurants and other eating places	65	13	91
4451	Grocery stores	30	15	254
7224	Drinking places (alcoholic beverages)	13	7	42
3114	Fruit and vegetable preserving	9	9	561
3117	Seafood product preparation and packaging	6	6	283
4244	Grocery and related product merchant wholesalers	5	5	151
3119	Other food manufacturing	4	4	312
1113	Fruit and tree nut farming	2	2	77
1119	Other crop farming	2	2	131
7223	Speciality food services (food trucks)	2	1	3
3112	Grain and oilseed milling	1	1	13
1122	Hog and pig farming	1	1	60
1151	Support activities for crop production	1	1	6
1112	Vegetable and melon farming	1	1	14
Unk	Unknown	2	2	1032
	Total	**276**	**200**	**30734**

(interactive map) and industry sector (e.g., food processing, restaurant, grocery stores), and number of cases/deaths[4].

4.2 RoBERTa Model Evaluation

Using our RoBERTa model enabled article classification at 88.4% accuracy with a 62.8% F1 score (Fig. 2). The combined RoBERTa-with-Food-NAICS model achieved the best performance (Fig. 2). Furthermore, RoBERTa-with-Food-NAICS flagged 1,112 articles as positive of the 67,078 total articles, which is less than the 4681 flagged by Food-NAICS-only (Fig. 2). However, the recall was lower for RoBERTa-with-Food-NAICS than for Food-NAICS only (Fig. 2, 79.2% vs. 91.7%).

Next, we extended our initial dataset of 67,078 news articles to include news articles through to the end of 2020, thus adding 17,170 new news articles. We used this expanded dataset of 84,248 articles to generate Fig. 3, which

[4] We developed a public dashboard of news article keywords of COVID-19 in food processing facilities from March 15-September 30, 2020. It is available at https://public.tableau.com/profile/charity.hilton\#!/vizhome/COVID-19NewsReportsaboutFoodSettings/COVID-19NewsMap.

shows the daily number of articles flagged as positive (i.e., containing mention of ≥1 COVID-19 cases in food establishments) by considering all articles, Food-NAICS-only, and combined RoBERTa-with-Food-NAICS. To generate Fig. 3, we applied our RoBERTa model trained on the original 67,078 articles to the expanded dataset, which spans from March 15, 2020 to December 31, 2020. When considering all articles, the maximum number of daily articles was 2315, which occurred on May 28, 2020, and the mean number of daily articles is 277 (Fig. 3). Food-NAICS-only has a maximum number of daily articles of 143 (on May 27) and a mean of 25 articles, while RoBERTa-with-Food-NAICS model has a maximum number of daily articles of 40 (May 29) and a mean of 4 (Fig. 3).

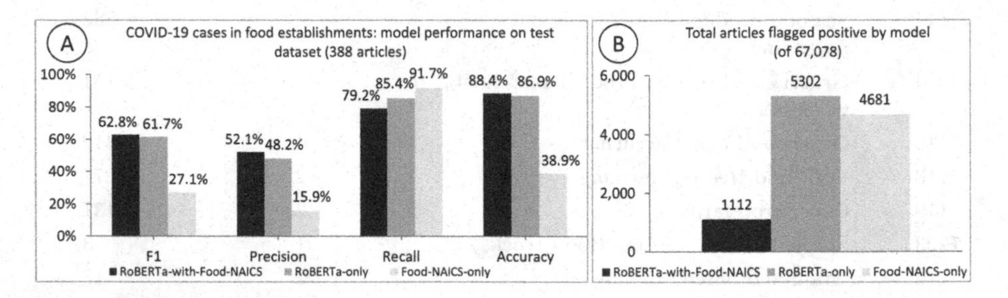

Fig. 2. In (A), we show the precision, recall, F1 score, and accuracy for 3 variations of our model on a held-out test dataset of 388 articles (48 positive examples and 340 negative). In (B), we apply our models to our overall dataset and display how many articles are flagged as containing mention of ≥1 COVID-19 cases in a food establishment.

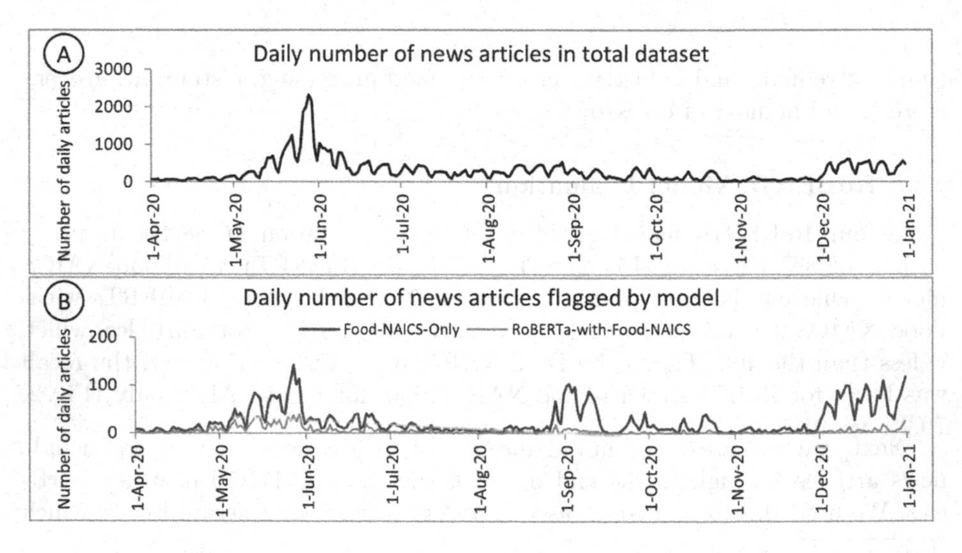

Fig. 3. In (A), we show the number of daily news articles without any NAICS or model-based filtering (i.e., the total number of articles). In (B), we show the number of daily news articles using a food NAICS filter only (blue) and a combined RoBERTa model output and food NAICS filter (orange). (Color figure online)

5 Discussion

5.1 RoBERTa Model and NAICS Codes

Because RoBERTa-with-Food-NAICS flagged a total of 1,112 articles and Food-NAICS-only flagged 4,681 articles, RoBERTa-with-Food-NAICS reduces the amount of human manual review needed for validation. The recall being lower for RoBERTa-with-Food-NAICS than for NAICS-only means that RoBERTa with NAICS misses more positive examples than NAICS-only; however, this is counterbalanced by the greater precision of RoBERTa with NAICS.

One limitation of the combined approach is that the NAICS code is not always available, in which case RoBERTa-only (i.e., without the food NAICS filter) could be used. Another limitation is the inability to differentiate between true transmission among workers in a food setting versus exposure within the community, particularly in geographic locations with sustained COVID-19 transmission. A third limitation is the likely differences in case definitions reported by various news sources. There were no efforts made to account for inaccuracies in reporting and all "news" sources were treated equally. Future efforts are needed to compare results from the proposed model to proven methods of public health surveillance relying on case investigation and contact tracing.

5.2 COVID-19 Surveillance

Some of the earliest reported cases of COVID-19 in food settings were identified in meat processing facilities in early April 2020 [14]. During that time, this web-scraping platform was in the early stages of development. In fact, a number of outbreaks had occurred within food settings by the time the tool was fully developed (see Fig. 3). Though we could not detect outbreaks using this systematic web-scraping approach during the early months of the COVID-19 pandemic, we were able to capture and quantify these outbreaks retroactively. This methodology can prove invaluable in addressing any time lapse between outbreak and response; during infectious disease outbreaks it takes significant time and effort to ramp up the public health response, yet news agencies often have the capacity to quickly collect and disseminate reports on cases and localized outbreaks. Because early detection of outbreaks is crucial for effective mitigation and successful public health coordination [33], a proactive approach in using this methodology and addressing that time lapse may be a useful tool to aiding in earlier outbreak detection, particularly for novel pathogens for which no transitional surveillance methods or reporting requirements exist.

This methodology could also be used to identify health equity issues within these worker populations during infectious disease outbreaks. Using an Occupational Health Equity framework (OHE), i.e., that social, economic, environmental, and structural factors impact work-related disparities in illness and disease in avoidable ways [6], this methodology could be used to identify the frequency of OHE-relevant mentions in news articles. By leveraging the by-the-minute reports of news agencies to compare outbreak data and OHE variables, we can aid in early identification of potential compounding social factors affecting disease transmission. Such early identification is vital; as we have seen during the COVID-19 pandemic, rates of infection and disease can be heavily impacted by the interactions between type of work and social variables (like race, ethnicity, and migration status). To properly address how social and structural vulnerabilities impact work-related illness and disease we must enhance our data collection through both innovative targeted efforts and expansion of current surveillance efforts. By developing new ways to incorporate race, ethnicity, work arrangement, and other variables, into OSH surveillance we can better understand how these variables interact to affect different worker populations [27].

Our findings improve our understanding of outbreaks in food settings and help identify health equity issues and health disparities among essential workers. Ideally this methodology may produce more focused and industry-specific public health interventions and provide meaningful feedback to partners, including federal agencies and policymakers.

6 Conclusion

By combining NLP-based web-scraping and a RoBERTa language model, our approach enables an efficient human-in-the-loop process to identify COVID-19 outbreaks in food establishments based on news article contents, which addresses the limitation of current COVID-19 surveillance largely underreporting occupational data. We found that combining the RoBERTa outputs with a NAICS code filter for food settings yielded better performance on our dataset than using only the RoBERTa outputs or only the NAICS code filter. Our approach could be applied to help predict geographic areas of concern for COVID-19 outbreaks based on early event-based reporting in news articles. Other future directions include extending the RoBERTa method to improve its performance without requiring the NAICS code (e.g., by gathering more human-labeled data and/or applying self-supervised learning) and applying our approach to other industries or news article datasets. Future studies are warranted for true validation of the surveillance model results against existing public health surveillance tools.

7 Disclaimer

The findings and conclusions in this report are those of the author(s) and do not necessarily represent the official position of the Centers for Disease Control and Prevention (CDC).

References

1. Coronavirus - outbreak reporting. www.michigan.gov/coronavirus/0,9753,7--406-98163_98173_102057---,00.html
2. COVID-19 outbreaks: department of health: state of Louisiana. https://ldh.la.gov/index.cfm/page/3997

3. Investigating and responding to COVID-19 cases in non-healthcare work settings. https://www.cdc.gov/coronavirus/2019-ncov/php/open-america/non-healthcare-work-settings.html
4. Places schema. https://docs.safegraph.com/docs/places-schema
5. What is a NAICS code and why do i need one? September 2018. https://www.naics.com/what-is-a-naics-code-why-do-i-need-one/
6. CDC - NIOSH program portfolio: occupational health equity: program description, December 2019. https://www.cdc.gov/niosh/programs/ohe/default.html
7. https://www.doh.wa.gov/Portals/1/Documents/1600/coronavirus/data-tables/StatewideCOVID-19OutbreakReport.pdf, December 2020
8. List of newspapers in the united states (2020). https://en.wikipedia.org/wiki/List_of_newspapers_in_the_United_States
9. List of radio stations in the united states (2020). https://en.wikipedia.org/wiki/Lists_of_radio_stations_in_the_United_States
10. List of television stations in the united states (2020). https://en.wikipedia.org/wiki/Lists_of_television_stations_in_the_United_States
11. Althubaiti, A.: Information bias in health research: definition, pitfalls, and adjustment methods. J. Multi. Healthc. **9**, 211 (2016)
12. Brown, T.B., et al.: Language models are few-shot learners. arXiv preprint arXiv:2005.14165 (2020)
13. Devlin, J., Chang, M.W., Lee, K., Toutanova, K.: BERT: pre-training of deep bidirectional transformers for language understanding. arXiv preprint arXiv:1810.04805 (2018)
14. Dyal, J.W.: COVID-19 among workers in meat and poultry processing facilities–19 states, April 2020. MMWR Morb. Mortal. Wkly Rep. **69** (2020)
15. Honnibal, M., Montani, I., Van Landeghem, S., Boyd, A.: spaCy: industrial-strength natural language processing in Python (2020). https://doi.org/10.5281/zenodo.1212303
16. Georgia Tech Research Institute: Clarity NLP (2018). https://github.com/ClarityNLP
17. Georgia Tech Research Institute: Clarity NLP documentation (2018). https://clarity-nlp.readthedocs.io/en/latest/
18. Kakimoto, K., Kamiya, H., Yamagishi, T., Matsui, T., Suzuki, M., Wakita, T.: Initial investigation of transmission of COVID-19 among crew members during quarantine of a cruise ship–Yokohama, Japan, February 2020
19. Kingma, D.P., Ba, J.: Adam: a method for stochastic optimization. arXiv preprint arXiv:1412.6980 (2014)
20. Krebs, C.: Guidance on the essential critical infrastructure workforce: ensuring community and national resilience in COVID-19 response. Cybersecurity and Infrastructure Security Agency (CISA) **5** (2020)
21. Liu, Y., et al.: RoBERTa: a robustly optimized BERT pretraining approach. arXiv preprint arXiv:1907.11692 (2019)
22. McMichael, T.M.: COVID-19 in a long-term care facility—king county, Washington, February 27–March 9, 2020. MMWR Morb. Mortal. Wkly Rep. **69**, 339 (2020)
23. Paszke, A., et al.: PyTorch: an imperative style, high-performance deep learning library. In: Advances in Neural Information Processing Systems, pp. 8026–8037 (2019)
24. Radford, A., Narasimhan, K., Salimans, T., Sutskever, I.: Improving language understanding by generative pre-training (2018)
25. Richardson, L.: Beautiful soup documentation, April 2007

26. Rodriguez-Lainz, A., et al.: Collection of data on race, ethnicity, language, and nativity by us public health surveillance and monitoring systems: gaps and opportunities. Public Health Rep. **133**(1), 45–54 (2018)
27. National Academies of Sciences, Engineering, and Medicine and others: A smarter national surveillance system for occupational safety and health in the 21st century. National Academies Press (2018)
28. Steinberg, J., et al.: COVID-19 outbreak among employees at a meat processing facility–South Dakota, March-April 2020. Morb. Mortal. Wkly Rep. **69**(31), 1015 (2020)
29. Sun, C., Qiu, X., Xu, Y., Huang, X.: How to fine-tune BERT for text classification? In: Sun, M., Huang, X., Ji, H., Liu, Z., Liu, Y. (eds.) CCL 2019. LNCS (LNAI), vol. 11856, pp. 194–206. Springer, Cham (2019). https://doi.org/10.1007/978-3-030-32381-3_16
30. Vaswani, A., et al.: Attention is all you need. In: Advances in Neural Information Processing Systems, pp. 5998–6008 (2017)
31. Wallace, M.: Public health response to COVID-19 cases in correctional and detention facilities—Louisiana, March–April 2020. MMWR Morb. Mortal. Wkly Rep. **69** (2020)
32. Waltenburg, M.A., et al.: Update: COVID-19 among workers in meat and poultry processing facilities–united states, April-May 2020. Morb. Mortal. Wkly Rep. **69**(27), 887 (2020)
33. Wilson, K., Brownstein, J.S.: Early detection of disease outbreaks using the internet. Cmaj **180**(8), 829–831 (2009)
34. Wolf, T., et al.: Transformers: state-of-the-art natural language processing. In: Proceedings of the 2020 Conference on Empirical Methods in Natural Language Processing: System Demonstrations, pp. 38–45 (2020)

Deep Kernel Learning for Mortality Prediction in the Face of Temporal Shift

Miguel Rios(✉) and Ameen Abu-Hanna

Department of Medical Informatics, Amsterdam UMC, University of Amsterdam, Amsterdam, The Netherlands
{m.a.riosgaona,a.abu-hanna}@amsterdamumc.nl

Abstract. Neural models, with their ability to provide novel representations, have shown promising results in prediction tasks in healthcare. However, patient demographics, medical technology, and quality of care change over time. This often leads to drop in the performance of neural models for prospective patients, especially in terms of their calibration. The deep kernel learning (DKL) framework may be robust to such changes as it combines neural models with Gaussian processes, which are aware of prediction uncertainty. Our hypothesis is that out-of-distribution test points will result in probabilities closer to the global mean and hence prevent overconfident predictions. This in turn, we hypothesise, will result in better calibration on prospective data.

This paper investigates DKL's behaviour when facing a temporal shift, which was naturally introduced when an information system that feeds a cohort database was changed. We compare DKL's performance to that of a neural baseline based on recurrent neural networks. We show that DKL indeed produced superior calibrated predictions. We also confirm that the DKL's predictions were indeed less sharp. In addition, DKL's discrimination ability was even improved: its AUC was 0.746 (±0.014 std), compared to 0.739 (±0.028 std) for the baseline. The paper demonstrated the importance of including uncertainty in neural computing, especially for their prospective use.

Keywords: Deep kernel learning · Temporal shift · Time series · Calibration · Gaussian process · Mortality prediction

1 Introduction

In the ICU, the prediction of in-hospital mortality is the task of providing probabilities for Intensive Care patients to die in the hospital, either in the ICU or after discharge to another ward. The (early) detection of such patients is relevant for clinical decision making. Mortality prediction models (MPMs) are often trained with large collections of electronic health records (EHR) that contain structured patient information such as demographics and physiological variables. MPMs based on deep learning are becoming prevalent in medical applications [18]. One reason for this is that NNs automatically derive *representations* for time series

A. Tucker et al. (Eds.): AIME 2021, LNAI 12721, pp. 199–208, 2021.
https://doi.org/10.1007/978-3-030-77211-6_22

data, which may provide predictive ability superior to that of standard regression models [9,21]. Specifically, neural models learn features from the input data by the incremental composition of simpler layers, resulting in complex representations for non-linear prediction models [1].

However, patient characteristics, medical technology, and clinical guidelines change over time, thus forming a challenge for the validity of MPMs for prospective patients, as these models were learned on historical data [14]. In particular, due to their flexibility, NNs have the ability to leverage on slight patterns in the data, but such patterns may not be stable over time and hence NN models may be sensitive to such temporal shifts causing a change (usually a drop) in performance [16]. For prediction models of a binary outcome, not only the discriminatory capability of the model may suffer, but especially its (mis)calibration. Calibration refers to the correspondence between the predicted probabilities and the true probabilities. The true probabilities are estimated on the test set by some measure of averaging the number of events for a set of patients. Performance drift has consequences for the task at hand, and the detrimental effects on benchmarking ICUs have been demonstrated [14]. One way to tackle this problem is to augment NNs with the notion of uncertainty: whenever the data distribution changes due to shift, the predictions should be more uncertain [13].

In contrast to NNs, The Gaussian process (GP) is a probabilistic framework for time series modelling that is able to increase model capacity with the amount of available data, and to produce uncertainty estimates. A GP characterises a distribution over possible functions that fit the input data. It is defined by a Gaussian function with a certain mean and, more importantly, a kernel function that captures the correlations between any two observations. The kernel encompasses the notion of uncertainty by performing a pairwise computation among all input data using some notion of similarity between the observations. The kernel can be viewed as providing a probability distribution over all possible models fitting the data.

The prediction models based on GPs successfully model time series data, incorporate confidence regions to predictions, and offer interpretability of the variables with the kernel function [20]. Moreover, the GP framework has been used to develop clinical prediction models [2,5]. In particular, Ghassemi et al. [8] use a multitask GP to model time series with physiological variables and clinical notes for mortality prediction. Directly relevant to our paper is the proposition in [23] to combine both NNs and GPs on a common framework of deep kernel learning (DKL). DKL leverages inductive biases from the NNs and from the non-parametric GPs.

In this paper, we investigate the behaviour of mortality prediction models based on DKL. In particular, we are interested in inspecting the robustness of the DKL model to a temporal shift. We also compare it to a strong NN-based baseline. Our hypothesis is that incorporation of uncertainty improves predictions. More specifically, we expect the DKL, when faced with uncertainty in the test set, to provide less extreme predictions that are closer to the global mean rather than providing overconfident predictions. In turn, the resultant prediction set would be less sharp than for the baseline model. Sharpness, which is also referred to refinement in weather forecast [15] measures the tendency of predictions to be close to 0 and 1. We therefore also compare the sharpness of both models but check that this does not come at the cost of discrimination. Finally, we also performed internal validation of the DKL model with all the population (i.e. no temporal shift) to understand whether the DKL's behaviour is specific to temporal validation.

Our main contribution in this paper is the introduction of a DKL model for in-hospital mortality prediction based on the first hours of an ICU stay in the context of temporal validation. The GP component in the DKL is shown to be robust to the shift in population and produces better calibrated predictions, without sacrificing discrimination. Our feature extraction is based on an open source benchmark [9] using the publicly available MIMIC-III [11] database. This facilitates the reproducibility of our results[1].

2 Deep Kernel Learning

The Gaussian Process [19] is a Bayesian non-parametric framework based on kernels for regression and classification. The set of functions that describes a given input data is possibly infinite and the GP assigns a probability to each one. For a dataset $\mathcal{X} = \{(\mathbf{x}_1, y_1), (\mathbf{x}_2, y_2), \ldots, (\mathbf{x}_n, y_n)\}$ where $\mathbf{x}$ is an input vector and y a corresponding output, we want to learn a function f that is inferred from a GP prior:

$$f(\mathbf{x}) \sim \mathrm{GP}(m(\mathbf{x}), k(\mathbf{x}, \mathbf{x}')) \tag{1a}$$

where $m(\mathbf{x})$ defines a mean (often set to 0) and $k(\mathbf{x}, \mathbf{x}')$ defines the covariance in the form of a kernel function. The kernel function models the covariance between all possible pairs $(\mathbf{x}, \mathbf{x}')$ and provides a measure of uncertainty. The choice of kernel determines properties of the function that we want to learn, usually this choice is based on background knowledge of the problem.

Wilson et al. [22] propose kernels based on deep learning architectures for GP regression. The DKL employs a GP with a base kernel as the last hidden layer of a NN. In other words, the DKL is a pipeline for learning complex NN features, and a distribution over functions that fit our input data. The base kernel $k(\mathbf{x}, \mathbf{x}' \mid \theta)$ with hyperparameters θ is parameterized by a non-linear function.

$$k(\mathbf{x}, \mathbf{x}' \mid \theta) \rightarrow k(g(\mathbf{x}, \omega), g(\mathbf{x}', \omega) \mid \theta, \omega), \tag{2a}$$

where $g(\mathbf{x}, \omega)$ is a NN architecture with weights ω. In addition, the DKL jointly learns the NN weights and kernel hyperparameters under the GP probabilistic framework. Learning a GP involves computing the kernel function, and finding the best kernel hyperparameters. The DKL optimises both the kernel hyperparameters and the NN weights, by maximising the marginal likelihood.

In Fig. 1, we define the architecture for extracting features $g(\mathbf{x}, \omega)$, $\mathbf{x}_i$ denotes the input vector in the ith element of $\mathcal{X}$.

[1] Code is available at: https://github.com/mriosb08/dkl-temporal-shift.git.

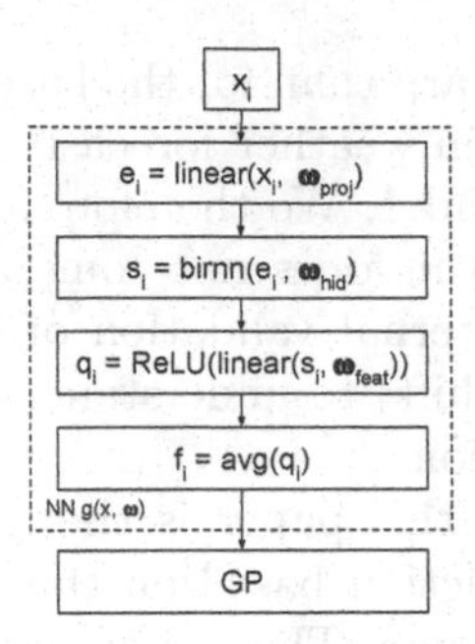

Fig. 1. NN architecture $g(\mathbf{x}, \omega)$ for extracting features $\mathbf{f}_i$ for the GP prediction layer.

The input features are first projected with an affine layer (linear(.)), then fed to a bidirectional LSTM (birnn(.)) [10] for encoding time series. Next the result goes through an affine layer with a non-linearity (ReLU(.)) that combines the hidden states of the bidirectional LSTM. Next the features $\mathbf{f}_i$ are summarised by averaging (avg(.)) and then fed to the GP layer.

3 Experiments

The Medical Information Mart for Intensive Care (MIMIC-III) database includes over 60,000 ICU stays across 40,000 critical care patients [11]. Harutyunyan et al. [9] propose a public benchmark and baselines based on MIMIC-III for modelling mortality, length of stay, physiologic decline, and phenotype classification. We use the benchmark for predicting in-hospital mortality based on the first 48 h of an ICU stay. The cohort excludes all ICU stays with unknown length-of-stay, patients under 18, multiple ICU stays, stays less than 48 h, and no observations during the first 48 h. The in-hospital mortality class is defined by comparing the date of death against hospital admissions and discharge times with a resulting mortality rate of 13.23%.

We use the benchmark to extract 17 input physiological variables (i.e. features), that are a subset of the Physionet challenge[2].

The benchmark [9] code processes the time series data with imputation of missing values with the previous hour, and normalisation from MIMIC-III. The normalisation of the features is performed by subtracting the mean and dividing by the standard deviation. The features also provide a binary mask for each variable indicating which time-step is imputed. All categorical variables are encoded using one-hot vectors (e.g. Glasgow coma scales). The final feature vector is formed by the concatenation of the clinical variables and the one-hot vectors with a total of 76 features. The clinical variables are shown in Table 1.

We use the architecture $g(.)$ as the baseline defined as: **BiLSTM**, which is based on a bidirectional LSTM for feature representation, and a linear prediction layer. We implement the **DKL** model with GPyTorch [7], with the following

[2] https://physionet.org/content/challenge-2012/1.0.0/.

Table 1. Clinical variables used in our experiments from MIMIC-III.

Variable
Capillary refill rate
Diastolic blood pressure
Fraction inspired oxygen
Glascow coma scale eye opening
Glascow coma scale motor response
Glascow coma scale total
Glascow coma scale verbal response
Glucose
Heart Rate
Height
Mean blood pressure
Oxygen saturation
Respiratory rate
Systolic blood pressure
Temperature
Weight
pH

components: the RBF kernel as the base kernel, feature extractor $g(.)$, and grid size 100 which is the number of inducing points used to approximate the GP for faster computations. The computation of the posterior distribution in the GP is expensive and several methods have been proposed to accelerate it by approximating it with a function over a set of inducing points [17,24]. In addition, we perform a simple ablation on the architecture by replacing the bidirectional LSTM with a LSTM for both models, baseline and DKL defined as: **LSTM**, and **DKL-LSTM**.

We use the following hyperparameters: optimiser Adam [12], learning rate 1e−3, epochs 30, encoder size 16, hidden size 16, batch size 100, dropout 0.3 applied after the linear layer. We perform model selection with the validation dataset based on AUC-ROC.

3.1 Temporal Shift: Strategy and Results

The MIMIC-III dataset includes data using the CareVue electronic patient record (EPR) system from 2001 to 2008. From 2008 to 2012 the MetaVision system was used instead. In the first experiment for inspecting temporal shift, we split the datataset into the CareVue period for training with 9,646 instances and 1,763 for validation (for tuning the hyper-parameters), and the data in the MetaVision period with 7,689 as the test set. We excluded patients present in both registries. This constitutes a temporal validation strategy in which the model is tested on data collected in the future relative to the data on which it has learned. This means that the model faces possible temporal shift due to changes that occur in time, and indeed possibly also due to the change of the EPR system that collects

Table 2. In-hospital mortality results with a temporal population shift over 10 runs ± one standard deviation. The training and validation datasets are on CareVue (2001–2008), and the test on MetaVision (2008–2012).

Model	Validation		Test	
	AUC-ROC	AUC-PR	AUC-ROC	AUC-PR
LSTM	0.838 ± 0.003	0.532 ± 0.006	0.693 ± 0.027	0.317 ± 0.037
BiLSTM	0.857 ± 0.002	0.572 ± 0.007	0.739 ± 0.028	$\mathbf{0.386} \pm 0.018$
DKL-LSTM	0.854 ± 0.002	0.562 ± 0.010	0.701 ± 0.033	0.327 ± 0.026
DKL	0.856 ± 0.002	0.569 ± 0.004	$\mathbf{0.746} \pm 0.014$	0.373 ± 0.018

the data that could have affected the workflow and/or the way of registration. Performance was measured in terms of: Discrimination, by the AUC-ROC; the balance between the positive predicted value and sensitivity, by the AUC-PR; the accuracy of predictions by the Brier score; and calibration by calibration graphs and the Cox recalibration approach [3] in which the observed outcome in the test set is regressed using logistic regression on the log odds of the predictions. If the predictions were perfectly calibrated then the linear predictor of this model would have an intercept of 0 and a slope of 1. We test deviations from these ideal value of 0 and 1, respectively. To test our hypothesis whether the DKL approach provides more conservative predictions due to uncertainty for areas in the test set, we measure the (un)sharpness of the predictions. We use the following measure of unsharpness: $\frac{\sum_1^N p_i(1-p_i)}{N}$ where p_i is the ith prediction and N is number of observations.

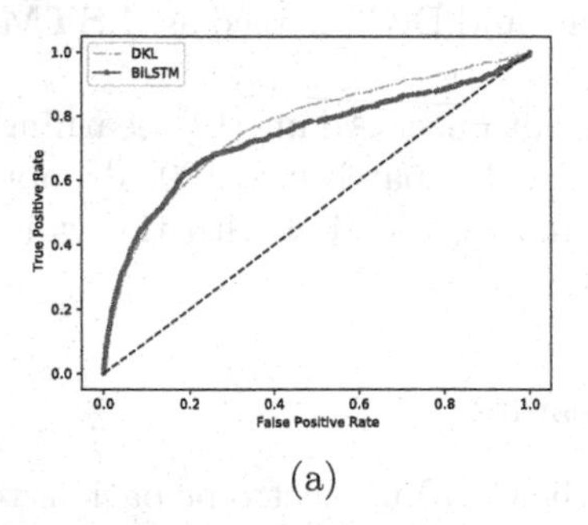

(a)

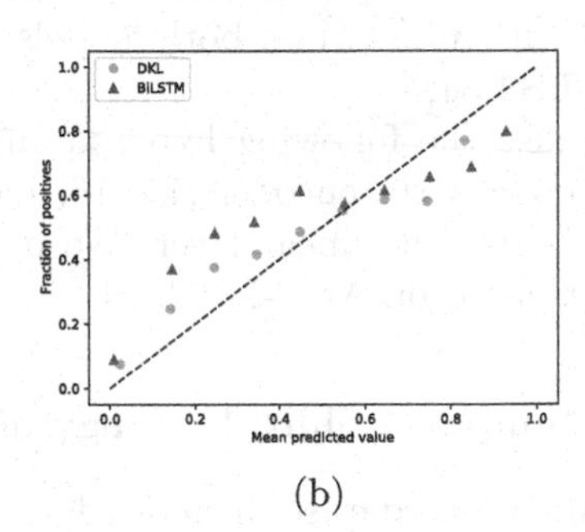

(b)

Fig. 2. Receiver operating characteristic curve (a) and calibration curve (b) for in-hospital mortality with temporal shift in population.

Table 2 shows the AUC-ROC and AUC-PR results for in-hospital mortality with a temporal shift in population. The baseline outperforms the DKL model on the validation (tuning) dataset for both metrics. On the test dataset, however, the DKL shows competitive performance on the AUC-ROC. We use the best

Table 3. In-hospital mortality results over 10 runs ± one standard deviation. Validation and test dataset from all sources (2001–2012).

Model	Validation		Test	
	AUC-ROC	AUC-PR	AUC-ROC	AUC-PR
LSTM	0.843 ± 0.003	0.513 ± 0.006	0.840 ± 0.005	0.434 ± 0.008
BiLSTM	0.858 ± 0.004	0.549 ± 0.010	**0.851** ± 0.004	**0.478** ± 0.016
DKL-LSTM	0.838 ± 0.002	0.485 ± 0.014	0.841 ± 0.003	0.425 ± 0.013
DKL	0.854 ± 0.004	0.536 ± 0.010	0.847 ± 0.005	0.454 ± 0.018

run from the validation based on the AUC-ROC for reporting the ROC and calibration curves. In addition, we select the best performing models from Table 2 based on AUC-ROC, namely BiLSMT and DKL, for comparing the calibration and ROC curves. The LSTM models consistently underperform compared to the bidirectional ones. Figure 2 shows the ROC and calibration curves for in-hospital mortality with a temporal shift. The Brier score for the DKL is 0.101 which is better that the 0.109 of the BiLSTM. The DKL outperforms the baseline and it shows better calibration.

In the Cox re-calibration on both models the BiLSTM had a calibration intercept of 1.965 (1.88, 2.049), and slope of 0.538 (0.5, 0.577) compared to the DKL's of 0.6615 (0.586, 0.734), 0.712 (0.652, 0.772). Although both models deviated significantly from the ideal values (of 0 and 1), the DKL showed significantly much better calibration. The DKL's predictions were also much less sharp: unsharpness of 0.061 for DKL versus 0.025 for BiLSTM.

3.2 Experiment 2: Internal Validation

We report the results with all the sources (2001–2012) for in-hospital mortality, with no shift in population. The training, validation and test datasets consisted of respectively 14, 681, 3, 222, and 3, 236 instances.

Table 3 shows the AUC-ROC and AUC-PR results for in-hospital mortality with all sources (2002–2012). The baseline outperforms the DKL model on the test dataset for both metrics the AUC-ROC, and AUC-PR. Figure 3 shows the ROC and calibration curves for in-hospital mortality with all sources. Both of our models perform similarly on the ROC curve. The Brier score for the DKL is 0.082 slightly better than the 0.084 of the BiLSTM.

In the Cox re-calibration the BiLSTM's calibration intercept was −0.358 (−0.49, −0.229), and slope 0.802 (0.726, 0.88); compared to the DKL's −0.066 (−0.185, 0.05), and 1.177 (1.062, 1.298). Unlike the BiLSTM the DKL showed no significant deviations from the ideal values of 0 and 1. The DKL was slightly more unsharp: 0.089 versus 0.081 for the BiLSTM.

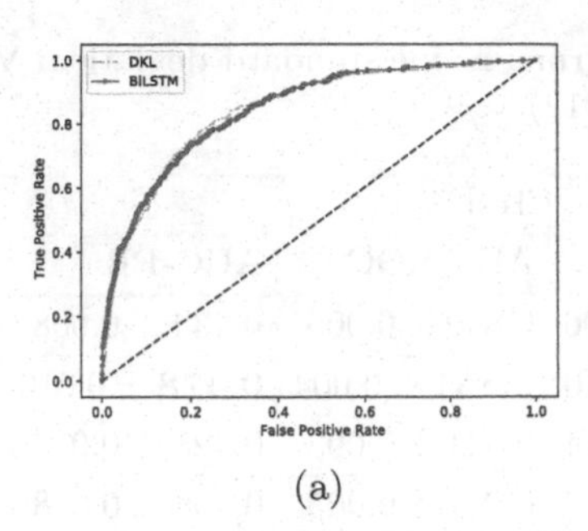

(a)

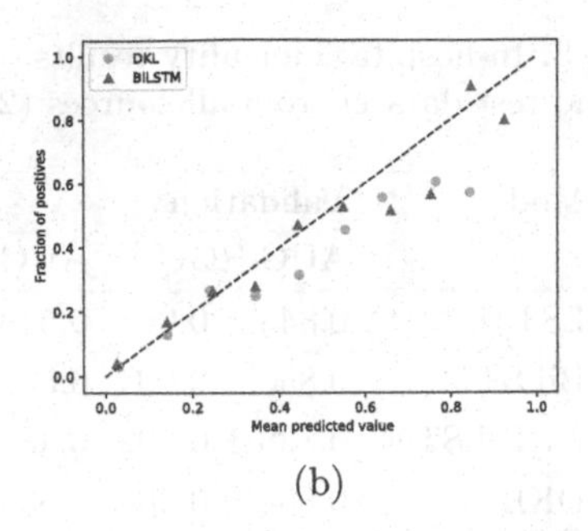

(b)

Fig. 3. Receiver operating characteristic curve (a) and calibration curve (b) for in-hospital mortality with all sources.

4 Related Work

Dürichen et al. [5] propose a multi-task GP that jointly models physiological variables for clinical time series. Cheng et al. [2] develop a real-time clinical prediction model based on a GP model. Aside from producing confidence regions in the predictions, the GP also scales to large patient databases, and produces interpretable relations across (clinical) variables. The interprtability is produced by inspecting the correlation across variables in the kernel function. Futoma et al. [6] propose a sepsis prediction model based on a pipeline with a GP that produces inputs for a NN classifier. The model takes into account uncertainty estimates and outperforms strong sepsis prediction baselines. On the other hand, our DKL model uses RNNs to model the time series physiological variables and feed the resulting features into the GP for prediction. Our work, however, is the first to investigate DKL in the context of temporal shift.

5 Conclusions and Future Work

We investigated the DKL framework for the task of in-hospital mortality prediction under a temporal shift in population. The DKL shows competitive performance compared to a strong NN baseline, as well as a better calibration. However, when the test dataset is in the same distribution as the training both models show similar results. The GP component does not degrade the overall performance, and in addition, it provides extra guarantees such as uncertainty estimates. By contrasting the two experiments and inspecting the sharpness of the predictions we can ascribe the improved performance on the test set to the robustness of the GP when facing uncertainty.

For future work, we will analyse different base kernels, evaluate the uncertainty estimate of the DKL, and use the framework described in [4] for better understanding of discrepancies in performance over time.

References

1. Bengio, Y.: Learning deep architectures for AI. Found. Trends Mach. Learn. **2**(1), 1–127 (2009)
2. Cheng, L.F., Darnell, G., Chivers, C., Draugelis, M.E., Li, K., Engelhardt, B.E.: Sparse multi-output Gaussian processes for medical time series prediction. BMC Med. Inform. Decis. Making **20**(152) (2020)
3. Cox, D.R.: Two further applications of a model for binary regression. Biometrika **45**, 562–565 (1958)
4. Debray, T.P.A., Vergouwe, Y., Koffijberg, H., Nieboer, D., Steyerberg, E.W., Moons, K.G.M.: A new framework to enhance the interpretation of external validation studies of clinical prediction models. J. Clin. Epidemiol **68**(3), 279–89 (2015)
5. Dürichen, R., Pimentel, M., Clifton, L., Schweikard, A., Clifton, D.: Multi-task Gaussian processes for multivariate physiological time-series analysis. IEEE Trans. BioMed. Eng. **62**, 314–322 (2014)
6. Futoma, J., Hariharan, S., Heller, K.: Learning to detect sepsis with a multitask Gaussian process RNN classifier. In: International Conference on Machine Learning. JMLR.org (2017)
7. Gardner, J.R., Pleiss, G., Bindel, D., Weinberger, K.Q., Wilson, A.G.: GPyTorch: blackbox matrix-matrix Gaussian process inference with GPU acceleration. CoRR abs/1809.11165 (2018)
8. Ghassemi, M., et al.: A multivariate timeseries modeling approach to severity of illness assessment and forecasting in ICU with sparse, heterogeneous clinical data. In: Proceedings of the Twenty-Ninth AAAI Conference on Artificial Intelligence, AAAI 2015, pp. 446–453. AAAI Press (2015)
9. Harutyunyan, H., Khachatrian, H., Kale, D.C., Ver Steeg, G., Galstyan, A.: Multitask learning and benchmarking with clinical time series data. Sci. Data **6**(1), 1–18 (2019)
10. Hochreiter, S., Schmidhuber, J.: Long short-term memory. Neural Comput. **9**(8), 1735–1780 (1997)
11. Johnson, A.E., et al.: MIMIC-III, a freely accessible critical care database. Sci. Data **3**, 160035 (2016)
12. Kingma, D.P., Ba, J.: Adam: a method for stochastic optimization (2014). Cite arxiv:1412.6980Comment: Published as a conference paper at the 3rd International Conference for Learning Representations, San Diego (2015)
13. Mackay, D.J.C.: Bayesian methods for adaptive models. Ph.D. thesis, USA (1992). uMI Order No. GAX92-32200
14. Minne, L., Eslami, S., de Keizer, N., de Jonge, E., de Rooij, S.E., Abu-Hanna, A.: Effect of changes over time in the performance of a customized SAPS-II model on the quality of care assessment. Intensive Care Med. **38**, 40–46 (2012)
15. Murphy, A., Winkler, R.: A general framework for forecast verification. Mon. Weather Rev. **115**, 1330–1338 (1987)
16. Nestor, B., et al.: Feature robustness in non-stationary health records: caveats to deployable model performance in common clinical machine learning tasks. In: Doshi-Velez, F., et al. (eds.) Proceedings of the 4th Machine Learning for Healthcare Conference. Proceedings of Machine Learning Research, Ann Arbor, Michigan, vol. 106, pp. 381–405. PMLR (2019)
17. Quiñonero Candela, J., Ramussen, C.E., Williams, C.K.I.: Approximation methods for Gaussian process regression. Technical report MSR-TR-2007-124 (2007)

18. Rajkomar, A., et al.: Scalable and accurate deep learning for electronic health records. CoRR abs/1801.07860 (2018)
19. Rasmussen, C.E., Williams, C.K.I.: Gaussian Processes for Machine Learning (Adaptive Computation and Machine Learning). The MIT Press, Cambridge (2005)
20. Roberts, S., Osborne, M., Ebden, M., Reece, S., Gibson, N., Aigrain, S.: Gaussian processes for timeseries modelling. Philos. Trans. R. Soc. (2012)
21. Shickel, B., Loftus, T.J., Ozrazgat-Baslanti, T., Ebadi, A., Bihorac, A., Rashidi, P.: DeepSOFA: a real-time continuous acuity score framework using deep learning. CoRR abs/1802.10238 (2018)
22. Wilson, A.G., Hu, Z., Salakhutdinov, R.R., Xing, E.P.: Stochastic variational deep kernel learning. In: Lee, D.D., Sugiyama, M., Luxburg, U.V., Guyon, I., Garnett, R. (eds.) Advances in Neural Information Processing Systems 29, pp. 2586–2594. Curran Associates, Inc. (2016)
23. Wilson, A.G., Hu, Z., Salakhutdinov, R., Xing, E.P.: Deep kernel learning. In: Gretton, A., Robert, C.C. (eds.) Proceedings of the 19th International Conference on Artificial Intelligence and Statistics. Proceedings of Machine Learning Research, Cadiz, Spain, vol. 51, pp. 370–378. PMLR (2016)
24. Wilson, A.G., Nickisch, H.: Kernel interpolation for scalable structured Gaussian processes (KISS-GP). In: International Conference on Machine Learning. JMLR.org (2015)

Model Evaluation Approaches for Human Activity Recognition from Time-Series Data

Lee B. Hinkle(✉) and Vangelis Metsis

Texas State University, San Marcos, TX 78666, USA
{leebhinkle,vmetsis}@txstate.edu

Abstract. There are many evaluation metrics and methods that can be used to quantify and predict a model's future performance on previously unknown data. In the area of Human Activity Recognition (HAR), the methodology used to determine the training, validation, and test data can have a significant impact on the reported accuracy. HAR data sets typically contain few test subjects with the data from each subject separated into fixed-length segments. Due to the potential leakage of subject-specific information into the training set, cross-validation techniques can yield erroneously high classification accuracy. In this work (Source code available at: https://github.com/imics-lab/model_evaluation_for_HAR.), we examine how variations in evaluation methods impact the reported classification accuracy of a 1D-CNN using two popular HAR data sets.

Keywords: Model evaluation · Time-series data · Deep learning · Human activity recognition · Data resampling · Cross-validation

1 Introduction

With the advent of inexpensive wearable sensors in recent years, Human Activity Recognition (HAR) has been a hot topic of research both for medical applications and in human-computer interaction in general. In HAR, the methodology used for model evaluation differs from other areas such as image recognition due to the sequential nature of the data sets. HAR data sets typically consist of accelerometer and gyroscopic data recorded using a smartphone or wrist-worn device. Movement patterns specific to given activities such as running, walking, and sitting are identified using classic machine learning or newer deep learning approaches. HAR data sets typically differ from image and natural language data sets because the number of subjects is usually quite small, typical ranges are from 5 to 50 [13], with each subject contributing multiple samples while performing a range of activities. Traditional cross-fold and train/test split techniques can result in subject data from the test group being included in the training set.

The goal of trained models is generalized performance which means the performance on independent test data [7]. In the case of HAR the ability of a

A. Tucker et al. (Eds.): AIME 2021, LNAI 12721, pp. 209–215, 2021.
https://doi.org/10.1007/978-3-030-77211-6_23

model to correctly classify activities for an unknown subject. The primary issue seen in many accompanying analyses is that samples from a given subject may be present in both the train and test groups. This work examines the impact of subject assignment on two data sets. The remainder of this section briefly describes three data sets, their evaluation method, and the reported accuracy to illustrate the multiple approaches found in the literature. Section 2 describes the two data sets and processing used in this evaluation.

An example of a popular data set and evaluation with subjects preallocated into train and test groups is the UCI-HAR data set [1] which contains acceleration data captured on a waist-worn smartphone. Subjects were randomly assigned: 21 in the training set and 9 in the test set. The accompanying analysis reports an accuracy of 96% for six activities. Another example of a model evaluation with preallocated subjects is [5] which contains Android-based Smartphone data from 100 subjects. The reported accuracy *without resampling* is 93.8% for eight different activities. The authors state "the signals of the training set and test set are collected by different volunteers." An example of hold-one-subject-out with individual results is [6] which uses multimodal motion data from the mHealth data set [2] and reports an average accuracy for 12 activities of 91.94%.

2 Materials and Methods

This section provides a brief overview of the two data sets used, the configuration of the 1D CNN, and the overall methodology.

The first data set used in this work is the MobiAct data set [12] which contains smartphone acquired raw accelerometer, gyroscope, and magnetometer data. 50 subjects were recorded performing nine types of activities of daily living (ADLs) and four types of falls. The accompanying analysis reports a best overall accuracy of 99.88% using 10-fold cross-validation. The authors state "we expect [the accuracy] to decrease when using leave-one-out cross-validation, which is a more realistic scenario." For this work the timestamp 'nanoseconds' and accelerometer data (accel_x/y/z) for the six Activities of Daily Living (ADL) were imported. The four types of falls, 'sit chair', 'car step in', and 'car step out' activities are not used as these are more events than activities. Gyro and magnetometer data are also not used for simplification. One second was discarded from the start/end of each record and the remaining data were segmented into 3-s windows. Prior works, including UCI-HAR [1] have used a 2.56-s window based on the mechanics and timing of human gait. This window length will yield multiple steps in each segment [3]. The six activity labels in y were one-hot-encoded.

The impact of the variable sampling rate and benefits of resampling were investigated using the MobiAct data. Sample timing is very consistent when using specialized equipment such as the BioRadio[1] or the Empatica E4 wristband[2]. However, when using a general-purpose device such as a smartphone preemption by other tasks results in a variation of timing between samples.

[1] https://www.glneurotech.com/products/bioradio/.
[2] https://www.empatica.com/research/e4/.

Figure 1a shows the delta time between data samples for a 30-s MobiAct walking segment. The Python's Pandas mean resampling method was used to resample and downsample the data.

The second data set used is the Smartphone Human Activity Recognition data set from the University of Milan Bicocca (UniMiB SHAR) [8] which contains both fall and ADL data from 30 subjects that have been preprocessed into 3-s samples. The subjects are not preallocated into train/test and the accompanying analysis reports results for both component and total acceleration using 5-fold cross-validation and hold-one-subject-out validation. The highest performing RNN classifier achieves an accuracy of 88.41% using component acceleration and 5-fold cross-validation. Each classifier showed a decrease in accuracy in the Leave-One-Subject-Out validation. The accuracy drops to 73.17% using Leave-One-Subject-Out and 72.67% using total acceleration. The authors state that human subjects perform tasks in unique ways. The UniMiB SHAR acceleration data were transformed into a 153 × 3 array and the total acceleration was calculated. The nine ADL class labels were one-hot-encoded.

A fixed 1D-CNN Keras [11] model shown to have good performance on time-sliced accelerometer data [4] was used for all experiments for consistency. Minimal tuning was performed, the primary change was increasing the convolution kernels to span one second of activity time. For a brief description of the layer functions with respect to time-series see [9]. The topology of the 1D-CNN is shown in Table 1.

Table 1. Keras sequential model 1D-CNN layers

Type	Input	Conv1D	Conv1D	Dropout	Max Pl	Flatten	Dense	Dense
Params	[60 × 1]	#f = 50 size = 1 s	#f = 50 size = 1 s	rate = 0.5	size = 2,		act = relu	act = softmax

All subject allocation experiments use total acceleration; MobiAct was resampled 20 Hz, UniMiB SHAR remains the 50 Hz. The next section describes how subjects were allocated to the training, validation, and test groups.

Allocation Using Stratification: While is easy to implement using the Scikit-learn [10] train_test_split method with stratification enabled a single subject's samples are likely to be present in each of the groups.

Allocation of Subjects by Attributes: The UCI-HAR data set preallocates subjects but the UniMIB SHAR and MobiACT data sets do not. To generate a baseline each subject was allocated to the train, validate, *or* test group in a 60%/20%/20% ratio. Assuming that height would affect the mechanics of motion more than weight for the ADLs, subjects were sorted by height and manually allocated. Swaps were made to preserve the male to female ratio and a mix of age and weight[3]. The subject allocation is shown in Table 2.

[3] Several MobiAct subjects did not complete all ADLs were dropped resulting in a non-contiguous subject list. E.g. there is no subject number 14.

Table 2. Subject numbers: attribute based assignment

Data Set	Training	Validation	Test
MobiAct	[2,4,5,9,10,16,18,20,23-28, 32,34-36,38,42,45-54,57]	[3,6,8,11,12,22, 37,40,43,56]	[7,19,21,25,29, 33,39,41,44,55]
UniMiB SHAR	[4–8,10-12,14,15,19-22,24]	[1,9,16,23,25,28]	[2,3,13,17,18,30]

Subject Aware Cross-Validation: Each subject was placed into the test group with the remaining subjects used for training and validation for hold-one-subject-out. The process was repeated with two, three, five, and ten subjects held out. To establish a range of possible results, the best and worst classified hold-one-out subjects were placed into min and max test groups.

3 Results and Discussion

Figure 1a shows the variation in sampling time for a walking sample. Figure 1b shows 5 Hz sampling results in reduced accuracy 10 Hz and above were largely the same. Reducing the sampling frequency significantly reduced the GPU-based[4] training time. This was even more pronounced when using CPU-based training where the 20 Hz data required just 4.5% of the training time required for 100 Hz data. For MobiAct the accuracy increased from 95.3% to 97.5% when using total acceleration and the attribute-based subject allocation with negligible impact on GPU training time. Table 3 shows that the accuracy when using stratification is extremely high at 99.3% (average of 10 runs, 200 epochs). Using the same model but with subjects allocated based by attribute, the accuracy drops to 96.9% for an error rate of 3.1% versus the stratified error of only 0.7%. The allocation of individual subject's data into both the train and test groups results in erroneously increased accuracy when using stratified split. The UniMiB SHAR data results show the same trend.

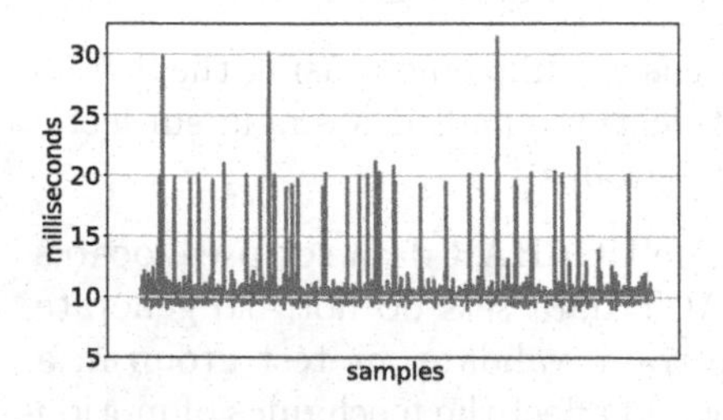

(a) Δ-time between samples

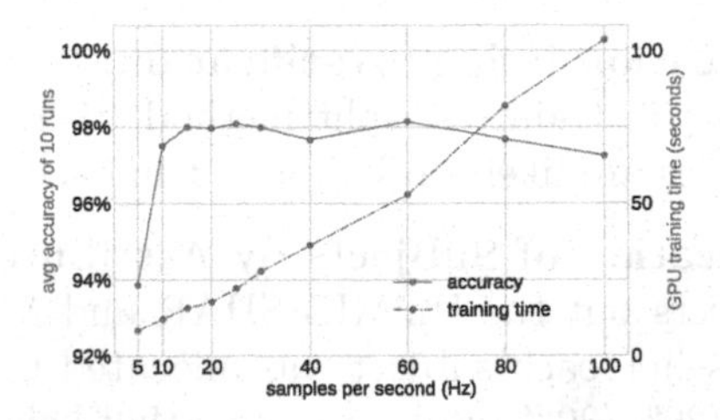

(b) Accuracy & training time vs. sampling rate

Fig. 1. Smartphone data sample fluctuation and impact of resample.

[4] GPU model Tesla P100-PCIE-16 GB at https://www.colab.research.google.com.

Figure 2 is a box plot of five runs for each subject and shows the large variation in accuracy among individual subjects. The overall by-subject cross-validation results are shown in Table 4.

Table 3. Stratified versus attribute-based subject split accuracy

Data set: Train/Validate/Test Split Method	Avg	Error	Delta
MobiAct: Stratified (incorrect)	99.3%	0.7%	–
MobiAct: Manual by Subject Attributes	96.9%	3.1%	2.4%
UniMiB SHAR: Stratified (incorrect)	93.9%	6.1%	–
UniMiB SHAR: Manual by Subject Attribute	92.3%	7.7%	1.6%

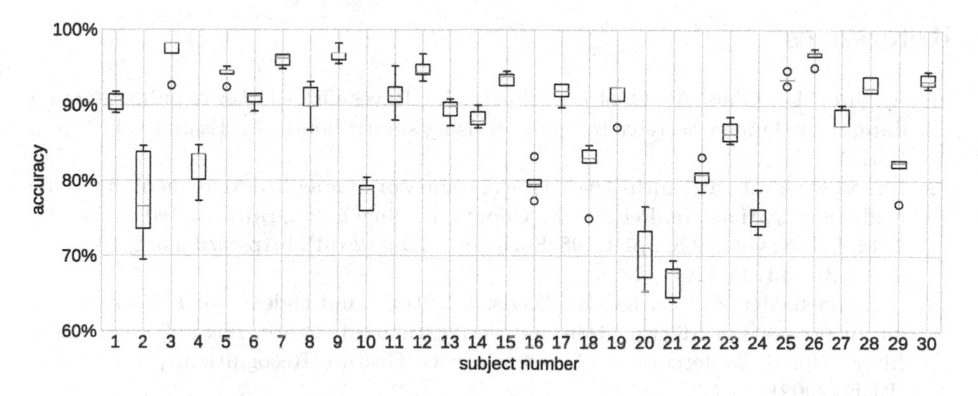

Fig. 2. Accuracy of each UniMiB Subject Tested Individually

Table 4. Accuracy based on X-fold and best/worst subjects. The delta between best/worst vs average accuracy narrows as more subjects are placed in a fold.

	MobiAct				UniMiB SHAR		
#subj/fold	All	Min	Max	#subj/fold	All	Min	Max
1	95%	78%	100%	1	87%	67%	97%
2	95%	84%	98%	2	86%	72%	94%
3	95%	89%	98%	3 (10-fold-CV)	85%	70%	93%
5 (10-fold-CV)	95%	92%	98%	6 (5-fold-CV)	86%	74%	91%
10 (5-fold-CV)	95%	92%	97%	–	–	–	–

4 Conclusion

In this work, we have shown that re-sampling smartphone acceleration data does not improve accuracy but downsampling can substantially reduce training time. This is important because consistent with prior work, stratified random allocation where samples from a single subject are present in both the training and testing groups generated higher accuracy than can be expected given an unknown subject. Hold-one-subject out is recommended but requires a train/test pass for each subject. We have shown that individual subject accuracies can vary greatly in a hold-one-out scenario and as the number of subjects in each fold increases the delta between possible min and max folds is reduced. Group-based 5-fold cross-validation can be used and closely matches the accuracy reported by averaging hold-one-subject-out.

References

1. Anguita, D., Ghio, A., Oneto, L., Parra, X., Reyes-Ortiz, J.L.: A public domain dataset for human activity recognition using smartphones. In: Esann, vol. 3, p. 3 (2013)
2. Banos, O., et al.: mHealthDroid: a novel framework for agile development of mobile health applications. In: Pecchia, L., Chen, L.L., Nugent, C., Bravo, J. (eds.) IWAAL 2014. LNCS, vol. 8868, pp. 91–98. Springer, Cham (2014). https://doi.org/10.1007/978-3-319-13105-4_14
3. BenAbdelkader, C., Cutler, R., Davis, L.: Stride and cadence as a biometric in automatic person identification and verification. In: Proceedings of Fifth IEEE International Conference on Automatic Face Gesture Recognition, pp. 372–377. IEEE (2002)
4. Brownlee, J.: 1D convolutional neural network models for human activity recognition, July 2020. https://machinelearningmastery.com/cnn-models-for-human-activity-recognition-time-series-classification/
5. Chen, Y., Xue, Y.: A deep learning approach to human activity recognition based on single accelerometer. In: 2015 IEEE International Conference on Systems, Man, and Cybernetics, pp. 1488–1492. IEEE (2015)
6. Ha, S., Choi, S.: Convolutional neural networks for human activity recognition using multiple accelerometer and gyroscope sensors. In: 2016 International Joint Conference on Neural Networks (IJCNN), pp. 381–388. IEEE (2016)
7. Hastie, T., Tibshirani, R., Friedman, J.: The Elements of Statistical Learning. SSS. Springer, New York (2009). https://doi.org/10.1007/978-0-387-84858-7. http://www-stat.stanford.edu/ tibs/ElemStatLearn/
8. Micucci, D., Mobilio, M., Napoletano, P.: UniMiB SHAR: a dataset for human activity recognition using acceleration data from smartphones. Appl. Sci. **7**(10), 1101 (2017)
9. Nils: Introduction to 1d convolutional neural networks in keras for time sequences. https://blog.goodaudience.com/introduction-to-1d-convolutional-neural-networks-in-keras-for-time-sequences-3a7ff801a2cf
10. Pedregosa, F., et al.: Scikit-learn: machine learning in Python. J. Mach. Learn. Res. **12**, 2825–2830 (2011)
11. François, C.: Keras. GitHub repository (2015). https://github.com/fchollet/keras

12. Vavoulas, G., Chatzaki, C., Malliotakis, T., Pediaditis, M., Tsiknakis, M.: The mobiact dataset: recognition of activities of daily living using smartphones. In: ICT4AgeingWell, pp. 143–151 (2016)
13. Wang, J., Chen, Y., Hao, S., Peng, X., Hu, L.: Deep learning for sensor-based activity recognition: a survey. Pattern Recogn. Lett. **119**, 3–11 (2019)

Unsupervised Learning

Unsupervised Learning to Subphenotype Heart Failure Patients from Electronic Health Records

Melanie Hackl[1], Suparno Datta[1,2](✉), Riccardo Miotto[2], and Erwin Bottinger[1,2]

[1] Digital Health Center, Hasso Plattner Institute, University of Potsdam, Potsdam, Germany

[2] Hasso Plattner Institute for Digital Health at Mount Sinai, Icahn School of Medicine at Mount Sinai, New York, USA

suparno.datta@hpi.de

Abstract. Heart failure (HF) is a deadly disease and its prevalence is slowly increasing. The sub-types of HF are currently mostly determined by the so-called ejection fraction (EF). In this work, we try to find novel subgroups of heart failure following a complete data-driven approach of clustering patients based on their electronic health records (EHRs). Using a validated phenotyping algorithm we were able to identify 14,334 adult patients with heart failure in our database. We derived embeddings of patients using two different strategies, one processing aggregated clinical features using principal component analysis (PCA) and uniform manifold approximation and projection (UMAP), and one where we learn embeddings from the sequence of medical events using a long short-term memory (LSTM) autoencoder. Then we evaluated different clustering strategies like k-means and agglomerative hierarchical to derive the most informative subtypes. The results were compared based on different metrics such as silhouette coefficient and so on and also based on comparing outcomes such as hospitalization, EF etc. between the clusters. In the most promising result, we were able to identify 3 subclusters using the aggregated data approach in combination with UMAP as dimension reduction method and k-means as cluster method. Patients in cluster 1 had the lowest number of hospital days and comorbidities, while patients in cluster 3 had a significantly higher number of hospital days together with a higher prevalence of comorbidities such as chronic kidney disease and atrial fibrillation. Patients in cluster 2 had a high prevalence of drug allergies in their medical history.

Keywords: Unsupervised learning · Electronic health records · Heart failure

M. Hackl and S. Datta—Contributed equally to this paper.

A. Tucker et al. (Eds.): AIME 2021, LNAI 12721, pp. 219–228, 2021.
https://doi.org/10.1007/978-3-030-77211-6_24

1 Introduction

Cardiovascular diseases are the number one cause of death from a global perspective [15]. Heart Failure (HF) is one of the major cardiovascular diseases. HF describes the medical condition, where the heart has no longer the original functional range of pumping blood [3]. In the US 6.2 million people are currently suffering from HF and 379,800 people died in 2018 because of HF [6]. HF is a long term illness that requires continuous treatment. The main treatments are lifestyle changes, such as more exercise or a healthier diet, drugs such as angiotensin-converting enzyme (ACE) inhibitors or beta blockers, and surgery such as bypass [13]. It is especially important here to identify patients with comorbidities, as treatments for these conditions could have a worsening effect on HF or, for example, could contribute to more hospital stays. To reduce the impact of HF it is crucial to get a better understanding of HF disease subtypes and with that provide more personalized treatment plans. Subgrouping or subtyping generally describes the process of clustering of a large group of data points, in this case patients, into different subgroups. The groups (or clusters) should then represent patients with similar features and can lead to new insights within a disease population. Patient subgrouping is an important tool to enable personalized medicine. In this context, electronic health records (EHR) can help to identify subgroups within a specific disease cohort and better understand comorbidities, demographics, and treatment patterns observed for that cohort [8].

Most of the works related to HF and machine learning are approaches to classify HF patients into the already known subgroups [2]. One study tried to subphenotype HF patients based on a clinical cohort study data [1]. To the best of our knowledge, no other studies have retrospectively and in a complete data-driven manner tried to subphenotype HF patients based on EHR data. In this work, we try to subphenotype HF patients based on data from a large EHR system. We adopt 2 different approaches to learn patient embeddings from the EHR data, one where we aggregate the different medical features over time, and a second one where we embed the sequences of medical events for each patient using a long short-term memory (LSTM) autoencoder. We experimented with broad range of dimensionality reduction and clustering techniques and present the best results we observed.

This work is detailed as follows: Sect. 2 describes the dataset and methods we used to subgroup HF patients. In Sect. 3 we convey our findings and results, followed by a discussion in Sect. 4 and Sect. 5 describing the implications of this work.

2 Methods

2.1 Data Source and Phenotyping Algorithm

The data used for this work was obtained from the Mount Sinai Health System (MSHS). The MSHS is an integrated health care system in the greater New York area. All the treatment data of the different facilities are combined in the so called Mount Sinai data warehouse (MSDW). MSDW contains clinical EHR data from over 3.4 million patients [7]. Since the Mount Sinai health system is an alliance of hospitals and outpatient facilities, the entire patient journey can be captured here. Therefore the data from MSDW usually provides a good representation of the patients current health status and history.

The main purpose of a phenotyping algorithms is to correctly identify patients exposed to a particular disease using different diagnostic codes, laboratory values and clinical free text notes a physician makes during an examination [9]. For creating our cohort of interest the HF phenotyping algorithm by Suzette J. Bielinski was used [19], was used. This algorithm is able to detect HF patients from EHR data and is also able to distinguish between the types of ejection fraction.

2.2 Features and Patient Embedding

The EHR data out of the MSDW provides different dimensions about a patients health journey like diagnosis medications, procedures, laboratory values and vital signs. All these dimensions were used to create a holistic view of the patient's health status. In order to avoid considering too many sparse features, individual concepts were selected from the dimensions based on their frequency of occurrence.

We chose different thresholds for extracting the features of each dimension. The threshold describes the fraction of patients that must have a certain code or dimension in their EHR, for that feature to be considered. For the dimensions medication and diagnosis we chose a threshold value of 20%, i.e. at least 20% of the patients must have that particular feature in their records, for it to be considered as a valid feature. For the dimensions procedures, vital signs and laboratory values we chose a threshold value of 80%, since they are usually more often present than the features of the other dimensions. The number of different type of features we considered is shown in the Fig. 1 inside the rectangle titled *Feature selection*. In total, our cohort consists of 14,334 patients and 284 features. The average age of onset of HF is 69 years. 7,828 (54.6%) are male and 6,506 (45.4%) are female. Next up we describe the two strategies we used to learn patient embeddings from the EHR data:

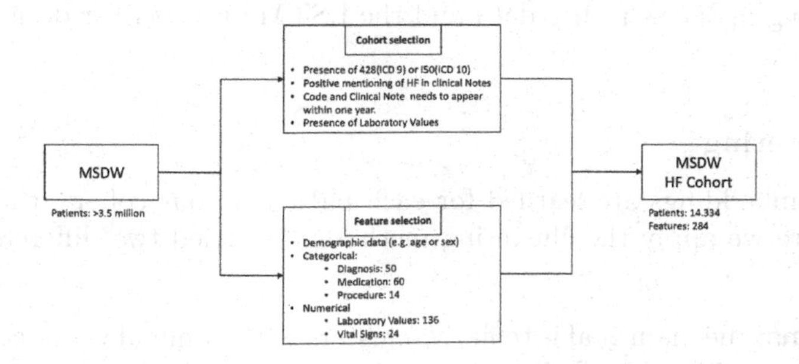

Fig. 1. Cohort and feature selection

Embeddings from Aggregated Data. In the first strategy we get the aggregated medical concepts from the different dimensions (excluding demographics) mentioned above. Since the dimensions diagnosis, procedures, and medication are categorical, we just notice if they were present or not, while for the numerical dimensions such as vital signs and laboratory values we took the minimum, maximum, and mean value a patient has during the observation period. Then the data was corrected for biologically infeasible outliers, scaled (using a min-max scaler) and finally the missing values were imputed using a simple mean imputation. We experimented with different imputation strategies such as K-Nearest Neighbour(K-NN), Multiple Imputation by Chained Equations(MICE) etc., but the results didn't seem to be much effected by the imputation strategy. For the aggregated data we used two strategies to reduce the dimension of the patient vector namely: Principal Component Analysis (PCA) and Uniform Manifold Approximation and Projection (UMAP). Having a large number of dimensions in the feature space can mean that the volume of that space is very large which can dramatically harm the performance of the clustering techniques to be applied later. This is usually known as "curse of dimensionality". PCA, one of the most popular dimension reduction technique, is used to compute the orthogonal projection of the data onto a lower dimensional linear space, known as the principal subspace, such that the variance of the projected data is maximized [4]. In contrast to PCA, UMAP is based on the topological mapping of the high-dimensional dataset. UMAP is primarily known as a visualization technique for high-dimensional data, but can also be used for dimensional reduction in general [11,18].

Embeddings from Sequential Data. In the second approach, a patient's embedding is learned from sequential data consisting of various medical events ordered chronologically. After arranging the events of that patient a "sentence" representing the patients journey is obtained. In order to encode the numerical features as a word in a sentence they are categorized in two categories, high and low, based on their value. In the next step each event (or "word" in the sentence) is replaced by a embedding. The embedding of the events were learned using the continuous bag of words (CBOW) model [12]. The sequence of embeddings were then fed into a LSTM autoencoder to ultimately get a vector per patient. The preprocessing of the sequence data and the LSTM autoencoder design is shown in Fig. 2.

2.3 Clustering

After the embeddings are learned for each patient in our cohort, the final step comes where we apply the clustering methods. We tried two different methods here

- **K-Means:** the main goal is to find k-clusters with minimal variance for grouping n data points [16] . To find the optimal value of k, the elbow method was used.

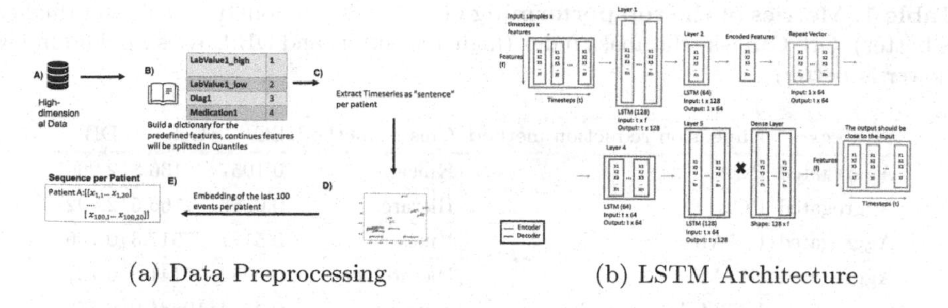

(a) Data Preprocessing (b) LSTM Architecture

Fig. 2. (a) displays the Entire preprocessing pipeline of the sequential data (b) shows the architecture of the LSTM Autoencoder.

- **Agglomerative-Hierarchical:** It is a "bottom up" approach, i.e. the algorithm starts at the individual data points and further groups them together hierarchically, resulting in a tree-like hierarchy, the so-called dendograms [14].

2.4 Evaluation

The resulting clusters were evaluated in different areas:

- **Performance of the Cluster Algorithm:** The clustering performance was evaluated with the Silhouette Coefficient [17], the Calinski-Harabasz Index [5] and the Davies-Bouldin Index [10].
- **Characteristics of the Clusters:** Insights such as the prevalence of different diagnoses or medications within the different clusters were compared. These insights can then be used to determine the characteristics of the clusters from a medical perspective.
- **Evaluation of Outcomes:** Days in hospital as inpatient and the ejection fraction measurement were compared between the different subgroups to see if there is a significant difference. It should be noted, ejection fraction and days in hospital were not used as input feature.

To determine if the difference between the different features and outcomes were significant between the clusters, chi-square (for categorical features and outcomes) and ANOVA (for numerical features and outcomes) tests were used.

3 Results

We have performed many different experiments with our clustering framework described in the methods section. Table 1 lists some of the top results from the two different embedding strategies and their corresponding evaluation metrics. In general the best results from the embeddings based on the aggregation strategy performed better than the embeddings based on the sequence strategy. UMAP performed markedly better than PCA as a dimensionality reduction technique. In the upcoming sections we take a deeper look into some of these clustering results.

Table 1. Metrics of the top performing clusters SC: Silhouette Coefficient (higher is better), CH: Calinski-Harabasz Index (higher is better) and DB:Davies Bouldin Index (lower is better).

Strategy	Dimension reduction method	Cluster method	SC	CH	DB
Aggregated	PCA	Kmeans	0.1057	2136.8	2.2607
Aggregated	PCA	Hierarchical	0.0856	1766.5	2.4927
Aggregated	UMAP	Kmeans	0.5187	27517.3	0.7061
Aggregated	UMAP	Hierarchical	0.5084	26141.8	0.7273
Sequential	LSTM-Autoencoder	Kmeans	0.4604	19881.9	0.7665
Sequential	LSTM-Autoencoder	Hierachical	0.4159	17089.6	0.8347

3.1 Results from the Aggregated Data Strategy

UMAP and K-means. The overall best results that we observed, were obtained using UMAP as dimension reduction method and k-means as the clustering technique. The data was initially reduced to 70 dimensions and clustered into 3 groups. We experimented with the dimension of the intermediate representation and found 70 to be most optimal. Figure 3 (part 2) shows the scatter plot of the 3 different clusters using a UMAP.

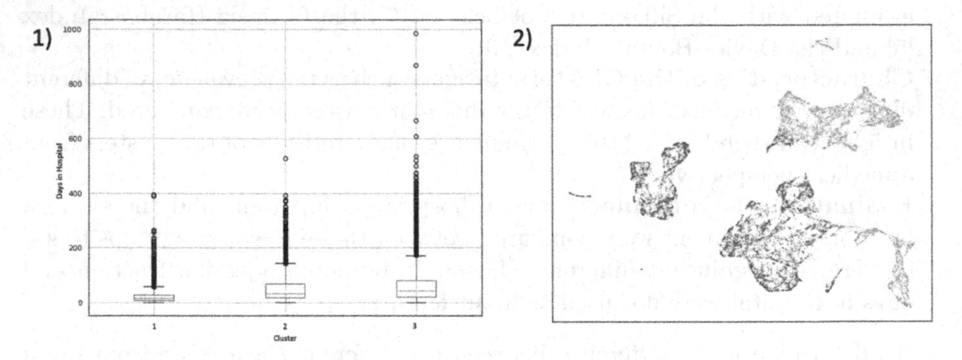

Fig. 3. Evaluation of the UMAP clusters 1: Distribution of days spent in hospital; 2: Scatter plot of clusters

In Table 2 the characteristics of the different clusters (list of features that were differing between the clusters the most) are shown. The Fig. 3 (part 1) shows the box plots of the hospitalization days per cluster. With a p-value of 2.79e-45 a significant difference in hospitalization rates among the clusters was observed. Though no significant differences in terms of ejection fraction was observed among the clusters. After investigating the clusters further the following characteristics were observed:

Cluster 1: The 6216 patients within cluster 1 have an average onset age of 69 years. This cluster has the lowest incidence of the comorbidites. Only for the diagnosis of primary hypertension and atrial fibrillation cluster 1 has a higher prevalence than cluster 2. Also for the cluster 1 beta blockers were prescribed more often. Compared to the other clusters, patients here spend lower number of days in the hospital. Hence, this can be described as the comparatively healthier cluster.

Cluster 2: The 2657 patients in cluster 2 have the oldest onset age with an average age of 72. Cluster 2 has the second highest values for all diagnoses except atrial fibrillation and hypertension, where it has the lowest prevalence. Patients in this cluster are most likely to have (72%) personal history of allergy to medical agents. This may contribute to the fact that patients in this cluster spend the second highest number of days in the hospital on average, 52 days. Besides that also the low prevalence of prescribed beta-blockers stands out.

Cluster 3: The average HF onset age of the 5461 patients in Cluster 3 is 68 years. Cluster 3 has the highest prevalence of comorbidites. The different comorbidities lead to a higher rate of hospitalization, 64 days on average.

Table 2. Characteristics of the UMAP clusters

Feature	Cluster 1	Cluster 2	Cluster 3
Onset Age	69	72	68
Male	55%	52%	56%
Female	45%	48%	44%
Patients	6216	2657	5461
Atrial fibrillation	0.222312045	0.10082769	**0.257330177**
Personal history of allergy to medicinal agents	0.284559418	**0.726862302**	0.437807321
Stroke hemorrhagic	0.047049313	0.060948081	**0.110362411**
Metoprolol succinate (beta-blocker)	0.263217462	0.0413845	**0.339464578**
Essential (primary) hypertension	0.216168149	0.010158014	**0.348752504**
Chronic kidney disease	0.296523848	0.380737397	**0.558550355**
Depression	0.176556184	0.267870579	**0.321981424**

3.2 Results from the Sequence Data Strategy

LSTM Autoencoder and K-means. The best performing clustering results from the experiments, where the sequence per patient was fed into the LSTM autoencoder, was achieved with an LSTM autoencoder (with two encoder and decoder layers of sizes 32 and 16 and the sigmoid activation function) in combination with the k-means clustering algorithm. Figure 4 (part 2) is showing the scatter plot of the 3 resulting clusters in 2-dimension using a UMAP.

Among the different clusters no big differences were observed in terms of the underlying features and are therefore not shown here. Figure 4 (part 1) shows

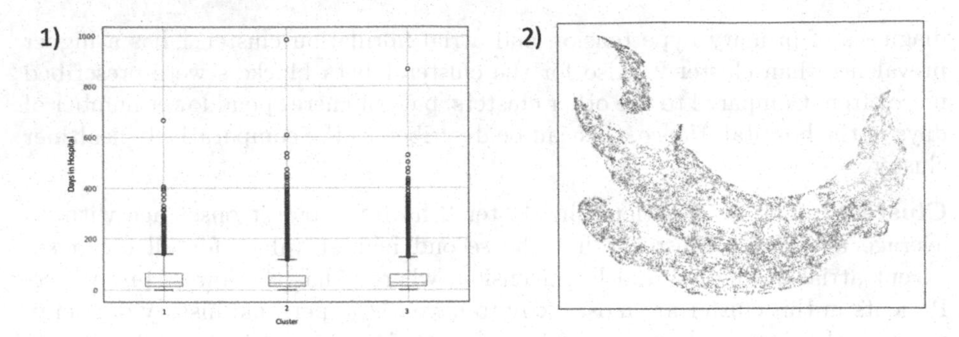

Fig. 4. Evaluation of the sequential clusters 1: Distribution of days spent in hospital; 2: Scatter plot of clusters

the box plots of the number of days spent by patients in hospital. No significant difference was observed here as well. As there was no significant difference in the features and in days spend in hospital, it was not possible to individually describe the different clusters here.

4 Discussion

The main objective of this work was to identify subphenotypes of HF patients using EHR data and an elaborate clustering framework. We managed to identify three subphenotypes that showed a significant difference in terms of prevalence of comorbidities and medications. We describe the significance of our work below.

4.1 Technical Significance

As a part of this work we developed a framework in python that manages to cluster specific cohort of patients from EHR data using different dimensionality reduction and clustering techniques. Due to it's modular design additional methods of clustering or dimensionality reduction can be added here with very little effort. Additionally the framework also allows to address the sequential nature of the data with the help of a LSTM autoencoder. To the best of our knowledge, our work is the first of it's kind in comparing aggregated and sequential data based approaches with EHR data for subphenotyping specific disease cohorts. We also show the UMAP approach can potentially outperform other methods like PCA for embedding high-dimensional EHR data.

4.2 Clinical Significance

To the best of our knowledge, our work is the first where we perform a retrospective data-driven subphenotyping of HF patients based on EHR data. Our work shows, it is possible to detect meaningful subphenotypes for HF patients with EHR data and unsupervised learning. We were able to clearly detect one cluster which had considerably higher number of comorbidites. We show the comorbidites can effect subsequent hospitalization rate. We also detected a cluster which showed significantly higher rate of personal history of allergy to medicinal agents. This group also showed a comparatively higher hospitalization rate compared to the healthier subgroup. Whether the more days spend it hospital was due to allergic reaction to medicinal agents needs to be investigated. Notably, when compared to [1], who performed subphenotyping of HF patients based on data from clinical trials, our clusters also differed in terms of incidence of diseases such as atrial fibrillation.

4.3 Limitations and Future Work

The main limitation of our work is it based on data from only one health system. We need to externally validate our models to see if similar clusters can be found in other EHR systems as well. Also, as the information of death related to HF was not available in our dataset we couldn't examine if the clusters also had different mortality rates. In future, we plan to integrate more modalities of data such as clinical notes and genetic data to investigate if we can enrich our clustering results with this added information.

5 Conclusion

In this work two different approaches were compared to identify subphenotypes of HF patients based on the EHR data from MSHS. The first approach was based on aggregating the medical concepts from the EHR data, and in the second approach we represented the patients as a sequence of medical events. Various dimensionality reduction and clustering techniques were employed to obtain the subgroups of patients. In general the aggregated data based approach performed better than the sequence data based approach. In our best results, we obtained 3 different clusters which varied significantly in terms of underlying medical features, number of hospitalizations and disease severity. Our best results were obtained with UMAP as a dimensionality reduction technique and k-means as the clustering technique. Our framework can help better understand varying subphenotypes in heterogeneous sub-populations and unlock patterns for EHR-based research in the realm of personalized medicine. In future we want to verify the various subgroups we found in the medical literature and with the support of clinicians.

References

1. Ahmad, T., et al.: Clinical implications of chronic heart failure phenotypes defined by cluster analysis. Technical report (2014)
2. Alonso-Betanzos, A., Bolón-Canedo, V., Heyndrickx, G.R., Kerkhof, P.L.: Exploring guidelines for classification of major heart failure subtypes by using machine learning. Clin. Med. Insights Cardiol. **9**, 57–71, March 2015. https://doi.org/10.4137/CMC.S18746

3. Association, A.H.: What is heart failure?—American Heart Association. https://www.heart.org/en/health-topics/heart-failure/what-is-heart-failure
4. Bishop, C.M.: Pattern Recognition and Machine Learning. Information Science and Statistics, Springer, New York (2006)
5. Cengizler, C., Kerem-Un, M.: Evaluation of Calinski-Harabasz criterion as fitness measure for genetic algorithm based segmentation of cervical cell nuclei. Br. J. Math. Comput. Sci. **22**(6), 1–13 (2017). https://doi.org/10.9734/bjmcs/2017/33729. https://www.journaljamcs.com/index.php/JAMCS/article/view/24229
6. Centers for Disease Control and Prevention: Heart Failure—cdc.gov. https://www.cdc.gov/heartdisease/heart_failure.htm
7. Glicksberg, B.S.: Platforms for multimodal health and clinical datasets. https://www.tele-task.de/lecture/video/7820/#t=1305
8. Landi, I., et al.: Deep representation learning of electronic health records to unlock patient stratification at scale. NPJ Digit. Med. **3**(1) (2020). https://doi.org/10.1038/s41746-020-0301-zhttp://arxiv.org/abs/2003.06516dx.doi.org/10.1038/s41746-020-0301-z
9. Liao, K.P., et al.: Development of phenotype algorithms using electronic medical records and incorporating natural language processing. BMJ (Online) **350** (2015). https://doi.org/10.1136/bmj.h1885. www.nlm.nih.gov/research/umls
10. Maulik, U., Bandyopadhyay, S.: Performance evaluation of some clustering algorithms and validity indices. IEEE Trans. Pattern Anal. Mach. Intell. **24**(12), 1650–1654 (2002). https://doi.org/10.1109/TPAMI.2002.1114856
11. Mcinnes, L., Healy, J., Melville, J.: UMAP: uniform manifold approximation and projection for dimension reduction. Technical report (2020)
12. Mikolov, T., Chen, K., Corrado, G., Dean, J.: Efficient estimation of word representations in vector space. Technical report. http://ronan.collobert.com/senna/
13. NHS: heart failure - treatment - NHS. https://www.nhs.uk/conditions/heart-failure/treatment/
14. Nithya, N.S., Duraiswamy, K., Gomathy, P.: A survey on clustering techniques in medical diagnosis. Int. J. Comput. Sci. Trends Technol. **2**, 17–23 (2013). www.ijcstjournal.org
15. Organization (WHO, W.H.O.: Cardiovascular diseases. https://www.who.int/health-topics/cardiovascular-diseases/#tab=tab_1
16. Oyelade, O.J., Oladipupo, O.O., Obagbuwa, I.C.: Application of k-means clustering algorithm for prediction of students' academic performance. Technical report 1 (2010). http://sites.google.com/site/ijcsis/
17. Rousseeuw, P.J.: Silhouettes: a graphical aid to the interpretation and validation of cluster analysis. J. Comput. Appl. Math. **20**(C), 53–65 (1987). https://doi.org/10.1016/0377-0427(87)90125-7
18. Sánchez-Rico, M., Alvarado, J.M.: A machine learning approach for studying the comorbidities of complex diagnoses. https://doi.org/10.3390/bs9120122. www.mdpi.com/journal/behavsci
19. Bielinski, S.J., (Mayo Clinic): Heart Failure (HF) with differentiation between preserved and reduced ejection fraction—PheKB. https://phekb.org/phenotype/heart-failure-hf-differentiation-between-preserved-and-reduced-ejection-fraction

Stratification of Parkinson's Disease Patients via Multi-view Clustering

Anita Valmarska[1,2(✉)], Nada Lavrač[2,3], and Marko Robnik-Šikonja[1]

[1] Faculty of Computer and Information Science, University of Ljubljana, Ljubljana, Slovenia
{anita.valmarska,marko.robnik}@fri.uni-lj.si
[2] Jožef Stefan Institute, Jamova 39, Ljubljana, Slovenia
nada.lavrac@ijs.si
[3] University of Nova Gorica, Vipavska 13, Nova Gorica, Slovenia

Abstract. Parkinson's disease is a neurodegenerative disease characterised by heterogeneity of the sets of symptoms patients experience and the trajectories of disease progression. The PPMI study includes patients' symptoms explaining different aspects of patients' life, i.e. motor, non-motor, and autonomic symptoms. This paper proposes a multi-view clustering approach for determining groups of Parkinson's disease patients from the PPMI study with distinct disease trajectories over 4 years. The proposed multi-view clustering approach searches groups of patients who share similar disease progression trajectories over multiple types of symptoms. We detected two groups of patients with different disease progression trajectories and significant differences in severity of motor, non-motor, and autonomic symptoms. On the other hand, while we did not detect any significant differences between the patients from the two groups based on their demographics, medications treatment or their disease types, we identified over-sensitivity to bright light as a possible early screening symptom for type of disease progression.

Keywords: Parkinson's disease · Multi-view clustering · Disease progression

1 Introduction

Parkinson's disease is the second most common neurodegenerative disease after Alzheimer's disease. Patients who have Parkinson's disease experience various symptoms whose severity can significantly affect the quality of life of both the patients and their families. Parkinson's disease is incurable, and the patients are provided only with symptomatic treatment. The treatment consists mainly of prescribing antiparkinson medications designed to treat patients' motor symptoms, which are the most characteristic symptoms of the Parkinson's disease.

Efforts have been made in the scientific community to connect the variability of symptoms to some underlying subtypes of Parkinson's disease. The definition

A. Tucker et al. (Eds.): AIME 2021, LNAI 12721, pp. 229–239, 2021.
https://doi.org/10.1007/978-3-030-77211-6_25

of actual subtypes of Parkinson's disease would lead to a better understanding of its underlying mechanisms. It could improve Parkison's disease patients' treatment and lead to a better design of clinical trials [3].

The commonly used subtype classification, conceptualising the heterogeneous motor manifestations of early Parkinson's disease, was proposed by Jankovic et al. [6]. This is the division of PD patients into tremor dominant (TD), postural instability and gait difficulty (PIGD), and the indeterminate subtype (Indeterminate). Based on the analysis of data from the DATATOP trial [10], the authors show that the patients classified into the TD and PIGD groups exhibit differences in their ability to perform activities of daily living, as well as differences concerning key non-motor symptoms, supporting the existence of the TD and PIGD clinical Parkinson's disease subtypes. The classification of Parkinson's disease patients in the TD/PIGD subtypes can be derived from the assessment of patients' symptoms severity using the well-established scale MDS-UPDRS [9][1]. However, the classification into PIGD/TD subtypes is usually done only at the beginning of the patient's diagnosis. It does not offer an insight into how the disease will progress, even though the classification of patients into TD/PIDG can change as the disease progresses. The patient's symptoms are affected by both the disease's natural progression and their symptomatic treatment [13].

The definition of consistent subtypes of Parkinson's disease is still an open issue and may contribute to better understanding of the disease. A consistent subtype would include patients with similar symptoms and trajectories of disease progression. In this work, we propose a novel multi-view clustering approach—clustering of data from multiple sources [1]—to detect groups of Parkinson's disease patients with similar trajectories of disease progression after four years of their involvement in the PPMI[2] study [7]. Our methodology addresses the heterogeneity of Parkinson's disease by considering the patients' motor, non-motor, and autonomic symptoms. We cluster the patients into groups based on the changes of symptoms severity that the patients have experienced at the beginning of the study and four years later. Our methodology detects two groups of patients with a different speed and severity of disease progression. A smaller group of Parkinson's disease patients experience more severe problems with their rigidity, postural instability, leg agility after four years. At that time, these patients also experience significantly worse motor experiences of daily living, non-motor symptoms, and autonomic symptoms.

The paper is structured as follows. Section 2 presents the data used in the analysis. Section 3 presents the proposed methodology and Sect. 4 outlines the experimental results. The conclusions and plans for further work are presented in Sect. 5.

[1] MDS-UPDRS is Movement Disorder Society sponsored revision of the Unified Parkinson's Disease Rating Scale.

[2] Parkinson's Progression Markers Initiative, https://www.ppmi-info.org/.

2 Data

In this paper, we use the data from the Parkinson's Progression Markers Initiative (PPMI) data collection [7]. The PPMI data collection records data for over 400 Parkinson's disease patients involved in the study for up to 5 years. During their involvement, the patients regularly (every 3–6 months) visit their assigned clinicians to assess their symptoms. In this way, the clinicians monitor the disease progression over time. The clinical data used in patient partitioning was gathered using several standardised questionnaires, briefly described below.

- MDS-UPDRS (Movement Disorder Society-sponsored revision of Unified Parkinson's Disease Rating Scale) [4] is the most widely used, four-part questionnaire addressing 'non-motor experiences of daily living' (Part I, subpart 1 and subpart 2), 'motor experiences of daily living' (Part II), 'motor examination' (Part III), and 'motor complications' (Part IV). It consists of 65 questions, each addressing a particular symptom. Each question is anchored with five responses that are linked to commonly accepted clinical terms, ranging from 0 = normal (patient's condition is normal, the symptom is not present) to 4 = severe (symptom is present and severely affects the normal and independent functioning of the patient); scores 1, 2, and 3 denote degrees of intermediate symptom severity.
- SCOPA-AUT (Scales for Outcomes in Parkinson's disease - Autonomic) [14] is a specific scale for assessing autonomic dysfunction in Parkinson's disease patients.

In the context of multi-view clustering, patients' data from the mentioned questionnaires represent separate views (NUPDRS1, NUPDRS1P, NUPDRS2P, NUPDRS3, SCOPA-AUT). Answers to the questions from each questionnaire form the vectors of attribute values. All considered questions have ordered values, where larger values suggest higher symptom severity and lower quality of life. In this work, we look only at the severity of patients' symptoms when they were admitted to the study and at the patients' 10th visit. We chose the 10th visit as it shows the status of patients after a slightly longer period (4 years), and a good portion (194 patients, 55 females) of the PPMI patients had records for this visit.

In addition to the clinical data, in order to establish significant differences of attribute values based on the assigned progression group, in the post-processing we also used the patients' demographics data, i.e. gender, years of education, patient's age at the time the patient was admitted to the PPMI study, patient's age when the first symptoms appeared, and the patient's age when the diagnosis was established. We also looked at the patients' therapy with antiparkinson medications data at the time of their admission to the PPMI study and at their 10th visit. We also report the statistical significance of the patients' subtypes (TD/PIGD) [6,9] at the start of the study and on the patients' 10th visit.

3 Multi-view Clustering Methodology

Our methodology for determining groups of Parkinson's disease patients with different disease progression trajectories consists of four steps, presented in Fig. 1.

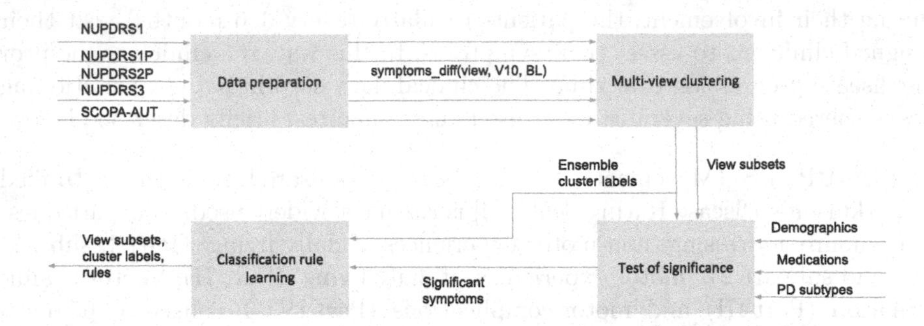

Fig. 1. The overall multi-view clustering methodology for detecting groups of Parkinson's disease patients with different trajectories of disease progression.

The first step is data preparation, where for each patient of the PPMI study with records of symptoms at the start of the study (V1) and 10th visit (V10), we calculate the difference of symptoms severity the patients had experienced at V10 and V1. We calculate these differences for each view separately. The i-th row of the j-th view represents the differences of the i-th patient's symptoms from the j-th view at V10 and V1.

In the second step, we perform multi-view clustering of the symptoms severity data. The multi-view clustering approach consists of three key steps: selecting the best clustering for each step separately, performing multi-view feature selection using the mvReliefF algorithm, and ensemble clustering of the best selected clusters from individual views. More details of the used multi-view clustering approach are provided below. The algorithm for multi-view clustering will be presented in a forthcoming paper from the authors.

In the third step, we performed statistical tests (the Kolmogorov-Smirnov test) to detect differences in the severity of symptoms for patients from the detected clusters based on their values at baseline and V10. In addition to the symptoms data used in the multi-view clustering, we also test differences in demographic and medications data, as well as the patients' assigned subtypes at V1 and V10.

In the fourth step, we perform rule learning to determine descriptions of the obtained clusters. The descriptive attributes for the rule learning algorithm of choice are the attributes (symptoms) from the third step of the methodology that showed a significant difference between the patients assigned to the obtained clusters. The cluster labels obtained from the second step of our methodology are used as class labels in the classification rule learning algorithm.

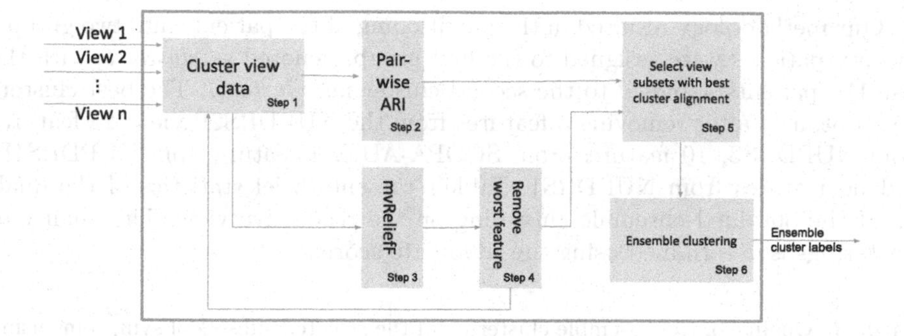

Fig. 2. Outline of the used multi-view clustering approach.

The outline of the multi-view classification algorithm is presented in Fig. 2. The methodology consists of six steps forming three key components. As the quality of clustering is highly dependent on the quality of the selected features and vice versa, the first two key components of the methodology are clustering of individual views and multi-view feature selection. The third key component of the multi-view clustering methodology is the ensemble clustering of the chosen best clusters per view based on their alignment.

In step 1 of the methodology, we choose the best clustering for each view. The decision is made based on an adaptation of the silhouette score [8]. In step 2, we record the pairwise cluster alignment using the adjusted rand index [5] and record the alignments as a time point. The third step evaluates features from all views using the mvReliefF approach. The mvReliefF algorithm is an adaptation of the ReliefF algorithm. It determines the neighbourhood of close hits (examples with the same class label) and close misses (examples with different class labels) based on their cluster labels in pairs of views. This algorithm selects features that are predictive of the cluster labels and coherent among the views. Based on the determined feature importance, in step 4, we remove the worst feature, i.e. the feature whose removal will lead to the biggest improvement of the alignment between the cluster labels of individual views. Steps 1–4 are performed in a loop until all views have only one remaining feature. In step 5, we select the time point with the best alignment between cluster labels of views. Based on the selected cluster labels from step 5, in step 6, we perform ensemble clustering using Python package *Cluster_Ensembles* [11].

4 Results

We applied our methodology to a dataset of 194 patients (55 females) from the PPMI study who already made their 10th visit to the clinician. The patients' average age on their first visit is 61.16 years (median value of 62 years), with the oldest patient being 84 years old, while the youngest patient was 33 years. The average duration of education of the involved patients is 15.83 years.

Our methodology resulted in the partitioning of the patients into two groups. Twenty patients were assigned to the first group, denoted as *cluster 0*, with the rest 174 patients assigned to the second cluster i.e. *cluster 1*. The best clusters were obtained after removing 5 features from the NUPDRS2P view, 24 features from NUPDRS3, 10 features from SCOPA-AUT, 1 feature from NUPDRS1P, and no features from NUPDRS1. Table 1 presents brief statistics of the quality of the obtained ensemble clustering on the respective view. The quality of clusterings is determined using the silhouette score.

Table 1. Quality of the ensemble clustering of the selected subsets of symptoms using the silhouette score. The subsets of symptoms are selected in the 5th step of the multi-view clustering approach (see Fig. 2). The ensemble cluster quality is the quality of ensemble clustering based on the selected feature subsets.

	View	Ensemble cluster quality
1	NUPDRS1	0.177
2	NUPDRS1P	0.140
3	NUPDRS2P	0.277
4	NUPDRS3	0.319
5	SCOPA-AUT	0.215

We used the Kolmogorov-Smirnov test to test the distribution equality of symptoms between patients assigned to *cluster 0* and *cluster 1*. For this purpose, we used the symptoms from the multi-view clustering step, i.e. the symptoms from the views NUPDRS1, NUPDRS1P, NUPDRS2P, NUPDRS3, and SCOPA-AUT and their respective sum attributes denoting the sum of patients' symptoms severity. For each of these attributes, we constructed two versions, the symptoms severity at the baseline, here denoted with the suffix *_V1* and the symptom severity at V10, denoted with the suffix *_V10*. From the motor symptoms of NUPDRS3, we also calculated more abstract symptoms, representing a type of motor symptoms. Table 2 presents the newly derived attributes and their underlying members. The value of the derived attributes is calculated either as *max* or *mean* value of the severity of underlying symptoms (here denoted with the suffixes *_max* and *_mean*. We are interested in potential differences of symptoms severity among patients assigned to either *cluster 0* or *cluster 1*. Significant differences at V1 can help early screening and division of patients based on their expected disease progression. Differences at V10 will reveal potential trajectories of the disease.

In addition to the clinical symptoms, in this post-processing step, we also included patients' demographic attributes (see Sect. 2), as well as data about the patients' antiparkinson medications treatment and their underlying subtypes at V1 and V10. Table 3 presents the symptoms/attributes with significant difference between patients from *cluster 0* and *cluster 1* at the significance level 0.1. Symptoms are ordered according to their p-value. Symptoms whose severity at

Table 2. Hierarchy of derived motor symptoms. The value of the derived symptoms was calculated as *mean* and *max* value of the severity of the respective questionnaire symptoms. All questionnaire symptoms belong to the NUPDRS3 view.

Derived symptom	Questionnaire symptoms
RIGIDITY	NP3RIGN, NP3RIGRU, NP3RIGLU, PN3RIGRL, NP3RIGLL
FTAPPING	NP3FTAPR, NP3FTAPL
HAND_MOVING	NP3HMOVR, NP3HMOVL
PRONATION/SUPINATION	NP3PRSPR, NP3PRSPL
TTAPPING	NP3TTAPR, NP3TTAPL
LEG_AGILITY	NP3LGAGR, NP3LGAGL
TREMOR	NP3PTRMR, NP3PTRML, NP3KTRMR, NP3KTRML, NP3RTARU, NP3RTALU, NP3RTARL, NP3RTALL, NP3RTALJ, NP3RTCON
POSTURAL TREMOR	NP3PTRMR, NP3PTRML
KINETIC TREMOR	NP3KTRMR, NP3KTRML
TREMOR AT REST	NP3RTARU, NP3RTALU, NP3RTARL, NP3RTALL

the baseline is significantly different between patients from *cluster 0*, and *cluster 1* are presented at the upper half of the table. For a more comprehensive presentation, we also present the description of the identified significant symptoms and their respective views (questionnaires).

The results did not reveal any significant differences regarding the patients' demographics or their treatment with antiparkinson medications. We detected a significant difference in symptoms severity at V1 for the SCAU19 symptom (over sensitivity to bright light) from the SCOPA-AUT questionnaire. As the disease progresses, the patients assigned to *cluster 0* experience more rapid and harsh disease progression, as manifested by all types of symptoms: motor, non-motor, and autonomic. Patients from *cluster 0* experience significantly more severe rigidity symptoms, freezing of gait, problems with their posture and postural instability, and consequently problems with their motor experiences of daily living (dressing, personal hygiene, hobbies, swallowing, etc.).

To describe the underlying structure of the obtained ensemble clusters, we used the ensemble cluster labels as class labels with the Ripper algorithm [2] for classification rule learning (WEKA implementation). The descriptive features were constructed by concatenating the resulting view subsets. This resulted in one classification rule, describing the patients assigned to *cluster 0* and a default rule describing the remaining patients. The number of true positive (TP) examples covered by the first rule is 15, and the number of false positives (FP) is 2. The obtained clusters can be described based on the value of patients sum of symptoms severity. Similarly to the results presented in [12], the patients with the worst symptoms severity are those whose sum of severity of motor symptoms NUPDRS3_SUM (NUPDRS3_SUM_V10 is NUPDRS3_SUM at V10)

Table 3. Statistics of statistically different symptoms of patients from *cluster 0* and *cluster 1*. The _V1 suffix (upper part of the table) denotes symptoms with significant difference in severity at V1, while the _V10 suffix (bottom part of the table) denotes symptoms with significant difference in severity at V10 (after 4 years). Symptoms are ordered according to the p-value.

Symptom	Description	Questionnaire	Mean value (STD)	Mean value cluster 0 (STD)	Mean value in cluster 1(STD)	p-value
SCAU19_V1	OVER SENSITIVITY TO BRIGHT LIGHT	SCOPA-AUT	0.387 (0.619)	0.75 (0.786)	0.345 (0.586)	0.0609
NUPDRS2P_SUM_V10	Sum of NUPDRS2P symptom severity	NUPDRS2P	10.876 (6.546)	19.750 (7.559)	9.856 (5.602)	0.0000
NP3PRSPL_V10	PRONATION-SUPINATION - LEFT HAND	NUPDRS3	1.000 (0.976)	2.250 (0.851)	0.856 (0.885)	0.0000
NP2DRES_V10	DRESSING	NUPDRS2P	0.918 (0.784)	1.900 (0.788)	0.805 (0.702)	0.0000
NP3HMOVL_V10	HAND MOVEMENTS - LEFT HAND	NUPDRS3	0.938 (0.931)	2.150 (0.813)	0.799 (0.840)	0.0000
NP2HYGN_V10	HYGIENE	NUPDRS2P	0.433 (0.574)	1.100 (0.788)	0.356 (0.492)	0.0001
NP3LGAGL_V10	LEG AGILITY - LEFT LEG	NUPDRS3	0.814 (0.868)	1.850 (0.875)	0.695 (0.786)	0.0001
NP2SALV_V10	SALIVA + DROOLING	NUPDRS2P	1.026 (1.176)	1.900 (1.021)	0.925 (1.153)	0.0006
SCAU11_V10	WEAK STREAM OF URINE	SCOPA-AUT	0.686 (0.971)	1.200 (0.768)	0.626 (0.976)	0.0011
NP2HOBB_V10	DOING HOBBIES AND OTHER ACTIVITIES	NUPDRS2P	0.866 (0.871)	1.850 (1.040)	0.753 (0.777)	0.0012
NUPDRS1P_SUM_V10	Sum of NUPDRS1P symptom severity	NUPDRS1P	6.964 (3.931)	10.450 (4.249)	6.563 (3.700)	0.0018
NP3PRSPR_V10	PRONATION-SUPINATION - RIGHT HAND	NUPDRS3	0.943 (0.889)	1.600 (1.046)	0.868 (0.840)	0.0025
NP1URIN_V10	URINARY PROBLEMS	NUPDRS1P	1.088 (0.943)	1.900 (1.119)	0.994 (0.877)	0.0027
SCAU2_V10	SALIVA DRIBBLED OUT OF MOUTH	SCOPA-AUT	0.655 (0.741)	1.200 (0.696)	0.592 (0.721)	0.0029
SCAU18_V10	EXCESSIVE PERSPIRATION DURING THE NIGHT	SCOPA-AUT	0.320 (0.558)	0.750 (0.639)	0.270 (0.528)	0.0032
NP3TTAPR_V10	TOE TAPPING - RIGHT FOOT	NUPDRS3	1.170 (0.931)	1.900 (1.021)	1.086 (0.886)	0.0041
NP2RISE_V10	GETTING OUT OF BED_CAR_OR DEEP CHAIR	NUPDRS2P	0.840 (0.795)	1.550 (0.999)	0.759 (0.729)	0.0047
NP3RISNG_V10	ARISING FROM CHAIR	NUPDRS3	0.299 (0.605)	1.000 (1.170)	0.218 (0.441)	0.0052
NP2SWAL_V10	CHEWING AND SWALLOWING	NUPDRS3	0.381 (0.674)	0.800 (0.696)	0.333 (0.657)	0.0055
NP3FTAPL_V10	FINGER TAPPING LEFT HAND	NUPDRS3	1.139 (0.980)	2.000 (0.973)	1.040 (0.933)	0.0075
SCAU19_V10	OVER SENSITIVITY TO BRIGHT LIGHT	SCOPA-AUT	0.541 (0.756)	1.100 (0.968)	0.477 (0.703)	0.0086
SCAU21_V10	TROUBLE TOLERATING HEAT	SCOPA-AUT	0.412 (0.709)	1.200 (1.196)	0.322 (0.569)	0.0092
NP3LGAGR_V10	LEG AGILITY - RIGHT LEG	NUPDRS3	0.727 (0.847)	1.500 (1.051)	0.638 (0.776)	0.0095
NP3RIGLL_V10	RIGIDITY - LLE	NUPDRS3	0.742 (0.867)	1.400 (1.095)	0.667 (0.807)	0.0098
NUPDRS1_SUM_V10	Sum of NUPDRS1 symptom severity	NUPDRS1	2.397 (2.641)	4.650 (3.787)	2.138 (2.357)	0.0105
SCAU15_V10	LIGHT-HEADED FOR SOME TIME AFTER STANDING UP	SCOPA-AUT	0.247 (0.499)	0.600 (0.598)	0.207 (0.472)	0.0116
NP1LTHD_V10	LIGHTHEADEDNESS ON STANDING	NUPDRS1P	0.505 (0.757)	1.200 (1.056)	0.425 (0.674)	0.0134
NP3HMOVR_V10	HAND MOVEMENTS - RIGHT HAND	NUPDRS3	0.918 (0.835)	1.550 (0.826)	0.845 (0.808)	0.0136
NP3RIGLU_V10	RIGIDITY - LUE	NUPDRS3	1.046 (0.889)	1.800 (0.834)	0.960 (0.856)	0.0176
NP2TURN_V10	TURNING IN BED	NUPDRS2P	0.619 (0.690)	1.350 (0.988)	0.534 (0.595)	0.0189
NP2EAT_V10	EATING TASKS	NUPDRS2P	0.696 (0.724)	1.300 (0.865)	0.626 (0.675)	0.0230
NP1DPRS_V10	DEPRESSED MOODS	NUPDRS1	0.438 (0.697)	0.900 (0.852)	0.385 (0.659)	0.0238

(*continued*)

Table 3. (*continued*)

Symptom	Description	Questionnaire	Mean value (STD)	Mean value cluster 0 (STD)	Mean value in cluster 1(STD)	p-value
SCAU17_V10	EXCESSIVE PERSPIRATION DURING THE DAY	SCOPA-AUT	0.314 (0.635)	0.850 (0.988)	0.253 (0.553)	0.0252
NP3SPCH_V10	SPEECH	NUPDRS3	0.722 (0.664)	1.300 (0.733)	0.655 (0.624)	0.0342
NP2WALK_V10	WALKING AND BALANCE	NUPDRS2P	0.768 (0.75)	1.450 (0.945)	0.690 (0.685)	0.0394
NP2FREZ_V10	FREEZING	NUPDRS2P	0.768 (0.75)	1.450 (0.945)	0.690 (0.685)	0.0394
SCAU1_V10	DIFFICULTY SWALLOWING OR CHOKING	SCOPA-AUT	0.361 (0.579)	0.900 (0.912)	0.299 (0.495)	0.0406
SCAU10_V10	AFTER PASSING URINE BLADDER NOT COMPLETELY EMPTY	SCOPA-AUT	0.680 (0.950)	1.050 (0.759)	0.638 (0.962)	0.0479
PN3RIGRL_V10	RIGIDITY - RLE	NUPDRS3	0.851 (0.866)	1.500 (0.946)	0.776 (0.827)	0.0498
NP3POSTR_V10	POSTURE	NUPDRS3	0.985 (0.890)	1.700 (0.979)	0.902 (0.844)	0.0540
NP3TTAPL_V10		NUPDRS3	1.253 (0.009)	2.000 (0.858)	1.167 (0.980)	0.0563
NP1COG_V10	COGNITIVE IMPAIRMENT	NUPDRS1	0.660 (0.886)	1.300 (1.490)	0.586 (0.761)	0.0593
NP3RIGN_V10	RIGIDITY - NECK	NUPDRS3	0.943 (0.894)	1.600 (1.046)	0.868 (0.846)	0.0593
SCAU14_V10	LIGHT-HEADEDNESS WHEN STANDING UP	SCOPA-AUT	0.433 (0.601)	0.800 (0.768)	0.391 (0.566)	0.0722
NP3PSTBL_V10	POSTURAL STABILITY	NUPDRS3	0.284 (0.806)	1.000 (1.451)	0.201 (0.654)	0.0768
NP1HALL_V10	HALLUCINATIONS AND PSYCHOSIS	NUPDRS1	0.180 (0.481)	0.600 (0.883)	0.132 (0.387)	0.0986

is greater or equal to 42. The severity of motor symptoms also influences the patients' motor experiences of daily living, as evident by NUPDRS2P_SUM, whose value is higher or equal to 17. Patients from *cluster 0* experience more rapid and severe disease progression, causing significant worsening of their motor symptoms—mostly rigidity and postural instability and consequently worsening of their motor experiences in daily living, as well as their non-motor and autonomic symptoms.

Rule for cluster 0: (NUPDRS2P_SUM_V10 >= 17) and (NUPDRS3_SUM_V10 >= 42) → CE_cluster = *cluster 0*, (TP: 15, FP: 2)
Default rule: → CE_cluster = *cluster 1* (TP: 179, FP: 7)

5 Conclusion

We proposed a multi-view clustering based methodology for determining groups of Parkinson's disease patients with different disease progression patterns. The proposed methodology detected the best clustering for the available data based on the clustering quality measure and performed multi-view feature selection to improve the clustering. We used data from the PPMI study, partitioned according to the difference in symptoms severity between their 10th visit and first visit.

The results revealed that a small group (20 patients) of patients differs from others concerning disease progression speed and severity. In the period from their first visit to their 10th visit, these patients have experienced significant worsening of their motor symptoms—mostly rigidity and postural instability, and consequently worsening of their motor experiences in daily living, as well as their non-motor and autonomic symptoms. There were no statistical differences between patients from both clusters based on their demographics, antiparkinson medications treatment, or Parkinson's disease subtypes (TD/PIGD). However, we were able to identify over-sensitivity to bright light at the first visit as a distinguishing symptom between patients involved in the identified clusters. Although a further discussion with clinicians is needed, this symptom can pose a possibility for screening patients at the time of their diagnosis and adapt their treatment based on the expected disease progression.

In future work, we will examine the significance of bio-markers and image data for the disease's progression. A potential biomarker associated with different trajectories of Parkinson's disease progression can be utilised for early screening of patients, prediction of disease course, and designing clinical trials addressing the needs of patients with more severe disease progression. Our future research will also investigate the point at which the trajectories of the two groups of patients start to diverge. We plan to test our methodology and our conclusions on other available data sets for Parkinson's disease patients.

Acknowledgment. The research was supported by the Slovenian Research Agency (research core funding programs P2-0209, P6-0411 and P2-0103) and the Slovenian Ministry of Education, Science and Sport (project R 2.1 - Public call for the promotion of researchers at the beginning of a career 2.1). Data used in the preparation of this article were obtained from the Parkinson's Progression Markers Initiative (PPMI) (www.ppmi-info.org/data). For up-to-date information on the study, visit www.ppmi-info.org.

References

1. Bickel, S., Scheffer, T.: Multi-view clustering. In: Proceedings of the 4th IEEE International Conference on Data Mining, ICDM 2004, pp. 19–26. IEEE Computer Society, Washington, DC, USA (2004)
2. Cohen, W.W.: Fast effective rule induction. In: Proceedings of the 12th International Conference on Machine Learning, pp. 115–123 (1995)
3. Fereshtehnejad, S.M., Postuma, R.B., Dagher, A., Zeighami, Y.: Clinical criteria for subtyping Parkinson's disease: biomarkers and longitudinal progression. Brain **140**(7), 1959–1976 (2017)
4. Goetz, C., et al.: Movement disorder society-sponsored revision of the unified Parkinson's disease rating scale (MDS-UPDRS): scale presentation and clinimetric testing results. Move. Disord. **23**(15), 2129–2170 (2008)
5. Hubert, L., Arabie, P.: Comparing partitions. J. Classif. **2**(1), 193–218 (1985)
6. Jankovic, J., et al.: Variable expression of Parkinson's disease - a base-line analysis of the DATATOP cohort. Neurology **40**(10), 1529–1529 (1990)
7. Marek, K., et al.: The Parkinson's Progression Markers Initiative (PPMI). Prog. Neurobiol. **95**(4), 629–635 (2011)
8. Rousseeuw, P.J.: Silhouettes: a graphical aid to the interpretation and validation of cluster analysis. J. Comput. Appl. Math. **20**, 53–65 (1987)

9. Stebbins, G.T., Goetz, C.G., Burn, D.J., Jankovic, J., Khoo, T.K., Tilley, B.C.: How to identify tremor dominant and postural instability/gait difficulty groups with the movement disorder society unified Parkinson's disease rating scale: comparison with the unified Parkinson's disease rating scale. Mov. Disord. **28**(5), 668–670 (2013)
10. Steering, D.: DATATOP: a multicenter controlled clinical trial in early Parkinson's disease. Arch. Neurol. **46**(10), 1052–1060 (1989)
11. Strehl, A., Ghosh, J.: Cluster ensembles–a knowledge reuse framework for combining multiple partitions. J. Mach. Learn. Res. **3**, 583–617 (2002)
12. Valmarska, A., Miljkovic, D., Lavrač, N., Robnik-Šikonja, M.: Analysis of medications change in Parkinson's disease progression data. J. Intell. Inf. Syst. **51**(2), 301–337 (2018)
13. Valmarska, A., Miljkovic, D., Robnik–Šikonja, M., Lavrač, N.: Connection between the Parkinson's disease subtypes and patients' symptoms progression. In: Riaño, D., Wilk, S., ten Teije, A. (eds.) AIME 2019. LNCS (LNAI), vol. 11526, pp. 263–268. Springer, Cham (2019). https://doi.org/10.1007/978-3-030-21642-9_32
14. Visser, M., Marinus, J., Stiggelbout, A.M., Van Hilten, J.J.: Assessment of autonomic dysfunction in Parkinson's disease: the SCOPA-AUT. Mov. Disord. **19**(11), 1306–1312 (2004)

Disentangled Hyperspherical Clustering for Sepsis Phenotyping

Cheng Cheng[1(✉)], Jason Kennedy[2], Christopher Seymour[2], and Jeremy C. Weiss[1]

[1] Carnegie Mellon University, Pittsburgh, PA 15213, USA
{ccheng2,jeremyweiss}@cmu.edu
[2] University of Pittsburgh, Pittsburgh, PA 15260, USA
{kennedyj4,seymourcw}@upmc.edu

Abstract. Sepsis is a heterogeneous disease. Clustering sepsis patients into homogeneous subgroups with characteristic phenotypes may help for studying the disease progression and for providing targeted therapies. Existing clustering methods use many or all input variables whereas clusters defined by few variables are preferred by clinicians investigating subgroup treatment. To address this gap, we propose a soft F-statistic loss that promotes disentangled clusters differentiating on a small subset of features. Empirical and qualitative results demonstrate our method excels at achieving the desired property against competing methods.

1 Introduction

Sepsis is defined by a physiologic malresponse to infection and often leads to adverse clinical outcomes including organ failure and death. Recent sepsis trials have targeted numerous inflammatory pathways, yet few new therapies have become standard of care [6]. A recurring story is the identification of a drug, test, or protocol first showing success in a trial then failure to replicate in subsequent studies due to population heterogeneity [10,11]. Recent work has focused on identifying sepsis phenotypes: clinically-measurable characteristics of patients meeting Sepsis-3 criteria [9]. Seymour et al. found that sepsis subgroups may share phenotypic characteristics and have recognizable pathways and propensities for the disease progression [9]. Previous studies used classical methods, *e.g.*, consensus k-means clustering, to identify sepsis phenotypes [4,9]. Although the resulting clusters differed on several attributes such as liver function tests, such distinctions were not built into the algorithmic design objective. While cluster means showed differences, the per-feature distributions overlapped considerably across clusters. Furthermore, responses to Seymour et al. [9] argue for further subgrouping based on feature collection postulated to more closely measure the underlying process in disease progression [7]. Since standard clustering methods summarize *all* of the separate processes, interpretations from the clusters has limited use for understanding and treatment of the disease.

A. Tucker et al. (Eds.): AIME 2021, LNAI 12721, pp. 240–245, 2021.
https://doi.org/10.1007/978-3-030-77211-6_26

We introduce a disentangled hyperspherical clustering (DHC) method with soft F-statistic loss which: 1. Optimizes for inter-cluster difference and intra-cluster similarity, aligning with the goal of creating homogeneous patient subgroups; 2. Encourages clusters to separate on limited embedding dimensions; 3. Has the disentangling *modularity* property, *i.e.* one dimension of the embedding corresponds to at most one dimension of the features. Therefore the embeddings can be mapped back to a small subset of features, identifying phenotypic differences between clusters, a property arguably central to the advancement of sepsis research. Code for our method can be found at DHC-repo.

Related Work. A seminal study on sepsis phenotyping used a consensus k-means clustering method [9]. Several recent works on clustering extended k-means with deep autoencoder (AE) because of its ability to model non-linear, high dimensional structures in data. Deep embedded clustering (DEC) [12] pretrained a denoising AE and refined a centroid-based clustering objective, while Improved DEC (IDEC) [3] further employed an undercomplete AE to preserve local structures. Deep clustering network (DCN) [13], on top of AE, optimized a k-means-based loss on the embedded space. Hyperspherical clustering (HC) [2], while also used an AE, enforced embeddings on a hypersphere where cluster membership could be directly obtained. Our method builds upon HC as its objective ensures that embeddings are learned for the direct purpose of clustering.

Disentangled representation learning (DRL) assumes that observed data is generated from a set of latent factors, and recovers these factors in embeddings. Although [5] shows that disentanglement is hard to achieve without supervision, useful properties of embeddings have been proposed and generalized into *modularity*, *compactness*, and *explicitness* [1,5,8]. We focus on modularity, which is used in DRL when an one-to-one mapping relation between an embedding dimension and a factor dimension is needed. Although we do not have latent factors as in DRL, we want that each dimension of our embedding to correspond to one feature, i.e. we want the *modularity* property but with respect to feature (not latent) variables. Note we refer to this modified modularity as a disentangling property but our setting is different from classic DRL. Inspired by the F-statistic loss [8] that optimizes for modularity and class homogeneity using limited dimensions, we derive a soft F-statistic loss promoting homogeneous and disentangling embeddings useful to create clusters separating on limited features.

2 Method

DHC consists of a HC network for clustering and a soft F-statistics loss for achieving disentangled clusters separated by limited features. We employ the HC in [2] which builds upon an AE. During training, an input instance $x \in \mathcal{R}^m$ is encoded into embedding $z \in \mathcal{R}^h$ and then decoded back into $\widehat{x} \in R^m$. From z a subnetwork learns clustering probability $c \in \mathcal{R}^k$. The loss term for training HC is defined as $\mathcal{L}_{HC}$. Full details of HC can be found in our Appendix.

2.1 Soft *F*-statistic

Assume we have two classes (α, β) and embedding z where $z_{k,i,j}$ denotes instance j of class i for dimension k. HC gives us probability $c_{i,j}$ of instance j to be clustered as class i. The soft F-statistic based on [8] for dimension k of embedding z with two class identities is:

$$s_k = \tilde{n} \frac{\sum_i n_i (\bar{z}_{k,i} - \bar{z}_k)^2}{\sum_{i,j} (c_{i,j} z_{k,i,j} - \bar{z}_{k,i})^2} \tag{1}$$

where $\bar{z}_{k,i} = \frac{1}{n_i} \sum_j c_{i,j} z_{k,i,j}$, $\bar{z}_k = \frac{1}{n_\alpha + n_\beta} \sum_{i,j} c_{i,j} z_{k,i,j}$, $n_\alpha = \sum_j c_{\alpha,j}$, $n_\beta = \sum_j c_{\beta,j}$, and $\tilde{n} = n_\alpha + n_\beta - 2$. Maximizing the clustering separation is equivalent to maximizing the CDF of the F-distribution, which can be approximated by the regularized incomplete beta function I as:

$$\Phi(k, \alpha, \beta) = \Pr(S < s_k | \mu_{k,\alpha} = \mu_{k,\beta}, \tilde{n}) = I(\frac{s}{s + \tilde{n}}, \frac{1}{2}, \frac{\tilde{n}}{2}) \tag{2}$$

For any class pair chosen from the overall class set C, we select a set $D_{\alpha,\beta}$ of top k features with top cluster separations (CDFs). Phenotypic groups of patients may differ on some characteristics yet remain the same on others. The joint loss function to optimize over all class pairs among top features is:

$$\mathcal{L}_F = - \sum_{k \in D} \sum_{\alpha,\beta \in C} \ln \Phi(\alpha, \beta, k) \tag{3}$$

Total Loss. The total trainable loss is defined as $\mathcal{L}_{total} = \mathcal{L}_{HC} + \lambda_F \mathcal{L}_F$.

2.2 Measurements

Modularity Score. Modularity measures how well that each embedding dimension could be mapped to one feature dimension. Soft F-statistic loss gives us clusters separating on top k embedding dimensions, and we want to map them to features for phenotyping. We extend the modularity score defined by [8]. Let m_{ij} denote the mutual information between embedding dimension i and feature dimension j. For each i, a template vector $\boldsymbol{t}_i$ denotes the idea modularity case, i.e. embedding dimension i encodes information of only one feature dimension:

$$t_{i,j} = \begin{cases} m_{ij} & \text{if } j = \text{argmax}_g(m_{ig}) \\ 0 & \text{otherwise} \end{cases}, \delta_i = \frac{\sum_j m_{ij} - t_{ij}}{\theta_i^2 (N-1)} \tag{4}$$

where δ_i is the dimension-wise deviation from the idea template and N is the number of features. Modularity is measured as the average of $1 - \delta_i$ over all i.

Predictive-Ability-From-Few-Features Analysis. Recall the soft F-statistic loss promotes cluster pairs to separate on k embedding dimensions, and the modularity property allows the mapping from each of the k embedding dimensions to one feature variable with which is has the highest mutual information. To further evaluate how well *these selected features* predict cluster membership, termed as predictive-ability-from-few-features property (PAFFF), we perform the following analysis: for DHC, we train a multilayer perceptron (MLP) to predict cluster membership based on the selected features for each cluster pair (train/test split is 80/20, see the Appendix for details). We calculate the average AUC score of the test set across all cluster pairs. For each of the clustering methods under comparison, a same number of input features are randomly selected to predict corresponding clusters for 100 random runs, and the average AUC is calculated.

Unsupervised Clustering Metrics. We use silhouette coefficient score (SS) with cosine similarity, Calinski-Harabasz Index (CHI) and Davies-Bouldin Index (DBI) as our evaluation metrics. Detailed formula can be found at scikit-learn.

3 Experiments

Data. Data was collected under the National Institutes of Health-funded Sepsis Endotyping in Emergency Care project , with 43,086 critical care patients meeting Sepsis-3 criteria. A complete data description can be found in the Appendix.

Training Details. Our network architecture followed [2]. We chose top **10** (k) dimensions in soft F-statistic, and **4** clusters to directly compare with the benchmark sepsis phenotypic clusters [9]. We compare with classic clustering methods including K-means++ (KM++), consensus K-means (CKM), Agglomerative clustering (AG), and Gaussian Mixture Models Clustering (GMM) as well as AE-based methods including DEC, IDEC, DCN and HC. Training details for both DHC and the baselines can be found in the Appendix.

4 Results and Discussions

Table 1 records the clustering results. Note that for PAFFF below the dashed line, DHC use features that are mapped back from the k distinguishing embedding dimensions identified by the soft F-statistic. We could do the mapping because DHC has a high modularity score of 0.95. Our DHC method outperforms all other algorithms in terms of SS, indicating good inter-cluster separation and intra-cluster cohesion. DCN and CKM outperform us in terms of CHI and DBI, and DBI could be low for us since we do not optimize for centroid. In general, our method has competitive unsupervised clustering power as we perform relatively well across the three metrics. In PAFFF, DHC has the highest AUC, meaning our limited features have the best predictive power for cluster membership. This property encourages us to build our phenotypes from these features, as clinicians may assess subgroup membership with only partial information collected.

Table 1. SS, CHI, and DBI (higher SS, higher CHI, and lower DBI indicates better clusters), and AUC from PAFFF analysis. Above the dashed line input features used in PAFFF are selected at random, and below, DHC use features mapped back from the embedding dimensions.

Model	SS	CHI	DBI	AUC
AG	0.001	1204	3.70	0.82
GMM	0.011	329	7.37	0.82
DEC	0.011	372	5.46	0.85
IDEC	0.039	1034	5.62	0.60
DCN	0.081	2107	**3.24**	0.73
KM++	0.080	1855	3.76	0.81
CKM	0.090	**2140**	3.30	0.81
HC	0.094	2024	3.39	0.84
DHC (ours)	**0.095**	2043	3.36	0.83
1-5 DHC (ours)	**0.095**	2043	3.36	**0.88**

Table 2. Cluster phenotypes. The number next to the variable indicates the number of other clusters on which this cluster differs.

Counts ratio	Feature differences	Cluster description
C1 11086 (0.26)	Bilirubin 2, BUN 2, **ESR 3**, **PaO2 3**, Sodium 2,	Healthier, low-risk group
C2 11181 (0.26)	Bilirubin 2, BUN 2, **HCO3 3**, Chloride 2, Glucose 2, sBP 2, Sodium 2	Acid-base imbalance; relative hyponatremia
C3 10691 (0.25)	Bilirubin 2, BUN 2, Glucose 2, Lactate 2, Sodium 2	Renal dysfunction; inflammatory; low lactate
C4 10128 (0.23)	Bilirubin 2, BUN 2, Chloride 2, GCS 2, Glucose 2, sBP 2, Sodium 2	Electrolyte derangement; advanced signs; high mortality

Sepsis Phenotyping. Table 2 shows our cluster phenotypes. We compare our phenotypes with the benchmark CKM phenotypes [9]. Both cluster indices are sorted based on increasing death ratios. We observe that (Appendix Fig. 1): 1. Our healthier, low-risk cluster 1 has the lowest median ESR compared to the rest clusters, not observable in CKM. 2. Our cluster 2 with relative hyponatremia has the lowest median sodium level compared to the rest, not observable in CKM. 3. Our cluster 3 with low lactate has the lowest median lactate compared to the rest, not observable in CKM. In general, although our phenotypes largely correspond to the CKM results, the median of selected features separate more across clusters in our method, revealing more distinct subgroup phenotypic difference.

References

1. Eastwood, C., Williams, C.K.I.: A framework for the quantitative evaluation of disentangled representations. In: ICLR (2018)
2. Fillmore, N., et al.: Hypersphere clustering to characterize healthcare providers using prescriptions and procedures from medicare claims data. In: AMIA (2019)
3. Guo, X., et al.: Improved deep embedded clustering with local structure preservation. In: IJCAI-17, pp. 1753–1759 (2017)
4. Knox, D.B., et al.: Phenotypic clusters within sepsis-associated multiple organ dysfunction syndrome. Intensive Care Med. **41**(5), 814–822 (2015)
5. Locatello, F., et al.: Challenging common assumptions in the unsupervised learning of disentangled representations. In: Proceedings of the 36th ICML (2019)
6. Marshall, J.C.: Why have clinical trials in sepsis failed? Trends Mol. Med. **20**(4), 195–203 (2014)
7. Moser, J., et al.: Identifying sepsis phenotypes. JAMA **322**(14), 1416–1416 (2019)

8. Ridgeway, K., Mozer, M.C.: Learning deep disentangled embeddings with the f-statistic loss. In: Advances in NIPS 31, pp. 185–194. Curran Associates, Inc. (2018)
9. Seymour, C.W., et al.: Derivation, validation, and potential treatment implications of novel clinical phenotypes for sepsis. JAMA **321**(20), 2003–2017 (2019)
10. Shankar-Hari, M., et al.: Developing a new definition and assessing new clinical criteria for septic shock. JAMA **315**(8), 775–787 (2016)
11. Talisa, V.B., et al.: Arguing for adaptive clinical trials in sepsis. Front. Immunol. **9**, 1502 (2018)
12. Xie, J., et al.: Unsupervised deep embedding for clustering analysis. In: ICML (2016)
13. Yang, B., et al.: Towards k-means-friendly spaces: simultaneous deep learning and clustering. CoRR abs/1610.04794 (2016)

Phenotypes for Resistant Bacteria Infections Using an Efficient Subgroup Discovery Algorithm

Antonio Lopez-Martinez-Carrasco[1(✉)], Jose M. Juarez[1], Manuel Campos[1,2], and Bernardo Canovas-Segura[1]

[1] AIKE Research Group (INTICO), University of Murcia, Murcia, Spain
{antoniolopezmc,jmjuarez,manuelcampos,bernardocs}@um.es
[2] Murcian Bio-Health Institute (IMIB-Arrixaca), Murcia, Spain

Abstract. The phenotyping process consists of selecting sets of patients of special interest and identifying their key characteristics. Subgroup Discovery (SD) is a suitable supervised approach for this task. In this work, we have proposed a two step process with an efficient SD algorithm (VLA4SD) for an exhaustive exploration of the search space with very effective prunes based on equivalence classes. We use the Coverage and the Incremental Response Rate quality measures to evaluate general and interesting subgroups. The suitability of our approach has been tested by identifying phenotypes of patients in the MIMIC-III open access database.

Keywords: Subgroup Discovery · Patient phenotype · Algorithm

1 Introduction

The phenotyping process consists of selecting sets of patients of special interest and identifying their key characteristics [4]. Descriptive Machine Learning techniques have proven to be useful as regards generating clinical hypotheses of this nature. However, from a practical perspective, clinicians' low level of confidence in such techniques is a limiting factor and some researchers are, therefore, currently, focusing on interpretability and patient traceability properties [2,3].

Subgroup Discovery (SD) methods are a suitable approach by which to tackle phenotyping. SD methods are supervised methods that obtain descriptions of subgroups of the population. These descriptions are simple explicit (i.e. legible and interpretable) relations of variable-value of the dataset with respect to a target variable of interest and a quality measure.

In this research, we propose a two step process for extracting phenotypes using an efficient SD algorithm (VLA4SD) based on the equivalence classes strategy. We study the suitability of our proposal as regards analyzing patients infected with different Staphylococcus and Enterococcus microorganisms treated with Vancomycin recorded in the MIMIC-III open access database [1]. We evaluate both the performance of the algorithm and the subgroups obtained with the Incremental Response Rate quality measure.

A. Tucker et al. (Eds.): AIME 2021, LNAI 12721, pp. 246–251, 2021.
https://doi.org/10.1007/978-3-030-77211-6_27

2 Methods

We first introduce the general definitions used in SD techniques. A *selector* e is a 3-tuple of the form '`<attribute name, operator, attribute value>`' which represents a property of a set of individuals in a dataset. A *pattern* p is a sequence '`[selector]+`' of non-repeated selectors that represents the characteristics of a set of individuals in a dataset. A *subgroup* s is a 2-tuple of the form '`<description, target variable>`' in which the description is a pattern and the target variable is a selector. Given a subgroup s and a dataset D, a *quality measure* $q(s, D) \in \mathbb{R}^+$ is a function that quantifies how interesting s is according to certain criteria in a dataset.

2.1 Algorithm Proposal

VLA4SD is an efficient algorithm for exhaustive exploration that uses the Equivalence Classes [5] search strategy. In this approach, the key element is a data structure denominated as the Vertical List (a.k.a. TID-List). A Vertical List vl contains a pattern (*vl.selectors*), a list with the IDs of the dataset instances covered by the pattern (*vl.instances*), and a quality (*vl.quality*). This strategy has two advantages: the operations to refine the subgroups and compute their quality are very efficient, and the algorithm to explore the equivalence classes is easily parallelizable. The VLA4SD algorithm consists of two methods: the main method (shown in Algorithm 1) and the search method (shown in Algorithm 2).

2.2 Subgroup Discovery Process for Phenotypes

A two step process for SD is proposed: (1) we use an exhaustive SD algorithm to extract general subgroups, and (2) we post-process and evaluate the subgroups obtained to select a set of subgroups with some properties. It is important to state that our interests lay in large subgroups and, therefore, we chose the Coverage quality measure since it focuses on the generality of the subgroups. In the second step, we obtain a set of consistent patterns that cover different parts of the dataset and to rank the subgroups according to their predictive value. To this end, we firstly run a tenfold cross stratified evaluation with the full dataset and only the subgroups present in all the folds are selected. For the ranking, we chose the Incremental Response Rate (IRR) measure from the marketing and uplift modeling field. As last step, we keep the most relevant and general subgroups by removing the refinements of the subgroups whose IRR is equal or lower than the one of its parent subgroup.

3 Experiments and Dataset

In order to allow a reproducible research, instead of using the private clinical database of the SITSUS research project (www.um.es/sitsus), we analyzed a clinical dataset generated from the MIMIC-III open access database [1]. The final

Algorithm 1. VLA4SD algorithm. Main method.

Input: D {dataset}, target {selector}, q {quality measure}, min_threshold {$\mathbb{R}$}

1: $\mathcal{P} := []$; $\mathcal{S} := []$; $\mathcal{E} := []$
2: Scan D in order to obtain the list of individual selectors (adding them to $\mathcal{E}$)
3: For each selector $e \in \mathcal{E}$, create a Vertical List vl, $\mathcal{P}.add(vl)$, and compute $e.quality := q(< [e], target >, D)$
4: For each instance $r \in D$, compute $r.quality := \sum_{selector\ e \in r} e.quality$
5: For each Vertical List $vl \in \mathcal{P}$, compute $vl.quality := \sum_{instance\ r \in vl} r.quality$
6: Sort $\mathcal{P}$ according to the quality of each $vl \in \mathcal{P}$
7: Prune the Vertical Lists in $\mathcal{P}$ such that $vl.quality < min\_threshold$
8: Create an empty matrix $\mathcal{M}$, where $\mathcal{M}[i, j] \in \mathbb{R}$ (i, j selectors acting as index)
9: **for each** $(vl_x, vl_y) \in \mathcal{P}$, with $x \geq y$ **do**
10: Generate a new Vertical List vl_{xy} such that:
$vl_{xy}.selectors := concatenate(vl_x.selectors, vl_y.selectors)$
$vl_{xy}.instances := vl_x.instances \cap vl_y.instances$
$vl_{xy}.quality := \sum_{instance\ r \in vl_{xy}} r.quality$
11: $\mathcal{M}[last(vl_x.selectors)][last(vl_y.selectors)] := vl_{xy}.quality$
12: **end for**
13: $\mathcal{F} := SEARCH(null, \mathcal{P}, min\_threshold, \mathcal{M})$
14: **for each** $vl \in \mathcal{F}$ **do**
15: $\mathcal{S}.add(s :=< [vl.selectors], target >)$ with quality $vl.quality$
16: **end for**
17: **return** $\mathcal{S}$

Algorithm 2. VLA4SD algorithm. `SEARCH` method.

Input: vl {Vertical List}, $\mathcal{P}$ {list of Vertical Lists}, min_threshold {$\mathbb{R}$}, $\mathcal{M}$ {Matrix}

1: $\mathcal{F} := []$
2: **for each** Vertical List $w \in \mathcal{P}$ **do**
3: **if** $w.quality \geq min\_threshold$ **then**
4: $\mathcal{F}.add(w)$
5: **end if**
6: **if** (w is not last in $\mathcal{P}$) and ($vl.quality + \mathcal{M}[last(w.selectors)][last(w.selectors)] \geq min\_threshold$) **then**
7: $\mathcal{V} := []$
8: **for each** Vertical List $z \in \mathcal{P}$, with z after w in $\mathcal{P}$ list **do**
9: **if** $\mathcal{M}[last(w.selectors)][last(z.selectors)] \geq min\_threshold$ **then**
10: Generate a new Vertical List new_vl such that:
11: $new\_vl.selectors := w.selectors.add(last(z.selectors))$
12: $new\_vl.instances := w.instances \cap z.instances$
13: $new\_vl.quality := \sum_{instance\ r \in new\_vl} r.quality$
14: $\mathcal{V}.add(new\_vl)$
15: **end if**
16: **end for**
17: **if** $\mathcal{V} \neq []$ **then**
18: $\mathcal{F} := concatenate(\mathcal{F}, SEARCH(w, \mathcal{V}, min\_threshold, \mathcal{M}))$
19: **end if**
20: **end if**
21: **end for**
22: **return** $\mathcal{F}$

Table 1. Set of relevant subgroups after post-processing.

#	Subgroup description	Size	IRR
0	culture_microorganism_name = Enteroc. SP., specimen_type= SWAB	788	0,681
1	microorganism = Enteroc. SP., icu_when_culture = SICU	609	0,611
2	microorganism = Enteroc. SP., service_when_culture = SURG	696	0,507
3	microorganism = Enteroc. SP., days_between_admission_and_first_ICU >0	623	0,458
4	admission_location = Transfer from Hosp/Extram, microorganism = Enteroc. SP.	424	0,441
5	microorganism = Enteroc. SP., patient_age = Adult, patient_gender = M	495	0,434
6	microorganism = Enteroc. SP., exitus = N discharge_location = Rehab/Distinct Part Hospital	516	0,406
7	microorganism = Enteroc. SP., icu_when_culture = NO ICU, patient_age = Adult	395	0,400
8	microorganism = Enteroc. SP., service_when_culture = MED	571	0,393
9	microorganism = Enteroc. SP., exitus = N, patient_age = Adult	687	0,364

mining view obtained had 9240 instances, 14 attributes and 129 selectors. By setting as target *'culture_susceptibility = Resistant'* and the minimum threshold 0.1 for the Coverage quality measure, we obtained 1432 subgroups.

4 Results and Discussion

With the process defined, we make two contributions. In the first place, the tenfold cross validation returns a set of consistent patterns, which is necessary for a small or medium size database. In the second place, the IRR quality measure focuses on the presence from the instances of the subgroup with the target variable (i.e., bacterium resistant to vancomycin) and on the absence from the instances without the target variable (i.e., bacterium sensitive to vancomycin). That is, the IRR promotes in the ranking the subgroups with a balance between positive predictive value and specificity.

The results obtained provided preliminary evidence that there are many interesting features and values related to antibiotic resistance. The top-10 subgroups shown in Table 1 illustrate that in the subgroups with the best quality: (1) only the microorganism 'Enteroc. SP.' appears, (2) only the hospital services 'SURG' and 'MED' appear, and (3) only include individuals in the surgical ICU 'SICU' and individuals that were not admitted in any ICU ('NO ICU').

Regarding the set of subgroups, the average number of selectors in the 1432 subgroups obtained is 3.5 and the average number of selectors in the top-10 subgroups is 2.4. These 10 subgroups have few descriptors and, together, cover

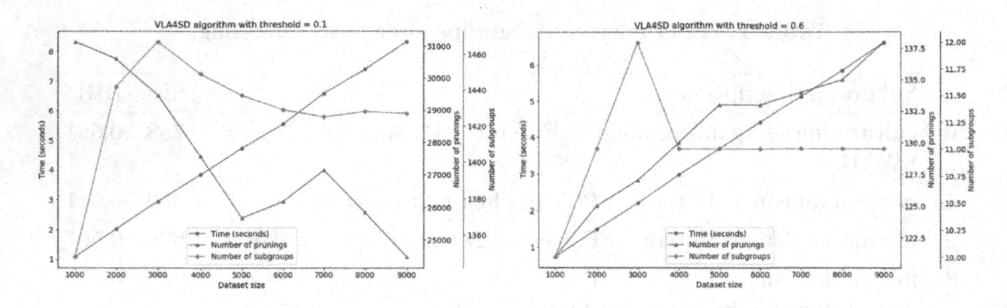

Fig. 1. Runtime, number of prunes and number of subgroups.

a good part of the population with resistance to vancomycin (the 76,98%, i.e. 1676 out of 2177).

Regarding the efficiency of the VLA4SD we can see in Fig. 1 the runtime, the number of prunes, and number of subgroups extracted using a coverage of 0.1 and 0.6 of the dataset for different dataset sizes.

5 Conclusions and Future Work

In this research, we propose a two step process based on Subgroup Discovery (SD) to extract phenotypes in a clinical database. We proposed an efficient exhaustive SD algorithm (VLA4SD) based on equivalence. In the experiments we show that the prunes included in VLA4SD are very efficient. VLA4SD shows a good execution time, and the preliminary results show a good scalability. However, a more comprehensive performance evaluation is required.

A contribution is the use the IRR as a quality measure that balances positive predictive value and specificity to rank the subgroups. The results obtained with the MIMIC-III database show some general and interesting subgroups related to antibiotic resistance using the coverage and IRR quality measures.

Acknowledgment. This work was partially funded by the SITSUS project (Ref: RTI2018-094832-B-I00), given by MCIU/AEI/FEDER, UE, and by a national grant (Ref:FPU18/02220) supported by the MCIU.

References

1. Johnson, A., et al.: MIMIC-III, a freely accessible critical care database. Sci. Data **3**, 160035 (2016)
2. Martinez-Carrasco, A.L., Juarez, J.M., Campos, M., Morales, A., Palacios, F., Lopez-Rodriguez, L.: Interpretable patient subgrouping using trace-based clustering. In: Riaño, D., Wilk, S., ten Teije, A. (eds.) Artificial Intelligence in Medicine (AIME 2019). Lecture Notes in Computer Science, vol. 11526, pp. 269–274. Springer, Cham (2019). https://doi.org/10.1007/978-3-030-21642-9_33
3. Stiglic, G., Kokol, P.: Discovering subgroups using descriptive models of adverse outcomes in medical care. Methods Inf. Med. **51**, 348–352 (2012)

4. Wojczynski, M.K., Tiwari, H.K.: Definition of Phenotype. Adv. Genet. **60**, 75–105 (2008)
5. Zaki, M.J., Parthasarathy, S., Ogihara, M., Li, W.: Parallel algorithms for discovery of association rules. Data Min. Knowl. Discov. **1**(4), 343–373 (1997)

Predicting Drug-Drug Interactions from Heterogeneous Data: An Embedding Approach

Devendra Singh Dhami[1,2](✉), Siwen Yan[1], Gautam Kunapuli[1,3], David Page[4], and Sriraam Natarajan[1]

[1] The University of Texas at Dallas, Richardson, USA
[2] Technical University of Darmstadt, Darmstadt, Germany
devendra.dhami@tu-darmstadt.de
[3] Verisk Analytics, Jersey City, USA
[4] Duke Univeristy, Durham, USA

Abstract. Most approaches for predicting drug-drug interactions (DDIs) have focused on text. We present the first work that uses multiple drug structure data - images, string representations and relationship representations. We exploit the recent advances in deep networks to integrate these varied sources of inputs in predicting DDIs. Our empirical evaluations clearly demonstrate the efficacy of combining heterogeneous data in predicting DDIs.

1 Introduction

ADEs account for as many as one-third of hospital-related complications, affect up to 2 million hospital stays annually, and prolong hospital stays by 2–5 d [5]. We focus on a specific problem of **drug-drug interactions** (DDIs), which are an important type of ADE and can potentially result in healthcare overload or even death. An ADE is characterized as a DDI when multiple medications are co-administered and cause an adverse effect on the patient. Predicting and discovering drug-drug interactions (DDIs) is an important problem and has been studied extensively both from medical and machine learning point of view. Identifying DDIs is an important task during drug design and testing, and several regulatory agencies require large controlled clinical trials before approval. Beyond their expense and time-consuming nature, it is impossible to discover all possible interactions during such clinical trials. This necessitates the need for computational methods for DDI prediction.

Our goal is to predict DDIs in large drug databases by exploiting heterogeneous data types of the drugs and identifying patterns in drug interaction behaviors. We take a *fresh and novel perspective on DDI prediction by seamlessly combining heterogeneous data representations* of the drug structures such as images, string representations and relations with other proteins. While in

D.S. Dhami and S. Yan—Equal contribution.

A. Tucker et al. (Eds.): AIME 2021, LNAI 12721, pp. 252–257, 2021.
https://doi.org/10.1007/978-3-030-77211-6_28

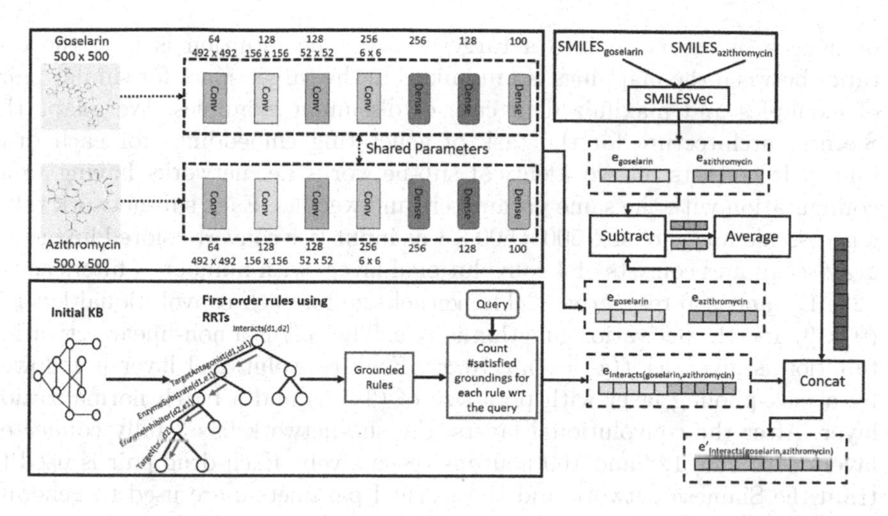

Fig. 1. Overview of architecture for predicting DDIs using heterogeneous data types.

principle, multi-view learning methods such as co-training [1] can be used, these methods assume that each view independently provides enough information for classification while we assume that each of these data source essentially provides a weak prediction of DDI. While it is possible to directly combine the data sources, standardization can be a major bottleneck. We take an embedding based approach to achieve the combination.

We make the following contributions: (1) we combine heterogeneous data types representing drug structures for DDI prediction. (2) we create embeddings to build a DDI prediction engine that can be integrated into a drug database seamlessly. (3) we show that using heterogeneous data types is more informative than using homogeneous data types.

Related Work: Previous approaches to DDI prediction have employed kernels on text data [11], multiple kernels on molecular properties [4], embeddings on a single source data [10] or Siamese GCNs on images [2]. Our work is *the first generalization of these multiple methods* where we consider multiple data sources including images and combine them seamlessly through embeddings.

2 Embeddings Using Heterogeneous Data Sources

We consider 3 different types of data, (1) images of drug structures, (2) SMILES (**S**implified **M**olecular **I**nput **L**ine **E**ntry **S**ystem) strings [12] representation of drug structures and (3) relational representation of various associations between the drugs and proteins (target, transporter and enzymes). Figure 1 shows the overall architecture of our approach. We now discuss the different components.

1. **Drug Structure Image Embeddings:** A discriminative approach for learning a similarity metric using a Siamese architecture [3] maps the input (pair

of images in our case) into a target space. The intuition is that the distance between the mappings is minimized in the target space for similar pairs of examples and maximized in case of dissimilar examples. We adapt the Siamese architecture for the task of generating embeddings for each drug image. It consists of two identical sub-networks i.e. networks having same configuration with the same parameters and weights. Each sub-network takes a gray-scale image of size $500 \times 500 \times 1$ as input (we convert colored images to gray-scale) and consists of 4 convolutional layers with number of filters as 64, 128, 128 and 256 respectively. The kernel size for each convolutional layer is (9×9) and the activation function is *relu*. The *relu* is a non-linear activation function is given as $f(x) = max(0, x)$. Each convolutional layer is followed by a max-pooling layer with pool size of (3×3) and a batch normalization layer. After the convolutional layers, the sub-network has 3 fully connected layers with 256, 128 and 100 neurons respectively. Each drug pair is used to train the Siamese network and the learned parameters are used to generate embeddings of dimension 100×1 for each drug image.
2. **Relational Data Embeddings:** DDIs can be considered as the characterization of the relationships between the drugs and the various proteins (enzymes, transporters etc.) using ADMET (absorption, distribution, metabolism, excretion and toxicity) features. A natural representation for such data is using first-order logic and the rules can then be induced. Using the given facts and the +ve and -ve examples, we learn a relational regression tree (RRT) where all the paths from the root to the leaves can be interpreted as first-order rules. The obtained first-order rules are first partially grounded with the query drug pairs and then completely grounded using the fact set. The number of satisfied groundings for each drug pair are then counted to obtain the final embeddings.
3. **SMILES Strings Embeddings:** SMILES strings represent the drug structure in form of a simple textual representation. We use the existing model of SMILESVec [13] which divides the SMILES string into several interacting sub-structures and then uses the word2vec method [8] to generate embeddings for these sub-structures. These embeddings are combined to generate the final embedding of the drugs.

The obtained embeddings need to be aggregated to generate a lower level representation. In the case of both image and SMILES strings embeddings (size 100×1), we hypothesize that more similar the structure of the drugs, higher is the probability of their interaction. To capture this similarity notion between both sets of embeddings, we use **subtraction** as the aggregation function to obtain 2 sets of embeddings for the image and SMILES strings data. These 2 sets are then averaged to obtain a single set of embeddings of size 100×1.

Each relational embedding represents the counts of the satisfied groundings of the query, in our case, $Interacts(d_1, d_2)$ i.e. the interaction between pair of drugs and is of the size 19×1 (19 is the number of first-order rules learned using the relational regression trees). The relational embeddings are concatenated with the combined embeddings obtained from the SMILES and image data to yield the final embedding size of 119×1 which can then be passes to a machine learning classifier. We choose a neural network since it is a universal approximator, can handle large number of features and also learns inherently aggregated latent features in the hidden layers. The over all architecure is presented in Fig. 1.

3 Empirical Evaluation

We aim to answer the following research question: Does using multiple data sources give an advantage over using a single data source? **Data set(s):** Our image data set consists of images of 373 drugs of size $500 \times 500 \times 3$ downloaded from the PubChem database[1] and converted to a grayscale format of size $500 \times 500 \times 1$. The images are then normalized by the maximum pixel value (i.e. 255). The SMILES strings of these drugs are obtained from PubChem and DrugBank[2]. For the relational data, we extract the different relations of the drugs with the proteins from DrugBank and convert it to a relational format with number of relations = 14 and the total number of facts = 5366. From the 373 drugs we create a total of 67, 360 drug interaction pairs excluding the reciprocal pairs (i.e. if drug d_1 interacts with drug d_2 then d_2 interacts with d_1 and are removed). From the 67, 360 drug pairs we obtain 19936 drug pairs that interact and 47424 drug pairs that do not.

Baselines: 7 baselines are considered. **Structural Similarity Index (SSIM)** is used for measuring perceptual similarity between images. **Autoencoders** are neural networks with an encoder and a decoder to extract features and then restore original images. **CASTER** [6] convert frequent substrings of SMILES strings to an embedded representation. The obtained latent features are then converted to linear coefficients for a decoder and predictor. **Siamese Neural Network with and without STNs** using contrastive loss. **RDN-Boost** [9] takes predictions of RRT to compute residues, and updates it with a new regression function fitted. **MLN-Boost** [7] boosts the undirected Markov logic networks (MLNs) using an approximation of likelihood.

Results: We optimize the Siamese network using the Adam optimizer with a learning rate of 5×10^{-5}, obtained using line search. We use the publicly available implementation[3] of SMILESVec method with default parameters. To learn the RRT, we use the publicly available software, BoostSRL[4], with the "-noBoost" parameter. For the classifier in our architecture we use a 4 hidden layer(s) neural network with hidden layer sizes 1000, 500, 200 and 50 with *relu* activation units and *Adam* optimizer. Table 1 shows the performance of our method.

To demonstrate the effectiveness of using heterogeneous data, we compare with methods that use homogeneous data. To that effect, the first 4 baselines consider the image data, CASTER uses the SMILES strings data and RDN-Boost and MLN-Boost use the relational data. The results show that combining

[1] https://pubchem.ncbi.nlm.nih.gov/.
[2] https://www.drugbank.ca/.
[3] https://github.com/hkmztrk/SMILESVecProteinRepresentation.
[4] https://starling.utdallas.edu/software/boostsrl/.

Table 1. Comparison of our method with baselines. The 1st 4 methods use images as input, CASTER uses SMILES strings and the next 2 use relational data.

Methods	Accuracy	Recall	Precision	F1 score
SSIM	0.519	0.487	0.304	0.374
Autoencoder	0.354	**0.911**	0.303	0.454
Siamese Network	0.837	0.780	0.705	0.741
Siamese Network + STN	0.823	0.825	0.661	0.734
CASTER	0.821	0.663	0.736	0.698
RDN-BOOST	0.773	0.832	0.413	0.552
MLN-BOOST	0.767	0.653	0.540	0.592
Our Method (agg=avg)	0.877	0.769	0.805	0.787
Our Method (agg=sub)	**0.884**	0.781	**0.818**	**0.799**
Our Method (with STN)	0.881	0.779	0.811	0.794

embeddings from heterogeneous data sources clearly outperform the methods using a single data source.

Conclusion: We considered the challenging task of predicting DDIs from multiple sources. To this effect, we combined multiple data sources and presented an architecture significantly outperforming strong baselines that learn from a single type of data. More rigorous evaluation using larger data sets is an interesting direction. Potentially identifying novel DDIs is an exciting future research. Allowing for domain expert's knowledge could significantly boost the performance of the architecture and can be achieved by considering the knowledge as constraints due to learning. Finally, understanding how it is possible to extract explanations of these interactions from the embeddings is an interesting direction.

Acknowledgements. We gratefully acknowledge DARPA Minerva award FA9550-19-1-0391. Any opinions, findings, and conclusion or recommendations expressed are those of the authors and do not necessarily reflect the view of the AFOSR, DARPA or the US government.

References

1. Blum, A., Mitchell, T.: Combining labeled and unlabeled data with co-training. In: COLT (1998)
2. Chen, X., Liu, X., Wu, J.: Drug-drug interaction prediction with graph representation learning. In: BIBM (2019)
3. Chopra, S., Hadsell, R., LeCun, Y., et al.: Learning a similarity metric discriminatively, with application to face verification. In: CVPR, vol. 1 (2005)
4. Dhami, D.S., Kunapuli, G., Das, M., Page, D., Natarajan, S.: Drug-drug interaction discovery: Kernel learning from heterogeneous similarities. Smart Health (2018)

5. DHHS: adverse events in hospitals: national incidence among medicare beneficiaries (2010). https://oig.hhs.gov/oei/reports/oei-06-09-00090.pdf
6. Huang, K., Xiao, C., Hoang, T.N., Glass, L.M., Sun, J.: Caster: predicting drug interactions with chemical substructure representation. AAAI (2020)
7. Khot, T., Natarajan, S., Kersting, K., Shavlik, J.: Learning Markov logic networks via functional gradient boosting. In: ICDM (2011)
8. Mikolov, T., Chen, K., Corrado, G., Dean, J.: Efficient estimation of word representations in vector space. arXiv preprint arXiv:1301.3781 (2013)
9. Natarajan, S., Khot, T., et al.: Gradient-based boosting for statistical relational learning: the relational dependency network case. Mach. Learn. **86**, 25–56 (2012)
10. Purkayastha, S., Mondal, I., et al.: Drug-drug interactions prediction based on drug embedding and graph auto-encoder. In: BIBE (2019)
11. Segura-Bedmar, I., Martinez, P., de Pablo-Sánchez, C.: Using a shallow linguistic kernel for drug-drug interaction extraction. J. Biomed. Inform **44**(5), 789–804 (2011)
12. Weininger, D.: SMILES, a chemical language and information system. 1.28 introduction to methodology and encoding rules. J. Chem. Inform. Comput. Sci. **28**, 31–36 (1988)
13. Öztürk, H., Ozkirimli, E., Özgür, A.: A novel methodology on distributed representations of proteins using their interacting ligands. Bioinformatics **34**, i295–i303 (2018)

Detection of Junctional Ectopic Tachycardia by Central Venous Pressure

Xin Tan[1(✉)], Yanwan Dai[1], Ahmed Imtiaz Humayun[1], Haoze Chen[1], Genevera I. Allen[1], and Parag N. Jain MD[2]

[1] Rice University, Houston, TX, USA
xin.tan@rice.edu

[2] Texas Children's Hospital, Houston, TX, USA
pnjain@texaschildrens.org

Abstract. Central venous pressure (CVP) is the blood pressure in the venae cavae, near the right atrium of the heart. This signal waveform is commonly collected in clinical settings, and yet there has been limited discussion of using this data for automatically detecting and monitoring arrhythmia and other cardiac events. In this paper, we introduce a signal processing and feature engineering pipeline for CVP waveform analysis. Through a case study on pediatric junctional ectopic tachycardia (JET), we show that our extracted CVP features reliably detect JET with comparable results to the more commonly used electrocardiogram (ECG) features. This machine learning pipeline can thus improve the clinical diagnosis and ICU monitoring of arrhythmia. It can also corroborate and complement the ECG-based diagnosis, especially when the ECG measurements are unavailable or corrupted.

Keywords: Arrythmia detection · Central venous pressure · Junctional ectopic tachycardia · Physiological signal feature extraction

1 Background and Introduction

JET is one of the most common types of tachyarrhythmia seen during early post-operative care [6] and is very dangerous and difficult to treat in an infant [7]. Currently, there is no automated bedside JET detection method that is available to clinicians, often leading to delay in diagnosis and subsequent provision of life-saving therapeutic interventions. Most of the current arrhythmia detection algorithms are based on electrocardiogram (ECG) waveform and result in a staggering number (∼72-99%) of false alarms [4]. Also, the absence of the P-wave of ECG is considered to be one of the primary morphological features of JET onset, while current methods cannot robustly detect and measure P-wave, which results in sub-optimal performance for ECG based classifiers.

Instead, the JET morphological features are more obvious for the CVP signal. The characteristics and amplitude of the CVP waveform components can change significantly with arrhythmia and tricuspid valve pathology [5,10]. Thus, CVP

A. Tucker et al. (Eds.): AIME 2021, LNAI 12721, pp. 258–262, 2021.
https://doi.org/10.1007/978-3-030-77211-6_29

provides valuable clinical information for arrhythmia diagnosis and automatic detection. Normal CVP waveform has 3 systolic components (c wave, x descent, v wave) and 2 diastolic components (y descent, a wave). During JET, the characteristics and amplitude of the CVP waveform components change significantly. A tall a wave, termed a cannon a wave is observed [5,10]. Figure 1a presents a processed and aligned median stack of CVP waveform. Upon JET onset, the fusion of a and c wave leads to an obvious cannon a wave. Although this morphological difference has been used extensively in clinical diagnosis, there has not been any published method of extracting CVP features for automatic arrhythmia detection. The major challenge is that the CVP waveform is very easily distorted by artifacts occurring through the water-filled, tubing transducer system and by respiration-induced cyclic changes. These artifacts and signal noises make CVP more difficult to analyze than the ECG signal. To solve this issue, we have developed a robust pipeline of removing these artifacts and extracting useful features to detect JET onset. We then compare the machine learning model based on extracted CVP features with the one based on gold standard ECG features [9,11].

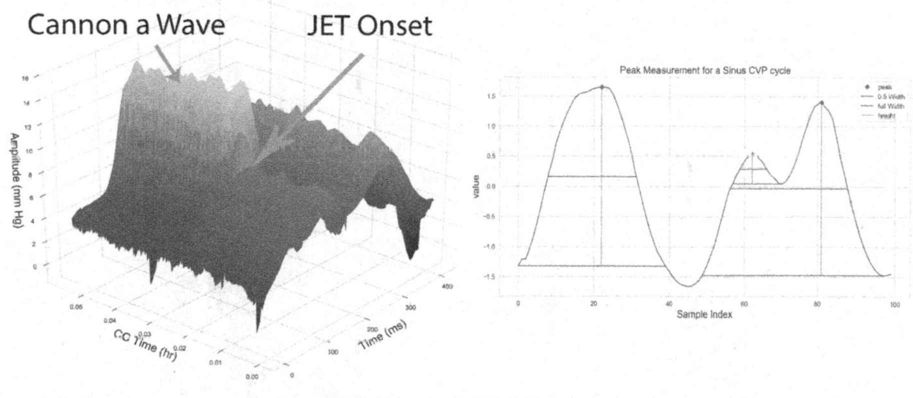

(a) Cannon a wave during JET Onset (b) Measure a, c, v waves for a single cycle

Fig. 1. CVP cycle morphology and measurement (Color figure online)

2 Methods

CVP data is very challenging to analyze because it is noisy and contains many artifacts, such as spikes and injection noise. We have developed an elaborate pipeline for CVP data preprocessing, which includes frequency filtering, spike removal, amplitude filter, median filtering, and dynamic alignment. Further details of all procedures can be found in an extended version of this paper [12]. These steps are able to recover true CVP waveform from very noisy data to facilitate reliable feature extraction.

As demonstrated by clinical study [5,10] and Fig. 1a, the primary morphological feature of JET onset lies in the a peak. We propose the following features extraction strategy to measure the characteristics of a peak and the overall CVP cycle waveform. We use four features to characterize the CVP a peak. Peak prominence measures how much a peak stands out from the surrounding baseline of the signal. In other words, it is the vertical distance between the peak and its lowest baseline (marked red in Fig. 1b). The peak height and width are the yellow and red lines in Fig. 1b. The width is identified as the distance between the detected endpoint in the peak, where the height is the distance from the peak to the baseline identified by the endpoint. The width of different relative heights can also be obtained. As shown in the graph, the green line represents the peak width at the 50% level of the peak. The peak slope is calculated as height/width.

Beyond features of the a wave, we introduce another 5 statistical features to characterize the overall shape of the waveform: waveform mean value, variance, and kurtosis, maximum value, and range. In Fig. 2, features in each boxplot display a clear separation between the sinus and JET group. These clinical-based features are able to characterize the morphological difference and provide reliable JET onset detection [5,10].

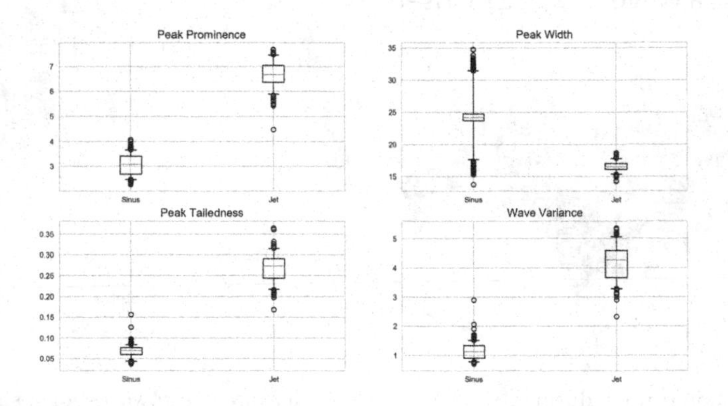

Fig. 2. Selected CVP features comparison

3 Experiment and Results

The data contains 23.3 h of signal with 6.3 h of JET and 17 h of sinus for 8 patients. Throughout the data, there are 4 channels available for the ECG signal and 2 channels for the CVP signal. We only select the channel that contains the best-quality signal to conduct feature engineering and subsequent classification experiment. We use a random forest model [1] with a maximum depth of 15 to classify JET vs sinus cardiac cycles from the 10 CVP features and 13 ECG features.

To demonstrate the effect of our proposed features, we compare them against 24 gold-standard ECG features that measure the temporal characteristics of the ventricular depolarization waves (QRS complex) [9,11]. They include the following: QRS complex widths, QS width, PR width, Peak Heights (P, R, Q, S), Peak Differences (PQ, RQ, RS), and normalized heart rate features, etc. These are well-validated features characterizing ECG waveform, and they have also been extensively utilized in ECG-based arrhythmia detectors [2,3,8,9].

We designed two experiments to demonstrate the effectiveness of the proposed CVP features. The first experiment conducts within-patient training and testing. The testing data and training data both come from the same patient with a 30%–70% split. The second experiment conducts cross-patient training and testing. In this experiment, the testing data comes from a single patient, and the training data comes from every other patient in the dataset. For each experiment, we report the sensitivity, specificity, and area under the curve (AUC) of the random forest model trained with CVP features, ECG features, and CVP + ECG features combined respectively. Note that we report the average feature importance of CVP and ECG features in the joint model in the extended version of this paper [12] (Table 1).

In both experiments, the model relying on CVP features alone achieves comparable performance with the model relying on ECG features. The within-patient experiment generally yields better performance than the cross-patient experiment. The reason is that each patient has other underlying diseases, which creates a morphological disparity in the CVP waveform. Despite the waveform disparity, the performance of the model relying on CVP features still matches the performance of the model relying on ECG features. As shown in the extended version of this paper, CVP features have higher importance scores than the ECG features in the joint model [12].

Table 1. Experiment result

Experiment	Within-patient performance								Cross-patient performance							
Test patient:	1	2	3	4	5	6	7	8	1	2	3	4	5	6	7	8
CVP features																
Sensitivity	1	0.99	1	1	0.53	1	0.95	1	1	0.38	0.98	0	0.26	0.99	0.93	0
Specificity	1	0.91	0.98	1	0.99	1	1	1	1	0.99	0.2	0.28	0.83	0.99	0.7	1
AUC	1	0.99	0.99	1	0.94	0.99	0.98	1	0.99	0.95	0.24	0.07	0.5	0.99	0.95	0.99
ECG features																
Sensitivity	1	1	1	1	0.79	0.98	0.98	1	1	0.17	0.89	0	0.3	0.97	0.93	0
Specificity	1	1	1	1	1	1	1	1	1	0.99	0.01	0.54	0.76	1	0.95	1
AUC	1	1	1	1	0.99	0.99	1	1	0.99	0.93	0.14	0.01	0.58	0.99	0.96	0.63
ECG + CVP																
Sensitivity	1	1	0.98	1	0.86	0.98	0.94	0.99	1	0.24	0.9	0	0.32	0.97	0.94	0
Specificity	1	1	1	1	1	1	1	1	1	0.98	0.01	0.36	0.77	1	0.91	1
AUC	1	1	1	1	0.98	0f.99	0.99	1	1	0.91	0.2	0.01	0.6	0.99	0.96	0.61

4 Conclusion

We have proposed a novel pipeline to process and extract features from Central Venous Pressure with a case study on the automatic detection of junctional ectopic tachycardia. The preprocessing pipeline and feature engineering pipeline provide a solution to remove complex artifacts in the CVP waveform and extract clinically valuable information. The within-patient and cross-patient experiments demonstrate that the CVP signal is as reliable as the ECG signal in detecting JET onset, and CVP features have higher importance scores than ECG features in the joint model. Thus, the quality and performance of arrhythmia detectors can be improved by incorporating CVP signals, and it can be particularly important when ECG signals are unavailable or contain major artifacts.

Acknowledgements. Genevera Allen acknowledges support from NSF DMS-1554821, NSF NeuroNex-1707400, and NIH 1R01GM140468.

References

1. Breiman, L.: Random forests. Mach. Learn. **45**(1), 5–32 (2001). https://doi.org/10.1023/a:1010933404324
2. Chang, C.C., Lin, C.J.: Libsvm. ACM Tran. Intelli. Syst. Technol. **2**, 1–27 (2011). https://doi.org/10.1145/1961189.1961199
3. deChazal, P., O'Dwyer, M., Reilly, R.: Automatic classification of heartbeats using ECG morphology and heartbeat interval features. IEEE Trans. Biomed. Eng. **51**, 1196–1206 (2004). https://doi.org/10.1109/tbme.2004.827359
4. Drew, B.J., et al.: Insights into the problem of alarm fatigue with physiologic monitor devices: a comprehensive observational study of consecutive intensive care unit patients. PLoS ONE **9**(10) (2014)
5. Fujita, Y., et al.: Central venous pulse pressure analysis using an R-synchronized pressure measurement system. J. Clin. Monit. Comput. **20**(6), 385 (2006). https://doi.org/10.1007/s10877-006-9035-y
6. Kabbani, M.S., Taweel, H.A., Kabbani, N., Ghamdi, S.A.: Critical arrythmia in postoperative cardiac children: recognition and management. Avicenna J. Med. **7**(3), 88–95 (2017)
7. Kylat, R.I., Samson, R.A.: Junctional ectopic tachycardia in infants and children. J. Arrhythm. **36**(1), 59–66 (2019)
8. Luz, E.J.D.S., Schwartz, W.R., Cámara-Chávez, G., Menotti, D.: ECG-based heartbeat classification for arrhythmia detection: a survey. Comput. Methods Programs Biomed. **127**, 144–164 (2016)
9. Mar, T., Zaunseder, S., Martinez, J.P., Llamedo, M., Poll, R.: Optimization of ECG classification by means of feature selection. IEEE Trans. Biomed. Eng. **58**, 2168–2177 (2011). https://doi.org/10.1109/tbme.2011.2113395
10. Pittman, J.A.L., Ping, J.S., Mark, J.B.: Arterial and central venous pressure monitoring. Int. Anesthesiol. Clin. **42**, 13–30 (2004)
11. Saenz-Cogollo, J.F., Agelli, M.: Investigating feature selection and random forests for inter-patient heartbeat classification. Algorithms **13**, 75 (2020)
12. Tan, X., Dai, Y., Humayum, A.I., Chen, H., Allen, G.I., Jain, P.: Detection of junctional ectopic tachycardia by central venous pressure. bioRxiv (2021). https://doi.org/10.1101/2021.04.02.438266

Planning and Decision Support

A Cautionary Tale on Using Covid-19 Data for Machine Learning

Diogo Nogueira-Leite[1], João Miguel Alves[1(✉)], Manuel Marques-Cruz[1], and Ricardo Cruz-Correia[1,2]

[1] Faculty of Medicine, University of Porto, 4200 Porto, Portugal
{up202002508,up199801351}@med.up.pt

[2] CINTESIS - Center for Health Technology and Services Research, 4200 Porto, Portugal

Abstract. *Introduction:* Good quality and real-time epidemiological COVID-19 data are paramount to fight this pandemic through statistical/machine-learning based decision-making support mechanisms.

Aims: Evaluate the resources available and used to gather COVID-19 epidemiological data by Portuguese health authorities from the onset of the pandemic until December 2020. The analysis laid on two main topics: (a) work processes at the Public Health Unit (PHU) level and (b) registry forms for epidemiological reporting and control procedures. Recommendations on requirements to overcome problems related to data integration and interoperability in order to build robust decision-making support mechanisms will also be produced.

Methods: For topic (a), we revised the Portuguese Directorate-General of Health (DGS) guidelines for data treatment. For topic (b), we analysed the forms used during first and second waves, while comparing them with DGS metadata provided to researchers.

Results: On topic (a), we detected the use of two complementary and non-interoperable systems. Further, the workflow does not seem to promote data quality and facilitates the occurrence of communication problems between health professionals. On topic (b), we found 27 deleted questions, 6 new questions, 1 displaced question, and 1 text modification between the 2 form versions.

Discussion: Both the workflow and data gathering methods are not the best suited for the generation of good quality data. They do not effectively support Public Health Professionals (PHP) nor provide the elements for posterior data analysis. The use of data by decision-making support mechanisms demands a careful planning of the data used to depict reality, and this condition is not met by the currently used forms.

Keywords: Data quality · Healthcare processes · Policy making · Public health surveillance · Workflow management

1 Introduction

COVID-19 has brought unprecedented pressure upon health systems across the world. After years of low investment, and of very little of it going to public health purposes (ca.

A. Tucker et al. (Eds.): AIME 2021, LNAI 12721, pp. 265–275, 2021.
https://doi.org/10.1007/978-3-030-77211-6_30

0,16% of GDP for Portugal in 2017 [1]), virtually all countries on Earth faced an enormous strain of their healthcare resources – infrastructural, technological, and human. However, few resources were, and remain, as scarce as trustworthy patient information. Overnight, and given the contagiousness and the fast spread of the disease, it was necessary to find effective ways to ensure timely verification, signalization, and communication that someone had been infected with COVID-19, so that appropriate measures could be taken.

COVID-19 is not only controlled by using all the precautionary measures we are aware of; it is, first and foremost, managed by gathering, analysing, and acting on accurate and reliable information [2–4]. The World Health Organization understands both roles as part of the Essential Public Health Operations [5], namely EPHO 2 (Monitoring and response to health hazards and emergencies) and EPHO 10 (Advancing public health research to inform policy and practice).

In general, to fulfil these operations, Portugal has a national epidemiological surveillance information system (SINAVE), in place since 2014, which dematerializes the mandatory notification of communicable diseases [6].

Regarding COVID-19, SINAVE includes a notification form and an epidemiological questionnaire for infection with SARS-CoV-2. As with other mandatory notifiable diseases, any physician in contact with a confirmed case of COVID-19 is legally obliged to notify the occurrence of transmissible diseases with mandatory reporting. The local public health authorities, and their respective Local Public Health Unit, responsible for the geographical area of the notified patient must then fill the epidemiological questionnaire, amidst the implementation of preventive and control measures.

Hence, local public health professionals, working in Public Health Units, are one of the firsts to come in contact with clinical data of COVID-19 cases and should ensure that these data are confirmed and corrected. Valuable information, such as epidemiological, should be added to the notification and epidemiological questionnaire forms.

If we wish to make sense of the potential wealth of data around COVID-19 we must first understand how they are gathered. As such, our aim is to evaluate the resources available and used to gather and manage COVID-19 epidemiological data by Portuguese health authorities from the onset of the pandemic until December 2020, focusing on describing how data for COVID-19 is expected to be collected and managed by public health professionals, including the questionnaires used for epidemiological surveillance.

2 Methods

This study was divided in two parts, namely (a) to study the work processes and (b) to study the registry forms used to collect the data.

2.1 Work Processes

We searched all relevant norms and guidelines from the Portuguese Directorate-General of Health, in order to shape an expected workflow for Public Health Units operating at a local level regarding the management of COVID-19 cases. Our main concern was to include all references to SINAVE or other platforms referred, as well as to all mandatory processes and professionals involved.

We then critically appraised this work method in light of the need for good quality data for decision making in healthcare.

2.2 Registry Forms

In March 2020, SINAVE included an initial notification form and epidemiological questionnaire for infection with SARS-CoV-2, that replaced the previous version referring to other coronaviruses, namely MERS-CoV. These forms were then updated in November 2020. We compare both versions with each other, searching for new insertions, deletions, displacements, and text modifications in each question of both documents. Simultaneously, we critically analysed some of the forms' fields regarding possible problems with the data obtained using those fields.

In April 2020, the Portuguese Directorate-General of Health (DGS) made a dataset updated weekly available for investigators upon request [7]. These data are, as informed by DGS, collected through SINAVE forms. We compared the variables included with the fields present in the forms. Results of this part were divided into three sections: *Notification Form, Epidemiological Questionnaire* and *Comparison with Metadata set*.

3 Results

3.1 Work Processes

All individuals presenting in primary care or in an emergency room with cough, fever, dyspnoea, anosmia or dysgeusia are considered suspect for infection with SARS-CoV-2 [8, 9]. All suspected cases must undergo laboratory testing and the physician doing the consultation must fill a notification form in SINAVE and introduce the patient in a second national platform named TraceCOVID [8]. After notified, the jurisdiction of the suspected case is transferred to the Public Health Unit (PHU) responsible for that geographical area. All SINAVE notifications must be answered, regardless of the laboratory test result [10]. This process, even for negative results, involves the filling of mandatory fields both in the notification form and the epidemiological questionnaire. Access to fill epidemiological questionnaire is limited as it is only available for Public Health Authorities (PHA), unless otherwise justified. Public Health Authorities are physicians with specialization in Public Health. Figure 1 represents the normal work processes in a PHU after the notification of a suspected case of infection with SARS-CoV-2.

A patient is considered a confirmed case of COVID-19 when a laboratory test comes back positive. Some confirmed cases can occur without the patient ever being suspected. In this instance, the physician or the laboratory first in contact with the confirmed case must notify the infection in SINAVE [8].

Then, local PHA, with the help of the professionals working in the local Public Health Unit (PHU), contact the confirmed case for the Epidemiological Questionnaire, identifying all contacts of the confirmed case. Those contacts must be introduced in both SINAVE (in a contact list spreadsheet) by the PHA and TraceCOVID by any PHU professional [11].

A risk assessment for all contacts is ensued by PHA and PHU professionals after this first contact, dividing by type of exposure in contacts with high-risk exposures and contacts with low-risk exposures. All high and low risk contacts must be informed of their risk evaluation and given specific preventive measures. These measures include reducing contacts for 14 days after exposure without need for isolation for low-risk exposures, and 14-day quarantine and symptom monitoring for high-risk exposures. Additionally, high risk individuals should be contacted daily or every other day by PHA or PHU professionals, and this contact must be registered in TraceCOVID. PHA must emit a Declaration of Prophylactic Isolation to all employed contacts or children of employed parents, to be sent to the patient as justification for work or school absence. Also, if deemed necessary by the PHA, which is in most cases of high-risk exposure, PHA should prescribe a laboratory test for infection with SARS-CoV-2 [11]. The confirmed case and all high-risk contacts must also be included in a nominal list to be sent to security forces and services, responsible to ensure that quarantine is being complied [12].

The European Centre for Disease Prevention and Control estimates that every epidemiological questionnaire takes between 45 min and 2 h, and a call to each contact takes between 3.5 and 20 min, and that every case has between 7–20 contacts, unless a lockdown is enforced, reducing that number to between 2–3 contacts [13]. This means that each new case would take an estimated minimum of 52 min and a maximum of 8 h and 40 min to complete all contact tracing activities, excluding the additional time dedicated to filling SINAVE and TraceCOVID forms and the daily or every other day call to high-risk contacts of previous confirmed cases.

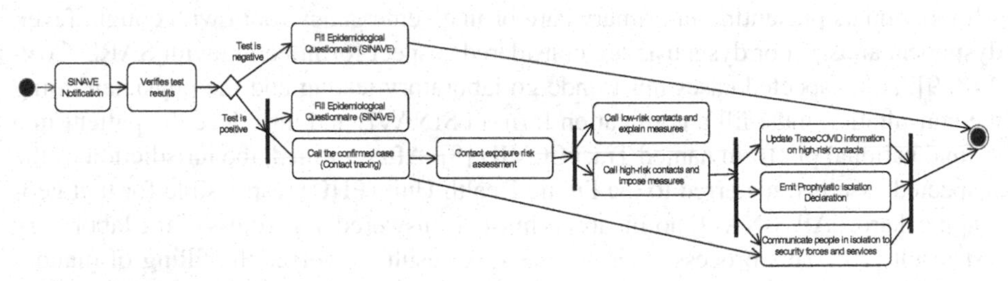

Fig. 1. Work processes in a Public Health Unit after notification of a suspected case of infection with SARS-CoV-2. All activities performed by PHP.

3.2 Registry Forms

The main differences between March and November versions of the forms are summarised in Table 1 and listed in Fig. 2. The variation in the number of fields was of –18 for the notification form and –9 for the epidemiological questionnaire.

Table 1. Summary of modifications in registry forms.

	Notification form	Epidemiological questionnaire
Questions existing only in original March form. Removed in November update	19	8
Questions not existing in original March form. Added to November update	3	3
Question existing in both versions but presented in different locations	1	-
Similar question in both versions, but with slightly different texts	1	-
Total number of fields:		
March 2020 (original version)	138	62
November 2020 (updated version)	120	53

Notification Form. The first noticeable change shows in the title of the document. Where it previously read "*Notification of Infection by new Coronavirus*", it now reads "*Notification of Infection by SARS-CoV-2/COVID-19*" in the November 2020 version – a clear statement on the purpose of this update. However, despite the expressed focus on Covid-19, a field "*Disease*" showing later in the report presents 2 unchanged dropdown options: i) "*Infection by MERS-CoV*", which seems to be out of the scope expected from the document's title, and ii) "*Infection by nCoV*", which despite being on scope of the documents, uses an outdated designation for SARS-CoV-2.

At the November 2020's update, the "*Clinical Presentation*" section had several symptoms removed: Pneumonia; Acute Respiratory Distress Syndrome; Convulsions; Pharyngeal exudate; Irritability/confusion; Abnormal pulmonary auscultation; Coma; Tachycardia; Neurocognitive disorder (including seizure); and Absence of an alternative diagnosis capable of fully explaining the disease. In replacement, new symptoms were included in the report list: Anosmia (loss of smell); and Dysgeusia/ageusia (gustatory dysfunction).

These changes reflect the reported list of symptoms recognized for Covid-19 by the European Centre for Disease Prevention and Control [14].

In the same section, the In-patient Hospitalization segment had questions on in-patient isolation removed. Additionally, a few listed symptoms have been removed as well: Chronic neurological or neuromuscular disease; Chronic neurological disorder; and Pregnancy Complications, including Acute renal failure, Heart failure, and Consumption coagulopathy.

Regarding the filling of symptoms-related fields, all questions are answerable using a drop list which reads "*Yes*", "*No*" and "*Unknown*". Since these fields are not mandatory, the fields could be left empty or answered with "*Unknown*", which would carry a myriad of different possible interpretations.

The "Clinical Manifestations" section has only two possible answers, besides "*Unknown*" and "*Not Applicable*": Moderate disease; and Severe disease.

The Portuguese Directorate-General of Health norms [8] state 4 levels of severity – the two previously referred, and "Mild" and "Critical", which are not selectable from the list. Therefore, the updated list of symptoms does not include all the symptoms that allow the physician to distinguish between the levels of disease severity.

In the "Laboratory Study" fields:

- A question was removed on whether a "biological sample was sent to the National Institute of Health (INSA)";
- A more specific question was added on whether the patient "has results of the laboratory test for qualitative detection of the SARS-CoV-2 antigen", including subfields for "Result", "Date of sample collection", and "Date of laboratory result";
- However, this new field regarding antigen testing has no option for not tested and includes various options that are not explained in the form, such as "No, awaits testing", "No, awaits result" and "No, pending".

No updates were made in "*Place of Occurrence*", "*Patient Identification*" and "*Physician Identification*" sections.

Our analysis identified a field which suffered a text modification while being kept at the same location. Whereas the original version read "*Lung X-ray with abnormalities*", the updated version reads "*Chest X-ray image with lesions compatible with SARS-CoV-2 infection*". In both cases, possible answers were the same: "*Yes*"/"*No*"/"*Unknown*". This text modification is worth of notice because it slightly alters the meaning of the question and, consequently, of the answer as well.

As the update results in a similar question and at the same position in the report, we cannot dismiss the possibility that the data codification for this field remained unchanged in the November 2020's update – an aspect that shall be taken in consideration for any data analysis which includes this field conveying data points from both input versions. Furthermore, the absence of a "*Not Applicable*" or "*Did not undergo Chest X-ray*" makes the interpretation of both "*No*" and "*Unknown*" answers difficult.

Furthermore, we noticed that the question on whether "*the patient is a health professional*" was displaced from its original location: whereas before it showed near the end of the report (last question of the "*Epidemiological Situation*" segment), it now appears in the beginning of the "*Notification*" section. This change to an earlier section of the report highlights the Health Authority's concern for the health condition of healthcare workers, particularly frontline workers. This is also revealed by this caption being one of the few that are mandatory to fill.

Along with this last field, the only other mandatory fields in the November 2020's update are "*Date of diagnosis*", "*In-patient hospitalization*", and "*Laboratory test performed*". Such low number of mandatory fields is certainly meant to obviate pressure in the report-filling process carried out by physicians, in a pandemic scenario where infection cases are constantly rising. Unfortunately, it comes at a cost: a potentially massive loss of data deemed dismissible by hectic professionals, that could otherwise provide invaluable information for a better management of COVID-19 outbreaks.

Epidemiological Questionnaire. As the previous document, the Epidemiological Questionnaire had its title amended to reflect a higher focus on Covid-19 specifically: from a previously broader designation "*Epidemiological Questionnaire of Infection by the new Coronavirus*", the updated title now reads "*SARS-CoV-2/COVID-19*".

The new version underwent deletions of 3 types of fields, specifically:

1. Fields with reference to MERS infection:
   a. "Recent stay (within 14 days before the onset of symptoms) in an area where MERS-CoV/nCoV infection has been reported or known"
   b. "Direct contact with dromedaries within 14 days before the onset of symptoms: exposure to the dromedary camel or food consumption (raw meat, unpasteurized milk, other) in an area where MERS-CoV infection has been reported or MERS-CoV in dromedaries is known"
2. Fields previously added in the beginning of Covid-19 outbreak:
   a. "Visits or work at live animal markets in Wuhan, China, 14 days before onset of symptoms"
   b. "Visits or work in live animal markets, 14 days before the onset of symptoms"
   c. "Contact with live animals, 14 days before the onset of symptoms"
3. Fields for unspecified coronaviruses:
   a. "Were the 3 amplification reactions directed to three different regions of the viral genome (gene E - 1st line screening test; RdRp gene (confirmatory test) and N gene - additional confirmatory test) carried out"
   b. "Were laboratory tests for pan-coronavirus carried out"

These deleted fields were not replaced by new ones. Notwithstanding, 3 new fields were included in different locations of the questionnaire. Not only these new fields are more specific for Covid-19, but also tackle relevant maters for tracking and understanding the spread of the infection:

- "[Patient is] Resident or worker in an institution with people in vulnerable situations and where there is reported transmission of COVID-19"
- "Unprotected laboratory exposure to biological material infected with SARS-CoV-2"
- "[Patient] Presents imaging changes, which meet the imaging criteria"

Deletions and adding were the only noticeable modifications to the epidemiological questionnaire. No displacements nor text modifications of previously existing questions could be found between both versions of this document.

These reported changes refer to the "*Epidemiological Enquiry*" and "*Conclusion*" sections of the questionnaire. Not surprisingly, no updates were made in "*Place of Occurrence*", "*Patient Identification*" and "*Physician Identification*" sections.

The first fillable field in the Epidemiological Questionnaire reads "*Diagnosis at the date of discharge*" and has no explanation to the type of answer wanted nor a list from where to choose, which leaves the physician to wonder how to fill this optional field.

Figure 2 portrays a comparative analysis of the fields' modifications between both versions of each registry form.

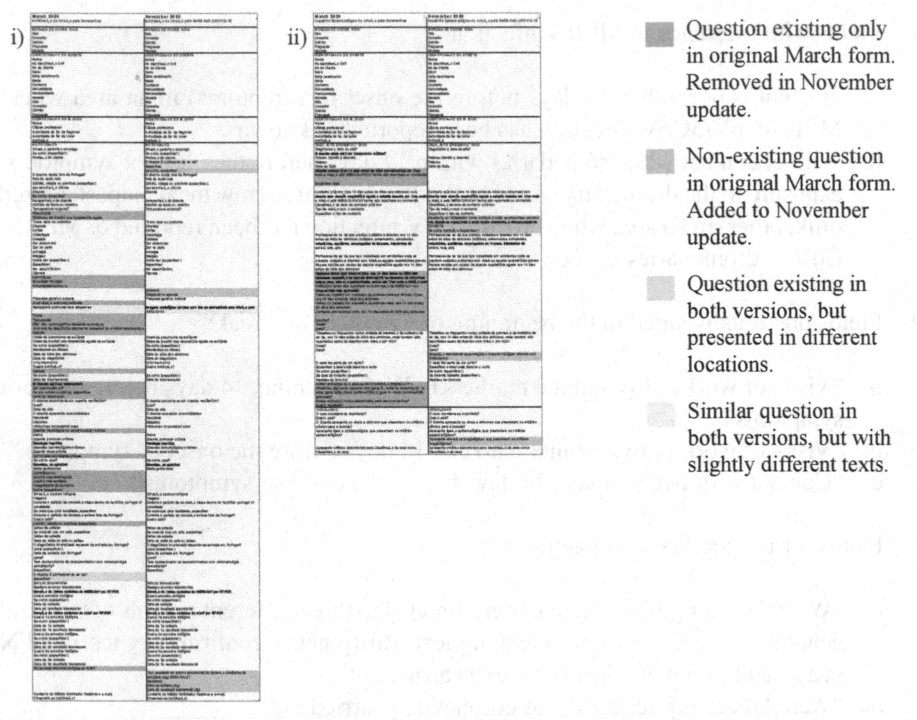
i)
ii)
Question existing only in original March form. Removed in November update.
Non-existing question in original March form. Added to November update.
Question existing in both versions, but presented in different locations.
Similar question in both versions, but with slightly different texts.

Fig. 2. Comparative analysis highlighting modifications between versions. Left-hand picture (i) represents the Notification Form; right-hand picture (ii) represents the Epidemiological Questionnaire. Each picture is vertically divided: left-hand panes represent original versions, dated March 2020; right-hand panes represent updated versions, dated November 2020.

Comparison with Metadata Set. The metadata set from the Portuguese Directorate-General of Health included 12 different variables.

Some of these variables are, to our best knowledge, collected in an automatic fashion and are likely correct, namely age and gender. *Probable place of infection*, classified in Portuguese NUTS 3 (municipalities) is collected through place of residence of the confirmed case, which unlikely describes accurately the most probable place of infection in all cases.

Other variables are likely the result of manual imputation on SINAVE forms. Admission to hospital and the respective date of hospitalisation are mandatory information to include in a notification. Nevertheless, if a confirmed case is not hospitalised when diagnosed, the answer to this field would have to be corrected afterwards, through manual search of the patient electronic health records by local or regional PHA, as there is no communication between these records and SINAVE.

Other manual imputed variables include date of first positive laboratory result, date of onset of disease, admission to intensive care unit, outcome of case, underlying conditions and need for respiratory support, collected through the notification form. These variables are, as previously discussed, not mandatory fields in the notification form, and if not filled by the physician who diagnoses the patient can only be corrected through manual search of the patient electronic health records by local or regional PHA, as with hospitalisation.

Costa-Santos 2020 [15] analysed the quality of the data of two requested datasets (one in April and other in August), and found added and deleted variables. The most important addition was date of diagnosis, which is a mandatory notification field and is likely correct, as it is the base for contact tracing and risk assessment operated by PHA and PHU professionals.

4 Discussion

4.1 Workflow Processes

The work processes described in the results section are likely to have been kept while the number of cases was relatively low, which might have been the case in the first wave of the pandemic in Portugal. For the first stage of the pandemic, however, the same processes may have been unsuited to handle the pressure resulting from the second wave. To allow for a comparison between stages, it took Portugal from March 16 until October 19 (217 days) to reach the 100.000 cases milestone; from October 19 until November 13 (25 days) to reach the following 100.000 cases, amounting to 204.664 in that day [16]. This brutal increase in contagion numbers forced PHU to adapt.

ECDC advised that in a situation with a large number of cases it would have to be necessary to reduce the intensity of interactions, replacing calling contacts with provision of initial information through other means, such as text messages. In this extreme scenario, only contacts reporting symptoms would receive follow-up phone calls [13]. This type of adaptations, whilst achieving productivity gains, is likely to do so at the expense of data quality and completeness, as the focus shifts from epidemiological surveillance to fast contact tracing.

Additionally, in the Portuguese scene, the redundancy of informatic platforms and of some processes (regarding the emission of a declaration per contact and the production of lists to security forces and services) are unlikely to have made up for a system which could effectively support PHU professionals in their contact tracing tasks, nor gather, curate and store data of sound quality which could be used by local, regional and national PHA to act on and tackle the pandemic.

Therefore, it is understandable that there was a need for various changes in the described workflow processes. Most importantly, efforts towards the case notification only after laboratory testing confirmation, the integration of the non-communicating systems (SINAVE and TraceCOVID) and the automation of processes leading to the emission of a declaration per contact and the production of lists to security forces and services should be prioritised. These changes would ensure that local PHA and PHU professionals could, even with an almost immeasurable increase in the number of confirmed cases, conciliate the primary mission of contact tracing with an effective epidemiological surveillance, leading to quality data supporting decision-making and investigation.

4.2 Registry Forms

The changes introduced in the registry forms likely increase the uncertainty around data quality and adds difficulty to data integration. For instance, the changes in symptoms observed in November 2020 make it plausible that symptoms not included as an option in March forms were then underreported and that the opposite is also true. Therefore, it is advisable to use caution when analysing data referring to symptoms obtained from SINAVE datasets.

Even though this uncertainty is more likely to occur when a field or a possible answer is deleted or a new one is added, changing the wording of a question without clear prior guidelines and keeping the same answers forcefully leads to different interpretations and, consequently, to answers with different meanings.

Additionally, we identified two problems with the forms as a whole: the lack of mandatory filling and the lack of clear indication of expected answers in some fields. A low number of mandatory fields, although meant to obviate pressure in the report-filling process carried out by physicians, comes at a cost: a potentially massive loss of data deemed dismissible, that could otherwise provide invaluable information. Both endanger the gathering of precious epidemiological information and make machine learning models harder to train, as it reduces the usable data for training and testing, without a clear outlined strategy to handle missing data.

After carefully analysing the DGS metadata set and comparing with the registry forms, there are some considerations that are worth noting. Firstly, the definition of the probable place of infection as the place of residence of the confirmed case has the potential to overlook important trends in infection starting in workplaces or related with tourism, with a potential to overestimate the infection rate of certain municipalities and underestimate in others, which directly affects decision-making.

Secondly, data regarding hospitalisation, as it is dependent on the need for admission upon notification or posterior correction, is likely to be underreported in the datasets assuming they are directly extracted from SINAVE.

Lastly, there is a risk of underreporting of other variables, namely date of first positive laboratory result, date of onset of disease, admission to intensive care unit, outcome of case, underlying conditions and need for respiratory support, mainly for two reasons: they are collected through the notification form, which means that these fields must be filled upon notification or await posterior correction; they are not mandatory fields.

Addressing all these concerns, or at least acknowledging them, is of paramount importance when conducting an investigation or building decision-making tools.

Acknowledgments. This work has been done under the scope of - and funded by - the PhD Program in Health Data Science of the Faculty of Medicine of the University of Porto, Portugal - heads.med.up.pt.

References

1. WHO Regional Office for Europe: Healthy, prosperous lives for all: the European Health Equity Status Report (2019). World Health Organization, Copenhagen (2019)
2. Morgan, O.: How decision makers can use quantitative approaches to guide outbreak responses. Philos. Trans. R. Soc. B Biol. Sci. **374**(1776), 20180365 (2019)
3. Yozwiak, N.L., Schaffner, S.F., Sabeti, P.C.: Data sharing: make outbreak research open access. Nature **518**(7540), 477–479 (2015)
4. Bo, X., et al.: Open access epidemiological data from the COVID-19 outbreak. Lancet Infect. Dis. **20**(5), 534 (2020)
5. WHO Regional Office for Europe: European Action Plan for Strengthening Public Health Capacities and Services. World Health Organization, Malta (2012)
6. Assembleia da República: Lei n.º 81/2009 de 21 de Agosto - Institui um sistema de vigilância em saúde pública, que identifica situações de risco, recolhe, actualiza, analisa e divulga os dados relativos a doenças transmissíveis e outros riscos em saúde pública, bem como prepara planos de contingência face a situações de emergência ou tão graves como de calamidade pública. Diário da República n.º 162/2009, Série I, pp. 5491–5495 (2009). Publication in Portuguese
7. Direção-Geral da Saúde: COVID-19 metadata (2020). https://covid19.min-saude.pt/wpcontent/uploads/2020/04/PT_COVID19_metadata-1.pdf. Accessed 13 Jan 2021. Publication in Portuguese
8. Direção-Geral da Saúde: Norma 004/2020 de 23/03/2020, COVID-19: Abordagem do Doente com Suspeita ou Confirmação de COVID-19. https://covid19.min-saude.pt/wp-content/uploads/2020/12/Norma-004_2020.pdf. Accessed 13 Jan 2021. First published 2020/03/23, updated 2020/10/14. Publication in Portuguese
9. Direção-Geral da Saúde: Norma 020/2020 de 09/11/2020, COVID-19: Definição de Caso de COVID-19. https://covid19.min-saude.pt/wp-content/uploads/2020/11/Norma_020_2020.pdf. Accessed 13 Jan 2021. First published 2020/11/09. Publication in Portuguese
10. Direção-Geral da Saúde: Despacho n.º 5855/2014. Ministério da Saúde, Diário da República n.º 85/2014, Série II, p. 11660 (2014). Publication in Portuguese
11. Direção-Geral da Saúde: Norma 015/2020 de 24/07/2020, COVID-19: Rastreio de Contactos. https://covid19.min-saude.pt/wp-content/uploads/2020/08/i026538.pdf. Accessed 13 Jan 2021. First published 2020/07/24. Publication in Portuguese
12. Presidência do Conselho de Ministros: Decreto n.º 2-A/2020 de 20 de Março. Diário da República n.º 57/2020, 1º Suplemento, Série I, pp. 11(5)–11(17) (2020). Publication in Portuguese
13. European Centre for Disease Prevention and Control: Contact tracing for COVID-19: current evidence, options for scale-up and an assessment of resources needed. European Centre for Disease Prevention and Control (2020)
14. European Centre for Disease Prevention and Control: Clinical characteristics of COVID-19. https://www.ecdc.europa.eu/en/covid-19/latest-evidence/clinical. Accessed 13 Jan 2021
15. Costa-Santos, C., Luísa Neves, A., Correia, R., Santos, P., Monteiro-Soares, M., Freitas, A., et al.: COVID-19 surveillance - a descriptive study on data quality issues (2020). https://doi.org/10.1101/2020.11.03.20225565
16. Associação Nacional de Médicos de Saúde Pública: COVID-19 - Mapa Epidemiológico Portugal. https://www.anmsp.pt/covid19-mapa. Accessed 13 Jan 2021

MitPlan 2.0: Enhanced Support for Multi-morbid Patient Management Using Planning

Martin Michalowski[1](✉), Malvika Rao[2], Szymon Wilk[3], Wojtek Michalowski[2], and Marc Carrier[4]

[1] University of Minnesota, Minneapolis, MN 55455, USA
martinm@umn.edu
[2] University of Ottawa, Ottawa, Canada
[3] Poznan University of Technology, Poznan, Poland
[4] The Ottawa Hospital Research Institute, Ottawa, Canada

Abstract. The complexity of patient care is growing due to an ageing population. As chronic illnesses become more common, the incidence of multi-morbidity increases. Generating disease management plans for multi-morbid patients requires the integration of multiple evidence-based interventions, represented as clinical practice guidelines (CPGs), that are designed to treat a single condition. Our previous work developed a mitigation framework called MitPlan that represented the generation of treatment as a planning problem. The framework used the Planning Domain Definition Language (PDDL) to represent clinical and patient information needed to identify and mitigate adverse interactions resulting from the concurrent application of multiple CPGs for a given patient encounter. In this paper we describe MitPlan 2.0 that supports shared decision-making by identifying a treatment plan optimized according to patient preferences, treatment cost, or perceived patient's adherence to medication. It mitigates adverse interactions using planning constructs, eliminating the need for procedural handling of adverse interactions and as such provides flexible and comprehensive decision support at the point of care. We demonstrate MitPlan 2.0's extended capabilities using synthetic scenarios approximating real-world clinical use cases and demonstrate its new capabilities within the context of atrial fibrillation.

Keywords: Clinical practice guidelines · Multi-morbidity · Planning

1 Introduction

Clinical practice guidelines (CPGs) are statements developed systematically from available evidence to assist practitioners in the management of a patient with a specific disease and their application improves quality of care and patient outcomes [1]. Yet their adoption in clinical practice is lacking and one of the major obstacles is the limited support for complex patients suffering from multi-morbidity [2]. Disease management of these

A. Tucker et al. (Eds.): AIME 2021, LNAI 12721, pp. 276–286, 2021.
https://doi.org/10.1007/978-3-030-77211-6_31

patients is multidimensional by nature, and apart from the clinical dimension needs to consider care settings and patient preferences to maximize patient outcomes.

In this work, we present a significant expansion of our previous planning-based approach to mitigation of adverse interactions between recommendations coming from different CPGs [3] that improves support for multi-morbid patient management by enriching the representation of patient and clinical information. This improvement allows us to provide support for shared decision making by explicitly capturing the multidimensional nature of treatment using a multivariate *objective function* customized to a specific patient encounter. We demonstrate how our improved framework, called MitPlan 2.0, builds upon and extends our previous work using both synthetic and clinical examples. We highlight its use across clinical settings, position it within the context of related work, and describe future work to realize our goal of integrating MitPlan 2.0 into a clinical decision support system used at the point of care.

2 MitPlan 2.0

In planning, a planner is given an initial state of the world, a set of desired goals, and a set of planning actions to find a sequence of planning actions that are guaranteed to generate a new state that satisfies the desired goal(s). Each planning action has a set of parameters, preconditions that must be true for the action to be taken, and effects resulting from its execution. These planning actions are also characterized by a duration, conditional effects, and a cost.

The original version of MitPlan (which we refer to as MitPlan 1.0), described in detail in [3], addressed the problem of mitigation by combining an algorithmic approach with the use of a planner. MitPlan 1.0 accepted as input patient data, patient preferences, the length of a planning horizon, and clinical goals and produced a safe management plan (i.e., where all adverse interactions were addressed), executed within the specified planning horizon. It detected adverse interactions using a combination of revision predicates and revision actions in its domain and mitigated these adverse interactions algorithmically and outside the planner. Specific to the revision operators in the MitPlan 1.0, the planner would terminate having identified what actions needed to be replaced/deleted/added to mitigate adverse interactions between different extended Actionable Graphs (AGs) [3] representing CPGs. The MitPlan 1.0 algorithm would then create a revised problem instance with new actions representing the required revisions and the planner would be applied to this instance. The process would be repeated until the management plan inferred by the planner required no further revisions. In addition to disjointed algorithmic and planning requirements for generating management scenarios, MitPlan 1.0 had very basic support for a single cost associated with the plan.

MitPlan 2.0 addresses these shortcomings by taking a fully planning-based approach, bringing the process of mitigation into the planning space via a new encoding of the planning problem in PDDL. We expand the extended AG to include the nodes derived from the underlying CPG and add to it all nodes introduced by revision operators that are possibly applicable to the AG. The extended AG encapsulates the contingencies introduced by revision operators, making them available to the planner much like patient preferences in MitPlan 1.0 [3].

Revision operators come from knowledge repository (KR). The representation of revision operators is flexible to include clinical actions found in many CPGs, clinical actions for a specific CPGs, and a combination of both. When applied to a specific AG, a revision operator is translated to a binary vector of length equal to the number of nodes in the AG, where each vector element indicates if a node is part of an adverse interaction addressed by the given revision operator. In this way a binary vector can represent multiple revision operators each with different costs. An adverse interaction is present only if a chosen path through the extended AG contains *all* the nodes flagged in a binary vector (if only a subset of nodes are contained, the adverse interaction is not present). If detected, the planner is forced to search for an alternate path in the AG, avoiding the adverse interaction. The binary vectors are automatically created for each planning problem instance, making their implementation scalable.

All nodes in the AG are associated with costs, and revision actions are designed to be costlier than the actions they (potentially) revise. The planner prioritizes paths through the AGs with no adverse interactions, finding a clinically feasible plan first and foremost, even if such a plan is more expensive than one that is clinically infeasible. If a path without executing a revision action does not exist, the planner chooses an alternate path containing revision actions that mitigate the adverse interactions and reach the goal nodes with no adverse interactions present. Because revision information is already built into the extended AG, the planner can optimize over various alternative paths, selecting the path associated with the lowest cost. As a simplification, MitPlan 2.0 implements cost minimization however utilities and their maximization are easily supported.

A revision action may itself introduce new adverse interactions, resulting in second-order adverse interactions. MitPlan 1.0 processed revisions sequentially and the plan depended on the order of revisions applied. In contrast, MitPlan 2.0 optimizes over all revision information in a single run and returns the optimal plan, if one exists, selecting the order of revisions to optimize the defined objective function.

MitPlan 1.0 treated patient preferences (e.g., preferred way of drug administration) as alternatives nodes in an AG [3] that are selected based on their dispreference costs (i.e., lower cost indicates a higher preference level). MitPlan 2.0 presents a natural extension of this approach, by unifying patient preferences and revisions, and modelling both as cost-based alternatives. At the same time, it considers more than one metric, each corresponding to a certain dimension, when looking for an optimal path. MitPlan 2.0 employs an objective function given as a weighted sum of selected cost metrics where weights indicate the importance of specific metrics. Possible metrics include various clinical (resources, specialists, capacity, etc.) and patient (financial, burden, preferences, adherence, etc.) indicators, thus it is possible to specify objective functions for a wide range of care settings (e.g., urban or rural), patient populations with unique attributes (e.g., health literacy, income, attitude towards treatment, etc.), and care planning approaches (physician-, nurse-, and patient-centered).

The overview of MitPlan 2.0 is given in Fig. 1(a) and the pseudocode illustrating its operation is given in Fig. 1(b). MitPlan 2.0 is invoked for a specific patient/physician encounter and takes patient data, patient preferences, a planning horizon, and an objective function as input. The objective function can be defined by both the physician and the patient in a shared decision-making process to combine patient- and the provider-oriented perspectives of management.

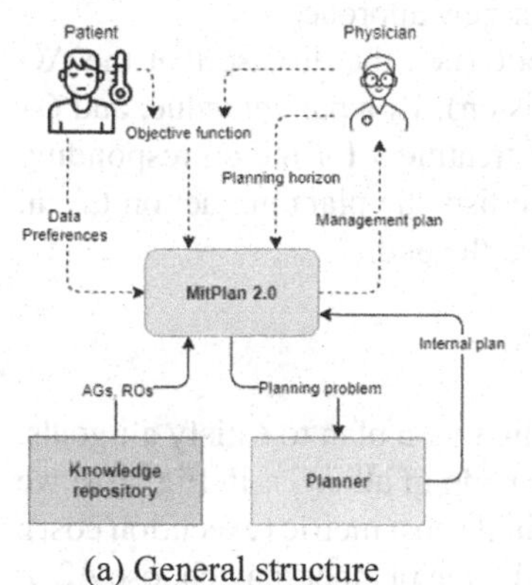

(a) General structure

input : data: patient data, prefs: patient preferences, horizon: planning horizon, objFunc: objective function for plan optimization
output: managementPlan: management plan
1 AGs := select from **knowledge repository** extended AGs representing CPGs used to manage the patient
2 ROs := select from **knowledge repository** revision operators possibly applicable to AGs
3 problem := create a planning problem using AGs, ROs, data, prefs, horizon and objFunc
4 internalPlan := apply **planner** to problem to find an optimal plan
5 managementPlan := post-process internalPlan to management plan
6 return managementPlan

(b) Operations (pseudo-code)

Fig. 1. Overview of MitPlan 2.0

MitPlan 2.0 starts by retrieving from the KR extended AGs that represent CPGs applied to manage conditions of the patient. It then identifies and retrieves from the KR revision operators (ROs) that are applicable to AGs selected in the previous step (see [3] for a more detailed explanation). Subsequently, it creates a planning problem in PDDL based on extracted AGs and ROs and provided input (patient data and preferences, planning horizon, and objective function). A planner solves the planning problem to obtain an internal plan optimized according to the provided objective function. Some of the actions in the internal plan are then filtered out as they represent implementation details of the planner (e.g., reaching a goal node) and as such are not relevant for patient management (e.g., performing a test, prescribing a drug). The management plan is finally presented to the physician and the patient.

MitPlan 2.0 significantly changed the management of revisions as stated above and the algorithmic part is no longer needed. We note that clinical quality of the final management plan depends on the completeness of the KR and the quality of available revision operators. If the KR is incomplete and does not contain an operator(s) addressing a certain interaction, then the resulting plan may be clinically unsafe. Moreover, if the KR contains a revision operator for a specific interaction that is not based on the most recent evidence, then the obtained plan may be clinically sub-optimal even though it optimized the objective function. The maintenance of the KR in knowledge driven CDSSs is an important part of their life cycle [4]. While it is beyond the scope of this work, we acknowledge it is critical for the clinical validity of MitPlan 2.0.

3 Illustrative Example

We illustrate the extensions introduced in MitPlan 2.0 through examples. We use synthetically generated examples, grounded in real-world applications and vetted by the physician on our team, to highlight how the embedding of costs (Sect. 3.1) and the new representation of revisions (Sect. 3.2) in the planning problem expands the capabilities of the planning framework. We use a subset of actions from the Atrial Fibrillation (Afib) CPG (Sect. 3.3) to show the exploration enabled by our new approach.

In each extended AG, *Dx* represents a context node (i.e., the disease that the AG represents), *Ax* a clinical action, *Tx* a clinical test (decision), *Vx* a patient value, and *Gx* a goal node representing the successful completion of treatment for the corresponding disease. For the MitPlan 2.0 generated plans, each line lists the planning action taken, the time step it is taken in at the start, and its duration at the end.

3.1 Cost Optimization

MitPlan 2.0 makes use of an objective function when finding a plan to satisfy all goals, where satisfying all goals means reaching all the goal node in all AGs. In Fig. 2(a) we illustrate the use of a simple objective function with a single cost metric (execution cost). Notice in this example that depending on the values of *V1* the planner may choose *A2* or *A5*. Suppose $V1 = 9$ and $V3 = 1$. While *A5* is costlier than *A2*, choosing *A2* will cause an adverse interaction and necessitate a revision. Because revisions are assigned a higher execution cost than original actions in the AGs (i.e., cost(*A7*) < cost(*newAction*)), the planner opts to traverse *A5* and *A6* to reach goal *G1* to minimize the overall cost of the plan. Figure 2(b) shows the generated plan.

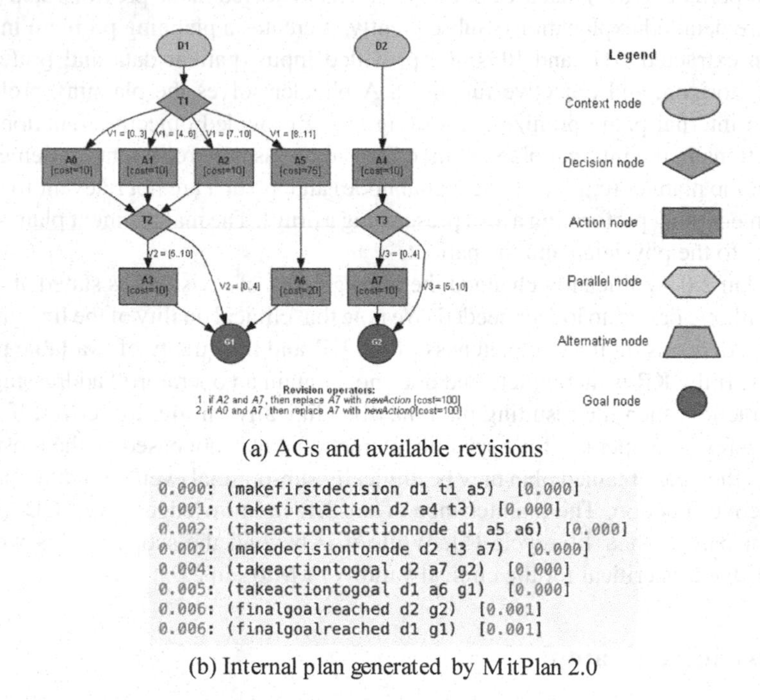

(a) AGs and available revisions

```
0.000: (makefirstdecision d1 t1 a5)  [0.000]
0.001: (takefirstaction d2 a4 t3)  [0.000]
0.002: (takeactiontoactionnode d1 a5 a6)  [0.000]
0.002: (makedecisiontonode d2 t3 a7)  [0.000]
0.004: (takeactiontogoal d2 a7 g2)  [0.000]
0.005: (takeactiontogoal d1 a6 g1)  [0.000]
0.006: (finalgoalreached d2 g2)  [0.001]
0.006: (finalgoalreached d1 g1)  [0.001]
```

(b) Internal plan generated by MitPlan 2.0

Fig. 2. Selection of minimum cost path

3.2 Revision Application

We demonstrated in Sect. 3.1 how execution costs of an action and optimization come into play when selecting actions. Here we build on that example to show how execution costs are used to select and prioritize revisions when adverse interactions are present. Consider the extended AG in Fig. 3(a) with patient values $V1 = 7$ and $V3 = 3$. In this case, an adverse interaction is unavoidable and one of the two available revisions must be taken. The two choices for replacing *A7* are *newAction*, which costs 100 units, and *newAction2*, which costs 50 units. Since *newAction2* is less expensive, the planner prioritizes *newAction2*, replacing *A7* with it. The generated plan is shown in Fig. 3(b). In MitPlan 1.0, the prioritization of revisions would have been done algorithmically and outside the planning process by revising a problem instance and passing it to the planner. Now the selection *and* application of revisions is embedded in the planner's search for an optimal plan.

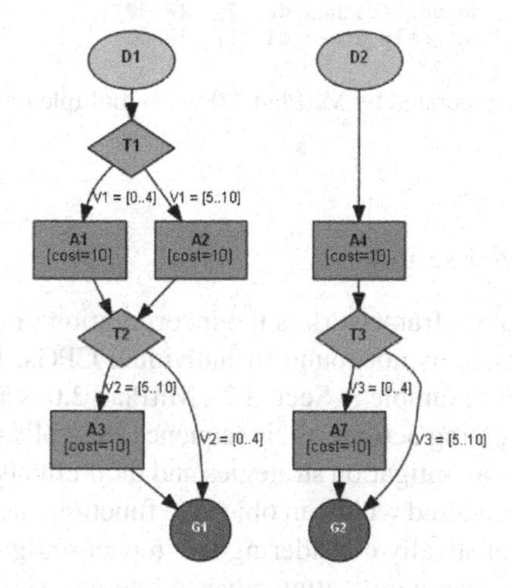

(a) AGs and available revisions

```
0.000: (makefirstdecision d1 t1 a2)  [0.000]
0.001: (takefirstaction d2 a4 t3)  [0.000]
0.002: (makedecisiontonode d2 t3 newaction2)  [0.000]
0.002: (takeactiontodecisionnode d1 a2 t2)  [0.000]
0.003: (makedecisiontonode d1 t2 a3)  [0.000]
0.004: (takeactiontogoal d2 newaction2 g2)  [0.000]
0.005: (takeactiontogoal d1 a3 g1)  [0.000]
0.006: (finalgoalreached d2 g2)  [0.001]
0.006: (finalgoalreached d1 g1)  [0.001]
```

(b) Internal plan generated by MitPlan 2.0

Fig. 3. Prioritization of revisions

We use the extended AG in Fig. 3(a) also to demonstrate how MitPlan 2.0 finds an optimal plan when multiple revisions are required (the second revision operator changed to *if A4 and A3 then replace A3 with newAction1* [$cost = 100$] for the sake of this example). In this problem instance, the patient values for *V1* and *V2* fall within the range [5 ... 10] and *V3* falls within the range [0 ... 4] , thus necessitating both revisions. Figure 4 shows the resulting plan that does just that. This case demonstrates that MitPlan 2.0 can apply any number of revisions necessary to find an optimal plan while tracking and accounting for any second-order effects from the applied revisions. This is a significant improvement over MitPlan 1.0 where each revision was applied individually by revising the problem instance and rerunning the planner.

```
0.000: (makefirstdecision d1 t1 a2)  [0.000]
0.001: (takefirstaction d2 a4 t3)  [0.000]
0.002: (makedecisiontonode d2 t3 newaction)  [0.000]
0.002: (takeactiontodecisionnode d1 a2 t2) [0.000]
0.003: (makedecisiontonode d1 t2 newaction1)  [0.000]
0.004: (takeactiontogoal d2 newaction g2)  [0.000]
0.005: (takeactiontogoal d1 newaction1 g1)  [0.000]
0.006: (finalgoalreached d2 g2)  [0.001]
0.006: (finalgoalreached d1 g1)  [0.001]
```

Fig. 4. Internal plan generated by MitPlan 2.0 when multiple revisions are required

3.3 Clinical Illustrative Example

At the core of our mitigation framework is the incorporation of external knowledge for mitigating adverse interactions not found in individual CPGs. By including revisions within the planning space (example in Sect. 3.2), MitPlan 2.0 is able to keep track of the application of these mitigating actions, their sequence of application, and their second-order effects. When clinical mitigation strategies and their effects are embedded into the planning process and considered within an objective function, the resulting management plan is constructed by holistically considering the impact mitigation strategies have on patient management. Iteratively mitigating adverse interactions, as done in the MitPlan 1.0 fails to account for second-order effects.

To demonstrate the power of using an objective function involving multiple metrices, consider the subset of the Afib AG, generated from the CPG, shown in Fig. 5 (full AG is presented in [3] and not reproduced here due to space limitations). In this example, an Afib patient with CHA_2SDS_2-VASc score greater or equal to 1 can be prescribed an anticoagulant such as Warfarin (*WARF*, dosage 5 mg daily) or one of the direct oral anticoagulants (DOACs) such as Dabigatram (*DABI*, dosage 110 or 150 mg twice daily). For a CHA_2SDS_2-VASc score less than 1, they are prescribed low dose aspirin (*ASP*).

MitPlan 2.0 can consider financial cost, patient's burden, or the "cost" of adherence, amongst others, during planning to generate different management scenarios. Let's first consider each independently. **Financial Cost**: the annual cost of anticoagulation treatment (at a dosage mentioned above) with Warfarin is about $55USD and with Dabigatram is about $1200USD. We therefore assign a financial cost of 55 to the WARF node and 1200 to the DABI node in the AG. When financial cost minimization is the optimization goal, treatment with Warfarin is the returned option (see the plan in Fig. 6(a)).

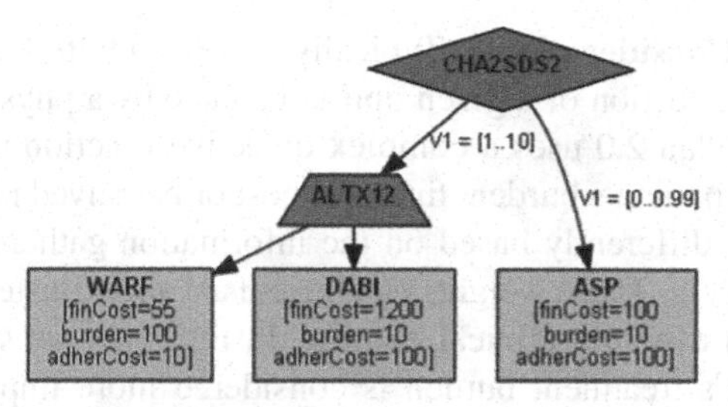

Fig. 5. Subset of the Afib AG from [3]

```
0.000: (makefirstdecision d cha2sds2 altx12)  [0.000]
0.001: (makedecisiontonode d altx12 warf)  [0.000]
0.002: (takeactiontogoal d warf g)  [0.000]
0.003: (finalgoalreached d g)  [0.001]
```

(a) Financial cost optimized plan

```
0.000: (makefirstdecision d cha2sds2 altx12)  [0.000]
0.001: (makedecisiontonode d altx12 dabi)  [0.000]
0.002: (takeactiontogoal d dabi g)  [0.000]
0.003: (finalgoalreached d g)  [0.001]
```

(b) Weighted cost optimized plan

Fig. 6. Afib generated plans for various cost considerations

Patient's Burden: A patient on Warfarin needs their international normalized ratio (INR) value checked as a stand-in to standardize the results of the prothrombin time test. Checking the INR value requires a blood test. When a patient is started on Warfarin, this test needs to be done weekly. When their INR value stabilizes, the frequency of the test falls between 2–4 weeks. Performing each blood test requires the patient to visit a laboratory service, imposing an additional treatment burden. On the other hand, a patient on Dabigatram (or any other DOAC) does not need a blood test to measure their INR value and does not incur any additional treatment burden. Using MitPlan 2.0, we assign a burden cost to WARF that is greater than the burden cost of DABI. When minimizing the burden is the optimization goal, Dabigatram is the action returned by MitPlan 2.0.

Adherence Likelihood: It is well documented in literature [5], that Afib patients on anticoagulation treatment poorly adhere to their prescribed medication. Considering that a patient on Dabigatram only takes pills and does not need to get a blood test, their only checkpoint for adherence is during an annual visit with their specialist. On the other hand, a patient on Warfarin needs to have regular blood tests in order to measure that their INR value is in the optimal range and any adjustment of medication dosage requires a visit with a specialist. Consequently, adherence to treatment with Warfarin is typically higher than treatment with Dabigatram due to regular consults with a specialist. As such, MitPlan 2.0 assigns a lower adherence cost to Warfarin than to Dabigatram. When the minimization of non-adherence is the desired goal, regardless of a patient's burden consideration, treatment with Warfarin is the returned option.

Multiple Weighted Considerations: Typically, there are multiple treatment options for multimorbidity. The selection of a given option is made by a physician in consultation with the patient. MitPlan 2.0 uses a complex objective function to represent multiple dimensions such as a patient's burden, financial cost or perceived medication adherence and weighs each one differently based on the information gathered during the shared decision-making process. These weights can be revised and adjusted to reflect changing patient's attitude and evolving clinical context. In Fig. 6(b) we show the plan where minimizing a patient's treatment burden is considered more important than both the financial cost of a treatment and the perceived patient's adherence using the multivariate objective function $0.2 * cost + 0.6 * burden + 0.2 * nonadherence$ (all weights are rescaled to the range [0, 1.0]). Note how this plan differs from the one in Fig. 6(a) when only financial cost is considered. The ability to customize a multivariate objective function to a specific patient and encounter represents a powerful tool supporting shared decision-making.

4 Discussion and Future Work

MitPlan 2.0 fully encapsulates the mitigation problem within a planning context with a unified approach to supporting preferences and revisions of CPGs' recommendations, simultaneously handling multiple revision operations, and optimizing across different metrics. Using various costs associated with nodes in the extended AG, MitPlan 2.0 finds management plans that are optimized according to a weighted multivariate objective function. This approach supports the combination of clinical dimensions with different treatment aspects such as financial cost, patient's burden, patient's perceived adherence to treatment, or cost of clinical resources required for treating a patient.

There are other approaches to mitigation of adverse interactions among multiple CPGs reported in the literature and below we briefly summarize the ones that are most closely related to MitPlan 2.0. META-GLARE [6] considers temporal characteristics of CPG actions during mitigation, employs goals to control the planning process, and it has been extended to model physician preferences [7]. However, no different types of preferences or costs are considered and there is no optimization over possible plans. Jafarpour et al. [8] propose an ontology-based framework for integrating multiple CPGs during execution time using policies. Some of these policies can optimize the use of clinical resources but other types of costs or preferences are not considered and there is no global optimization of generated plans.

A multi-agent planning (MAP) framework [9] is used to automatically generate several candidate management plans, evaluate them according to predefined patient- and institution-related metrics, and select the optimal one that minimizes the overall cost. MAP requires a more complex computational framework with multiple agents, and it assumes that secondary knowledge representing revision operators is embedded in CPGs. This assumption makes it more difficult to maintain than MitPlan 2.0, which uses an external knowledge repository for the revision operators. Finally, Kogan et al. [10] propose a goal-driven mitigation framework that uses standard representations (PROforma and HL7 FHIR), relies on existing knowledge sources, operates on different levels of abstraction, and generates explanations for proposed mitigations. Yet, it does not consider patient preferences and other costs to prioritize candidate plans as the final selection of a management plan is made by the clinician.

We are pursuing several directions for future work. The use of hard and soft constraints, that is, constraints that must be satisfied in any solution (e.g., a hypertension drug must be administered) and those that are optional (e.g., a blood test applied only if lab resources are available), respectfully, will add an additional level of personalization. We are also exploring how to generate management plans from a partial satisfaction of a subset of defined goals. Finally, we are studying the use of a stochastic model of mitigation. As the representational and functional complexity of the CPGs and in turn the planning problem increases, it will be necessary to shift from a deterministic representation towards one that supports probabilities tied to the execution of actions. In the real-world, medication may not be taken by a patient or may not have the intended effects, test results may be inconclusive or inaccurate, and future test results assumed for the sake of planning could return unanticipated values.

Acknowledgements. We thank Andrew Coles and Amanda Coles for their clarifications regarding PDDL and OPTIC and the reviewers for their helpful comments.

References

1. Goud, R., et al.: The effect of computerized decision support on barriers to guideline implementation: a qualitative study in outpatient cardiac rehabilitation. Int. J. Med. Inform. **79**(6), 430–437 (2010)
2. Peleg, M.: Computer-interpretable clinical guidelines: a methodological review. J. Biomed. Inform. **46**(4), 744–763 (2013)
3. Michalowski, M., Wilk, S., Michalowski, W., Carrier, M.: MitPlan: a planning approach to mitigating concurrently applied clinical practice guidelines. Artif. Intell. Med. **112**, 102002 (2021)
4. Sutton, R.T., Pincock, D., Baumgart, D.C., et al.: An overview of clinical decision support systems: benefits, risks, and strategies for success. npj Digit. Med. **3**, 17 (2020)
5. Raparelli, V., Proietti, M., Cangemi, R., Lip, G.Y., Lane, D.A., Basili, S.: Adherence to oral anticoagulant therapy in patients with atrial fibrillation. Focus on non-vitamin K antagonist oral anticoagulants. Thromb. Haemost. **117**(2), 209–218 (2017)
6. Bottrighi, A., Piovesan, L., Terenziani, P.: Coping with "exceptional" patients in META-GLARE. In: Cliquet Jr., A., et al. (eds.) BIOSTEC 2018. CCIS, vol. 1024, pp. 298–325. Springer, Cham (2019). https://doi.org/10.1007/978-3-030-29196-9_16
7. Terenziani, P., Andolina, A.: Considering temporal preferences and probabilities in guideline interaction analysis. In: Riaño, D., Wilk, S., ten Teije, A. (eds.) AIME 2019. LNCS (LNAI), vol. 11526, pp. 120–124. Springer, Cham (2019). https://doi.org/10.1007/978-3-030-21642-9_16
8. Jafarpour, B., Abidi, S.R., Woensel, W.V., Abidi, S.S.R.: Execution-time integration of clinical practice guidelines to provide decision support for comorbid conditions. Artif. Intell. Med. **94**, 117–137 (2019)

9. Fdez-Olivares, J., Onaindia, E., Castillo, L., Jordan, J., Cozar, J.: Personalized conciliation of clinical guidelines for comorbid patients through multi-agent planning. In: Proceedings of Artificial Intelligence in Medicine (AIME 2019), pp. 167–186 (2019)
10. Kogan, A., Peleg, M., Tu, S.W., Allon, R., Khaitov, N., Hochberg, I.: Towards a goal-oriented methodology for clinical-guideline-based management recommendations for patients with multimorbidity: GoCom and its preliminary evaluation. J. Biomed. Inform., **112**, 103587 (2020)

Explanations in Digital Health: The Case of Supporting People Lifestyles

Milene Santos Teixeira[1,2], Ivan Donadello[3], and Mauro Dragoni[2(✉)]

[1] University of Trento, Trento, Italy
m.santosteixeira@unitn.it
[2] Fondazione Bruno Kessler, Trento, Italy
dragoni@fbk.eu
[3] Free University of Bolzano, Bolzano, Italy
ivan.donadello@unibz.it, donadello@fbk.eu

Abstract. Systems that aim at supporting users on behavior change are expected to implement strategies that can both motivate and gain the users' trust, like the use of human understandable justifications for system's decisions. While the literature has dedicated great effort on providing accurate system's decisions, less focus has been given on addressing the problem of explaining to the user the reasons for a decision. This work presents a SPARQL-based reasoner enabling explainability on systems thought for supporting users in following healthy lifestyles. Our results demonstrate that users that received such information were able to reduce unhealthy behaviors over time.

1 Introduction

Explainable Artificial Intelligence (XAI) [12] exploits the challenges of providing transparent evidence for intelligent systems outcomes. XAI has gained visibility in recent years given the increasing number of automated systems that raise questions such as whether we can trust their decisions without any human interference. In fact, the benefits that could be provided by AI systems are frequently limited by the lack of transparency and understanding on the rationals behind their decisions. This lack of explainability prevents the adoption of these systems in real world scenarios.

Explainability is a critical requirement for the healthcare domain [7]. AI systems for health require human-understandable explanations that are aimed at increasing the trust and consequent acceptance by their final users, i.e., physicians and patients. As a matter of fact, an early research on expert systems has revealed that explaining the system decisions to the physicians was considered the most desired feature for a medical diagnostic system [10]. Health systems are required to be comprehensible [14] and, in addition to its outcome (e.g. recommendation), provide symbols (e.g. words) that enable the generation of user-driven explanations that help to understand the reasons behind the system's conclusion.

In this work, we provide a SPARQL-based reasoner to be integrated to systems that aim at supporting users to follow health behaviors. The proposed reasoner identifies violations in the user expected behavior and extracts the relevant information to be used either with natural language or visualization models in the generation of human

A. Tucker et al. (Eds.): AIME 2021, LNAI 12721, pp. 287–292, 2021.
https://doi.org/10.1007/978-3-030-77211-6_32

understandable explanations. Our intention is to support the generation of explanations that describe the expected behavior, relying on the user own data (violations) to justify why the system is presenting this information. To showcase the proposed approach in action, we developed a mobile application that relies on the state-of-the-art HeLiS ontology[1] [6], which formalizes concepts related to diet and physical activity for a healthy lifestyle.

2 Related Work

Rule-based and ontology-based systems have long supported digital health domains like food [2], drug [13] and diagnostic or health treatment [1]. While these systems are able to generate efficient recommendations, not many of them have addressed the challenge of justifying such recommendations. On the other hand, some approaches have justified their outcomes and exploited explainability in different ways. An example is the work of [11] that aims at personalizing the result of searches about health information according to the needs and profile of the individual (specified in an ontology). Their approach provides brief explanations that mention the individual's own information to justify why the resulting information is presented. Visualization techniques, like the *rainbow boxes* used by [8], that provides decision support in antibiotic treatment, are also used to support ontology based approaches on the generation of human understandable explanations. In some other works, like [4], that provided an ontology based framework for risk assessment, semantic explainability is obtained through ontology interpretation. That is, the approach is explainable in the sense that any user can navigate the ontology for inspection. In [3], instead, benefits from semantic explainability to provide semantic explanations that inform which elements (e.g. behavior, actions) may cause undesired outcomes and, therefore, should be changed in the smoking cessation domain.

Our work tackles the challenge of providing explanations that are not a mere description of the inference process performed by the system. The strategy proposed in this paper goes a step forward by translating the output of the designed SPARQL-based reasoner into a natural language description of the undesired events detected by the system. This way, we aim to reduce the barriers between the intelligent machinery and both domain experts and users of the final applications in understanding why a system inferred specific facts or provided specific suggestions.

3 The Reasoning-Based Explainable Approach

In this Section, we describe a SPARQL-based reasoner that exploits the HeLiS ontology [6] in order to identify users' behaviors that may lead to undesired situations. Examples are dietary intakes that do not meet the rules of a healthy lifestyle defined by domain experts. The detection of an inconsistency triggers the population of the knowledge base with an individual of type `Violation`. Violations are aimed to support the generation of explainable feedbacks to the user.

[1] http://w3id.org/helis.

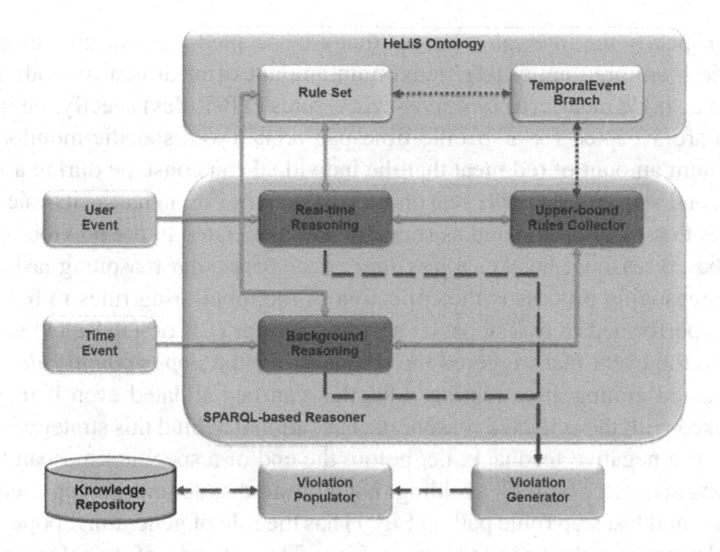

Fig. 1. The overall picture of the online reasoning process. (Color figure online)

The proposed SPARQL-based reasoner relies on the architecture implemented in RDFpro [5]. RDFpro is a reasoner that provides out-of-the-box OWL 2 RL reasoning and supports the fixed point evaluation of `INSERT... WHERE...` SPARQL-like entailment rules that leverage the full expressivity of SPARQL (e.g., `GROUP BY` aggregation, negation via `FILTER NOT EXISTS`). RDFpro has been used for the following reasons. First, through the architecture of RDFpro, it is possible to convert custom methods into reasoning operations able to (i) perform mathematical calculations on users' data and (ii) exploit real-time data obtained from external sources, without the need of storing this data in the knowledge repository. Second, this reasoner has been reported as efficient and suitable for real-time scenarios [5].

In our approach, reasoning is first conducted *offline* by taking into account the *static* part of the ontology (monitoring rules, food, nutrients, activities) with the aim of materializing the ontology deductive closure, based on OWL 2 RL and some additional pre-processing rules that identify the most specific types of each individual defined in the static part of the HeLiS ontology ABox. Whereas, *online* reasoning is triggered (i) every time the user adds or modifies a data package in the knowledge base or (ii) at the end of a specific timespan (e.g. end of the day) modeled as a HeLiS concept. The former triggers the *real-time reasoning* task that is responsible, mainly, for analyzing the single data package provided by the user. The latter triggers the *background reasoning* that is in charge of processing the data packages provided within the considered timespan. During online reasoning, the user data is merged with the closed ontology and the deductive closure of the rules is computed. Resulting `Violation` individuals (if any) and their RDF descriptions are then stored back in the knowledge base.

In Fig. 1, we illustrate the *online* stage of the reasoning process. The green path is the first step to be carried out. Its role is to gather the rules to validate, according to what triggered the reasoner. Two types of monitoring rules are specified: (i) *event-based* rules

(EB-Rules) specify the interval of values that are accepted for a specific event related to a specific monitored entity (e.g. maximum amount of meat that the individual can consume in a single meal); (ii) *timespan-based* rules (TB-Rules) specify the interval of values that are accepted for a specific timespan related to a specific monitored entity (e.g. maximum amount of red meat that the individual can consume during a week).

The second step of the *online* reasoning (red path in Fig. 1) has as its role to gather further rules that can be validated as semantically associated to the previous ones. The SPARQL-based reasoner invokes a *real-time* or a *background* reasoning task. The first step of the reasoning process is the collection of the monitoring rules to be validated. This task is performed in two steps: (i) a group of rules (EB or TB-Rules) is collected according to the event that triggered the reasoner; (ii) the *upper-bound rules collector* task is executed aiming at extracting rules that can be validated even if they are not directly linked with the activated reasoning. The rational behind this strategy is the early generation of a negative feedback, i.e. before the end of a specific timespan (e.g. `Day isSubTimespanOf Week`), avoiding that undesired situations get reported too late.

The third and last step (blue path in Fig. 1) has the role of generating, populating and storing violations into the knowledge repository. The outcome of the inference activity is a set of structured data packages that encapsulate the information about the detected undesired behaviors, i.e., violations. Each data package corresponds to an instance of the `Violation` concept and it is stored within the knowledge repository. The generation and the population of each instance of type `Violation` is performed in two separated steps. First, the `Violation` is generated as result of the reasoning activity and all information inferred by the SPARQL-based reasoner is stored into it. Second, accessory information are integrated into the `Violation` instance to provide extra details that can support the explanation of the detected violation. Examples of accessory information are references to other concepts of the ontology or the number of times that the specific rule has been violated. This way, each `Violation` instance is a self-contained object that includes all information needed to generate a feedback for the user and for statistics purposes.

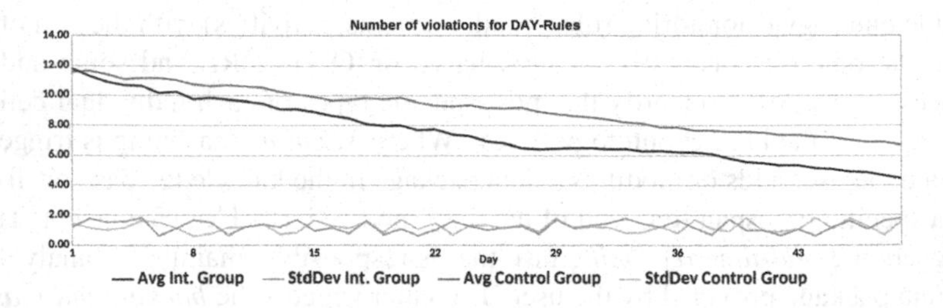

Fig. 2. Variation of the (average) number of detected daily violations within the experiments time span. The intervention group present a more effective decay of the violations. (Color figure online)

4 Evaluation

In this Section, we report our evaluation of the proposed reasoner. With the aim of testing our approach in a real scenario, we developed a mobile application for monitoring food intake and physical activities. A template based sentence generator [9] was integrated to deliver natural language sentences from violation instances. Our goal was to measure the effectiveness of the explanations generated by our platform by observing the evolution of the number of violations generated. The system was used during a period of seven weeks by 120 individuals inside our institution. We split the users in two groups: (i) the *intervention group* (92 users) received the explainable messages generated by using the template system; (ii) the *control group* (28 users) received only canned text messages notifying when a rule was violated (e.g. 'Today you had too much (300 ml of maximum 200 ml) fruit juice').

In Fig. 2, we show the variation in the number of daily violations for the two groups. The blue and green lines represent the average number of violations for the intervention and control group respectively. Whereas, the red and orange lines represent the relative standard deviations. Rules are verified at the end of each day by analyzing the food information provided by each user. The increasing trend of the gap between the blue and green lines demonstrates the positive impact of the explanation sent to users when an undesired event occurred. Indeed, we can observe a drop of 62.20% of the violations for the intervention group with respect to a drop of 42.33% of the violations for the control group. By considering the standard deviation lines, we can appreciate how both lines remain contained within low bounds without the presence of outliers.

We are also interested in the time spent by our system to be effective. Figure 2 shows us that the two groups tend to diverge at a certain point during the experiments time span. We measure when the two groups start to diverge with a statistical significance. We observed Table 1 reports this value along with the p-values and average number of violations for both the intervention and control group.

Table 1. Starting point of the project time where the persuasion system takes effect with statistical significance.

Starting day/week	p-value	Violations intervention group	Violations control group
19th day	0.011	8.09 ± 2.88	9.82 ± 2.85

5 Conclusion

In this paper, we have employed SPARQL based reasoning to identify possible violations on a user's expected behavior. This information is exploited to provide explanations that justify system's recommendations with the final aim of supporting behavior change. To test our reasoner, we developed a mobile application to monitor food intake. The application was tested by 120 users and our results show that users that received explainable messages were able to reduce significantly the number of violations over time. We are aware that, in order to effectively support behavior change, further strategies have to be taken into account such as persuasiveness and argumentation. These aspects are out of scope of this work, but are already being exploited by us as part of the *Key to Health* project. Future work includes the integration of our solution into a conversational agent that, through a multi-turn dialogue, tries to understand the user situation, being able to infer new knowledge from the answers given by the user and, then, provides an explainable recommendation.

References

1. Alharbi, R.F., Berri, J., El-Masri, S.: Ontology based clinical decision support system for diabetes diagnostic. In: 2015 Science and Information Conference (SAI), pp. 597–602. IEEE (2015)
2. Baek, J.W., Kim, J.C., Chun, J., Chung, K.: Hybrid clustering based health decision-making for improving dietary habits. Technol. Health Care (Prepr.), 1–14 (2019)
3. Brenas, J.H., Shaban-Nejad, A.: Health intervention evaluation using semantic explainability and causal reasoning. IEEE Access **8**, 9942–9952 (2020)
4. Cattelani, L., Chesani, F., Palmerini, L., Palumbo, P., Chiari, L., Bandinelli, S.: A rule-based framework for risk assessment in the health domain. Int. J. Approx. Reason. **119**, 242–259 (2020)
5. Corcoglioniti, F., Rospocher, M., Mostarda, M., Amadori, M.: Processing billions of RDF triples on a single machine using streaming and sorting. In: ACM SAC, pp. 368–375 (2015)
6. Dragoni, M., Bailoni, T., Maimone, R., Eccher, C.: HeLiS: an ontology for supporting healthy lifestyles. In: Vrandečić, D., et al. (eds.) ISWC 2018. LNCS, vol. 11137, pp. 53–69. Springer, Cham (2018). https://doi.org/10.1007/978-3-030-00668-6_4
7. Holzinger, A., Biemann, C., Pattichis, C.S., Kell, D.B.: What do we need to build explainable ai systems for the medical domain? arXiv preprint arXiv:1712.09923 (2017)
8. Lamy, J.B., Sedki, K., Tsopra, R.: Explainable decision support through the learning and visualization of preferences from a formal ontology of antibiotic treatments. J. Biomed. Inform. **104**, 103407 (2020)
9. Maimone, R., Guerini, M., Dragoni, M., Bailoni, T., Eccher, C.: PerKApp: a general purpose persuasion architecture for healthy lifestyles. J. Biomed. Inform. **82**, 70–87 (2018). https://doi.org/10.1016/j.jbi.2018.04.010
10. Moore, J.D., Swartout, W.R.: Explanation in expert systemss: a survey. Technical report. University of Southern California Marina Del Rey Information Sciences Inst. (1988)
11. Nguyen, V.L.: An ontology-based health self-education framework to facilitate the patient-health practitioner collaboration in healthcare. Ph.D. thesis, Queensland University of Technology (2018)
12. Samek, W., Wiegand, T., Müller, K.R.: Explainable artificial intelligence: understanding, visualizing and interpreting deep learning models. arXiv preprint arXiv:1708.08296 (2017)
13. Shang, J., Ma, T., Xiao, C., Sun, J.: Pre-training of graph augmented transformers for medication recommendation (2019)
14. Vellido, A.: The importance of interpretability and visualization in machine learning for applications in medicine and health care. Neural Comput. Appl. **32**(24), 18069–18083 (2019). https://doi.org/10.1007/s00521-019-04051-w

Predicting Medical Interventions from Vital Parameters: Towards a Decision Support System for Remote Patient Monitoring

Kordian Gontarska[1,2,3](✉), Weronika Wrazen[1], Jossekin Beilharz[1,3], Robert Schmid[1,3], Lauritz Thamsen[2], and Andreas Polze[1]

[1] Hasso Plattner Institute, University of Potsdam, Potsdam, Germany
{kordian.gontarska,weronika.wrazen,jossekin.beilharz,robert.schmid, andreas.polze}@hpi.de

[2] Technische Universität Berlin, Berlin, Germany
lauritz.thamsen@tu-berlin.de

[3] Charité – Universitätsmedizin Berlin, corporate member of Freie Universität Berlin, Humboldt-Universität zu Berlin, and Berlin Institute of Health, Berlin, Germany

Abstract. Cardiovascular diseases and heart failures in particular are the main cause of non-communicable disease mortality in the world. Constant patient monitoring enables better medical treatment as it allows practitioners to react on time and provide the appropriate treatment. Telemedicine can provide constant remote monitoring so patients can stay in their homes, only requiring medical sensing equipment and network connections. A limiting factor for telemedical centers is the amount of patients that can be monitored simultaneously. We aim to increase this amount by implementing a decision support system. This paper investigates a machine learning model to estimate a risk score based on patient vital parameters that allows sorting all cases every day to help practitioners focus their limited capacities on the most severe cases. The model we propose reaches an AUCROC of 0.84, whereas the baseline rule-based model reaches an AUCROC of 0.73. Our results indicate that the usage of deep learning to improve the efficiency of telemedical centers is feasible. This way more patients could benefit from better health-care through remote monitoring.

Keywords: Telemedicine · Decision Support System · Remote Patient Monitoring · Machine Learning · Heart Failure

1 Introduction

According to the World Health Organization, cardiovascular diseases (CVDs) are the main cause of a non-communicable disease mortality in the world [10]. It is important to detect a patient's critical condition early to enable a timely

A. Tucker et al. (Eds.): AIME 2021, LNAI 12721, pp. 293–297, 2021.
https://doi.org/10.1007/978-3-030-77211-6_33

intervention. One way to ensure this is to monitor patients remotely in their homes from telemedical centers (TMCs). Modern technology makes it possible to provide patients with monitoring services even in areas without comprehensive medical infrastructures. In recent years, it was shown that telemedical interventions reduce the mortality in patients with heart failures [1,4].

This paper is a part of the Telemed5000 project and follows our previous work on clinical decision support systems for heart failure which was a part of the Fontane project in collaboration with Charité, Berlin [3,6]. Our aim is to scale up the TMC capacity to ensure that up to 5,000 patients will be cared for in the future utilizing Artificial Intelligence (AI).

In this paper we describe the development and evaluation of a machine learning model for the prediction of the daily per-patient risk of being in a medically critical condition. The patients are sorted by this estimated risk, enabling the TMC to concentrate on the most severe cases. To accomplish this we use a database with daily vital parameters from the TIM-HF 2 study [4].

2 Materials and Methods

In this research we used the Telemedical Interventional Management in Heart Failure II (TIM-HF2) database, which was created by Charité, Berlin during the Fontane project [4]. The trial has been conducted in Germany between 2013 and 2018. TIM-HF2 included 1,538 patients (773 usual care) whose stage of heart failure is classed II or III according to the New York Health Association (NYHA) classification. Additionally they were admitted to a hospital at most 12 months prior to the study due to heart failure and had a left ventricular ejection fraction (LVEF) of $< 45\%$. The dataset includes daily measurements performed by the patients themselves using a weight scale, a blood pressure monitor, a pulse oximeter, a small ECG device and a tablet-like device for the self-reporting of their well-being. In total the unprocessed dataset consists of records from 763 patients out of which 100 died before the end of the study (66 within 7 days after their last measurement). The database also contains clinical events like 387 endpoint-adjudicated hospitalizations and 4,329 interventions performed by the TMC. Patients were asked to participate for one year.

We included the following features into the data: age, weight, blood pressure, oxygen saturation, gender, diabetes, the NYHA class, several symptoms and signs of heart failure (e.g. AV Block, LBBB), automatically extracted data from ECG (heart rate, sinus rhythm, ventricular tachycardia, atrial fibrillation), self-assessed state of health, weight difference (1, 3 and 8 days difference), social variables (e.g. living alone, anxiety). The binary predictor variable is a union of TMC's intervention, hospitalization or death events. Missing values were linearly imputed for up to 2 consecutive missing days, the rest got dropped. The positive class forms only approximately 2% of the dataset, thus we oversampled observations from the minority class in the training set to balance the classes. The dataset was split into three sets: train, validation, and test in a proportion 4:1:1 respectively. The distribution of samples and events per patient was preserved across all sets. Each patient was assigned to exactly one set. To evaluate model performance we took the following metrics into consideration: Receiver Operating Characteristic (ROC) curve, area under ROC curve (AUCROC), Precision - Recall curve, and area under PR curve (AUCPR).

We investigate different deep neural network (DNN) models and compare them to a rule-based baseline. The rule-based model (RB) is based off the TIM-HF 2 study [4]. The rules are heart related and consist of engineered features and thresholds defined by an expert group at the Charité [5]. All DNN models had an output layer with *a sigmoid activation* function, *binary cross-entropy* as a loss function, and *Adam* as the optimization algorithm. We tested between 2 and 5 hidden layers with 5 to 150 neurons in each. Additionally we tested linear, sigmoid, and ReLU activation functions, and dropout rates between 0 and 0.5.

3 Results

Figure 1a shows the ROC curves for the selected DNN and the rule-based model. The DNN outperforms the rule-based model having better sensitivity at any specificity and an AUCROC of 0.84 as compared to 0.73. Figure 1b shows the Precision/Recall curve for both models. The DNN outperforms the rule-based model in precision at any recall rates. The plots in Fig. 2 show the distributions of the predicted risk-scores for both classes, as predicted by the DNN and rule-based models. It can be seen that the DNN model performs better in detecting events than the rule-based model, as there is a clearer cut between the distributions. The final DNN model was trained for 453 epochs using a batch size of 4096, and a learning rate of 0.001. It has 3 hidden layers with 35, 20, and 35 neurons respectively. All neurons in the hidden layers use ReLU as their activation function and have dropout rates of 0.25, 0.15, and 0.3. The patient's self assessment, weight differences, pulse-rate, and complaints had the highest impact on the models decision making.

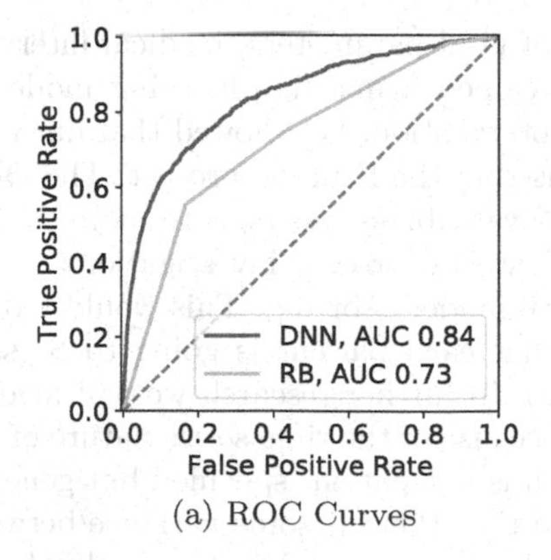

(a) ROC Curves

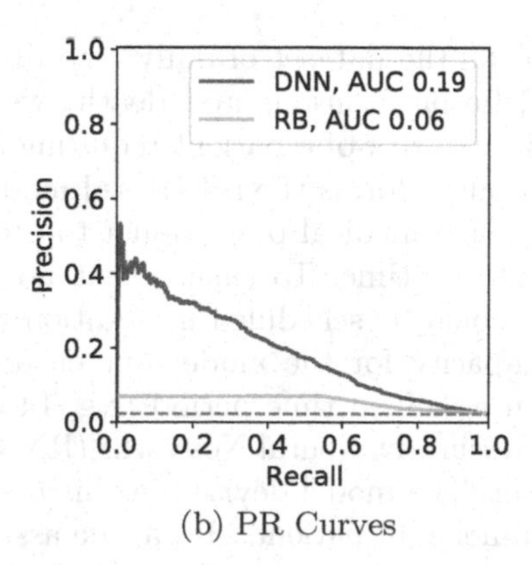

(b) PR Curves

Fig. 1. The figures show the (a) ROC curves and (b) PR curves for both the DNN and the rule-based model. The dashed lines represent what performance a purely random classifier would achieve.

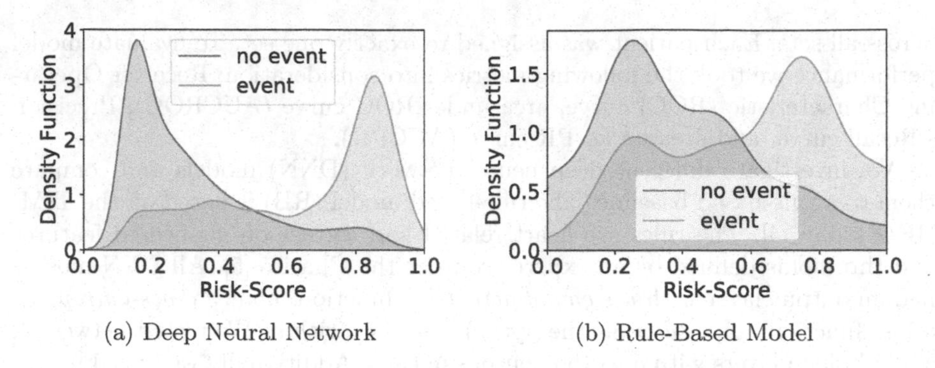

(a) Deep Neural Network (b) Rule-Based Model

Fig. 2. The depicted plots show the distribution of the predicted risk-scores in the test set, separated by the true label.

4 Related Work

Decision Support Systems (DSSs) have been used in the medical field since the seventies [8]. Seto et al. applied a rule-based model to monitor patients with heart failure [7]. A rule-based model was implemented for the Fontane project, which had to prioritize patients based on their daily vital parameters [4,6]. Groccia et al. proposed a linear Support Vector Machine (SVM) model which predicts major cardiovascular worsening events for patients with heart failure [2]. Stehlik et al. studied the potential efficiency of noninvasive remote monitoring in predicting heart failure re-hospitalizations [9]. Heinze et al. proposed a Hybrid AI model as an improvement for the rule-based model in Fontane [3]. The hybrid model consists of a Neural Network (NN) with one hidden layer and two rules which were handcrafted by medical experts.

5 Conclusion

Based on the dataset of daily recordings of vital parameters, medical interventions, hospitalizations and deaths we developed a machine learning model to predict the risk of a patient requiring an intervention. We showed that our approach outperforms the rule-based model used in the Fontane project. The DNN may help a medical practitioner to provide valuable assessment to more critical patients on time. To ensure that no patient is overseen, investigations in the TMC could be scheduled in addition to the model's sorting. This would reduce the capacity for the model but ensures that each patient is going to be seen within a defined time period (e.g. 14 days). In further research we will investigate Recurrent Neural Networks (RNN), because of the time series nature of the dataset. The model devised in this research is not patient specified but generalized among all patients. It can be assumed that there is some variance between the patients which could be used to adapt the model to each patient individually and thus boosting its performance.

Acknowledgment. We thank Prof. Dr. med. Friedrich Köhler and his team for the provisioning of the dataset and Prof. Dr. med. Alexander Meyer for helping and supporting us in analyzing the dataset. We are grateful to Alexander Acker for the fruitful discussions, Volker Möller for his help with the dataset, and Boris Pfahringer for his valuable feedback on our evaluation. This research and the Telemed5000 project have been supported by the Federal Ministry for Economic Affairs and Energy of Germany as part of the program "Smart Data" (project number 01MD19014C).

References

1. Ekeland, A., Bowes, A., Flottorp, S.: Effectiveness of telemedicine: a systematic review of reviews. Int. J. Med. Inf. **79** (2010)
2. Groccia, M., Lofaro, D., Guido, R., Conforti, D., Sciacqua, A.: Predictive models for risk assessment of worsening events in chronic heart failure patients. In: CinC 2018 (2018)
3. Heinze, T., Wierschke, R., Schacht, A., von Löwis, M.: A hybrid artificial intelligence system for assistance in remote monitoring of heart patients. In: Corchado, E., Kurzyński, M., Woźniak, M. (eds.) HAIS 2011. LNCS (LNAI), vol. 6679, pp. 413–420. Springer, Heidelberg (2011). https://doi.org/10.1007/978-3-642-21222-2_50
4. Koehler, F., et al.: Efficacy of telemedical interventional management in patients with heart failure (TIM-HF2): a randomised, controlled, parallel-group, unmasked trial. The Lancet (2018)
5. Koehler, F., et al.: Telemedical interventional management in heart failure II (TIM-HF2), a randomised, controlled trial investigating the impact of telemedicine on unplanned cardiovascular hospitalisations and mortality in heart failure patients: study design and description: tim-hf2: study design. Eur. J. Heart Fail. **20** (2018)
6. Polze, A., Tröger, P., Hentschel, U., Heinze, T.: A scalable, self-adaptive architecture for remote patient monitoring. In: ISORCW 2010 (2010)
7. Seto, E., Leonard, K., Cafazzo, J., Barnsley, J., Masino, C., Ross, H.: Developing healthcare rule-based expert systems: case study of a heart failure telemonitoring system. Int. J. Med. Inf. **81** (2012)
8. Shortliffe, E.H., Davis, R., Axline, S.G., Buchanan, B.G., Green, C.C., Cohen, S.N.: Computer-based consultations in clinical therapeutics: explanation and rule acquisition capabilities of the mycin system. Comput. Biomed. Res. **8** (1975)
9. Stehlik, J., et al.: Continuous wearable monitoring analytics predict heart failure hospitalization: the link-HF multicenter study. Circ. Heart Fail. **13** (2020)
10. WHO: World health statistics 2020: monitoring health for the sdgs, sustainable development goals (2020)

CAncer PAtients Better Life Experience (CAPABLE) First Proof-of-Concept Demonstration

Enea Parimbelli[1(✉)], Matteo Gabetta[2], Giordano Lanzola[1], Francesca Polce[1], Szymon Wilk[3], David Glasspool[4], Alexandra Kogan[5], Roy Leizer[5], Vitali Gisko[6], Nicole Veggiotti[1], Silvia Panzarasa[1], Rowdy de Groot[7], Manuel Ottaviano[8], Lucia Sacchi[1], Ronald Cornet[7], Mor Peleg[5], and Silvana Quaglini[1]

[1] University of Pavia, Pavia, Italy
[2] Biomeris s.r.l., Pavia, Italy
[3] Poznan University of Technology, Poznan, Poland
[4] Deontics Ltd., London, England
[5] University of Haifa, Haifa, Israel
[6] Bitsens JSC, Vilnius, Lithuania
[7] AMC - Academic Medical Center, University of Amsterdam, Amsterdam, Netherlands
[8] Universidad Politecnica de Madrid, UPM, Madrid, Spain

Abstract. The CAncer PAtient Better Life Experience (CAPABLE) project combines the most advanced technologies for data and knowledge management with a socio-psychological approach, to develop a coaching system for improving the quality of life of cancer patients managed at home. The team includes complementary expertise in data- and knowledge-driven AI, data integration, telemedicine and decision support. The time is right to fully exploit Artificial Intelligence for cancer care and bring the benefits right to patients' homes. CAPABLE relies on predictive models based on both retrospective and prospective data, integrated with computer interpretable guidelines and made available to oncologists. CAPABLE's Virtual Coach component identifies unexpected needs and provides patient-specific decision support and lifestyle guidance to improve mental and physical wellbeing of patients. The demo, designed around a use-case scenario developed with clinicians involved in the project, addresses the ESMO Diarrhea guideline. It revolves around a prototypical fictional patient named Maria. Maria, 66, is affected by renal cell carcinoma and moderate insomnia. The demo follows Maria during the first three days of using the CAPABLE system. This allows the audience to understand the scope and innovation behind this AI-based decision-support and coaching system that personalizes lifestyle and medication interventions to patients, their carer and clinicians.

Keywords: Cancer · Side effects · Personalization · Coaching · Guideline · FAIR · FHIR · OMOP · AI · mHealth

A. Tucker et al. (Eds.): AIME 2021, LNAI 12721, pp. 298–303, 2021.
https://doi.org/10.1007/978-3-030-77211-6_34

1 Introduction

After the primary intervention, most cancer patients are managed at home, facing long-term treatments or sequelae, making the disease comparable to a chronic condition [1]. Despite their benefit, strong therapeutic regimens often cause toxicity, severely impairing quality of life. This may decrease adherence to treatment, thus compromising therapeutic efficacy. Also due to age-related multimorbidity, patients and their caregivers develop emotional, educational and social needs [1].

In 2019, a consortium comprising 5 universities across Europe and Israel, 3 small-medium enterprises, 1 large company, 2 hospitals and 1 patient association, was funded by the EC Horizon 2020 tender on "Big data and artificial intelligence for monitoring health status and quality of life after the cancer treatment". As a result, the CAncer PAtient Better Life Experience (CAPABLE)[1] project started Jan 1st 2020, with the objective to combine the most advanced technologies for data and knowledge management with a sound socio-psychological approach in order to develop a coaching system for improving the quality of life of cancer patients managed at home. The project addresses EU priorities such as shifting care from hospitals to home to face scarcity of healthcare resources, facilitating patients' re-integration in the society and promoting an effective, novel cancer care model for all EU citizens. The time is right to fully exploit Artificial Intelligence (AI) and Big Data for cancer care and bring them to patients' home. In this paper, we present the first proof-of-concept (POC) of the CAPABLE system, developed during the first 12 months of the project.

2 Methods

2.1 Consortium and Expertise

University of Pavia (UNIPV) is the project coordinator, with the "M. Stefanelli" Biomedical Informatics Laboratory group. UNIPV is also home to the Centre for Health Technologies (CHT)[2], and the European Centre for Law, Science and new Technologies (ECLT)[3], helping the consortium to tackle the medico-legal issues related to the application of IT and AI in medicine. The project leverages a strong collaboration between universities and SMEs. University of Haifa (UoH) has leading expertise in knowledge representation for decision support and for data integration; in CAPABLE they are focusing on representation and algorithms for a) planning conflict-free treatment plans for multimorbidity patients and b) bridging the semantic gap between clinical abstractions and retrieval of raw data. Deontics ltd (DEON) developed a computer-interpretable guideline (CIG) editor and enactment engine for the PROforma CIG formalism, adapting it to the project needs. Amsterdam Medical Center (AMC) has a wide experience on standards for medical data representation while Biomeris s.r.l. (BIOM) integrates all the data collected in CAPABLE exploiting their experience in data-warehousing. Poznan University of Technology (PUT) is responsible for the patient coaching development

[1] The CAPABLE project has received funding from the European Union's Horizon 2020 research programme under grant agreement No 875052. www.capable-project.eu.

[2] cht.unipv.it.

[3] www.unipv-lawtech.eu.

while Bitsens JCV (BIT) provides the final user interfaces. The large company involved is IBM, providing the necessary skills to exploit the latest developments in AI. Data is collected in retrospective studies and in the CAPABLE prospective clinical studies by Istituti Clinico-Scientifici "Maugeri" (ICSM) and Netherlands Cancer Institute (NKI), two leading hospitals for cancer treatment in Italy and the Netherlands. Associazione Italiana Malati di Cancro AIMAC, an important patient association from Italy networked with other associations and the European Patient Cancer Coalition contributes to maintain a patient-centred approach along the whole project execution.

2.2 Iterative Development of the First POC and Its Components

The first POC was developed in the project, following an iterative development process that started July 1st 2020, culminating in the production of deliverable 4.1 [2] in Dec 2020. Figure 1 highlights such a process and its sub-iterations. Details of the scope of each iteration are provided in the following, along with the POC architecture.

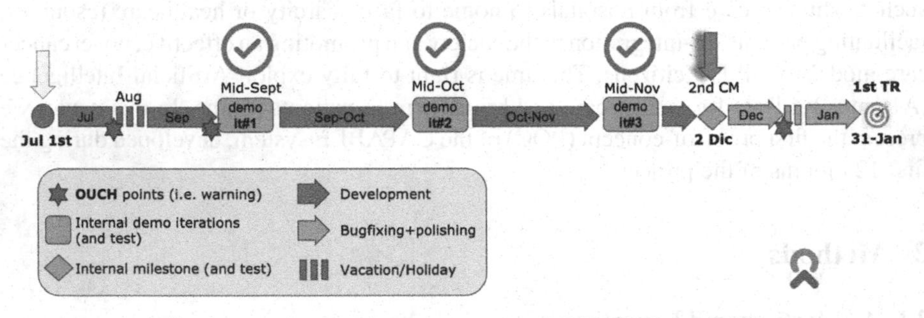

Fig. 1. Iterative development process of CAPABLE first POC.

Table 1 presents the CAPABLE system components that are part of the 1st POC, along with their main functionalities and responsible partner in the consortium. Figure 2 presents the scoped-down architecture of the 1st CAPABLE POC.

2.3 Data Model and FAIR Principles

The infrastructure is intended to be FAIR, i.e., findable, accessible, interoperable, and reusable. This is realized by adhering to existing standards as much as possible, i.e., OMOP CDM as the persistence model for the data, and HL7 FHIR for inter-component communication. This will be complemented by advertising metadata, including characteristics of the stored data, used vocabularies, and characterization of the included population [3].

2.4 AI

CAPABLE relies on diversified AI techniques, including knowledge- and data-driven approaches to provide comprehensive decision support to both patients and clinicians thorough Virtual Coach and Physician DSS, respectively. It employs complex CIGs, with

Table 1. CAPABLE first POC components.

Component (ABBR)	Role	Responsible partner	It#
Data Platform (DP)	Storing and providing patient-level data	UNIPV	It#1
Case Manager (CM)	Managing events related to Data Platform and providing notifications to other components	UNIPV	It#1
Physician DSS (DSS)	Providing guideline-based decision support for clinicians when managing cancer patients	DEON	It#2
Knowledge-Data Ontology Mapper (KDOM)	Using ontology mapping classes to define clinical abstractions in terms of raw data and FHIR queries	UoH	It#2
GoCom Multimorbidity controller (GOCO)	Checking for possible adverse interactions between clinical tasks for multimorbid patients and resolving them	UoH	It#2
Virtual Coach (VC)	Providing coaching support combining clinical and non-clinical recommendations to cancer patients at home	PUT	It#2
Deontics Engine (DE)	Executing computer-interpretable clinical practice guidelines (CIGs) defined using the PROforma language	DEON	It#2
Patient app	Providing user interface for patients	BIT	It#3
Physician dashboard	Providing user interface for physicians	BIT	It#3

physician- and patient-oriented components, represented in PROforma and executed by DE to provide evidence-based recommendations. CAPABLE also includes two classes of data-driven models – exploratory models and prediction models that are derived from multi-modal data (clinical data, patient reported outcomes, readings from environmental and wearable sensors). Exploratory models provide a concise summary of analyzed data and will be used for indirect decision support through infographics and other visual tools. Prediction models provide patient-specific recommendations, thus offering direct decision support. They are further divided into population and personal models – the former are derived from cohorts of patients suffering from the same type of cancer, while the latter are constructed for individual patients.

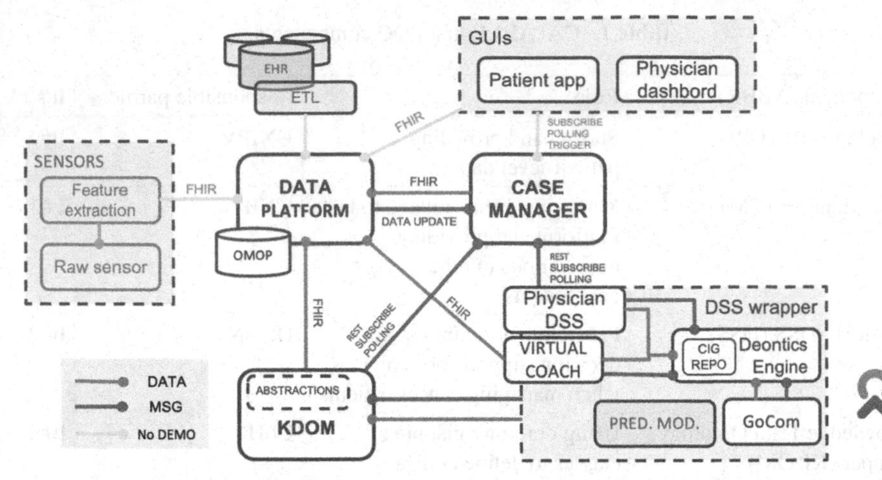

Fig. 2. General architecture of the CAPABLE system. Components and interactions that are out-of-scope for the 1st POC have been grayed-out.

Current population models have been derived from available retrospective data provided by NKI and ICSM and they aim predicting the survival and response to treatment. They will be further refined based on prospective data. Personalized models are aimed at facilitating application of non-clinical lifestyle interventions (so called *capsules*), such as breathing exercises, meditation or physical activity. Preliminary personal models have been constructed from available retrospective benchmark data (e.g., WESAD [4]) and will be further refined with CAPABLE prospective data.

3 Results

A recording of the POC demonstration was produced on December 9th 2020, reflecting what was shown during the CAPABLE consortium meeting held on December 2nd. Deliverable 4.1, focusing on the 1st POC is publicly available [2]. A second POC is scheduled for project M18. Late-breaking results may be available to be presented live at AIME2021 in June.

4 Conclusion

With its AI-enabled components, CAPABLE is more than a personalized tool for improving quality of life, but rather an advance for the AI in medicine research community. This first POC is being extended by the project consortium and CAPABLE will start its clinical studies, testing the system with real-world cancer patients, in Jan 2023.

References

1. Boele, F., Harley, C., Pini, S., Kenyon, L., Daffu-O'Reilly, A., Velikova, G.: Cancer as a chronic illness: support needs and experiences. BMJ Support. Palliative Care (2019). https://doi.org/10.1136/bmjspcare-2019-001882

2. Gabetta, M., Parimbelli, E.: CAPABLE D4.1: 1st Iteration of the Platform Proof Of Concept (2020). https://doi.org/10.5281/zenodo.4540457
3. Parimbelli, E., Cornet, R., Gabetta, M., Tibollo, V., Bottalico, B., et al.: CAPABLE D1.2: Data Management Plan (2020). https://doi.org/10.5281/zenodo.3970580
4. Schmidt, P., Reiss, A., Duerichen, R., Marberger, C., Van Laerhoven, K.: Introducing WESAD, a multimodal dataset for wearable stress and affect detection. In: 20th ACM International Conference on Multimodal Interaction, New York, NY, USA, pp 400–408 (2018). https://doi.org/10.1145/3242969.3242985

Deep Learning

Sensitivity and Specificity Evaluation of Deep Learning Models for Detection of Pneumoperitoneum on Chest Radiographs

Manu Goyal[1](✉), Judith Austin-Strohbehn[2], Sean J. Sun[2], Karen Rodriguez[2], Jessica M. Sin[2], Yvonne Y. Cheung[2], and Saeed Hassanpour[1,3]

[1] Department of Biomedical Data Science, Dartmouth College, Hanover, NH, USA
manu.goyal@dartmouth.edu
[2] Department of Radiology, Dartmouth-Hitchcock Medical Center, Lebanon, NH, USA
[3] Departments of Biomedical Data Science, Computer Science, and Epidemiology, Dartmouth College, Hanover, NH, USA

Abstract. Deep learning has great potential to assist with detecting and triaging critical findings such as pneumoperitoneum on medical images. To be clinically useful, the performance of this technology still needs to be validated for generalizability across different types of imaging systems. This retrospective study included 1,287 chest X-ray images of patients who underwent initial chest radiography at 13 different hospitals between 2011 and 2019. State-of-the-art deep learning models were trained on a subset of this dataset, and the automated classification performance was evaluated on the rest of the dataset by measuring the AUC, sensitivity, and specificity. All deep learning models performed well for identifying radiographs with pneumoperitoneum, while DenseNet161 achieved the highest AUC of 95.7%, Specificity of 89.9%, and Sensitivity of 91.6%. The DenseNet161 model was able to accurately classify radiographs from different imaging systems (Accuracy of 90.8%), while it was trained on images captured from a specific imaging system from a single institution. This result suggests the generalizability of our model for learning salient features in chest X-ray images to detect pneumoperitoneum, independent of the imaging system. If verified in clinical settings, this model could assist practitioners with the diagnosis and management of patients with this urgent condition.

Keywords: Pneumoperitoneum · Deep learning · Chest X-ray · Sensitivity · Specificity

1 Introduction

In recent years, advances in deep learning have presented new opportunities to assist and improve clinical diagnosis involving different medical imaging modalities such as magnetic resonance imaging (MRI), X-ray, Ultrasound, computed tomography (CT), and positron emission tomography (PET) [1–5]. Chest X-rays (CXR) are commonly used as an important imaging tool to screen patients for a number of diseases. In recent

A. Tucker et al. (Eds.): AIME 2021, LNAI 12721, pp. 307–317, 2021.
https://doi.org/10.1007/978-3-030-77211-6_35

studies, deep learning has provided end-to-end proof of concept for models achieving radiologist-level performance in the detection of different clinical findings on CXRs [6, 7]. In doing so, deep learning has helped physicians to prioritize urgent medical cases and focus their attention during clinical diagnosis.

Pneumoperitoneum is a critical clinical finding that requires immediate surgical attention [8, 9]. Although abdominal radiographs and CT scans are standard modalities for the detection of pneumoperitoneum, CXRs are often an initial exam that is ordered in the emergency room setting. Therefore, pneumoperitoneum is often detected on initial CXRs, before additional imaging, such as CT exams, are ordered. Free air in the abdomen is most visible on CXRs of patients in the standing position. Because gas ascends to the highest point in the abdomen, free air accumulates beneath the domes of the diaphragm in the standing or upright position. Therefore, CXR is one of the most sensitive modalities to detect pneumoperitoneum [10]. Solis et al. showed that performing abdominal CT exams can delay surgery, without providing any measurable benefit over a CXR for the diagnosis of pneumoperitoneum [11]. A few examples of positive pneumoperitoneum CXR images are shown in Fig. 1.

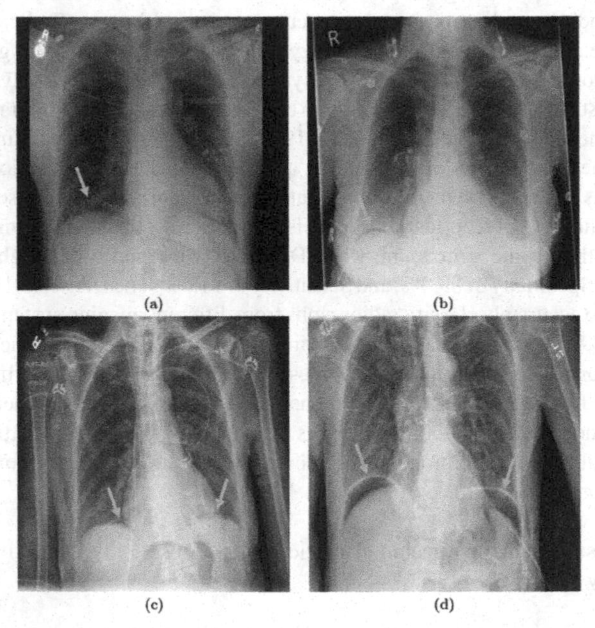

Fig. 1. Four examples of pneumoperitoneum positive cases in chest X-ray images in our dataset. The yellow arrow indicates the presence of Pneumoperitoneum. (Color figure online)

Despite the recent success of deep learning models in detecting disease on CXRs, it has been found that these models can be highly sensitive to the types of systems used for the training dataset. For instance, Marcus et al. [12] argued a deep learning model trained on standard CXR images captured by a particular imaging system in a fixed location may not perform as well on portable CXR images. This is because the trained deep learning model has to deal with variabilities in patterns and characteristics found in CXR images across different imaging systems, rather than variability and differences in chest anatomy and morphology intrinsic to the disease itself. In this study, we developed state-of-the-art deep learning models to detect pneumoperitoneum on CXR images and evaluated the sensitivity and specificity of these models on a diverse dataset assembled from different types of X-ray imaging systems from various hospitals to demonstrate the generalizability of our approach. The purpose of the deep learning tool is to assist radiologist readers with prioritizing interpretations of the most urgent exams and help them to reach a prompt, correct diagnosis.

2 Materials and Methods

2.1 Pneumoperitoneum Dataset and Expert Annotations

The pneumoperitoneum dataset consisted of 1,287 CXR images (from 1,124 patients) and was collected using Montage (Montage Healthcare Solutions, Philadelphia, PA) search functionality from the database of a tertiary academic hospital and several community hospitals serving rural populations between March 2011 and September 2019. The inclusion and exclusion criteria for this study is demonstrated in Fig. 2. This dataset is nearly balanced with 634 pneumoperitoneum positive cases and 673 pneumoperitoneum negative cases. The pneumoperitoneum negative cases consist of both normal and other conditions (such as pneumothorax, pneumonia, atelectasis, etc.). A brief description of this dataset is presented in Table 1. All CXR images in this dataset were retrieved in DICOM format. The resolution of CXR images in our dataset ranges from 1728 × 1645 pixels to 4280 × 3520 pixels. All CXR images from the academic hospital were taken with Philips imaging systems, whereas CXR images from other community hospitals were taken with imaging systems from various manufacturers (Philips, Fujifilm, Siemens, Kodak, Konica Minolta). Further details of the total number of CXR images specified by the imaging system manufacturer are shown in Fig. 3.

Although several images from the same patient were included, they were not identical in terms of positioning and appearance. Since our goal in this study is to detect pneumoperitoneum per exam, multiple images for a patient do not alter the findings. Furthermore, we kept all the images from same patient in one partition (training, validation, testing). This dataset is nearly balanced, with 634 pneumoperitoneum positive cases and 673 pneumoperitoneum negative cases.

We needed high-quality expert annotations indicating ground truth pneumoperitoneum labels (i.e., positive or negative) for each CXR image in our dataset to develop and evaluate our model. The expert annotations in our study were generated by four radiologists from the main academic hospital campus. To produce the ground truth labels, the CXR images were equally divided among two radiologists for annotation. Then, the other two radiologists independently reviewed all the ground truth labels generated by the previous radiologists for accuracy. Any disagreements among annotators were resolved by further review and discussion among all radiologists. We tested the consistency of expert annotation between two radiologists on 177 randomly selected test cases consisting of 45 positive and 132 negative cases. Out of 177 tested cases, there was one on which the radiologists disagreed about the presence of pneumoperitoneum. That case was negative for pneumoperitoneum, and the disagreement was resolved after further discussion among the radiologists. Radiologist 1 has over 30 years of general radiology experience; Radiologist 2 has over 7 years of general radiology experience; Radiologist 3 has 20 years of abdominal imaging experience; and Radiologist 4 is a 4th-year radiology resident.

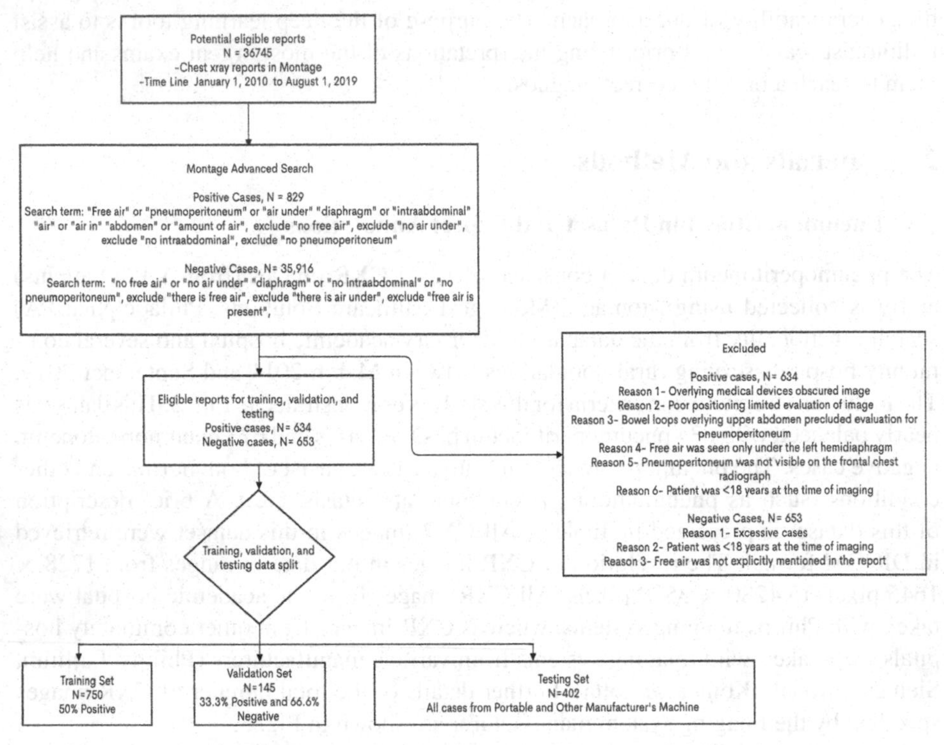

Fig. 2. Details of inclusion and exclusion criteria of this study.

2.2 Training, Validation and Test Split

Our study has two objectives: 1) to train and evaluate the performance of common deep learning architectures on our CXR image dataset for classification of pneumoperitoneum status, and 2) to analyse the sensitivity and specificity of these models based on different characteristics of the radiographs. Therefore, as shown in Table 1, we partitioned this dataset into training, validation, and test datasets. For the training and validation datasets, we only used the CXR images with the most common characteristics in the dataset, i.e., images taken by fixed X-ray machines at the main academic hospital (420 positive cases and 465 negative cases: Fig. 4). The training dataset consisted of 750 CXR images (375 positive and 375 negative), whereas the validation dataset consisted of 135 CXR images (45 positive and 90 negative). The cases in the training and validation datasets were randomly selected. In contrast, our test dataset consisted of 402 CXR images (214 positive and 188 negative) with images from different manufacturers and with both fixed and portable characteristics. Therefore, our test dataset was suitable to perform sensitivity and specificity analysis for the different deep learning models. Of note, in our data split, we ensured that CXR images from the same patient stayed in the same partition (training, validation, and testing datasets) to avoid any biases.

Table 1. Characteristics of our dataset stratified by pneumoperitoneum positive and negative cases.

Characteristic	Full dataset	Pneumoperitoneum positive	Pneumoperitoneum negative
No. of CXRs	1,287	634	653
Sex	Male - 697 Female - 590	Male - 344 Female - 290	Male - 353 Female - 303
Age (standard deviation)	Male - 61.44 ± 17.62 Female - 62.03 ± 17.96	Male - 61.06 ± 17.44 Female - 65.09 ± 16.04	Male - 61.82 ± 17.79 Female - 58.98 ± 19.89
Technique: Anteroposterior (AP)/ Posteroanterior (PA)	AP - 554 PA - 733	AP - 251 PA – 383	AP - 304 PA - 429
Imaging system type	Fixed - 969 Portable - 318	Fixed - 451 Portable - 183	Fixed - 518 Portable - 135
Institution/hospitals	Academic hospital - 1,061 Others - 226	Academic hospital - 545 Others – 89	Academic hospital - 516 Others - 137
Imaging system Manufacturer	Philips - 1,145 Others - 142	Philips - 576 Others - 58	Philips - 569 Others - 84

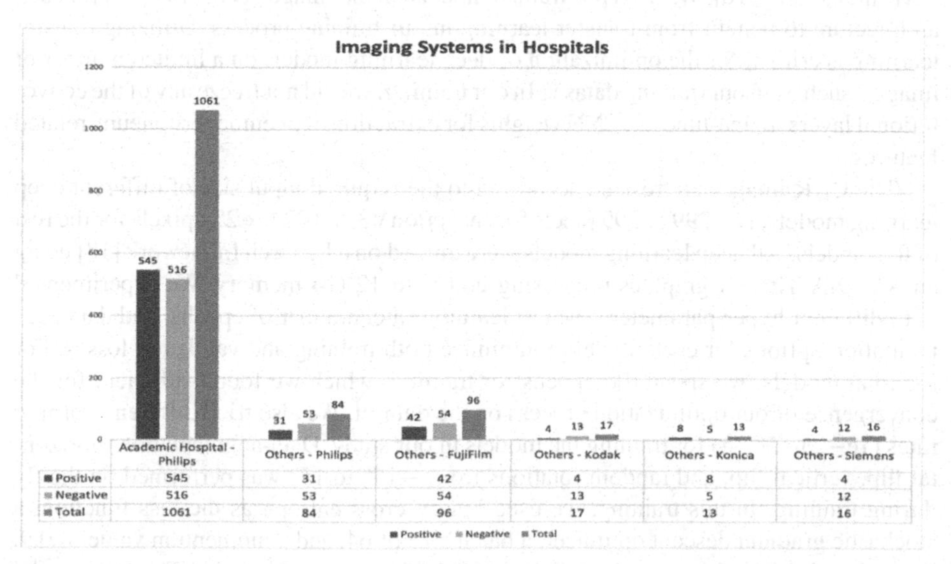

Fig. 3. Number of CXRs stratified in our dataset by their corresponding imaging system manufacturer.

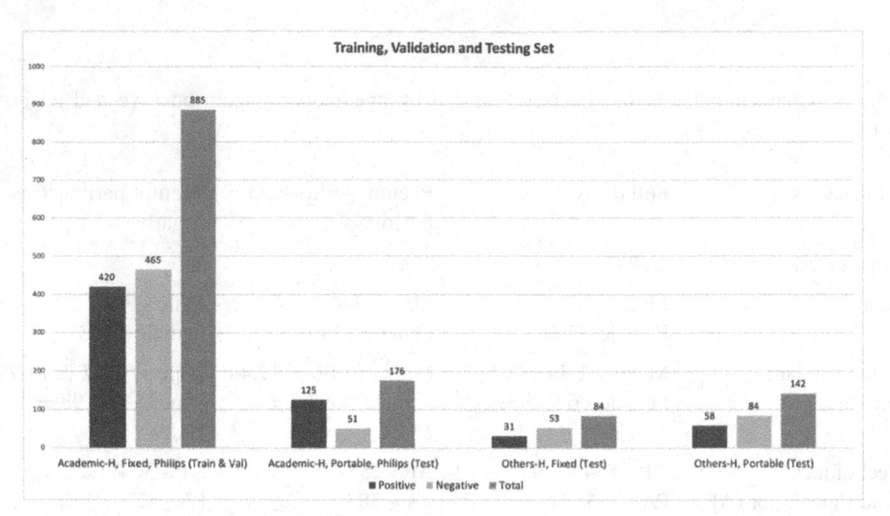

Fig. 4. Dataset split into training, validation, and test datasets. CXR images, taken by fixed Philips X-ray machines at the academic hospital, were used in the training and validation datasets, whereas all CXR images from portable and other imaging system manufacturers were used in the test dataset.

2.3 Deep Learning Methods

We utilized four different state-of-the-art deep learning architectures (ResNet50, DenseNet161, InceptionV3, ResNeXt101) for the detection of pneumoperitoneum on CXR images [13–16]. We used pre-trained models on the ImageNet dataset [17] for each architecture to benefit from transfer learning in our training process. Utilizing transfer learning is critical for the optimization of deep learning models on a limited number of images, such as in our training dataset. In our training, we did not freeze any of the convolutional layers to fine-tune the CNN weights for extraction of pneumoperitoneum-related features.

The CXR images are resized according to the required input size of different deep learning models, i.e., 299 × 299 pixels for InceptionV3 and 224 × 224 pixels for the rest of the models. All deep learning models were trained on a PyTorch framework [18] using an NVIDIA Titan X graphics processing unit with 12 GB memory. We experimented with different hyper-parameters such as learning rate, number of epochs, and data augmentation options for each model to minimize both training and validation losses. For the final models, we spent 100 epochs for training, which we found sufficient for the convergence of our optimization process on the dataset. We also tried different learning rates (1e−2 to 5e−4) for training the models in our study. Data-augmentation (horizontal flip, vertical flip, and random rotations from −15° to 15° was performed on the fly during training. In this training, we used binary cross-entropy as the loss function, a stochastic gradient descent optimizer, a batch size of 64, and a momentum value of 0.9. We reduced the learning rate by a factor of 0.1 after every 25 epochs. The final model was selected based on minimum validation loss during training.

3 Results

We evaluated the different deep learning models for the binary classification of pneumoperitoneum using our test dataset. In Table 2, we report sensitivity, specificity, accuracy, F1-score, and area under the receiver operating characteristic curve (AUC) as our evaluation metrics. These metrics are considered reliable measures for assessing the quality of machine learning models.

All deep learning models, particularly DenseNet161 and ResNeXt101, performed well for the binary classification of pneumoperitoneum. DenseNet161 achieved the highest accuracy of 0.908, whereas ResNeXt101 (0.905), ResNet101 (0.902), and InceptionV3 (0.883) performed slightly worse. For sensitivity, DenseNet161 (0.916) again outperformed ResNet101, InceptionV3, and ResNeXt101 by a margin of 0.43, 0.75, and 0.51, respectively. On the contrary, DenseNet161 achieved the lowest score of 0.899 for specificity, whereas ResNeXt101 performed best in this category, with a score of 0.952. The AUC score is considered to be a stable performance metric for evaluating machine learning approaches. ResNeXt101 and DenseNet161 achieved 0.951 and 0.957, respectively, for AUC. The ROC curves for all deep learning models are shown in Fig. 5.

Table 2. The performance measures of various deep learning models for binary classification of pneumoperitoneum.

Method	Sensitivity	Specificity	Accuracy	Precision	F-1 Score	AUC
InceptionV3	0.841	0.931	0.883	0.932	0.884	0.938
ResNet101	0.873	0.936	0.902	0.937	0.906	0.946
ResNeXt101	0.865	0.952	0.905	0.953	0.907	0.951
DenseNet161	0.916	0.899	0.908	0.911	0.913	0.957

3.1 Model Visualization and Error Analysis

We used the Grad-CAM algorithm [19], which uses pneumoperitoneum specific gradient information flowing into the final convolutional layer of the DenseNet161 deep learning model to mark the regions of interest on the CXR images that heavily influenced the outcomes of our model. Examples of Grad-CAM activations on randomly selected true positive cases of pneumoperitoneum are shown in Fig. 6. This visualization produces localization maps of the regions of interest for CXR images and can provide an explanation for the final diagnostic decisions of the deep learning models. The red coloring indicates the most important regions for the ultimate decision of the model on CXR images. We also applied the Grad-CAM algorithm on randomly selected false-positive cases of pneumoperitoneum, which were incorrectly identified by our DenseNet161 model. We found that false-positive cases, CXRs without pneumoperitoneum, were most frequently due to air in the stomach or small bowel below the left diaphragm, or lucency in the lungs above the diaphragm, as shown in Fig. 6.

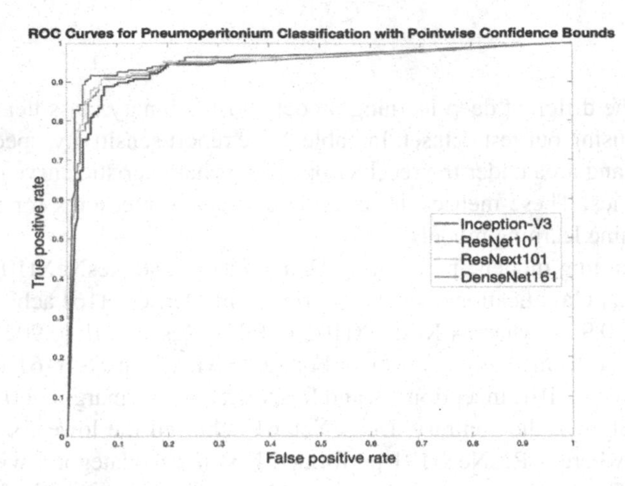

Fig. 5. ROC curve for all deep learning models trained on test dataset.

3.2 Discussion

In this study, we used various common deep learning architectures to develop a model for binary classification of pneumoperitoneum on chest X-ray exams. Pneumoperitoneum, also known as free air, is the abnormal presence of air in the peritoneal cavity. In this experiment, we developed our deep learning models using training and validation datasets that only consisted of CXR images from Philips fixed imaging systems, whereas, in the test dataset, we used CXR images from portable imaging systems or from other imaging system manufacturers (Siemens, Kodak, etc.). Our experiment showed that deep learning models trained on data from a fixed imaging system from a single institution performed well on heterogeneous data from other institutions. Particularly, our deep learning models in this study achieved a high specificity and sensitivity on our diverse test dataset overall.

The Grad-CAM algorithm showed that our models accurately identify the correct anatomic area and features on the CXR images for the detection of pneumoperitoneum. Radiologists can generally identify and interpret the urgency of the findings based on chart review of patients' medical history. However, the goal of this deep learning approach for pneumoperitoneum detection is to identify and triage possible urgent cases for interpretation rather than replacing the need for radiologist interpretation. A major application for our deep learning algorithm is to screen and triage all imaging with critical findings for expedited interpretation and patient care. Particularly, when the reading list is long, such deep learning approaches can assist with prioritizing urgent exams, especially when the finding is not tagged as STAT (immediate) priority. In addition, when there are many other findings in chest X-rays, subtle pneumoperitoneum cases can be missed. The proposed deep learning model can help radiologists by drawing attention to those cases.

This Study has Several Limitations: First, our study would benefit from further validation on a larger external test dataset and a prospective clinical trial, which we will

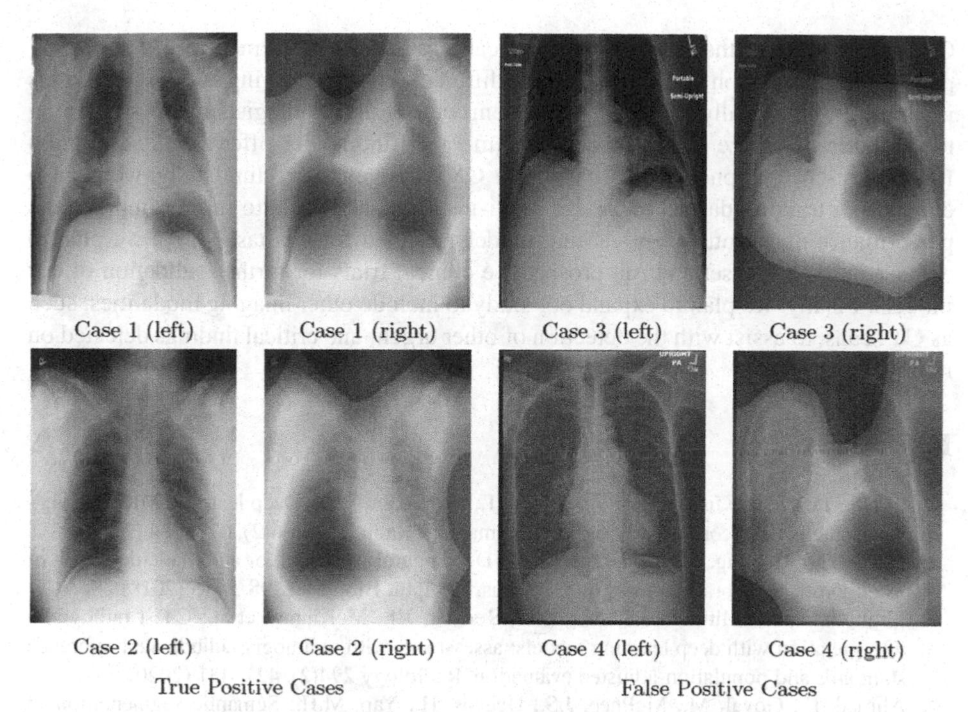

Fig. 6. Examples of Grad-CAM Activation of true-positive (1 and 2) and false-positive (3 and 4) cases by DenseNet161: left image of each case is original image whereas right image is activations by Grad-CAM method. The red coloring indicates a highly weighted region of interest. In the top true positive case (1), the pneumoperitoneum (free air) is beneath both the right and left diaphragm, and the heat map correctly marks both sides. In the bottom case (2), free air is only beneath under the right diaphragm, and the heat map identifies it on the correct side. Whereas, the false positive cases (3 and 4) in these examples are due to air in bowel below the left diaphragm in the upper images and air in stomach below the left diaphragm in the lower images, without pneumoperitoneum in either case. (Color figure online)

pursue as future work. Second, our pipeline is focused on distinguishing pneumoperitoneum negative and positive cases and does not recognize other urgent or critical findings such as pneumothorax or pneumonia on chest X-rays, and such findings could require patients to seek immediate medical or surgical attention. As future work, we plan to include other urgent findings in the next version of our model and will evaluate it in a multi-class classification setting. Finally, although we used Grad-CAM in this study to visualize the regions of interest in our classification, we plan to develop and evaluate a precise detection and a segmentation model to localize pneumoperitoneum on CXR images.

4 Conclusion

In summary, this study evaluated the generalizability of deep learning models across different image characteristics for the detection of pneumoperitoneum on CXR images.

Our results showed that end-to-end deep learning models performed well in detecting pneumoperitoneum on CXR images from different types of imaging systems at various institutions. If clinically validated, this system could assist radiologists as a pre-screening tool to help prioritize chest X-rays with emergent findings or offer a second opinion for the presence of pneumoperitoneum on CXR images. For future study, we plan to expand our training dataset to a large multi-institutional dataset to further improve the performance of various deep learning models selected for this task. Also, we plan to expand our test dataset and run prospective clinical trials for further validation of our models. Finally, we plan to expand our study to include other imaging modalities, such as CT scans, to assist with the detection of other urgent and critical findings detected on radiology exams.

References

1. Liu, F., Jang, H., Kijowski, R., Bradshaw, T., McMillan, A.B.: Deep learning MR imaging–based attenuation correction for PET/MR imaging. Radiology **286**(2), 676–684 (2018)
2. Tomita, N., Cheung, Y.Y., Hassanpour, S.: Deep neural networks for automatic detection of osteoporotic vertebral fractures on CT scans. Comput. Biol. Med. **98**, 8–15 (2018)
3. Majkowska, A., Mittal, S., Steiner, D.F., Reicher, J.J., McKinney, et al.: Chest radiograph interpretation with deep learning models: assessment with radiologist-adjudicated reference standards and population-adjusted evaluation. Radiology **294**(2), 421–431 (2020)
4. Ahmad, E., Goyal, M., McPhee, J.S., Degens, H., Yap, M.H.: Semantic segmentation of human thigh quadriceps muscle in magnetic resonance images. arXiv preprint arXiv:1801.00415 (2018)
5. Yap, M.H., et al.: End-to-end breast ultrasound lesions recognition with a deep learning approach. In: Medical Imaging 2018: Biomedical Applications in Molecular, Structural, and Functional Imaging, vol. 10578, p. 1057819. International Society for Optics and Photonics (2018)
6. Rajpurkar, P., et al.: ChexNet: radiologist-level pneumonia detection on chest x-rays with deep learning. arXiv preprint arXiv:1711.05225 (2017)
7. Wang, X., Peng, Y., Lu, L., Lu, Z., Bagheri, M., Summers, R.M.: Chestx-ray8: hospital-scale chest x-ray database and benchmarks on weakly-supervised classification and localization of common thorax diseases. In: Proceedings of the IEEE Conference on Computer Vision and Pattern Recognition pp. 2097–2106 (2017)
8. Stapakis, J.C., Thickman, D.: Diagnosis of pneumoperitoneum: abdominal CT vs. upright chest film. J. Comput. Assist. Tomogr. **16**(5), 713–716 (1992)
9. Woodring, J.H., Heiser, M.J.: Detection of pneumoperitoneum on chest radiographs: comparison of upright lateral and posteroanterior projections. AJR Am. J. Roentgenol. **165**(1), 45–47 (1995)
10. Chen, S.C., Yen, Z.S., Wang, H.P., Lin, F.Y., Hsu, C.Y., Chen, W.J.: Ultrasonography is superior to plain radiography in the diagnosis of pneumoperitoneum. Br. J. Surg. **89**(3), 351–354 (2002)
11. Solis, C.V., Chang, Y., De Moya, M.A., Velmahos, G.C., Fagenholz, P.J.: Free air on plain film: do we need a computed tomography too? J. Emerg. Trauma Shock **7**(1), 3 (2014)
12. Marcus, G., Little, M.A.: Advancing AI in health care: it's all about trust, 23 October 2019. https://www.statnews.com/2019/10/23/advancing-ai-health-care-trust/
13. Szegedy, C., Vanhoucke, V., Ioffe, S., Shlens, J., Wojna, Z.: Rethinking the inception architecture for computer vision. In: Proceedings of the IEEE Conference on Computer Vision and Pattern Recognition, pp. 2818–2826 (2016)

14. He, K., Zhang, X., Ren, S., Sun, J.: Deep residual learning for image recognition. In: Proceedings of the IEEE Conference on Computer Vision and Pattern Recognition, pp. 770–778 (2016)
15. Iandola, F., Moskewicz, M., Karayev, S., Girshick, R., Darrell, T., Keutzer, K.: DenseNet: implementing efficient convnet descriptor pyramids. arXiv preprint arXiv:1404.1869 (2014)
16. Chen, Y., Li, J., Xiao, H., Jin, X., Yan, S., Feng, J.: Dual path networks. In: Advances in neural Information Processing Systems, pp. 4467–4475 (2017)
17. Deng, J., Dong, W., Socher, R., Li, L.J., Li, K., Fei-Fei, L.: ImageNet: a large-scale hierarchical image database. In: 2009 IEEE Conference on Computer Vision and Pattern Recognition, pp. 248–255. IEEE, June 2009
18. Paszke, A., et al.: PyTorch: an imperative style, high-performance deep learning library. In: Advances in Neural Information Processing Systems, pp. 8024–8035 (2019)
19. Selvaraju, R.R., Cogswell, M., Das, A., Vedantam, R., Parikh, D., Batra, D.: Grad-cam: Visual explanations from deep networks via gradient-based localization. In: Proceedings of the IEEE International Conference on Computer Vision, pp. 618–626 (2017)

An Application of Recurrent Neural Networks for Estimating the Prognosis of COVID-19 Patients in Northern Italy

Mattia Chiari(✉), Alfonso E. Gerevini(✉), Matteo Olivato(✉), Luca Putelli(✉), Nicholas Rossetti(✉), and Ivan Serina(✉)

Università degli Studi di Brescia, Via Branze 38, Brescia, Italy
{m.chiari017,alfonso.gerevini,m.olivato,l.putelli002,n.rossetti004, ivan.serina}@unibs.it

Abstract. Hospital overloads and limited healthcare resources (ICU beds, ventilators, etc.) are fundamental issues related to the outbreak of the COVID-19 pandemic. Machine learning techniques can help the hospitals to recognise in advance the patients at risk of death, and consequently to allocate their resources in a more efficient way. In this paper we present a tool based on Recurrent Neural Networks to predict the risk of death for hospitalised patients with COVID-19. The features used in our predictive models consist of demographics information, several laboratory tests, and a score that indicates the severity of the pulmonary damage observed by chest X-ray exams. The networks were trained and tested using data of 2000 patients hospitalised in Lombardy, the region most affected by COVID-19 in Italy. The experimental results show good performance in solving the addressed task.

Keywords: COVID-19 · Recurrent Neural Networks · Clinical data

1 Introduction

Machine learning can offer powerful tools for fighting the spread of COVID-19, which reached more than 100 million cases worldwide at January 2021 [8]. Since March 2020, we have been working in collaboration with *Spedali Civili di Brescia*, one of the hospitals that has been treating more COVID-19 patients in Italy, on developing tools that support doctors in estimating in advance the prognosis of hospitalised patients [11]. This information can be helpful to the hospital staff for better managing limited resources such as ICU beds, ventilators and personnel.

Starting from the raw data of more than 2,000 patients, we designed a dataset for representing the course of the disease of each patient during the hospitalisation. This dataset contains the status of each patient for each day of the first 20 days from the beginning of the hospitalisation. These data include demographic information such as sex and age, several laboratory test results and a value that measures the severity of the pulmonary condition using chest X-ray [7].

A. Tucker et al. (Eds.): AIME 2021, LNAI 12721, pp. 318–328, 2021.
https://doi.org/10.1007/978-3-030-77211-6_36

For our task, we implemented an architecture based on Recurrent Neural Networks (RNNs) and Attention Mechanism. In order to find the best configuration, we conducted a hyperparameter search with the explicit goal of minimising the false negatives, i.e., those patients who are at high risk of death but the model considers them not at risk. The optimised networks are evaluated in terms of F2-Score (a metric strongly influenced by false negatives) and ROC-AUC, which is a standard measurement for medical analyses. Our results are promising, obtaining a score of 73.2% in terms of F2 and 81.4% in terms of ROC-AUC. In the following, after a brief discussion of related work, we describe our datasets; then we present our models and their experimental evaluation, and finally we give conclusions and mention future work.

2 Related Work

The work of Bullock et al. [8] presents an overview of recent studies on the use of Artificial Intelligence against COVID-19 for drug discovery and development, testing and diagnosis (especially via X-Ray imaging), tracking and epidemiology, and patient outcome prediction. Given the recency of the pandemic outbreak, many of these studies are still preliminary works, without an in-depth description of the developed techniques (often they are only pre-printed and not properly peer-reviewed).

A first study about prognosis prediction is presented in [22]. The authors of this work train a XGBoost model using lab tests, symptoms and some epidemiological information for predicting the mortality risk at the 10th day after the lab test findings. The work in [1] proposes a Feed-forward Neural Network model for the mortality prediction that is trained and tested using data of approximately 400 patients in the United Kingdom. This study does not consider any lab test performed during the hospitalisation, which is a major difference with our work; instead it uses information obtained when a patient is admitted in hospital (such as symptoms, demographics and smoking history). In [3], the authors investigate the tasks of predicting mortality with no time constraints, or within either 14 days or 30 days after the diagnosis using the LASSO algorithm. In [11], we described an approach based on ensemble of Decision Trees for estimating the prognosis of COVID-19 patients at different times during their hospitalisation using only demographic information and lab tests. In this work, starting from the same raw data, we present a completely different dataset and use RNNs for better monitoring the progression of the disease.

RNNs are widely used in healthcare and proved to be very effective in various contexts including diagnosis and mortality prediction. RNNs with Gated Recurrent Units were used in [6] to analyse dead patients' medical records in order to predict the life expectancy to help doctors in end-of-life decision making. In [16], a Long Short Term Memory (LSTM) neural network was trained over electronic health records (EHR) to predict mortality in rare diseases cases. Nonetheless, to the best to our knowledge, RNNs were used to predict the prognosis of COVID-19 patients only in [9], which exploits RNNs and Convolutional Neural Networks for training a model using sequences of X-Ray images (which are not available in our dataset), and predict the evolution of the observed lung pathology. Another major difference with our work is that [9] does not consider any lab test.

3 Available Data

Our data derives from 2015 patients who were hospitalised in Spedali Civili di Brescia from February to April 2020. For each of these patients, the specific features that were made available to us are: age and sex, the values and dates of several lab tests (i.e. PCR, LDH, Ferritin, Troponin-T, WBC, D-dimer, Fibrinogen, Lymphocites and Neutrophils), the scores (each one from 0 to 18), assigned by the physicians, assessing the severity of the pulmonary conditions resulting from the X-ray exams [7], the values and dates of the throat-swab exams for COVID-19, and the final outcome of the hospitalisation at the end of the stay, which is the classification value of our application (i.e. in-hospital death or released alive). We have no further information about symptoms, their timing, comorbidities, generic health conditions, admission in ICU or clinical treatment. In our cohort 18.4% of the patients are deceased and 81.6% were released or transferred to a rehabilitation centre.

When applying machine learning to real-world data, there are some non-trivial practical issues to deal with. In our case, one of such issues is that the length of the hospitalisation period can sensibly differ from one patient to another (from few days to two months). Subsequently, the number and the frequency of performed lab tests and relative findings significantly varies among the considered set of patients (from only three to hundreds). For many patients, this results in a substatial presence of missing values, i.e. we are not able to fully know the patients' conditions at a given day. In fact, on average, lab tests PCR, WBC, Lymphocites and Neutrophils are performed once every 2 days, LDH, Ferritin and Fibrinogen once every 4 or 5 days, while RX, D-dimer and Troponin-T once a week or even less frequently. If, for a given patient, a test is not performed at a considered day, we store a missing value for the corresponding lab result.

Moreover, an examination of the data available for our cohort of patients revealed that their prognostic risk is influenced by multiple factors. For instance, the number of the patients currently hospitalised, which impacts on the availability of ICU beds, and the increase of the clinical knowledge. Therefore, studying the death rate over time, we discovered that for patients hospitalised during the most critical weeks of the pandemic (February and March 2020) this value is about twice than the value obtained during the period were the pandemic was stabilising (April 2020). Similarly, also the average length of stay in the hospital varies, from 14 days in April 2020 down to only 7 days in March 2020.

In machine learning, this variation of data distributions over time is known as *concept drift* [10,20]. In a first attempt to deal with the observed concept drift, we used a classical method, which consists of training the algorithm using only a subset of samples depending on the data distribution that we are considering [10,21], dividing our set of patients into two groups [11]. However, this has the drawback of significantly reducing the data for training. Thus we decided to follow a different approach, using the full dataset with an additional feature that helps the learning algorithm to discriminate if a patient was hospitalised during the most critical pandemic phase or not. This allows to exploit more training data, achieving better performance. The new feature, called **death rate**, intends to provide an indicator of the status of the pandemic emergency at a given day (when the feature is evaluated), and it is defined as the average death rate computed considering seven days preceding such a day. Specifically, the death rate feature is the ratio of all the patients who died over all the patients discharged (dead or alive) over the considered 7-days period.

4 Description of the Datasets

In order to apply Recurrent Neural Networks to our dataset, we have to design a representation of the patients' hospitalisation over time. This can be done by building a matrix $M[l, e]$ of real numbers for each of our patients, where l is the length (number of days) of the hospitalisation and e is the number of the considered features, i.e. the patient's demographic information, the death rate and all the lab tests available in our dataset. Therefore, in this representation the value obtained for test j in the i-th day of hospitalisation is stored in $M[i, j]$. If test j has not been performed on that day, we indicate this missing value with -1 (i.e., $M[i, j] = -1$). Considering a set of n patients, the input of our RNN is a tensor with dimension $n \times l \times e$.

Given that the length of the hospitalisation can sensibly differ from one patient to another, as explained in Sect. 3, we set $l = 20$ considering in this way at most the first 20 days of hospitalisation for each patient. If a patient stays in the hospital for less than 20 days, the remaining period is filled by vectors of zero values with length e, called *padding*. The value of l is chosen taking into account the learning problems related to RNNs. In fact, we need a sequence that is long enough to allow the recurrent model to derive temporal dependencies and to capture the evolution of the patients' conditions over time. However, choosing a value for l that is too large, as for example the longest hospitalisation in our dataset (which is over 50 days), would require adding a considerable amount of padding, and this would have a negative impact on the performance. As a compromise, we chose $l = 20$ which is a value slightly higher than the average length of stay in April 2020. We call this representation the **Complete Dataset**.

Unfortunately, the Complete Dataset is not appropriate for the purpose of providing a prognosis estimation in advance. According to our data, more than 70% of considered patients stay in hospital for less than 20 days. Therefore, using the previous data representation, for these patients the RNN would provide a prediction considering their entire history (until the day of the release or eventual decease). Thus, in order to build a model that can provide an estimation of the prognosis in advance (with respect to the 20-days of hospital stay), we adopted a "cutting strategy" to generate a variant of the dataset called the **Cut Dataset**. Specifically, given a patient who is hospitalised for d days, our cutting strategy is to consider only the tests performed in the first d_p days of hospitalisation, where $d_p = p * d$ and p is a random fraction between 0.3 and 0.9. This fraction is chosen randomly in order to simulate the clinical usage of our system. In fact, our goal is to create a deep learning model that can be used for estimating the prediction of a patient at almost every day (with enough data) during his or her hospitalisation. Therefore, our system has to obtain good performance analysing a patient during its first period (with $p < 0.4$) of hospitalisation, in the middle ($0.4 \leq p \leq 0.6$) but also in the end ($p > 0.6$). With this strategy, the days between d_p and d are not provided to our model and they are replaced by padding. If $d_p > l$ (which is unlikely in our dataset) only the first 20 days of hospitalisation are considered.

The number of training instances is fundamental for the performance of machine learning and deep learning systems. Given the relatively limited amount of available data in our application, we have designed a method for the dataset augmentation that consists in adding artificial instances obtained by varying the those in the Complete Dataset as follows. Given a patient who is hospitalised for d days, instead of applying the cutting strategy only once, it is applied r times, obtaining several d_p^i with $i \in [1, r]$. In our case, we consider $r = 3$; we call this representation the **Augmented Dataset**. With this technique, the available data for a single patient are used multiple times in order to help the Recurrent Neural Network to better learn how the disease progresses, what are the most important lab tests, and how they are related with each other. It is important to point out that this technique is applied only to the training and validation sets. In fact, the trained models are finally evaluated on a test set where each patient is represented only once and only considering the (lab and X-ray) tests performed in the first d_p days of the hospitalisation with a random probability p between 30% and 90%. For all our datasets, the test set used for evaluating our models consists of 20% of all patients randomly selected (with stratified sampling). These patients are the same for all our datasets.

5 Recurrent Neural Network Model

A Recurrent Neural Network (RNN) is a neural network model for processing sequential data. Gated Recurrent Unit (GRU) networks and Long Short Term Memory Networks (LSTM) are two types of RNNs which deal with typical issues of the standard version of RNNs, such as vanishing gradient and long term dependencies, obtaining better performance than standard RNNs [17]. As shown in [12], even though usually LSTMs are better than GRUs, the latter can achieve better performance on smaller datasets.

5.1 Gated Recurrent Units and Attention Mechanism

Given $x_1, x_2 \ldots x_l \in \mathbb{R}^e$ representing the status (as a vector of demographic information and lab test features) of a patient for l days of hospitalisation (padding included), a GRU layer is formed by l cells. Each cell processes the t-th vector also taking into account the computation made by the previous cell (h_{t-1}) using a neural network layer with *tanh* activation function: $n_t = tanh(W_n[h_{t-1}, x_t]+b_n)$.

While in RNN n_t is simply passed to the next cell, in the GRU network we have also the *update gate* z_t, which is a layer with sigmoid activation function, that decides which part of the information contained in n_t and h_{t-1} has to be preserved, and which part can be forgotten: $z_t = \sigma(W_z[h_{t-1}, x_t] + b_z)$ where $W_z, W_n \in \mathbb{R}^{(n+e)\times n}$ are weight matrices, $b_z, b_n \in \mathbb{R}^n$ are bias vectors, and n is an hyperparameter that corresponds to the number of neurons in the GRU layer. The weight matrices and bias vectors are shared by all the cells. Finally, the output of the GRU cell, h_t is calculated as: $h_t = (1 - z_t) * n_t + z_t * h_{t-1}$.

In most implementations, the GRU includes a further gate (the *reset gate*) which has the possibility to erase some parts of h_{t-1}.

The attention mechanism [5] takes each h_t into consideration, and computes a set of weights α_i associated to the contribution of each day for the prognosis estimation:

$$u_t = tanh(W_a h_t + b_a),\ \alpha_t = softmax(v^T u_t) = exp(v^T u_t)/\sum_{k=1}^{n} exp(v^T u_k)$$

where $W_a \in \mathbb{R}^{n\times n}$, $b_a \in \mathbb{R}^n$ and $v \in \mathbb{R}^n$ are trainable parameters of the attention mechanism. Given that the weights α_t are calculated using the *softmax* function, it is guaranteed that each α_t is between 0 and 1, and the sum of them is 1. The attention mechanism outputs a final representation of the patient's history, also called the *context vector*, as the average of each h_t weighted by α_t.

Table 1. Hyperparameters space used for our Recurrent Neural Model search and tuning. Scaler is set to *MaxAbsScaler* and the activation function is set to *tanh*.

Hyperparameter	Type	Interval
Total hidden units	Integer	[0, 150]
Layers	Integer	[1, 5]
Dropout rate	Real	[0, 0.6]
Recurrent dropout rate	Real	[0, 0.6]
Batch normalisation	Categorical	[True, False]
Activity regularisation	Categorical	[True, False]
Batch size	Integer	[32, 40, 48, 56, 64]

5.2 Loss Function and Tuning of the Hyperparameters

Typically, a feed-forward neural network is trained using a standard loss function such as Log-Loss or Binary Cross Entropy. In our case these loss functions did not provide adequate results, especially in terms of number of false negatives, which are the most critical errors in our application context. As proposed in [19], the F-measure can be used as loss function in order to improve the performances in high unbalanced problems. Moreover, other works [13,14] use *F-β* losses with different β values in order to improve the performance specifically for unbalanced and difficult tasks in the medical domain.

Therefore, for our ANN models, we adopted a loss function based on the F-2 score metric defined as follows: $F\text{-}\beta_{loss} = 1 - \frac{(1+\beta^2)\cdot Precision\cdot Recall}{\beta^2 * Precision + Recall + \varepsilon}$, with $\beta = 2$, and $\varepsilon = 10^{-7}$ used to avoid zero-division errors.

Since the performance of a RNN is deeply influenced by its hyperparameters, such as the number of neurons or the recurrent dropout, we performed a hyperparameter search to generate a highly-performing architecture. In order to optimise the RNN architecture, we adopted the Bayesian-optimisation approach via the Optuna framework [2] with 128 search iterations. The hyperparameters search is performed using tenfold cross-validation on the training set of each dataset at different days of hospitalisation. For each randomly selected combination Σ of hyperparameter values:

1. The training set of each dataset is partitioned into k folds with $k = 10$.
2. For each fold of the cross validation, the performance of the algorithm using Σ is evaluated in terms of F2-score.
3. The overall evaluation score of the k-fold cross validation for a configuration Σ of the hyperparameters is computed by averaging the scores obtained for each fold.

Finally, the hyperparameter configuration with the best overall score is selected.

It is well known that the standardisation and scaling of input data can have a huge impact on the stability and performance of Neural Networks [4,18]. Therefore, we considered the scalers provided by the Scikit-Learn package and conducted some preliminary experiments for finding the most suitable one for our task, which is the *MaxAbsScaler*. In the same initial experiments, we have found that GRU units have much better performance with respect to standard RNNs or LSTM neural networks. We have also considered hyperparameters for the initialiser, number of layers, number of units (or neurons) for each layer, batch size, dropout rates, Batch Normalisation and Activity Regularisation. The considered ranges of values for these hyperparameters are shown in Table 1.

Finally, we used Adam [15] as optimiser with the default values for the parameters β_1, β_2 and ϵ, while the starting learning rate is set to 10^{-4}.

Table 2. Hyperparameters for the tested RNN models. **B.S.** is the batch size, **B.N.** stands for batch normalisation, **Layers** is the number of hidden layers, **Act Reg** is the activity regularisation and **Rec Dropout** is the recurrent dropout rate. The dropout rate was set very close to 0 by the optimizer for all the datasets.

Model	Layers	Neurons	B.S.	B.N.	Act reg	Rec dropout
Cut	2	66;34	32	False	False	0.57;0.28
Attention	1	81	32	True	True	0.26
Augmented	1	65	64	True	False	0.49

Table 3. Performance of our best RNN models in terms of F2-score, ROC-AUC score and accuracy over the test sets.

Dataset	F2-Score	ROC-AUC	Accuracy
Cut	66.81	74.33	64.55
Attention	68.13	75.8	66.93
Augmented	73.17	81.39	77.87

6 Experimental Results

In this section, we evaluate the performance of the RNN models over different datasets. Our system is implemented using the Scikit-Learn library for Python, and the experimental tests were conducted using an Intel(R) Xeon(R) Gold 6140M CPU @ 2.30 GHz. The performance of the models with the relative optimised hyperparameters are evaluated using the test set in terms of both F2 and ROC-AUC scores and Accuracy. First of all, we tested the performances using a model trained on the Complete Dataset, that contains the whole history of each patient. With this model we obtain a F2-score of 68.81% and a ROC-AUC of 75.95%. However, since this model is trained and tested using all the available data for patients who stayed in hospital less than 20 days (more than 70% of total patients), this model is not suitable to be used in an hospital, and we consider it as a reference point for a comparison and evaluation of the performance of the other models. Moreover, this model is tested on a different test set which does not implement the cutting strategy, and for this reason it shouldn't be compared directly to the other models. We evaluated the following models:

- *Cut model*: a model trained on the Cut dataset. This Dataset implements the cutting strategy that reduces the available data by cutting the length of stay according to a random probability between 30% and 90%.
- *Attention model*: a model including the Attention Mechanism (Sect. 5.1) and trained on the Cut Dataset.
- *Augmented model*: a model trained on the Augmented Dataset (the augmentation procedure is described in Sect. 4) along with the cutting strategy.

In Table 2, we show the hyperparameters of each tested model. In Table 3 and in Fig. 1 we show their results on the classification task for predicting the prognosis of hospitalised patients. The performances obtained by the Cut model are slightly worse compared to the ones obtained by the Complete model (66.81% F2-score and 74.33% ROC-AUC). This was expected because, by applying the cutting strategy to our dataset, we remove some of the exams data from the last days of hospitalisation. On the other hand, through the application of the of the cutting strategy we can simulate an average use case of our model where we want to estimate the outcome of the patient some days in advance (w.r.t. the last days of hospitalisation).

The Attention model performs better that the Cut model, although the improvement is not very significant (68.13% F2-score and 75.8% ROC-AUC).

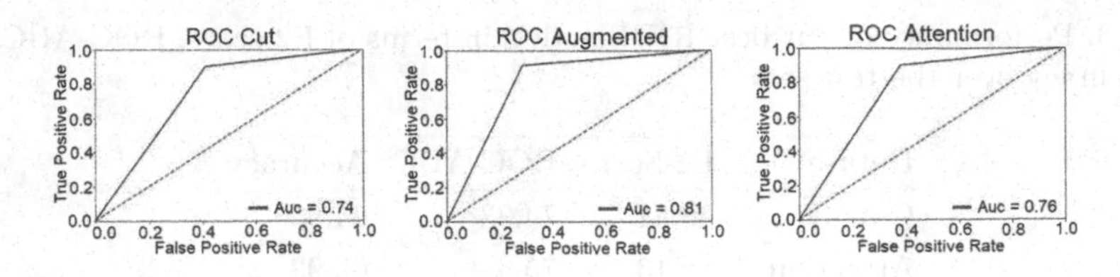

Fig. 1. Receiving operating characteristic curve for the best cut, attention and augmented models.

By using both the cutting strategy and the augmentation technique for our dataset, we can significantly improve the performances of our model thanks to the increased size of the training set: the Augmented model obtained a F2 score of 73.17% and a ROC-AUC of 81.39%. These results outperform the Complete model performances by more than 4 percentage points, which is impressive.

We also trained a model implementing the attention layer with the Augmented Dataset, but its performance is not satisfactory. The resulting model has 5 hidden layers, a high number of neurons per layer (121 on the first layer), and the same recurrent dropout rate for each layer. The number of layers is unusually high considering that most of the RNNs described in literature have at most 2 recurrent layers. The poor model quality and inferior average performances over the test set (65.55% of F2-score and 76.82% of ROC-AUC) led us to conclude that this model has substantial overfitting issues.

In previous work [11], starting from the same raw data, we used ensembles of Decision Trees (DT) for estimating the prognosis at different specific days of the hospitalisation (either 2, 4, 6, 8, 10 days or the last day). We built a different model for each considered day, obtaining performances between 68.4 and 76.5 in terms of F2-Score, and between 79.3 and 86.3 in terms of ROC-AUC. However, these models cannot be directly compared with the RNN-based approach here presented. In fact, while each DT-based model is specialised for patients at certain specific day of hospitalization, using RNNs we generated a *single* model that works considering hospitalisation periods of different length. Furthermore, some of the features that we introduced in the datasets for the DT-based models are automatically learned by our RNN models.

7 Conclusions and Future Work

We have presented a system for monitoring and evaluating the prognosis of COVID-19 patients focusing on the mortality risk. We built our predictive system using Recurrent Neural Networks trained using datasets engineered from lab tests and X-ray data of more than 2000 patients in an hospital in Northern Italy that was severely hit by COVID-19.

An experimental analysis shows that our best performing model achieves good predictive results (F2-score 73% and ROC-AUC 81%). In particular, we obtain the best results simulating a real use case of the system by a cutting strategy that reduces the available data to focus on earlier days of the hospitalizations, and improving the predictive performances by a powerful technique for dataset augmentation. Thanks to the dataset augmentation, the resulting model outperforms even the model trained on the Complete Dataset.

We think that there is room to improve our results by considering additional information about the patients that was not available to us, such as patient comorbidities, clinical treatments and administered drugs. Moreover, usually Recurrent Neural Network models require a large amount of data to achieve high performances. Given the limited size of our dataset, our approach is very promising. We intend to extend the dataset with more data by gathering more recent data and patients hospitalised during the second wave of the pandemic. This will further improve our results and validation of the proposed models.

References

1. Abdulaal, A., Patel, A., Charani, E., Denny, S., Mughal, N., Moore, L.: Prognostic modeling of COVID-19 using artificial intelligence in the united kingdom: model development and validation. J. Med. Internet. Res. **22**(8) (2020)
2. Akiba, T., Sano, S., Yanase, T., Ohta, T., Koyama, M.: Optuna: a next-generation hyperparameter optimization framework. In: Proceedings of the 25th ACM SIGKDD International Conference on Knowledge Discovery & Data Mining (2019)
3. An, C., Lim, H., Kim, D.W., Chang, J.H., Choi, Y.J., Kim, S.W.: Machine learning prediction for mortality of patients diagnosed with COVID-19: a nationwide Korean cohort study. Sci. Rep. **10**(1), 1–11 (2020)
4. Anysz, H., Zbiciak, A., Ibadov, N.: The influence of input data standardization method on prediction accuracy of artificial neural networks. Procedia Eng. **153**, 66–70 (2016)
5. Bahdanau, D., Cho, K., Bengio, Y.: Neural machine translation by jointly learning to align and translate. In: 3rd International Conference on Learning Representations, ICLR 2015, San Diego, CA, USA, 7–9 May 2015 (2015)
6. Beeksma, M., Verberne, S., van der Bosch, A., Das, E., Hendrickx, I., Groenewoud, S.: Predicting life expectancy with a LSTM neural network using electronic medical records. BMC Med. Inf. Decis. Making (2019)
7. Borghesi, A., Maroldi, R.: COVID-19 outbreak in Italy: experimental chest X-ray scoring system for quantifying and monitoring disease progression. La Radiol. Med. **125**(5), 509–513 (2020). https://doi.org/10.1007/s11547-020-01200-3
8. Bullock, J., Luccioni, A., Pham, K.H., Lam, C.S.N., Luengo-Oroz, M.A.: Mapping the landscape of artificial intelligence applications against COVID-19. J. Artif. Intell. Res. **69**, 807–845 (2020)
9. Fakhfakh, M., Bouaziz, B., Gargouri, F., Chaari, L.: Prognet: Covid-19 prognosis using recurrent and convolutional neural networks. medRxiv (2020)
10. Gama, J.A., Žliobaitundefined, I., Bifet, A., Pechenizkiy, M., Bouchachia, A.: A survey on concept drift adaptation. ACM Comput. Surv. **46**(4) (2014)
11. Gerevini, A.E., Maroldi, R., Olivato, M., Putelli, L., Serina, I.: Prognosis prediction in COVID-19 patients from lab tests and x-ray data through randomized decision trees. In: Proceedings of the 5th International Workshop on Knowledge Discovery in Healthcare Data. CEUR Workshop Proceedings, vol. 2675 (2020)

12. Gruber, N., Jockisch, A.: Are GRU cells more specific and LSTM cells more sensitive in motive classification of text? Front. Artif. Intell. **3**, 40 (2020)
13. He, H., Sun, X.: F-score driven max margin neural network for named entity recognition in Chinese social media. arXiv preprint arXiv:1611.04234 (2016)
14. Kawahara, J., Hamarneh, G.: Fully convolutional networks to detect clinical dermoscopic features. In: International Skin Imaging Collaboration (ISIC) 2017 Challenge at the International Symposium on Biomedical Imaging (ISBI) (2017)
15. Kingma, D.P., Ba, J.: Adam: a method for stochastic optimization (2017)
16. Liu, L., et al.: Multi-task learning via adaptation to similar tasks for mortality prediction of diverse rare diseases. arXiv preprint arXiv:2004.05318 (2020)
17. Mangal, S., Joshi, P., Modak, R.: LSTM vs. GRU vs. bidirectional RNN for script generation. CoRR abs/1908.04332 (2019). http://arxiv.org/abs/1908.04332
18. Mazzatorta, P., Benfenati, E., Neagu, D., Gini, G.: The importance of scaling in data mining for toxicity prediction. J. Chem. Inf. Comput. Sci. **42**(5), 1250–1255 (2002)
19. Pastor-Pellicer, J., Zamora-Martínez, F., España-Boquera, S., Castro-Bleda, M.J.: F-measure as the error function to train neural networks. In: Rojas, I., Joya, G., Gabestany, J. (eds.) IWANN 2013. LNCS, vol. 7902, pp. 376–384. Springer, Heidelberg (2013). https://doi.org/10.1007/978-3-642-38679-4_37
20. Quionero-Candela, J., Sugiyama, M., Schwaighofer, A., Lawrence, N.D.: Dataset Shift in Machine Learning. The MIT Press, Cambridge (2009)
21. Rakitianskaia, A.S., Engelbrecht, A.P.: Training feedforward neural networks with dynamic particle swarm optimisation. Swarm Intell. **6**(3), 233–270 (2012)
22. Yan, L., et al.: An interpretable mortality prediction model for COVID-19 patients. Nat. Mach. Intell. **2**, 1–6 (2020)

Recurrent Neural Network to Predict Renal Function Impairment in Diabetic Patients via Longitudinal Routine Check-up Data

Enrico Longato[1], Gian Paolo Fadini[2], Giovanni Sparacino[1], Angelo Avogaro[2], and Barbara Di Camillo[1](✉)

[1] Department of Information Engineering, University of Padova, 35131 Padua, Italy
{enrico.longato,gianni}@dei.unipd.it, barbara.dicamillo@unipd.it
[2] Department of Medicine, University of Padova, 35131 Padua, Italy
{gianpaolo.fadini,angelo.avogaro}@unipd.it

Abstract. People affected by diabetes are at a high risk of developing diabetic nephropathy, which, in turn, is the leading cause of end-stage chronic kidney disease worldwide. Predicting the onset of renal complications as early as possible, when kidney function is still intact, is of paramount importance for therapy selection due to existence of a class of antidiabetic agents (SGLT2 inhibitors) with known nephroprotective properties.

In the present work, we study the anthropometric and laboratory data of 28,955 diabetic patients followed for a median of 6.6 years (IQR 4.7–7.8) by 14 Italian diabetes outpatient clinics. We develop a deep learning model, based on the incorporation of variable-length longitudinal baseline data via recurrent layers, to predict the onset of impaired kidney function (KDOQI stage $\geq$ 3). We adopt a multi-label output-coding system to address the irregularity and sparsity in the sampling of endpoints induced by the real-life structure of the data.

Using the cumulative/dynamic AUROC with respect to a variable prediction horizon of 1 to 7 years, we compare the proposed model against the predictor of imminent deterioration of kidney function used in clinical practice, i.e., the estimated glomerular filtration rate (eGFR), and a set of year-specific logistic regressions trained on a single baseline visit.

The proposed deep learning model generally outperforms both benchmarks, especially in the medium-to-long term, with AUROC ranging from 0.841 to 0.895. Supplementary analyses confirm the effective encoding of sequence data within the network.

Keywords: Diabetes · Kidney disease · Predictive modelling · Recurrent neural network · Routine clinical data

1 Introduction

People affected by diabetes, a chronic, incurable disease characterised by elevated blood glucose concentration levels, often experience a broad range of macro- and micro-vascular complications. Among the latter, diabetic nephropathy is the leading cause

A. Tucker et al. (Eds.): AIME 2021, LNAI 12721, pp. 329–337, 2021.
https://doi.org/10.1007/978-3-030-77211-6_37

of end-stage chronic kidney disease (CKD) worldwide [1]. Indeed, it is estimated that the prevalence of CKD among people with diabetes may be as high as double that in the general population [2]. Key intervention targets include improvements in glycaemic control, blood pressure, and lipid profile, which, combined with frequent monitoring via routine check-ups, appropriate therapeutic choices, and positive lifestyle changes, have been shown to delay the onset and slow the progression of diabetic nephropathy [3, 4]. Recently, a novel class of antidiabetic agents known as sodium-glucose cotransporter 2 inhibitors (SGLT2is) have demonstrated marked nephroprotective properties in diabetic patients with pre-existing albuminuria or reduced estimated glomerular filtration rate (eGFR) [5–8]. However, as a much greater number of diabetic patients with preserved kidney function would need to be treated with SGLT2is to prevent even a single case of nephropathy [5, 9], suboptimal resource allocation remains a concern, and there is no clear indication for specific CKD-preventing therapies in subjects at non-immediate risk.

In light of these considerations, it is apparent that early prediction of impaired renal function is a crucial target with notable ramifications not only on individual quality of life, but also on resource allocation with respect to the early identification of potential candidates for innovative anti-CKD therapy. Recent research in this direction has highlighted that machine learning models based on routinely acquired real-world data have a great potential as tools to aid in the prediction of future CKD [10, 11]. Oftentimes, however, data collection objectives for clinical practice and model development do not align. This is the case, e.g., of routine check-up visits, where different batteries of laboratory tests are usually performed at a physician's discretion, resulting, on the one hand, in the potentially advantageous acquisition of additional longitudinal information, but, on the other, in incomplete or sparsely sampled data points, which might render baseline definition and outcome adjudication more difficult.

Taking into account this inherent divergence of purposes, in the present work, we develop a deep learning model based on recurrent neural networks to predict the onset of impaired renal function using the routine check-up data of 28,955 patients, acquired in 14 Italian diabetes outpatient clinics. In doing so, we address two main challenges related to model development with this type of data: 1) the incorporation of longitudinal baseline data in the form of the sequence of anthropometric and laboratory information collected during a series of past visits; and 2) the highly irregular sampling of endpoints that is ill-suited to traditional methods.

2 Prediction Target and Study Population

2.1 Prediction Target: Impaired Kidney Function on the KDOQI Scale

The prediction target was the onset of impaired kidney function, i.e., stage ≥ 3 on the Kidney Disease Outcomes Quality Initiative (KDOQI) scale [12]. As only stages ≥ 3 meet the criteria for CKD, we will refer to "CKD onset" and "impaired kidney function onset" interchangeably. Operatively, in terms of outcomes, we distinguished between subjects with preserved renal function, i.e., eGFR ≥ 60 (KDOQI stages 1 and 2) and those with eGFR < 60 (KDOQI stages 3a, 3b, 4, and 5).

2.2 Study Population and Dataset Split

The primary source for the present study was a multi-centre database comprising the data of 28,955 subjects treated at 14 diabetes outpatient clinics in the Veneto region between 1st January 2010 and 14th May 2019 (median observation time: 6.6 years; IQR 4.7–7.8). For each subject, a number of routine check-up visits, recorded with an irregular (on average, yearly) sampling rate, were available. At each visit, demographic, anthropometric, and laboratory data were collected as part of the subjects' regularly scheduled monitoring sessions. The complete list of variables comprised sex, age, diabetes duration, body-mass index (BMI), systolic and diastolic blood pressures, fasting glucose, glycated haemoglobin (HbA1c), total and HDL cholesterol levels, triglycerides, aspartate transaminase (AST), alanine transaminase (ALT), creatinine, and eGFR for a total of 15 variables (14 dynamic, 1 static). All subjects met the following inclusion criteria.

1. At least three visits with known eGFR (at least two to serve as a sequential input, and at least one more to determine the output).
2. At least two consecutive visits with eGFR ≥ 60.
3. No evidence of CKD at database entry.

We split the total cohort of 28,955 patients into a training, validation, and test sets, comprising, respectively, 80% (23,164), 10% (2,895), and 10% (2,896) of the subjects.

Table 1. Population characteristics. Continuous quantities are expressed as mean ± standard deviation, other quantities as counts. BP: blood pressure.

	Training	Validation	Test
Sample size	23,164	2,895	2,896
Male sex	13,913	1,747	1,754
Age (years)	66.3 ± 11.7	66.2 ± 11.6	66.4 ± 11.6
Diabetes duration (years)	9.4 ± 8.3	9.2 ± 8.2	9.5 ± 8.2
BMI (kg/m^2)	29.5 ± 5.4	29.5 ± 5.2	29.5 ± 5.2
Systolic BP (mmHg)	140.1 ± 18.9	139.9 ± 19.1	140.2 ± 18.7
Diastolic BP (mmHg)	79.6 ± 9.8	79.5 ± 10.1	79.6 ± 10.0
Fasting glucose (mg/dL)	144.7 ± 45.3	144.4 ± 42.9	145.5 ± 45.6
HbA1c (%)	7.2 ± 1.2	7.2 ± 1.2	7.2 ± 1.2
Total cholesterol (mg/dL)	175.5 ± 38.4	177.0 ± 38.8	173.9 ± 37.7
HDL cholesterol (mg/dL)	52.6 ± 15.3	52.7 ± 14.9	51.9 ± 14.8
Triglycerides (mg/dL)	125.2 ± 71.0	125.7 ± 73.0	125.8 ± 72.0
AST (IU/L)	23.1 ± 12.7	23.1 ± 12.0	22.8 ± 11.0
ALT (IU/L)	24.9 ± 16.7	24.9 ± 15.5	25.0 ± 16.0
Creatinine (mg/dL)	0.8 ± 0.2	0.8 ± 0.2	0.8 ± 0.2
eGFR (mL/min/1.73m^2)	86.1 ± 13.8	86.3 ± 13.6	86.1 ± 13.5

We define the most recent visit in the baseline sequence, i.e., the latest one before the start of the follow-up period, as the end-of-baseline (EOB) visit. The EOB visit is the one that would be used for prediction in the absence of sequence data. Table 1 summarises the characteristics of the study population at the EOB visit (see Sect. 3.1). The average subject had a 60% chance of being male, was 66 years old, had had diabetes for 9 years, a BMI of 29.5, a blood pressure of 140/80 mmHg, a fasting glucose of 145 mg/dL, and an HbA1c of 7.2%. The average eGFR was 86.1 mL/min/1.73 m^2.

Missing data were present (except for sex, age, and diabetes duration), but their proportion was small, i.e., <3.5% at the EOB visit.

3 Methods

3.1 Input Data Preparation

The input data preparation process was guided by our stated objective of incorporating longitudinal baseline data into the model development pipeline. In summary, we identified each patient via a multidimensional sequence of data corresponding to a variable-length sequence of baseline routine check-ups, and a single static feature, i.e., sex. The minimum number of baseline visits was 2, as per the inclusion criteria in Sect. 2.2, thus avoiding the degenerate case of 1-visit sequences. The actual number was subject-specific, i.e., between 2 and the minimum between: a) the number of available visits minus one (at least one was needed for the outcome, as per inclusion criterion 1); b) the number of consecutive outcome-free visits; and c) an arbitrary threshold of 6.

We formatted each subject's baseline data according to model requirements (see Sect. 3.3), thus obtaining a 14-variable × 6-visit padded matrix and a scalar value (technically, a 1-dimensional vector) encoding the static sex variable. Missing data in the matrix were set to "0" if they were, in fact, missing in the original dataset, whereas "-1" was the masking indicator to distinguish between informative and padded portions of the variable-length sequence. Additionally, to aid in data description and benchmarking, we created a static version of the dataset comprising only each subject's (unmodified) EOB visit and the "sex" variable.

3.2 Output Coding

The irregular and relatively sparce sampling rate induced by the real-life configuration of the data source prevented us from encoding outcome occurrence via the typical (event indicator, censoring time) tuples used in survival analysis. Indeed, survival analysis requires that exact information on outcome occurrence be known and that there be no gaps in the observation of follow-up. On the contrary, here, outcome information was only available via inspection of the eGFR values collected during each follow-up visit, meaning that status changes between two visits were inherently unknowable, and so was the exact time or reason for right censoring.

To overcome this limitation, we cast the problem of predicting impaired kidney function at different prediction horizons as a multi-label classification problem with a 7-dimensional output. Each of the 7 elements of the outcome vector, say j, encoded the answer to the question "Was there evidence of CKD onset by the end of the j-th year?" Hence, if an eGFR < 60 was recorded between the start of follow-up and the end of the j-th yeah, the j-th element of the outcome vector was equal to 1; if there was evidence of eGFR ≥ 60 after the end of the j-th year (but no evidence to the contrary before then) the j-th element of the outcome vector was equal 0; in all other cases, patient status was unknown and the j-th element was encoded as "NA" (note that this may happen both in the "natural" case of right censoring and due to gaps in eGFR sampling).

Table 2 shows the absolute frequencies of patient status across the 7 time points. We observe an expected, progressive inversion of the ratio between 1s and 0s as the prediction horizon moves forward into the future: as time goes on, follow-up visits that confirm undeteriorated renal function become rarer and rarer, whereas the number of CKD onsets accumulates. Predictably, "NA" values start appearing immediately after the start of follow-up, demonstrating the presence of subjects for whom endpoint information is temporarily unclear in addition to truly right-censored subjects.

Table 2. Outcome distribution at each prediction horizon (PH). 1: CKD onset within the year, 0: reportedly CKD free at the end of the year, NA: unknown status (% of right censored).

PH	Training			Validation			Test		
	0	1	NA	0	1	NA	0	1	NA
1 year	20,228	1,004	1,932 (0%)	2,545	117	233 (0%)	2,532	125	239 (0%)
2 years	13,971	2,570	6,623 (83%)	1,763	301	831 (82%)	1,720	331	845 (83%)
3 years	9,337	3,521	10,306 (93%)	1,195	418	1,282 (93%)	1,134	457	1,305 (94%)
4 years	5,723	4,107	13,334 (97%)	730	494	1,671 (96%)	684	533	1,679 (98%)
5 years	3,258	4,451	15,455 (99%)	404	550	1,941 (98%)	376	561	1,959 (99%)
6 years	1,572	4,634	16,958 (99%)	184	581	2,130 (99%)	172	579	2,145 (99%)
7 years	410	4,734	18,020 (100%)	56	592	2,247 (100%)	31	592	2,273 (100%)

3.3 Model Architecture and Development

Using a typical train/validate/test scheme, we developed a deep learning model based on the incorporation of longitudinal baseline data via a recurrent layer. Operatively, we optimised the network's weights on the training set, selected the best combination of hyperparameters via the validation set, and evaluated performance on the previously unseen test set. We carried out weight estimation via the ADAM optimiser with a fixed learning rate of 0.0005 for a maximum of 100 epochs. The cost function was a modified version of the binary cross-entropy where "NA" labels did not contribute to weight update via back-propagation (this is done, e.g., by artificially setting the missing prediction to the currently predicted value, resulting in a null contribution to the gradient).

As shown in Fig. 1, the proposed neural network initially handles sequence data via a recurrent layer, namely a gated recurrent unit (GRU) [13]. The objective, here, is encoding the variable-length multi-dimensional sequence as a fixed-length vector that can be concatenated with the static "sex" variable. In this portion of the network, the hyperparameters were the number of GRU units (16, 32, or 48), and the dropout fractions related to the inputs and recurrent connections (possible values for both: 0, 0.05, 0.1, 0.2, 0.3, 0.5). The result of this dynamic-to-static encoding step is then concatenated with the static "sex" variable and sent to a cascade of fully connected layers. The hyperparameters at this stage were the number of post-concatenation, pre-output layers (2 or 3) and their dimensions (valid combinations: {64, 32}, {48, 24}, {32, 16}, {16, 8}, {64, 32, 16}, {48, 24, 12}, {32, 16, 8}, {16, 8, 4}). The fully connected cascade ends on the 7-dimensional output layer. Finally, to obtain a more robust scalar score for each prediction horizon, we implemented a cumulative summation step such that each prediction at j years was the sum of the first j output neurons.

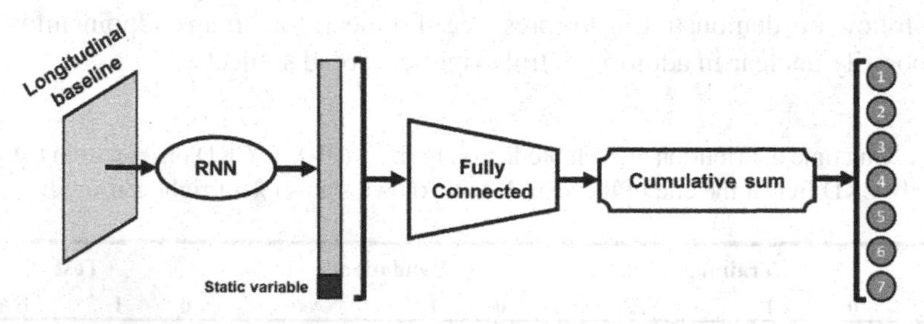

Fig. 1. High level overview of the network's architecture.

We carried out the hyperparameter selection phase in two steps. First, for each hyperparameter combination, we selected the set of weights that minimised the binary cross-entropy on the validation set, thus obtaining a set of 864 candidate models. Second, we computed the cumulative/dynamic areas under the receiver-operating characteristic curve (AUROC) [14] at 1 to 7 years on the validation set, and ranked all 864 candidates according to their predictive ability at each prediction horizon. The final model was the one with the minimum year-wise median rank.

3.4 Performance Evaluation and Secondary Analyses

In our primary performance analysis, we evaluated the discrimination power of the proposed model on the unseen test set via our target metrics, i.e., the seven AUROCs corresponding to the 1- to 7-year prediction horizons.

Our first secondary analysis was meant to challenge the hypothesis that the deep learning model was effectively encoding the sequence of visits comprising the longitudinal baseline. Hence, we measured the model's prediction ability on a modified version of the test set where we artificially inverted the order of the visits comprising each subject's longitudinal baseline.

In another secondary analysis, we compared the proposed model to a trivial model returning the eGFR collected at the time of the EOB visit, the known predictor of imminent renal function deterioration used in clinical practice [15].

In a third set of secondary analyses, we compared the proposed model with a battery of year-specific logistic regressions trained with the full EOB visit as the input and with each year's status (whenever available) as the output. The minority of missing values was imputed via mean imputation.

In all analyses, we estimated 95% confidence intervals (CIs) via the DeLong estimator [16], and assessed statistical significance at the 0.05 level.

4 Results

The hyperparameter selection process resulted in the identification of the optimal architecture as the one having 32 GRU units with standard and recurrent drop-out fractions of 0.05 and 0.1, and three fully connected layers of sizes 16, 8, and 4.

As shown in the second column of Table 3, model performance was satisfactory across the board (AUROC always > 0.84), and particularly promising in the medium term, where it ranged from year 5's 0.853 (CI. 0.828–0.878) to year 7's 0.895 (CI: 0.852–0.937). The performance comparison with the artificially inverted version of the test set (first secondary analysis) strongly suggests that the model's good behaviour was at least in part attributable to a fruitful encoding of temporal relationships between the longitudinal baseline's visits. This is apparent from the substantially (and significantly, except at the 7-year mark) diminished performance of the model when confronted with improperly ordered sequences (third column of Table 3). Had order been irrelevant, we would have expected to see a negligible difference.

As expected, the proposed model always outperformed EOB eGFR in terms of discrimination power. Interestingly, however, the AUROC difference at the 1-year mark (0.15) was only nominally greater than 0, suggesting that eGFR alone might be a sufficiently effective predictor of imminent deterioration in renal function, while additional information should be collected for longer-term prediction.

The comparison with the battery of year-specific logistic regressions (third secondary analysis, fourth column of Table 3) also yielded encouraging results. Except at the 1-year prediction horizon, the proposed model always outperformed logistic regression, and exhibited the most notable and statistically significant performance gains at 4, 5, and 6 years (respectively, AUROC 0.844 vs. 0.829, 0.853 vs. 0.830, and 0.874 vs. 0.839). Regrettably, despite the proposed model's excellent AUROC of 0.895 (CI: 0.852–0.937), the 7-year comparison was underpowered and failed to detect statistically significant differences. Overall, it appears that the inclusion of longitudinal baseline data, possibly combined with increased model capacity and with the simultaneous learning from different prediction horizons (via the proposed multi-label coding scheme), was beneficial to long-term prediction. While, under the current experimental framework, it is difficult to disentangle the contributions of these factors, it is notable that our deep learning model, i.e., a single, one-size-fits-all model, was able to compete with and generally outperform individual models specifically trained on the expected outcome distributions observed at each prediction horizon (recall the inversion of the 1:0 ratio shown in Table 2).

Table 3. Cumulative dynamic AUROC on the test set (95% CIs). PH: prediction horizon; DL inverted: deep learning model after sequence inversion. Statistical significance vs. deep learning model marked with*

PH	Deep learning	DL inverted	eGFR only	Logistic regression
1 year	0.844 (0.815–0.874)	0.712 (0.672–0.751)*	0.829 (0.797–0.860)	0.858 (0.830–0.885)
2 years	0.841 (0.819–0.862)	0.693 (0.665–0.721)*	0.819 (0.796–0.843)*	0.834 (0.811–0.856)
3 years	0.841 (0.820–0.861)	0.701 (0.675–0.728)*	0.820 (0.798–0.842)*	0.838 (0.817–0.859)
4 years	0.844 (0.822–0.866)	0.735 (0.707–0.763)*	0.803 (0.778–0.827)*	0.829 (0.806–0.852)*
5 years	0.853 (0.828–0.878)	0.765 (0.733–0.797)*	0.797 (0.767–0.826)*	0.830 (0.803–0.856)*
6 years	0.874 (0.846–0.903)	0.832 (0.796–0.869)*	0.798 (0.760–0.836)*	0.839 (0.807–0.871)*
7 years	0.895 (0.852–0.937)	0.880 (0.817–0.943)	0.797 (0.724–0.871)*	0.882 (0.833–0.932)

5 Discussion and Conclusions

An early prediction of CKD onset in people affected by diabetes but whose renal function is still satisfactory could be extremely useful in reconciliating therapeutic intervention with patient needs and resource allocation constraints. Motivated by previously reported, promising results obtained using machine learning and real-world data [10], in the present work we demonstrated the feasibility and potential benefit of developing a predictive model of impaired kidney function (KDOQI stage 3) using deep learning to integrate longitudinal information on routine check-ups. Thus, we obtained a well-performing model that yielded AUROC values between 0.841 (1-year prediction horizon) and 0.895 (7-year prediction horizon), generally and often significantly outperforming the tested benchmarks.

From a methodological perspective, our study showcases a fruitful approach to utilise routine data whose natural format is suboptimal for traditional survival analysis or classification approaches. Indeed, at variance with most similar models [17, 18], which attempt to recreate the clinical trial setting by predicting a well-behaved outcome via one-shot baseline data, here, we embraced the longitudinal vocation of routine diabetes check-ups by incorporating a sequence of past visits via a recurrent layer, and offset the inconsistent sampling scheme of the CKD endpoint using a multi-label framework and an opportunely modified cost function.

The main limitation of our study was the impossibility of disentangling the contributions to performance improvement of 1) adding baseline sequence data (although we showed that sequence order was effectively encoded by the model), 2) increasing model capacity with respect to traditional techniques such as logistic regression, and 3) casting the problem as a multi-label task. Future research will revolve around the systematic testing of the proposed architecture (or variants thereof, e.g., using different recurrent units, such as LSTMs [19]) against a stronger set of literature and custom-made benchmarks to determine the key factors in achieving high discrimination ability.

Acknowledgments. This work was partly supported by MIUR (Italian Ministry for Education) under the initiative "Departments of Excellence" (Law 232/2016).

References

1. Koye, D.N., Magliano, D.J., Nelson, R.G., Pavkov, M.E.: The global epidemiology of diabetes and kidney disease. Adv. Chron. Kidney Dis. **25**, 121–132 (2018). https://doi.org/10.1053/j.ackd.2017.10.011
2. Ene-Iordache, B., et al.: Chronic kidney disease and cardiovascular risk in six regions of the world (ISN-KDDC): a cross-sectional study. Lancet Global Health **4**, e307–e319 (2016). https://doi.org/10.1016/S2214-109X(16)00071-1
3. Lin, Y.-C., Chang, Y.-H., Yang, S.-Y., Wu, K.-D., Chu, T.-S.: Update of pathophysiology and management of diabetic kidney disease. J. Formosan Med. Assoc. **117**, 662–675 (2018). https://doi.org/10.1016/j.jfma.2018.02.007
4. Andrésdóttir, G., et al.: Improved survival and renal prognosis of patients with type 2 diabetes and nephropathy with improved control of risk factors. Diab. Care **37**, 1660–1667 (2014). https://doi.org/10.2337/dc13-2036
5. Tuttle, K.R., et al.: SGLT2 inhibition for CKD and cardiovascular disease in type 2 diabetes: report of a scientific workshop sponsored by the national kidney foundation. Diabetes **70**, 1–16 (2021). https://doi.org/10.2337/dbi20-0040
6. Perkovic, V., et al.: Canagliflozin and renal outcomes in type 2 diabetes and nephropathy. N. Engl. J. Med. (2019). https://doi.org/10.1056/NEJMoa1811744
7. Wanner, C., et al.: Empagliflozin and progression of kidney disease in type 2 diabetes. N. Engl. J. Med. **375**, 323–334 (2016). https://doi.org/10.1056/NEJMoa1515920
8. Neal, B., et al.: Canagliflozin and cardiovascular and renal events in type 2 diabetes. N. Engl. J. Med. **377**, 644–657 (2017). https://doi.org/10.1056/NEJMoa1611925
9. Yin, W.L., Bain, S.C., Min, T.: The effect of glucagon-like peptide-1 receptor agonists on renal outcomes in type 2 diabetes. Diab. Ther. **11**, 835–844 (2020). https://doi.org/10.1007/s13300-020-00798-x
10. Ravizza, S., et al.: Predicting the early risk of chronic kidney disease in patients with diabetes using real-world data. Nat. Med. **25**, 57–59 (2019). https://doi.org/10.1038/s41591-018-0239-8
11. Yang, C., Kong, G., Wang, L., Zhang, L., Zhao, M.-H.: Big data in nephrology: Are we ready for the change? Nephrology **24**, 1097–1102 (2019). https://doi.org/10.1111/nep.13636
12. National Kidney Foundation: K/DOQI clinical practice guidelines for chronic kidney disease: evaluation, classification, and stratification. Am. J. Kidney Dis. **39**, S1–266 (2002)
13. Gal, Y., Ghahramani, Z.: A theoretically grounded application of dropout in recurrent neural networks. arXiv:1512.05287 [stat]. (2016)
14. Bansal, A., Heagerty, P.J.: A tutorial on evaluating the time-varying discrimination accuracy of survival models used in dynamic decision making. Med. Decis. Making **38**, 904–916 (2018). https://doi.org/10.1177/0272989X18801312
15. Mayer, G., et al.: Systems biology-derived biomarkers to predict progression of renal function decline in type 2 diabetes. Diab. Care **40**, 391–397 (2017). https://doi.org/10.2337/dc16-2202
16. DeLong, E.R., DeLong, D.M., Clarke-Pearson, D.L.: Comparing the areas under two or more correlated receiver operating characteristic curves: a nonparametric approach. Biometrics **44**, 837–845 (1988). https://doi.org/10.2307/2531595
17. Dagliati, A., et al.: Machine learning methods to predict diabetes complications. J. Diab. Sci Technol. **12**, 295–302 (2018). https://doi.org/10.1177/1932296817706375
18. Retnakaran, R., Cull, C.A., Thorne, K.I., Adler, A.I., Holman, R.R.: Risk factors for renal dysfunction in type 2 diabetes: U.K. Prospective diabetes study 74. Diabetes **55**, 1832–1839 (2006). https://doi.org/10.2337/db05-1620
19. Gers, F.A., Schmidhuber, J., Cummins, F.: Learning to forget: continual prediction with LSTM. Neural Comput. **12**, 2451–2471 (1999)

Counterfactual Explanations for Survival Prediction of Cardiovascular ICU Patients

Zhendong Wang(✉), Isak Samsten, and Panagiotis Papapetrou

Stockholm University, Stockholm, Sweden
{zhendong.wang,samsten,panagiotis}@dsv.su.se

Abstract. In recent years, machine learning methods have been rapidly implemented in the medical domain. However, current state-of-the-art methods usually produce opaque, black-box models. To address the lack of model transparency, substantial attention has been given to develop interpretable machine learning methods. In the medical domain, counterfactuals can provide example-based explanations for predictions, and show practitioners the modifications required to change a prediction from an undesired to a desired state. In this paper, we propose a counterfactual explanation solution for predicting the survival of cardiovascular ICU patients, by representing their electronic health record as a sequence of medical events, and generating counterfactuals by adopting and employing a text style-transfer technique. Experimental results on the MIMIC-III dataset strongly suggest that text style-transfer methods can be effectively adapted for the problem of counterfactual explanations in healthcare applications and can achieve competitive performance in terms of counterfactual validity, BLEU-4 and local outlier metrics.

Keywords: Counterfactual explanations · Survival prediction · Explainable models · Deep learning

1 Introduction

Machine learning models have recently demonstrated high utility and applicability in the medical domain, and have been proven successful for various supervised and unsupervised learning tasks [8]. A contributing factor to this is the availability of rich medical data sources, such as electronic healthcare records (EHRs). For example, unstructured clinical notes have been employed in text classification for classifying diagnosis codes [8], while laboratory tests and patient demographic variables have been adopted to predicting daily sepsis, myocardial infarction (MI), and vancomycin antibiotic administration [11].

Due to the temporal nature of EHR data, it is common to model clinical events as event sequences so as to efficiently address prediction tasks, such adverse drug event (ADE) detection [2], by applying sequence classification models [18]. Nonetheless, many machine learning algorithms produce models that are often considered as black-boxes, inhibiting their use in application areas where

A. Tucker et al. (Eds.): AIME 2021, LNAI 12721, pp. 338–348, 2021.
https://doi.org/10.1007/978-3-030-77211-6_38

transparency is important for trust and reliability [15], since it is often not possible for practitioners to interpret and understand the predictions and the inherent model structure if the model is opaque. Model interpretability is even more important for critical application domains, such as healthcare, where such models are used for medical decision making [20]. There are several types and definitions of model interpretability (see [14]). In this paper, our focus is on *counterfactual explanations* [14] for Intensive Care Unit (ICU) patients suffering from cardiovascular disease. Counterfactuals refer to an example-based approach that can provide actionability in order to change the minimum set of specific features to achieve a desired outcome. For example, given a classifier and a test example for which the classifier has predicted a particular class label, a counterfactual is a modified version (e.g., a minimum-cost conversion) of that example so that the classifier switches its decision to an alternative class label.

Example. Consider a classification model for predicting the survival of an ICU patient. If for a given patient the model has a negative outcome (e.g., the patient will die), it is critical for the medical practitioner to understand whether there exists any set of treatment changes that if applied to the patient, it would lead to a positive outcome. Figure 1 illustrates such counterfactual, recommending the removal of "epinephrine" (*orange* colour) from the patient treatment and the insertion (*green* colour) of "procainamide" and "amiodarone".

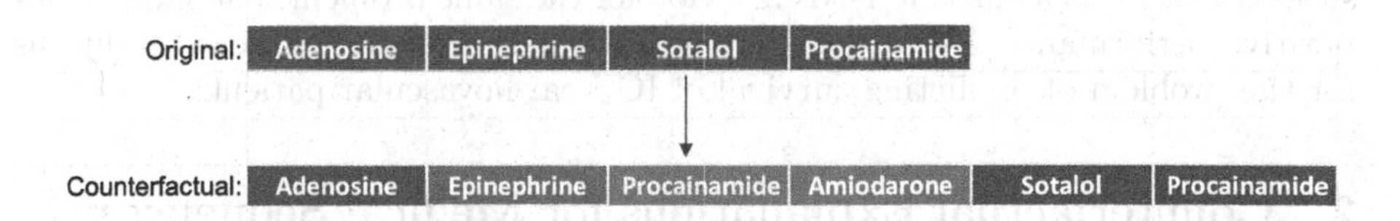

Fig. 1. An example of a counterfactual, suggesting the removal (*orange* colour) of "epinephrine" (*orange* colour) from the patient treatment and the insertion (*green* colour) of "procainamide" and "amiodarone". (Color figure online)

Related Work. Machine learning has been widely applied in predicting the outcomes of ICU patients. Hsieh et al. [7] employ a fuzzy neural network model for predicting the ICU survival of critically ill patients, while hybrid machine learning approaches [1] have also been applied for the same problem. However, none of these black-box methods is interpretable. Interpretable machine learning techniques have been adopted by many researchers recently. Especially, interpretable deep learning has drawn attention to different healthcare applications. For example, Caicedo-Torres and Gutierrez [4] applied a convolutional neural network (CNN) alongside coalitional game theory to produce interpretable visualizations for ICU mortality prediction using the MIMIC dataset. Furthermore, various types of medical events were investigated in addressing prediction problems in the healthcare domain. For instance, Esteban et al. [6] applied neural networks on embeddings of medical events (e.g., ordered tests, lab results and diagnoses) in predicting sequences of clinical events. The concept of counterfactuals was first presented by Wachter et al. [23] followed by several formulations for different data domains and classifiers, such as Shapelet-based classifiers for time series data [12] or variational auto-encoders for categorical and continuous features as well as images [21]. However, there is a lack of counterfactual explanation applications on event sequences, such as medical event sequences. Text style-transfer has been widely studied in recent years. For example, the Delete-Retrieve-Generate framework [13] explicitly disentangles content and style from a given sentence using two RNN-based models for generating the desired output text style. This model has also been extended to include an attention mechanism with a Transformer [19]. Style-transfer has been successfully implemented in different application areas, such as transferring medical notes from the expert level to the layman level or converting non-polite sentences to polite ones without adjusting the original meaning [5]. However, so far the focus has been on textual data and language applications. To the best of our knowledge, there has yet been no formulation of counterfactuals for event sequences using text style-transfer techniques, and specifically for the medical domain.

Contributions. In this paper, we formulate the problem of counterfactual explanations for medical event sequences, and provide a solution that exploits text style-transfer techniques. More concretely, our contributions include: (1) a counterfactual explanations solution for sequential data that is based on text style-transfer; (2) a baseline 1-NN solution for the same problem that shows competitive performance; (3) an experimental evaluation of the proposed solutions for the problem of predicting survival of ICU cardiovascular patients.

2 Counterfactual Explanations for Medical Sequences

We formulate the problem of counterfactual explanations for medical event sequences, followed by a description of our two solutions.

2.1 Problem Formulation

Let $\mathcal{E} = \{(x_1, y_1), ..., (x_m, y_m)\}$ be an EHR dataset of m patients, with each $x_i = e_1, \ldots, e_d$ being a sequence of d medical events for patient i coupled with a binary target class $y_i \in \mathcal{Y}$. We let $\mathcal{Y}$ define the set of possible class labels. In this paper, we assume a binary classification problem (positive and negative outcomes), i.e., $\mathcal{Y} = \{\text{'+'}, \text{'−'}\}$. Each medical event e_j can be either a medication or a medical procedure code. Finally, $f(\cdot)$ defines an opaque classification model (e.g., an LSTM). The problem studied in this paper is defined as follows.

Problem 1. **Counterfactual Explanations for Medical Sequences.** Given a trained classifier $f(\cdot)$ and a medical event sequence x with $f(x) = \text{'−'}$, we want to define a function $g(\cdot)$ that identifies the changes needed to modify x into another sequence x' so as to convert a negative prediction into a positive one, i.e.,

$$g : x \rightarrow x',\ s.t.\ f(x) = \text{`}-\text{'}\ and\ f(x') = \text{`}+\text{'},$$

such that an objective function $L(\cdot)$ is optimized.

For example, given a classifier f trained on a set of EHRs of ICU patients, g identifies the changes needed to be applied to an ICU patient x with an undesired prediction '−' (e.g., *death*) to a desired one '+' (e.g., *survival*).

2.2 Style-Transfer Counterfactual Explanations

Our first solution to the Problem 1 is to the adoption and adaptation of the Delete-Retrieve-Generate (DRG) framework, which was originally proposed for text style-transfer by Li et al. [13]. In its original formulation, DRG consists of three components, that correspond to the operations applied for generating the counterfactuals: Delete, Retrieve, and Generate. Using these components, we employ two different solutions (also in accordance to Li et al. [13]: (1) `Delete-Only` (Algorithm 1 with r = **False**), which includes the Delete and Generate components, (2) `DeleteAndRetrieve` (Algorithm 1 with r = **True**) containing all three components. The details of each component are described below.

Delete. For this operation, we consider n-grams of medical events. In our implementation, an n-gram is a contiguous sequence of n medical events from a given sample. Let S_y define a collection of event sequences sampled from $\mathcal{E}$, such that they all belong to class y. Given a target class $y \in \mathcal{Y}$, we define the *salience* of an n-gram u with respect to y by its relative frequency in S_y:

$$s(u, y) = \frac{\#(u, S_y) + \lambda}{\sum_{y' \in \mathcal{Y}, y' \neq y} \#(u, S_{y'}) + \lambda}, \tag{1}$$

where λ is the smoothing parameter and $\#(u, S_y)$ stands for the frequency of n-gram u in S_y. We consider u to be an attribute marker for class y if $s(u, y)$ is larger than a threshold γ. The main outcome of this operation is to delete a set of attribute markers (n-grams) from an input sequence x. Let $a(x, y)$ denote the set of attribute markers for class y in a sample x and $c(x, y)$ be the remaining sequence of medical events after removing $a(x, y)$ from sample x.

Retrieve. Our goal is to retrieve a set of attribute markers from x_{ret} of the target class y' based on the smallest sequence distance, i.e.,

$$x_{ret} = \underset{x^* \in S_{y'}}{\arg\min}\, d(c(x, y), c(x^*, y')), \tag{2}$$

where $d(\cdot)$ is any distance metric. We instantiate $d(\cdot)$ as the TF-IDF weighted overlap score between the input sequence and any other sequence in $S_{y'}$ [13].

Algorithm 1: The DRG method for counterfactual explanation

Input : A sequence of medical events x and boolean r
Output: A modified sequence x' with desired target class y'

```
1  if s(u, y) > γ then
2  |   Delete u from x
3  |   c(x, y) ← the remaining events
4  if r is True then
5  |   Retrieve x_ret using Eq. 2
6  |   e1 ← Encode(c(x, y), RNN1)
7  |   e2 ← Encode(a(x_ret, y'), RNN2)
8  |   x' ← Decode(e1 ⊕ e2, RNN3)
9  else
10 |   e1 ← Encode(c(x, y), RNN1)
11 |   x' ← Decode(e1, RNN3)
```

Table 1. Description of the dataset used in the experiments.

Dataset	No. samples
Training data (*survive*/*die*)	2,818/1,221
Validation data (*survive*/*die*)	200/200

Generate. We adopt the encoder-decoder structure, where RNN encodes a sequence of events into a fixed-length representation and another RNN decoder decodes the latent representations back into an output sequence of events. First, the remaining sequence $c(x, y)$ is embedded into a vector using the RNN_1 encoder. In `DeleteOnly`, the concatenation of the final hidden state with a learned embedding for the desired target y' gets fed into the RNN_3 decoder to generate a new sample x'. `DeleteAndRetrieve` employs an additional RNN_2 to encode the sequence of retrieved attribute markers $a(x_{ret}, y')$ from Eq. 1. This vector and the encoded embedding vector from the remaining sequence are concatenated into the decoder RNN_3 to generate a new sample x'. In both cases, x' is the counterfactual for the original sample x.

The goal of `DeleteOnly` is to reconstruct the medical event sequences by maximizing the following objective function: $L(\theta) = \sum_{(x,y)\in S} \log p(x \mid c(x, y), y; \theta)$. Similarly, `DeleteAndRetrieve` employs denoising to address the mismatch problem that $a(x, y)$ and $c(x, y)$ are originally from the same sequence when reconstructing x. We apply noise to replace attribute marker $a(x, y)$ with another randomly selected attribute marker of the same target class y and edit it using event-level edit distance of 1 with probability 0.1, denoted as $a'(x, y)$. As such, the objective of `DeleteAndRetrieve` is to maximize

$$L(\theta) = \sum_{(x,y)\in S} \log p(x \mid c(x, y), a'(x, y); \theta) \ . \tag{3}$$

2.3 Nearest Neighbour Counterfactual Explanations

As a baseline solution, we adopt a 1-NN approach (similar to the one presented for time series counterfactuals [12]). The counterfactual x' of x is defined as the 1-NN of x in the training examples of the opposite class. More formally,

$$x' = \underset{x^* \in S_{y'}}{\arg\min}\, L(x, x^*), \tag{4}$$

where $L(\cdot)$ is the hamming distance (that is minimized), $S_{y'}$ represents the collection of all desired outcome samples (y' ='+' in this case). Given a 1-NN classifier, Eq. 4 is guaranteed to return a *valid* counterfactual [12]. However, this guarantee does not hold for other classification models, e.g. the LSTM model.

3 Experiments

We first present our experimental setup followed by our empirical investigation and experimental results.

3.1 Experimental Setup

We evaluate the proposed methods on data extracted from the MIMIC-III dataset [9], a collection of EHRs from over 40,000 ICU patients at the Beth Israel Deaconess Medical Center, collected between 2001 and 2012. In this paper we select drug events, medical procedures, and diagnosis codes, and represent each patient as a historical sequence of events, limiting the selection to the last 12 months of patients visits. Moreover, to limit the scope our experiments, we focus on patients that have been diagnosed with cardiovascular diseases (corresponding to ICD-9 codes: 393–398, 410–414 and 420–429, representing chronic rheumatic, ischemic and other forms of heart disease separately).

To limit the impact of high frequency drugs prescriptions (e.g., electrolytes or paracetamol) and procedures, as well as very infrequent events, we filter the patient history by removing any event that has an overall frequency of over 4.1% or appears less than 6 times. Moreover, since our methods are agnostic to the drug dosage, we remove any consecutive and identical drug events. We reduce the impact of the skewed distribution of drug events by limiting the analysis to patients with fewer than 50 but more than 3 drug events. Finally, we concatenate the drug events and procedure events and pad shorter sequences with $\emptyset$ so that all sequences have uniform length, and record the label of each patient as *survive* or *die*. The number of positive (*survive*) and negative patients in the training set is 2,818 and 1,221, respectively (see Table 1).

Implementation. We compute counterfactual explanations for a 2-layer bidirectional LSTM model used for survival predictions. To improve model performance, we employ early stopping to monitor the validation accuracy. The model has one 128-dimensional embedding layer, 2 bidirectional LSTM layers with 64 hidden units, and one output layer with a sigmoid activation function. For the implementation of the DRG framework, we follow the implementation of the method as proposed by Pryzant et al. [17]. In our approach, both `DeleteOnly` and `DeleteAndRetrieve` are trained using the *Adam* optimizer with a mini-batch size of 128 samples. For both methods we use one 128-dimensional embedding layer and a bidirectional LSTM layer with 512 hidden units for both the encoders and decoders. Furthermore, we set the dropout rate to 0.2, the learning rate to 0.0003, and each model is trained for 300 epochs. Additionally, in Eq. 1 we set the threshold γ for filtering salience scores to 15, and the smoothing parameter λ in Eq. 1 is set to 0.5.

Evaluation Metrics. We evaluate the quality of the counterfactuals using: the fraction of valid counterfactuals (CFs), the local outlier factor, and the cumulative 4-gram BLEU score. More concretely, valid CFs is defined as the fraction of valid counterfactuals of the desired class [22]. The local outlier factor (LOF) measures the closeness of the counterfactuals to the training data distribution [3]. Finally, the BLEU score measures the fraction of common n-grams between the counterfactual and the original sequence [16]. In our case, we employ the cumulative 4-gram BLEU score (BLEU-4) to measure 1-gram through 4-gram individual BLEU scores, taking the uniformly weighted mean as the final score. Finally, we set the smoothing function to $\epsilon = 0.1$ if there are no n-gram overlaps.

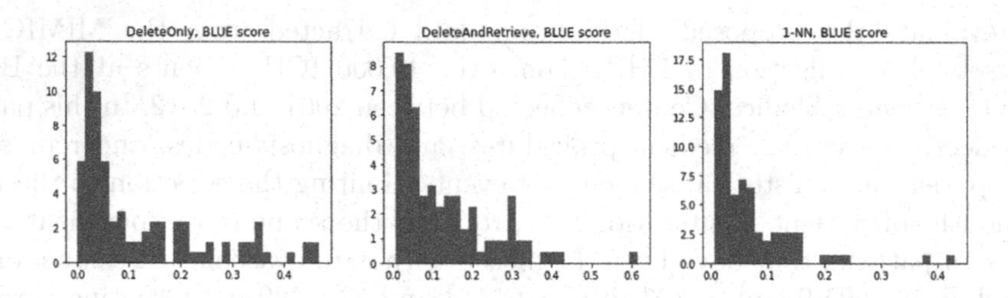

Fig. 2. Sentence-level BLEU-4 scores for the generated counterfactual explanations compared to the original input, grouped by `DeleteOnly`, `DeleteAndRetrieve`, and `1-NN`.

Table 2. Summary of evaluation metrics for the three approaches. The best score for each metric is highlighted in bold.

	Valid CFs	LOF	BLEU-4
`DeleteOnly`	0.5455	0.0364	0.1186
`DeleteAndRetrieve`	0.4909	**0.0182**	**0.1237**
`1-NN`	**0.7818**	0.0364	0.0662

3.2 Empirical Investigation

The LSTM model for survival prediction, obtained an accuracy of 64% on the independent validation data (with a training error of 19%). Among the 400 validation samples, 110 of them were predicted as negative. On those samples, we applied DeleteOnly, DeleteAndRetrieve and 1-NN to generate counterfactuals. Table 2 summarizes the empirical scores for the three methods. In Table 2, we observe that DeleteOnly achieved a CF of 54.55%, while 1-NN received the highest CF (78.18%) and DeleteAndRetrieve the lowest (49.09%). Similarly, the LOF score for DeleteOnly, DeleteAndRetrieve, and 1-NN was 0.0364, 0.0182, and 0.0364, respectively. In contrast, the reference LOF score on the validation data was 0.075, suggesting that the number of unusual (i.e., novel) counterfactuals was lower than the number of unusual samples in the validation data, indicating that all methods generate trustworthy explanations.

The BLEU-4 scores of DeleteAndRetrieve (0.1237) and DeleteOnly (0.1186) were higher than that of 1-NN (0.0662). As such, the generated counterfactuals from these methods contain more event sequence overlaps than 1-NN. Given the higher BLEU-4 score, one conclusion is that DeleteAndRetrieve and DeleteOnly produce more compact explanations. Moreover, we performed a detailed comparison concerning sentence-level BLEU-4 scores using histograms for DeleteOnly, DeleteAndRetrieve and 1-NN individually. In Fig. 2, we observe that 1-NN received sentence-level BLEU scores between 0 and 0.1 (only two of them achieved 0.4). In comparison, DeleteOnly and DeleteAndRetrieve produced slightly more samples with scores ranging from 0.1 and 0.4, which indicates that both methods can generate more sequentially relevant counterfactuals.

Finally, we investigated the differences between the original samples and the counterfactuals in terms of total, drug event, and procedure counts. Figure 4 depicts a comparison of subtraction of these three event counts for DeleteOnly, DeleteAndRetrieve and 1-NN methods, respectively. The x-axis and y-axis represent the subtraction of event counts and probability density of bin counts separately.

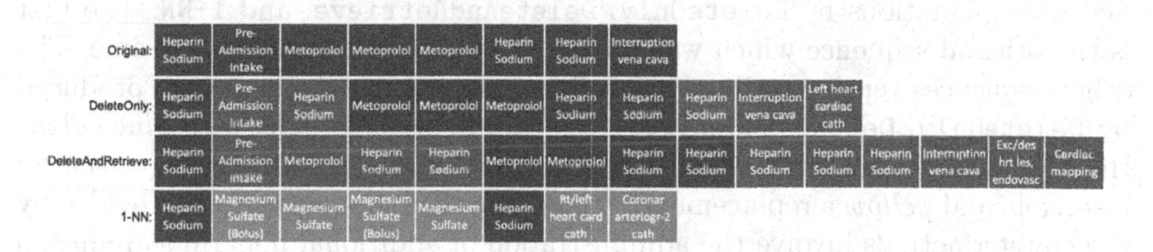

Fig. 3. Examples of generated counterfactual explanations by each algorithm. Modified events are highlighted with different colours: *orange* suggests a deletion, *green* an insertion, and *yellow* a substitution, while *blue* indicates an unchanged event. (Color figure online).

It can be observed that DeleteOnly and DeleteAndRetrieve have a maximum of ≈60 modifications, compared to 1-NN with ≈30 modifications. Conversely, DeleteAndRetrieve has the the fewest procedure modifications (from −5 to 3), while both DeleteOnly and 1-NN both have a broader range of procedure modifications. Considering drug events, the distribution differences are similar to the total event differences among the three methods. One reason for this is the increased frequency of drug events in these patients.

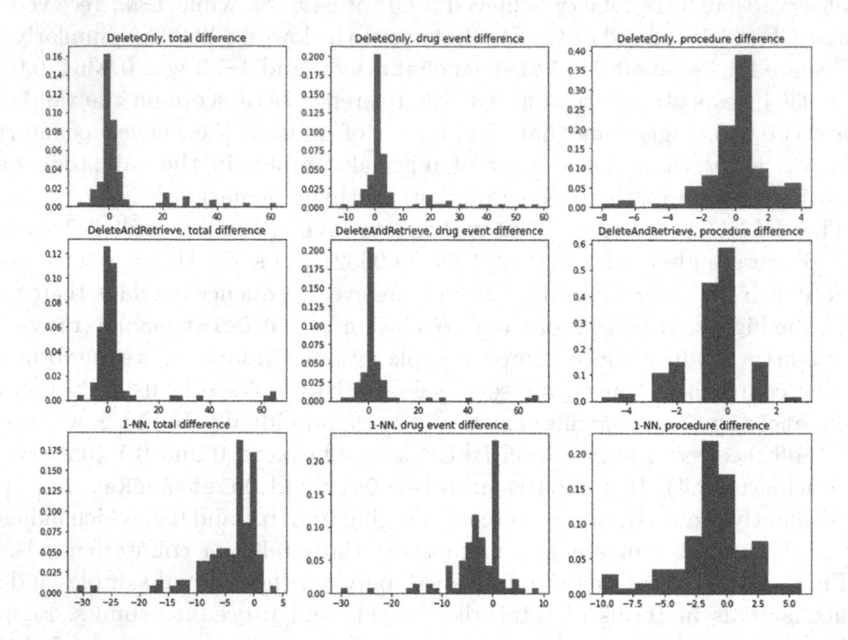

Fig. 4. Histograms of event count differences between original samples and counterfactual explanations, grouped by: total counts, drugs and procedure counts.

Example Counterfactuals. Figure 3 presents examples of generated counterfactual explanations by `DeleteOnly`, `DeleteAndRetrieve`, and `1-NN`. The first is the original sequence which was predicted as having a negative outcome. The other sequences represent, in order, valid counterfactual explanations produced by `DeleteOnly`, `DeleteAndRetrieve`, and `1-NN`. We highlight with *blue* colour the events that remain unchanged, while *orange* indicates a deletion, *green* an insertion, and *yellow* a replacement. We observe that the suggestions provided by the counterfactuals involve the administration of additional heparin sodium (an anticoagulant, blood thinner) as well as additional electrolytes (such as magnesium). This is in accordance with recent studies indicating that increased morbidity and mortality among critically ill ICU patients is associated with fluid and electrolyte imbalances [10]. Our preliminary investigation of these findings suggests that the proposed counterfactuals hold some medical relevance, which will be more thoroughly assessed by medical experts in our future work.

4 Conclusions

We proposed a counterfactual solution for medical event sequences that leverages techniques from text style-transfer. Our approach achieves a reasonable fraction of valid CFs in generating counterfactuals on medical sequence data, while outperforming an NN-based baseline in terms of BLEU-4 core. Moreover, our approach produces LOF scores that are lower than the validation result, implying that all the generated counterfactuals are considered close to the original samples. One limitation of our evaluation is that there were no humans involved in the approach. For future work, we intend to extend to involve medical practitioners into further assessing the medical relevance of the counterfactuals as well as integrating expert knowledge directly into their construction procedure. In addition, we plan to explore other counterfactual methods for an intensive comparison. Besides, we aim to integrate other possible features of patients to construct more meaningful event sequences in future experiments. For reproducibility, the code is publicly available at our GitHub repository[1].

Acknowledgments. This work was supported in part the EXTREMUM collaborative project of the Digital Futures framework.

References

1. Ahmad, F.S., et al.: A hybrid machine learning framework to predict mortality in paralytic ileus patients using electronic health records (EHRs). J. Ambient Intell. Hum. Comput. **30**, 1 (2020)
2. Bagattini, F., Karlsson, I., Rebane, J., Papapetrou, P.: A classification framework for exploiting sparse multi-variate temporal features with application to adverse drug event detection in medical records. BMC Med. Inform. Decis. Mak. **19**(1), 7:1–7:20 (2019)
3. Breunig, M.M., Kriegel, H.P., Ng, R.T., Sander, J.: LOF: identifying density-based local outliers. In: Proceedings of SIGMOD international conference on Management of data, pp. 93–104. New York (2000)
4. Caicedo-Torres, W., Gutierrez, J.: ISeeU: visually interpretable deep learning for mortality prediction inside the ICU. Biomed. Inform. **98**, 103269 (2019)
5. Cao, Y., Shui, R., Pan, L., Kan, M.Y., Liu, Z., Chua, T.S.: Expertise style transfer: a new task towards better communication between experts and laymen. In: ACL, pp. 1061–1071 (2020)
6. Esteban, C., Schmidt, D., Krompaß, D., Tresp, V.: Predicting sequences of clinical events by using a personalized temporal latent embedding model. In: International Conference on Healthcare Informatics, pp. 130–139 (2015)
7. Hsieh, Y.Z., Su, M.C., Wang, C.H., Wang, P.C.: Prediction of survival of ICU patients using computational intelligence. Comput. Biol. Med. **47**, 13–19 (2014)
8. Huang, J., Osorio, C., Sy, L.W.: An empirical evaluation of deep learning for ICD-9 code assignment using MIMIC-III clinical notes. Comput. Meth. Prog. Biomed. **177**, 141-153 (2019)
9. Johnson, A.E.W.: MIMIC-III, a freely accessible critical care database. Sci. Data **3**(1), 160035 (2016)
10. Lee, J.W.: Fluid and Electrolyte Disturbances in Critically Ill Patients. Electrolyte Blood Press (2010)
11. Kaji, D.A.: An attention based deep learning model of clinical events in the intensive care unit. PLOS ONE **14**(2), e0211057 (2019)

[1] https://github.com/zhendong3wang/counterfactuals-for-event-sequences.

12. Karlsson, I., Rebane, J., Papapetrou, P., Gionis, A.: Explainable time series tweaking via irreversible and reversible temporal transformations. In: 2018 IEEE International Conference on Data Mining (ICDM), pp. 207–216 (2018)
13. Li, J., Jia, R., He, H., Liang, P.: Delete, retrieve, generate: a simple approach to sentiment and style transfer. In: NAACL-HLT (2018)
14. Molnar, C.: Interpretable Machine Learning - A Guide for Making Black Box Models Explainable (2019)
15. Papernot, N., McDaniel, P., Goodfellow, I., Jha, S., Celik, Z.B., Swami, A.: Practical Black-Box Attacks against Machine Learning (2017). arXiv:1602.02697
16. Papineni, K., Roukos, S., Ward, T., Zhu, W.J.: BLEU: a method for automatic evaluation of machine translation. In: Proceedings of ACL, pp. 311–318. Philadelphia, Pennsylvania, USA (2002)
17. Pryzant, R., Richard, D.M., Dass, N., Kurohashi, S., Jurafsky, D., Yang, D.: Automatically neutralizing subjective bias in text. In: Association for the Advancement of Artificial Intelligence (AAAI) (2020)
18. Rebane, J., Samsten, I., Papapetrou, P.: Exploiting complex medical data with interpretable deep learning for adverse drug event prediction. Artif. Intell. Med. **109**, 101942 (2020)
19. Sudhakar, A., Upadhyay, B., Maheswaran, A.: "Transforming" delete, retrieve, generate approach for controlled text style transfer. In: Proceedings of EMNLP-IJCNLP, pp. 3269–3279. Hong Kong, China (2019)
20. Tonekaboni, S., Joshi, S., McCradden, M.D., Goldenberg, A.: What clinicians want: contextualizing explainable machine learning for clinical end use. In: Machine Learning for Healthcare Conference, pp. 359–380 (2019)
21. Van Looveren, A., Klaise, J.: Interpretable Counterfactual Explanations Guided by Prototypes (2020). arXiv:1907.02584
22. Verma, S., Dickerson, J., Hines, K.: Counterfactual Explanations for Machine Learning: A Review (2020). arXiv:2010.10596
23. Wachter, S., Mittelstadt, B., Russell, C.: Counterfactual Explanations Without Opening the Black Box: Automated Decisions and the GDPR. Technical report, Social Science Research Network (2017)

Improving the Performance of Melanoma Detection in Dermoscopy Images Using Deep CNN Features

Himanshu K. Gajera[1](✉), Mukesh A. Zaveri[1], and Deepak Ranjan Nayak[2]

[1] Department of Computer Engineering, Sardar Vallabhbhai National Institute of Technology, Surat, India
{ds19co001,mazaveri}@coed.svnit.ac.in

[2] Department of CSE, Malaviya National Institute of Technology, Jaipur, India
drnayak.cse@mnit.ac.in

Abstract. Deep learning based automated approaches mainly based on convolution neural networks (CNN) has recently brought significant attention to diagnose skin cancers (melanoma) from dermoscopic images. However, learning efficient features from these models has been challenging due to unavailability of ample amount of data. To address this problem, in this paper, we propose an improved automated system that derives visual features from a contemporary pre-trained deep CNN model (MobileNet) to identify melanoma from dermoscopic images. Further, skin lesion classification is performed using a set of classifiers. The method introduces boundary localization and cropping that helps in generating more relevant features. Our proposed method has been validated on PH^2 dataset for the classification of non-melanoma and melanoma cases. The experimental results reveal that the suggested approach obtained promising performance compared to state-of-the-art methods.

Keywords: Skin cancer · Melanoma · CNN · Deep features

1 Introduction

Melanoma is the most deadly form of skin cancer and has risen rapidly across the globe. Hence, early diagnosis of melanoma is of great significance for timely treatment and healthy living. This disease unexpectedly causes signs on normal skin with a dark mole and an irregular border [1]. Figure 1 shows sample dermoscopic images from melanoma and non-melanoma categories.

Several algorithms were already reported for the automatic diagnosis of melanoma using dermoscopic images in the last few years. A comprehensive analysis of different feature extraction and classification techniques studies over the past decade have been presented in [2,3]. Most of the earlier automated

A. Tucker et al. (Eds.): AIME 2021, LNAI 12721, pp. 349–354, 2021.
https://doi.org/10.1007/978-3-030-77211-6_39

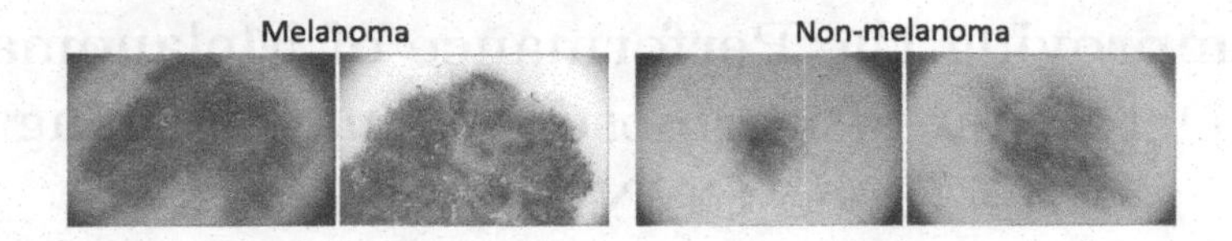

Fig. 1. Sample dermoscopic images of skin lesion

approaches are based on manual hand-crafted feature engineering and classification. On the other side, deep learning (DL) based methods, in particular convolution neural networks (CNN) facilitate automatic learning of hierarchical features directly from the images [4]. Artificial neural networks (ANN) were employed in [5] for skin lesion classification. The method used in [6] explored a CNN model for feature extraction. The VGG network based deep features with different classifiers was explored in [7]. Recently, in [8], a classification system based on sparse auto-encoder and SVM was introduced.

In general, DL algorithms learn effective features in the presence of an ample amount of data. Nevertheless, most skin cancer datasets are of limited size. Further, the features derived directly from original images in most of the existing CNN models induce a weak visual representation of lesion regions. To address the aforesaid issues, we propose an effective classification system for identification of melanoma cases from the dermoscopic images in which a contemporary deep CNN architecture pre-trained with a large dataset has been used to derive visual features and the final classification has been performed by a set of classifiers. Our contributions are two-fold: First, we extract effective features by using skin lesion boundary localization and a deep CNN model. Second, we investigate the potency of MobileNet and a set of classifiers over a publicly available dataset that outperforms compare to other existing schemes.

2 Proposed Methodology

The structure of the proposed method is depicted in Fig. 2. Our proposed method consists of three major phases: (1) boundary localization, (2) feature extraction, and (3) classification.

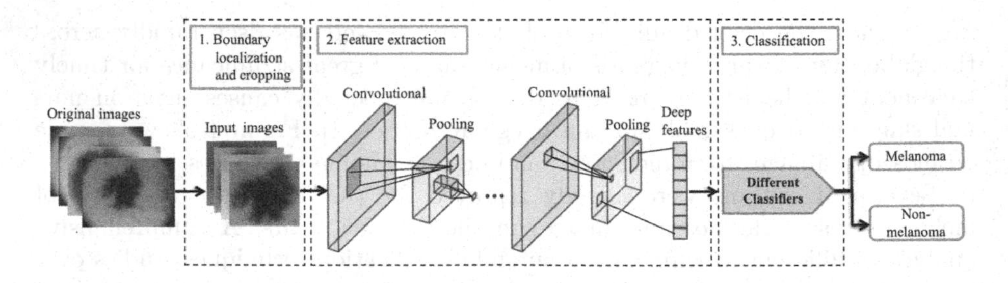

Fig. 2. Proposed method architecture

2.1 Boundary Localization

Dermoscopic images may be affected by noise because of electronic dermatoscope and other event occurrences that influence the originality of images. Cropping only salient region of images may reduce artifacts and noises present in the input images and thereby, facilitating the extraction of valuable features. An example of boundary localization of a sample dermoscopy image is depicted in Fig. 3.

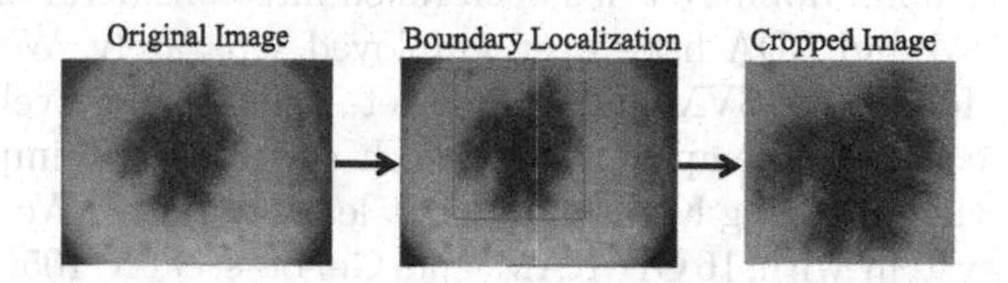

Fig. 3. Illustration of boundary localization process

After performing boundary localization and cropping, images of different dimensions have resulted. Therefore, image resizing has been performed in the present study. Then, the normalization has been applied over the resultant images that ensure each feature value within a uniform and limited range. Moreover, it establishes numerical stability in a CNN model.

2.2 Feature Extraction

CNNs have been found noteworthy in a wide range of image centric applications and the features derived from these models have also been proved significant in large scale image classification tasks. In our approach, we adopt MobileNet [9] architecture which is pre-trained with ImageNet dataset to extract in-depth visual features from the skin lesion. The concept of transfer learning has been used as there is no sufficient amount of data available in the considered dataset for training. MobileNet [9] is a simplified CNN system that utilizes depthwise separable convolution to construct a compact deep convolution neural network and offers an effective model for mobile and vision applications. We extract 1024 features from the global average pooling layer.

2.3 Classification

To perform the classification, SVM with linear kernel and linear discriminant analysis (LDA) have been separately applied. The advantages of employing linear classifiers are simplicity and computational attractiveness.

3 Experimental Setting and Results

3.1 Experimental Setting

1) Dataset: PH^2 [10] dataset has been used to validate the proposed approach, which includes 200 dermoscopy images of skin lesions: 80 normal nevi,

80 atypical nevi and 40 melanoma cases. The samples were classified into two groups, melanoma and non-melanoma, by combining the two forms of normal and atypical nevi. The dataset has been randomly split into 70:30 ratio for training and testing, respectively.

2) Implementation and evaluation measures: The goal of this study is to demonstrate the benefit of boundary localization and extracted visual features from skin images for accurate classification of melanoma. As a baseline feature extraction method, MobileNet has been taken into consideration and for classification, SVM and LDA have been employed separately. We labeled these methods as MobileNet+SVM and MobileNet+LDA, respectively. The parameters have been chosen empirically for each classifier. We implemented our proposed method by using Keras and Scikit-learn libraries. We ran our experiment on a system with 16 GB RAM and GeForce GTX 1050Ti GPU. The performance metrics used for evaluation include accuracy (Acc), specificity (Spec), precision (Pre), sensitivity (Sen) and F1-score (F1).

3.2 Experiments with Proposed Method

To confirm the feasibility of the proposed method, we thoroughly assessed its efficiency on PH^2 dataset by employing boundary localization, cropping, and image normalization methods as described in Sect. 2. The performance of MobileNet has been tested with SVM and LDA classifiers separately and the results with and without preprocessing are tabulated in Table 1. Further, our method has been assessed using 10-fold cross-validation strategy that achieved an accuracy of 95.50% and 94.00% for SVM and LDA classifier respectively.

Table 1. Classification results of the proposed method

Classifier	Input images	Acc (%)	Pre (%)	Sen (%)	Spec (%)	F1 (%)
MobileNet + SVM	With preprocessing	95.00	90.90	83.33	97.91	86.95
MobileNet + LDA		95.00	90.90	83.33	97.91	86.95
MobileNet + SVM	Without preprocessing	85.00	66.66	50.00	93.75	54.14
MobileNet + LDA		91.66	100.00	58.33	100.00	73.68

3.3 Comparison with State-of-the-Art Methods

Table 2 summarizes the performance comparison of the proposed approach with other automated melanoma classification methods on PH^2 dataset. Our proposed method achieved a higher accuracy of 95% and specificity of 97.91% when compare to all other models. Compared to other existing studies, our method included boundary localization, cropping of only interested regions of skin lesion, and image normalization. The combined benefit of all these techniques prompted to higher performance. The aforementioned results confirmed that the combination of MobileNet based features and SVM/LDA classifier lead to better performance.

Table 2. Comparison with existing methods using PH^2 dataset

Method	Year	Acc (%)	Sen (%)	Spec (%)
Joint Reverse Classification (JRC) [11]	2016	92.00	87.50	93.13
ANN [5]	2017	82.00	85.71	81.25
AlexNet [6]	2018	93.00	86.00	94.00
Bag Tree Ensemble Classifier [12]	2019	93.50	96.00	93.00
Autoencoder + SVM [8]	2020	94.00	90.00	66.66
MobileNet + SVM (**Proposed**)		**95.00**	**83.33**	**97.91**
MobileNet + LDA (**Proposed**)		**95.00**	**83.33**	**97.91**

4 Conclusion

In this paper, we proposed an improved classification system for melanoma diagnosis in dermoscopic images. We introduced boundary localization of skin lesions and cropped only the fascinating region. The resultant images were then fed to MobileNet architecture for deriving a set of discriminative features. The derived features were finally supplied to classifiers like SVM and LDA to perform melanoma classification. The experimental results on the PH^2 dataset indicated that MobileNet with SVM and LDA classifiers obtained a higher performance compared to the state-of-the-art methods. This method is hence highly effective in identifying melanoma from skin cancer images. The future research directions related to this study include the following: Introduction of more efficient classifiers with different combinations of CNN models and development of ensemble CNN models for further performance improvement. The performance of the proposed method could be verified over large and diverse skin cancer datasets.

References

1. Negin, B.P., Riedel, E., Oliveria, S.A., Berwick, M., Coit, D.G., Brady, M.S.: Symptoms and signs of primary melanoma: important indicators of breslow depth. Cancer **98**(2), 344–348 (2003)
2. Barata, C., Celebi, M.E., Marques, J.S.: A survey of feature extraction in dermoscopy image analysis of skin cancer. IEEE J. Biomed. Health Inform. **23**(3), 1096–1109 (2018)
3. Gupta, A., Thakur, S., Rana, A.: Study of melanoma detection and classification techniques. In: 8th International Conference on Reliability, Infocom Technologies and Optimization (ICRITO), pp. 1345–1350. IEEE (2020)
4. Nayak, D.R., Dash, R., Majhi, B.: Automated diagnosis of multi-class brain abnormalities using MRI images: a deep convolutional neural network based method. Pattern Recogn. Lett. **138**, 385–391 (2020)
5. Pathan, S., Siddalingaswamy, P., Lakshmi, L., Prabhu, K.G.: Classification of benign and malignant melanocytic lesions: a cad tool. In: International Conference on Advances in Computing, Communications and Informatics (ICACCI), pp. 1308–1312. IEEE (2017)

6. Salido, J.A.A., Ruiz, C.: Using deep learning to detect melanoma in dermoscopy images. Int. J. Mach. Learn. Comput. **8**(1), 61–68 (2018)
7. Maia, L.B., Lima, A., Pereira, R.M.P., Junior, G.B., de Almeida, J.D.S., de Paiva, A.C.: Evaluation of melanoma diagnosis using deep features. In: 25th International Conference on Systems, Signals and Image Processing (IWSSIP), pp. 1–4. IEEE (2018)
8. Zghal, N.S., Kallel, I.K.: An effective approach for the diagnosis of melanoma using the sparse auto-encoder for features detection and the SVM for classification. In: 5th International Conference on Advanced Technologies for Signal and Image Processing (ATSIP), pp. 1–6. IEEE (2020)
9. Howard, A.G., et al.: MobileNets: efficient convolutional neural networks for mobile vision applications. ArXiv Preprint ArXiv:1704.04861 (2017)
10. Mendonça, T., Ferreira, P.M., Marques, J.S., Marcal, A.R., Rozeira, J.: PH2-a dermoscopic image database for research and benchmarking. In: 35th Annual International Conference of the Engineering in Medicine and Biology Society (EMBC), pp. 5437–5440. IEEE (2013)
11. Bi, L., Kim, J., Ahn, E., Feng, D., Fulham, M.: Automatic melanoma detection via multi-scale lesion-biased representation and joint reverse classification. In: 13th International Symposium on Biomedical Imaging (ISBI), pp. 1055–1058. IEEE (2016)
12. Lynn, N.C., War, N.: Melanoma classification on dermoscopy skin images using bag tree ensemble classifier. In: International Conference on Advanced Information Technologies (ICAIT), pp. 120–125. IEEE (2019)

Mobile Aided System of Deep-Learning Based Cataract Grading from Fundus Images

Yaroub Elloumi[1,2,3](✉)

[1] Medical Technology and Image Processing Laboratory, Faculty of Medicine, University of Monastir, Monastir, Tunisia
yaroub.elloumi@esiee.fr
[2] LIGM, Univ Gustave Eiffel, CNRS, ESIEE Paris, 77454 Marne-La-Vallée, France
[3] ISITComHammam-Sousse, University of Sousse, Sousse, Tunisia

Abstract. The cataract is an ocular disease which requires early detection to avoid reaching a higher severity level. However, a worldwide deficiency of ophthalmologists and medical imaging devices is registered, which prevents early cataract detection. Our main objective is to propose a high performance method of cataract grading with a lower computational processing to be suitable for mobile devices. The main contribution consists in extracting features through a transfer-learned and fine-tuned MobileNet-V2 model, and deducing the cataract grade using a random forest classifier. The evaluation is conducted using a dataset of 590 fundus images, where 91.43% sensitivity, 89.58% specificity, 90.68% accuracy and 92.75% precision are achieved. In addition, the method implemented into a smartphone requires an average execution time of 1.41 s. The method implementation as an app into a smartphone associated to an optical lens for retina capturing, presents a mobile-aided-grading system that facilitates diagnosing the cataract disease.

Keywords: Cataract · Deep learning · MobileNet-V2 · Random forest · M-health

1 Introduction

The cataract is an ocular disease where the eye lens, initially transparent, becomes cloudy due to protein accumulation [1, 2]. This pathology is the main cause for half of the blind worldwide, where their number may achieve 40 million in 2025. It is always diagnosed using the ophthalmoscopy technique where the blurriness of the retina components is similar to the protein accumulation and hence to the cataract severity, where the mild, moderate and severe grade are respectively shown in Fig. 1.

A higher difficulty has been noticed in recognizing cataract, despite that significantly affects the life quality by imposing activity limits. Therefore, it is highly recommended a periodical diagnosis to avoid achieving advanced stages [1–3]. However, worldwide deficiencies of ophthalmologists and medical imaging devices are observed. Elsewhere, the cataract affects a population that exceeds 50 yrs old, having a limited mobility. Consequently, a delay of early cataract diagnosis is registered.

This work was supported by the PHC-UTIQUE 19G1408 Research program.

A. Tucker et al. (Eds.): AIME 2021, LNAI 12721, pp. 355–360, 2021.
https://doi.org/10.1007/978-3-030-77211-6_40

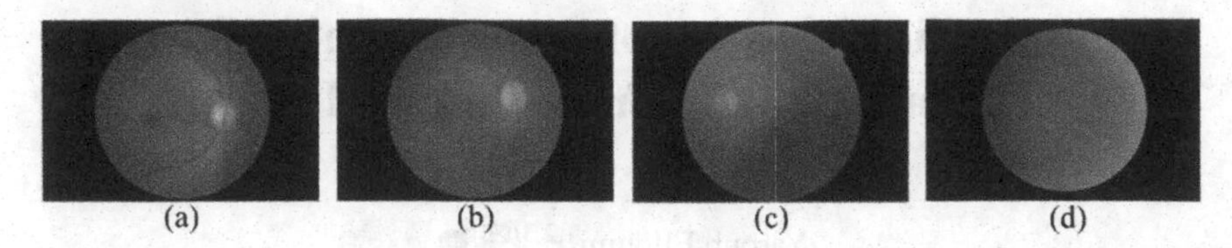

Fig. 1. Retinal images of cataract grades. (a) Non-cataract. (b) Mild. (c) Moderate. (d) Severe

Actually, several lenses have been recently proposed which can be snapped onto mobile devices to ensure capturing fundus images [5]. The provided fundus images are easily readable and have an acceptable quality with respect to the ones captured by classical ophthalmoscopy. For this purpose, our main idea is to suggest a novel method for cataract grading. The challenge is to perform reliable grading from fundus images using lower complexity processing to be suitable for mobile devices. For such need, we perform cataract grading through the MobileNet-V2 deep learning architecture associated to the random forest classifier. The implementation of the method into a smartphone as an app leads to a Mobile-Aided-Grading (MAG) system for the cataract disease. The paper is organized as follows. Section 2 describes the propounded method for cataract grading. The method evaluation is presented in Sect. 3. The implementation of the MAG system is detailed in Sect. 4, followed by a conclusion in Sect. 5.

2 Novel Method for Cataract Grading

2.1 Preprocessing and Data Augmentation

The fundus images are resized to (224 × 224 × 3) as required for MobileNet-V2. In addition, a cataract-affected fundus image appears with a blurred region with a different contrast than the retina background. Furthermore, handheld capturing with a smartphone leads to a light leakage, and so to a randomly increased contrast in image sections. To resolve this problem, we apply the Contrast Limited Adaptive Histogram Equalization (CLAHE) approach to enhance the fundus image contrast. Thereafter, a mask is applied to encompass the retina from its background. Elsewhere, we put forward data augmentation where the fundus images are rotated, zoomed, shifted and flipped in both vertical and horizontal direction [6]. Both preprocessing and data augmentation are depicted in the black dotted squares of Fig. 2.

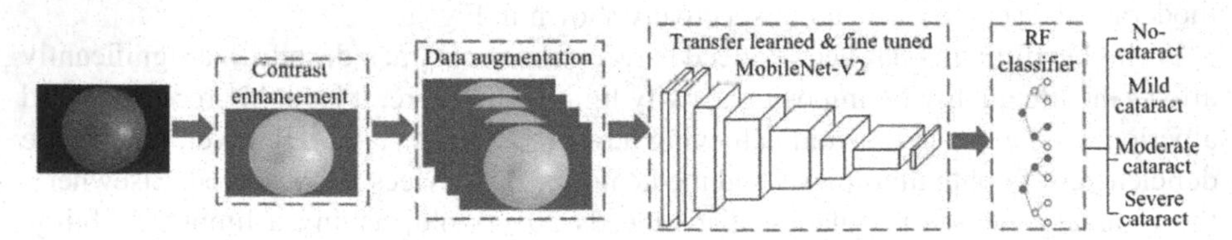

Fig. 2. Processing flowchart of the proposed cataract grading method

2.2 Fine Tuning and Transfer Learning

The preprocessed image is provided to a MobileNet-V2 architecture to generate feature map. This architecture is a convolutional neural network, which is composed of 17 bottleneck residual blocks followed by a global average pooling layer. The limited size of images, as the case of cataract-affected fundus images, avoids converging the model weight, thus not achieving higher performance grading. For this purpose, we adopt the transfer learning method where a MobileNet-V2 model, trained with the "ImageNet" dataset containing 1000 categories, is used in our method [7]. The initial classification layer is replaced by a set of three layers. The first one consists in a fully connected dense layer with the ReLU function where the feature map size is reduced to 256. The second layer applies a dropout function to initiate the classification. The last one is a dense layer with the Softmax function which provides a vector of probability features, where the size is identified experimentally as mentioned in Sect. 3.2. To enhance the weight convergence, we proceed to fine-tune the model, which allows updating its weights during training with a learning rate of 0.0001. In addition, the optimal weight with respect to accuracy is saved after each epoch, in order to accelerate the model convergence while avoiding overfitting [7]. Training is done into 150 epochs using the "Adam" optimizer and the "categorical-cross entropy" loss function.

2.3 Classification for Cataract Grading with Random Forest

Our goal is to achieve a higher performant classification even with a reduced dataset size while having a lower complexity to be suitable for a mobile implementation. The classifier retrieves a feature vector from the deep learning model, which is composed of four probability values with respect to the four cataract grades. It is noticed that some grades have a higher dependency with specific features, while other grades correspond to a significant variation in a whole feature sub-set. To handle the feature correlation, the Random Forest (RF) is used to ensure the classification into cataract grading. It is considered as a set of decision trees where each one is formed by a feature subset, randomly selected [19]. The RF has two main parameters, which are the number of trees in the forest and the maximum depth of the trees. We perform an experimental study, as detailed in Sect. 3.2, where the RF is employed with a varied depth to identify the one achieving better detection performances.

3 Experimental Results

3.1 Dataset and Evaluation Metrics

Two public databases have recently uploaded in Kaggle which are the "Cataract Dataset" and "Ocular Disease Recognition (ODiR)" respectively containing 100 and 293 fundus images affected by different cataract stages. To guarantee a reliable evaluation, a dataset of 590 fundus images is build which are split into 220 non-cataract, 65 mild stage, 145 moderate stage and 160 severe stage fundus images. The images of each grade are randomly divided into five subsets where the three first ones are used for training, the fourth is dedicated for validation, while the last is used for testing. The experimental evaluation is conducted by referring to four metrics which are sensibility (Sens), the specificity (Spec), the accuracy (Acc) and the precision (Prec), respectively computed as indicated in Eqs. (1–4).

$$\text{Sens} = \frac{TP}{TP + FN} \tag{1}$$

$$\text{Spec} = \frac{TN}{TN + FP} \tag{2}$$

$$\text{Acc} = \frac{TP + TN}{TP + TN + FP + FN} \tag{3}$$

$$\text{Prec} = \frac{TP}{TP + FP} \tag{4}$$

where TP, TN, FP and FN are respectively True Positive, True Negative, False Positive and False Negative cataract-detected images.

3.2 Cataract Grading Performance

A first experimentation is performed where the size of feature vector is varied. It was deduced that a vector of four feature allows achieving a better classification performance. In a second experimentation, we conduct an experimentation of four different evaluations, where the tree depth is varied from 1 to 4, which corresponds to the number of probability features taken into account. The maximal number of trees is fixed to 300 in order to guarantee achieving better feature combinations. The metrics for the four evaluations are illustrated in Table 1, where it is deduced that using a depth of 4 allows achieving the highest performances.

Table 1. Performance of cataract grading

Arch.	MobileNet-V2			MobileNet-V2 & Random Forest			
Training parameters	Size of feature vector			Tree depth			
	4	8	12	1	2	3	4
Sens (%)	87.50	85.94	83.08	93.22	92.19	91.18	**91.43**
Spec (%)	77.78	74.07	75.47	78.57	79.63	86.00	**89.58**
Acc (%)	83.05	80.51	79.66	86.09	86.44	88.98	**90.68**
Prec (%)	82.35	79.71	80.60	82.09	84.29	89.86	**92.75**

Then, we evaluate our methods with respect to the existing ones by comparing the whole accuracy and sensitivity and specificity of each grade, where the values are shown in Table 2. We notice that our method achieves the highest accuracy and realizes better sensitivity performances in the first, third and fourth grades.

Table 2. Cataract grading performance of related methods

Methods of cataract grading	Non-cataract		Mild		Moderate		Severe		Acc
	Sen	Spec	Sen	Spec	Sen	Spec	Sen	Spec	
Yang et al. [4]	89.3	90.4	79.5	87.9	74.6	**96.7**	75.0	**98.9**	84.5
Cao et al. [2]	93.53	**95.44**	**82.17**	**93.83**	73.27	94.55	89.66	97.37	85.98
Song et al. [3]	- -	- -	- -	- -	- -	- -	- -	- -	88.6
Zhou et al. [1]	- -	- -	- -	- -	- -	- -	- -	- -	89.23
Our method	**97.73**	93.48	61.54	80.00	**89.66**	86.67	**93.75**	93.75	**90.68**

4 Mobile-Aided-Grading System for Cataract Disease

The whole method is coded using the python language where MobileNet-V2 is implemented using the "Keras" API. The cloud service "google Colab" is employed for training and testing the deep learning model which contains 2,257,984 parameters, where 2,223,872 are updated through fine tuning. To run the MobileNet-V2 model into a smartphone, the "TFLiteConverter" class of the public "tf.lite" API is utilized to convert the trained "TensorFlow" model into a "TensorFlowLite" model. The whole method is coded with the JAVA language and implemented as an android app. To employ the model, the "TensorFlow Lite Task" library is implemented, which contains easy-to-use methods to create apps using "TensorFlowLite" model. Then, the "TensorFlow Lite Android Support" library is used, which allows managing input data and interpreting the provided output. Image processing and the RF classifier grading is performed using The "Open Source Computer Vision (OpenCV)" library which is compiled through Android Native Development Kit (NDK) [8].

The whole method is run as an app into a "Samsung Galaxy A31" smartphone having an octa-core processor (2 x 2 GHz & 6 × 1.95 GHz) and 4 Go RAM. To evaluate the execution time, three fundus images are randomly selected from each grade, where the average execution time of the whole method is about 1.41 s.

5 Conclusion

We have suggested a novel method for cataract grading, where the MobileNet-V2 architecture has been used for feature extraction, while the RF classifier has been performed for grading. The experimental evaluation has proved that our proposed method can achieve higher performance grading with respect to recent related work. In addition, the lower complexity has allowed performing grading under 2 s when run on a smartphone. Those grading and computational performances have resulted a MAG system for the cataract disease, where its mobility and lower cost boost the early cataract diagnosis. In our future work, we aim to enhance the grading performance by extracting other features that reflect the cataract disease. Subsequently, an ensemble learning principle can be put forward to perform accurate grading. In addition, we will be interested in extending the mobile system to detect other ocular pathologies such as the diabetic retinopathy and the glaucoma.

References

1. Zhou, Y., Li, G., Li, H.: Automatic cataract classification using deep neural network with discrete state transition. IEEE Trans. Med. Imaging **39**(2), 436–446 (2020). https://doi.org/10.1109/tmi.2019.2928229
2. Cao, L., Li, H., Zhang, Y., Xu, L., Zhang, L.: Hierarchical method for cataract grading based on retinal images using improved Haar wavelet (2019). arXiv:1904.01261v1
3. Song, W., Cao, Y., Qiao, Z., Wang, Q., Yang, J.: An improved semi-supervised learning method on cataract fundus image classification. In: 2019 IEEE 43rd Annual Computer Software and Applications Conference (COMPSAC), vol. 2, pp. 362–367 (2019). https://doi.org/10.1109/compsac.2019.10233
4. Yang, J.-J., et al.: Exploiting ensemble learning for automatic cataract detection and grading. Comput. Meth. and Prog. Biomed. **124**, 45–57 (2016). https://doi.org/10.1016/j.cmpb.2015.10.007
5. Akil, M., Elloumi, Y.: Detection of retinal abnormalities using smartphone-captured fundus images: a survey. In: Real-Time Image Processing and Deep Learning 2019, p. 21. Baltimore, United States (2019). https://doi.org/10.1117/12.2519094
6. Blaiech, A.G., Mansour, A., Kerkeni, A., Bedoui, M.H., Ben Abdallah, A.: Impact of enhancement for coronary artery segmentation based on deep learning neural network. In: 9th Iberian Conference on Pattern Recognition and Image Analysis, IbPRIA 2019, 1–4 July 2019. https://doi.org/10.1007/978-3-030-31321-0_23
7. Boudegga, H., Elloumi, Y., Akil, M., Bedoui, M.H., Kachouri, R., Ben Abdallah, A.: Fast and efficient retinal blood vessel segmentation method based on deep learning network. Comput. Med. Imaging Graph. (2021). https://doi.org/10.1016/j.compmedimag.2021.101902
8. Boukadida, R., Elloumi, Y., Akil, M., Bedoui, M.H.: Mobile-aided screening system for proliferative diabetic retinopathy. Int. J. Imaging Syst. Technol. https://doi.org/10.1002/ima.22547

Uncertainty Estimation in SARS-CoV-2 B-Cell Epitope Prediction for Vaccine Development

Bhargab Ghoshal[1], Biraja Ghoshal[2(✉)], Stephen Swift[2], and Allan Tucker[2]

[1] Queen Elizabeth's School, Barnet, London, UK
[2] Department of Computer Science, Brunel University, London, UK
Biraja.Ghoshal@brunel.ac.uk

Abstract. B-cell epitopes play a key role in stimulating B-cells, triggering the primary immune response which results in antibody production as well as the establishment of long-term immunity in the form of memory cells. Consequently, being able to accurately predict appropriate linear B-cell epitope regions would pave the way for the development of new protein-based vaccines. Knowing how much confidence there is in a prediction is also essential for gaining clinicians' trust in the technology. In this article, we propose a calibrated uncertainty estimation in deep learning to approximate variational Bayesian inference using MC-DropWeights to predict epitope regions using the data from the immune epitope database. Having applied this onto SARS-CoV-2, it can more reliably predict B-cell epitopes than standard methods. This will be able to identify safe and effective vaccine candidates to combat Covid-19.

Keywords: Covid-19 · Vaccine development · Dropweights · Epitope prediction · Deep learning · Uncertainty estimation · B-cell epitopes

1 Introduction

Adaptive immunity is orchestrated by lymphocytes. B-cells recognise antigens using the membrane bound immunoglobulins, which are the B-cell receptors (BCR). When antigens bind onto them, it results in a cascade of reactions, which concludes in the proliferation of B-cells and differentiation into plasma cells, that secrete antibodies, and memory B-cells. [1]. Immunological memory is the ability of the immune system to respond more rapidly and effectively to pathogens that have been encountered previously. With the COVD-19 pandemic, safe and effective vaccines are very desirable in controlling the transmission of the virus and thus limiting its effects on people, especially those who are vulnerable. B-cell epitope prediction is necessary as identifying epitopes highlights potential protein-based vaccine candidates.

B-cell epitopes are the sections of the antigen which interact with the BCR. They are generally classified into two categories: linear and conformational. Linear epitopes consist of consecutive peptides in the antigen's polypeptide chain

A. Tucker et al. (Eds.): AIME 2021, LNAI 12721, pp. 361–366, 2021.
https://doi.org/10.1007/978-3-030-77211-6_41

that are on the exterior of the antigens i.e. solvent-exposed). Conformational epitopes are solvent-exposed peptides that are discontinuous in the peptide sequence. Although around 90% of epitopes are discontinuous, linear B-cell epitopes can be readily used as candidates for a vaccine [2].

Early models that attempted to predict linear epitopes were based on simple characteristics. For instance, Hopp and Wood [3,4] considered the hydrophilic nature of some peptides and used it to make calculations, assuming that hydrophilic regions were mainly on the antigen surface and thus acted as epitopes. However, later research [5] showed that the proportions of hydrophilic and hydrophobic residues on protein surfaces are similar. Other characteristics, such as polypeptide flexibility, surface accessibility and β-turn tendencies have also been used. However, Blythe and Flower [6] showed that in the prediction of B-cell epitopes there is no correlation between the propensity profile and the presence of linear epitopes when qualities, or propensity scales, of amino acids are analysed. As a result, machine-learning based methods are used instead. These algorithms are trained to distinguish B-cell epitopes from residues that are not epitopes. Currently, a popular method is using BepiPred [2].

Quantifying uncertainty in the prediction of B-cell epitope of the protein SARS-CoV-2 regions can provide a measure for a model's confidence in its prediction. Providing an uncertainty measure could also improve subsequent steps in design and development of vaccines, providing clinicians an estimate on the likelihood of success, if a certain epitope were to be selected as a vaccine candidate. Bayesian Neural Networks (BNNs) provide a natural and principled way of modelling uncertainty in deep learning [7–11]. BNNs can be approximated by incorporating dropweights into the neural network to capture uncertainty in deep learning [12–14].

In this paper we present a natural way to quantify uncertainty in B-cell epitope prediction using Bayesian Neural Networks (BNNs) with dropweights, by decomposing predictive uncertainty into two parts: aleatoric and epistemic uncertainty. In order to produce suitable vaccines, various possible epitopes are considered and tested for efficacy and safety. We demonstrate that the proposed epitope prediction achieves better prediction accuracy compared with the existing method BepiPred2.0, by implementing it in the experiments on the immune epitope database (IEDB), a public database of immune epitopes [15]. We also demonstrate that the estimated uncertainty provides a better and more useful insight for epitope prediction.

2 Cost-Sensitive Calibrated Uncertainty

A Bayesian Neural Network (BNN) is a neural network with a prior distribution on its weights, which is robust to over-fitting (i.e. regularisation). Bayesian decision theory is a framework for making optimal decisions under uncertainty based on maximising expected utility over a model posterior. Exact inference is analytically intractable, and hence Variational Bayesian Inference (VBI) has been applied instead to approximate inference. While performing the inference, it calibrates the posterior approximation to maximise the expected utility including maximising the accuracy [14,16]. We leveraged a plug-in estimate of entropy and Jackknife resampling method to calculate bias-corrected uncertainty [13]. This approach addresses the issues with overconfidence and providing well-calibrated quantification of predictive uncertainty, giving us a way to model "when the machine does not know".

3 Experiment

We have used the publicly available dataset provided from The Immune Epitope Database (IEDB) and UniProt [15]. This contains two data files:

- B-Cell: The number of records is 14387 for all combinations of 14362 peptides and 757 proteins.
- SARS: The number of records is 520.

Datasets consists of information of protein and peptide: parent protein ID, parent protein sequence, start position of peptide, end position of peptide, peptide sequence, Isoelectric point, Aromaticity, Stability, Chou and Fasman β-turn prediction, Emini surface accessibility scale, Kolaskar and Tongaonkar antigenicity scale, and Parker hydrophilicity. Each peptide has a different sequence and is part of the parent protein sequence. Parameters have been correlated with the location of continuous epitopes. We use sequence length instead of sequence data for prediction.

4 Experimental Results

4.1 Model Performance

On average, Variational Bayesian Inference (VBI) improves the prediction accuracy of epitope regions [17]. The Confusion Matrices in Fig. 1 summarise the prediction accuracy of our implemented models.

4.2 Distribution of Uncertainty Estimates

We measured the aleatoric uncertainty and epistemic uncertainty associated with the predictive probabilities of the VBI by keeping dropweights on during test time. Figure 2 shows Kernel Density Estimation with a Gaussian Kernel is used to plot the output posterior distributions for all of the test data, grouped by correct and incorrect predictions.

4.3 The Contribution of Uncertainty Thresholds in Predictive Probabilities

We altered the uncertainty threshold (UT) in the range [0, 1], then computed and plotted the predictive accuracy of the evaluation metrics as in Fig. 3. As the uncertainty in data increases, the predictive accuracy decreases. Analysis shows

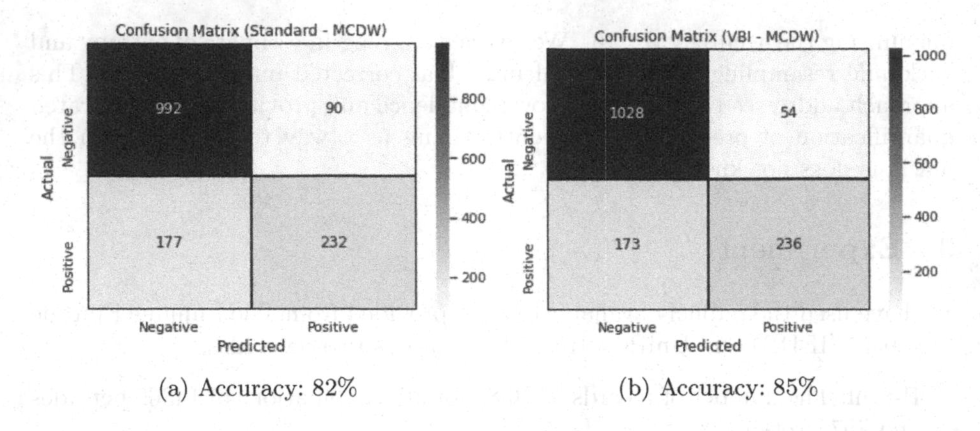

(a) Accuracy: 82% (b) Accuracy: 85%

Fig. 1. Confusion matrix.

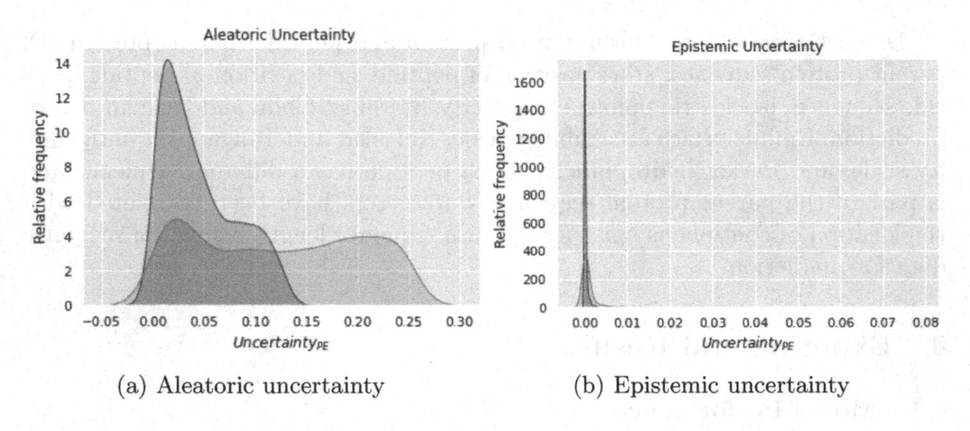

(a) Aleatoric uncertainty (b) Epistemic uncertainty

Fig. 2. The Posterior Distribution of normalised aleatoric (a) and epistemic (b) uncertainty values of the correct (green) and incorrect (red) class predictions. It shows that model uncertainty is higher for incorrect predictions. Therefore, it stands to reason to refer the uncertain samples to experts to improve the overall performance of the collaborative efforts of man and machine in prediction. Kernel density estimation with a Gaussian Kernel is used to plot the output posterior distributions. (Color figure online)

that the epistemic uncertainty threshold has a very little impact on predictive accuracy, whereas aleatoric uncertainty has a significant impact on the predictive accuracy. Therefore, unless the quality of B-cell processing data improves at the time of preparation, it is very likely that accuracy of results will not be sufficient, which may limit the efficacy of subsequent vaccine candidates. However, the aleatoric uncertainty may also provide an insight on the relative success rates of selected epitopes, if they were to be tested as vaccine candidates.

By accurately predicting B-cell epitopes, we can acutely determine the parts of the antigen that interact with B-cell receptors. Consequently, antibodies specific to the epitopes are synthesised by effector B-cells, which confer primary immune response. Following the primary immune response, some B-cells differentiate into memory B-cells. Memory is not dependent on repeated exposure to infection, and is established by populations of the memory cells, that persist regardless of the presence of antigens. Upon re-exposure to the same antigen, a secondary immune response will occur. The activation of memory B-cells is similar to that of naïve B-cells; however, it is more efficient. BCR of memory B-cells have a greater affinity to the antigens, so memory B-cells are stimulated more efficiently. Furthermore, memory B-cells can act as antigen-presenting cells for the activation of naive helper T cells, removing the need for these T cells to be activated by dendritic cells. Proliferation of memory B-cells results in plasma cells that have a greater affinity and are of diverse types. As a result, the secondary immune response is more successful in overcoming the pathogen. Inducing a secondary immune response is thus desired and is the overall aim of vaccines [18], especially for SARS-CoV-2.

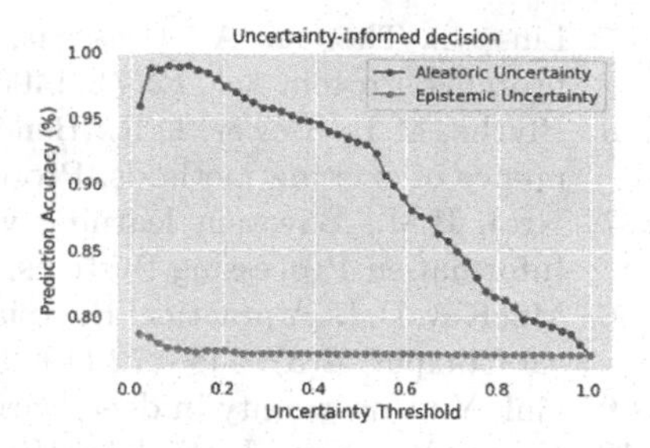

Fig. 3. Predictive accuracy (%) for different values of the uncertainty threshold.

5 Conclusion and Future Work

We demonstrated that SARS-CoV-2 B-cell epitope prediction with uncertainty information from Bayesian Neural Networks approximated using MC-dropweights provide a more accurate and reliable method than currently used methods and so can be harnessed to be able to identify potential vaccine candidates more successfully. Further research would include the extension of the ideas above to represent better uncertainty estimates in mRNA sequence analysis to identify potential mRNA sequences in the SARS-CoV-2 genome which would serve as suitable candidates for mRNA-based vaccines.

References

1. Janeway, C.A., Travers, P., Walport, M.: Immunobiology: the immune system in health and disease. Garland Publishing Inc. (1999)
2. Sanchez-Trincado, J.L., Gomez-Perosanz, M., Reche, P.A.: Fundamentals and methods for t-and b-cell epitope prediction. J. Immunol. Res. **2017** (2017)
3. Hopp, T.P., Woods, K.R.: Prediction of protein antigenic determinants from amino acid sequences. Proc. Natl. Acad. Sci. **78**(6), 3824–3828 (1981)
4. Hopp, T.P., Woods, K.R.: A computer program for predicting protein antigenic determinants. Mol. Immunol. **20**(4), 483–489 (1983)

5. Lins, L., Thomas, A., Brasseur, R.: Analysis of accessible surface of residues in proteins. Protein Sci. **12**(7), 1406–1417 (2003)
6. Blythe, M.J., Flower, D.R.: Benchmarking b cell epitope prediction: underperformance of existing methods. Protein Sci. **14**(1), 246–248 (2005)
7. Neal, R.M.: Bayesian learning via stochastic dynamics. In: Advances in Neural Information Processing Systems, pp. 475–482 (1993)
8. MacKay, D.J.: A practical Bayesian framework for backpropagation networks. Neural Comput. **4**(3), 448–472 (1992)
9. Gal, Y.: Uncertainty in deep learning, PhD thesis, University of Cambridge (2016)
10. Kwon, Y., Won, J.-H., Kim, B.J., Paik, M.C.: Uncertainty quantification using Bayesian neural networks in classification: application to ischemic stroke lesion segmentation. Med. Imaging Deep Learn. Conf. (2018)
11. Blundell, C., Cornebise, J., Kavukcuoglu, K., Wierstra, D.: Weight uncertainty in neural networks, arXiv preprint arXiv:1505.05424 (2015)
12. Gal, Y., Ghahramani, Z.: Dropout as a Bayesian approximation: Representing model uncertainty in deep learning. In: 33rd International Conference on Machine Learning, ICML 2016, vol. 3, pp. 1651–1660 (2016)
13. Ghoshal, B., Tucker, A., Sanghera, B., Lup Wong, W.: Estimating uncertainty in deep learning for reporting confidence to clinicians in medical image segmentation and diseases detection, Computational Intelligence (2019)
14. Ghoshal, B., Tucker, A.: Estimating uncertainty and interpretability in deep learning for coronavirus (covid-19) detection, arXiv preprint arXiv:2003.10769 (2020)
15. COVID-19/SARS B-cell Epitope Prediction (2019). https://www.kaggle.com/c/data-science-bowl2018
16. Ghoshal, B., Tucker, A.: On calibrated model uncertainty in deep learning. In: The European Conference on Machine Learning (ECML PKDD 2020) (2020)
17. Nomi, T., et al.: Epitope prediction of antigen protein using attention-based LSTM network, BioRxiv (2020)
18. Mak, T., Saunders, M., Jett, B.: B cell development, activation and effector functions, primer to the immune response. 2nd ed: Academic Cell, pp. 111–42 (2014)

Attention-Based Explanation in a Deep Learning Model For Classifying Radiology Reports

Luca Putelli[1,2(✉)], Alfonso E. Gerevini[1], Alberto Lavelli[2], Roberto Maroldi[1], and Ivan Serina[1]

[1] Università Degli Studi di Brescia, Brescia, Italy
{l.putelli002,alfonso.gerevini,roberto.maroldi,ivan.serina}@unibs.it
[2] Fondazione Bruno Kessler, Povo (TN), Italy
{lputelli,lavelli}@fbk.eu

Abstract. Although deep learning techniques have obtained remarkable results in clinical text analysis, the delicacy of this application domain requires also that these models can be easily understood by the hospital staff. The attention mechanism, which assigns numerical weights representing the contribution of each word to the predictive task, can be exploited for identifying the textual evidence the prediction is based on. In this paper, we investigate the explainability of an attention-based classification model for radiology reports collected from an Italian hospital. The identified explanations are compared with a set of manual annotations made by the domain experts in order to analyze the usefulness of the attention mechanism in our context.

1 Introduction and Background

Hospitals collect a huge amount of clinical narrative texts containing very significant information that can be used to improve the efficacy and quality of patients care. While Natural Language Processing and Deep Learning techniques have been proved to be very effective for extracting information from clinical texts, the use of such information in the clinical environment requires also an explanation. Highlighting the most important part of the text which *explains* the decision made by the model can assure the physicians that the system is not biased and is correct also from a medical point of view.

Our previous work regarding the analysis of radiology reports relied on the annotation of relevant snippets and on machine learning techniques [2]. A deep learning based system, which exploits also a higher number of reports, was introduced in [6] and greatly improved our results. This system is based on Long Short Term Neural Networks, which are particularly suited for processing sequential data like natural language sentences, and on the Attention Mechanism [1], which computes a weight for each word representing its contribution in the addressed task. Although in some works the weights are visualized highlighting the words with the highest weights and presenting them as the *explanation* of the inner

A. Tucker et al. (Eds.): AIME 2021, LNAI 12721, pp. 367–372, 2021.
https://doi.org/10.1007/978-3-030-77211-6_42

working of the model [5], there is much debate around the capability of the attention weights to provide an effective explanation.

In particular, in the works of Jain et al. [4] and Serrano and Smith [7], the authors show how it is possible to generate two different or even counterfactual explanations that nonetheless yield to the same prediction, undermining the use of attention weights for understanding the behaviour of a model. However, their claims are criticized in [9], where an alternative way to build an adversary distribution of the attention weights that alters the predictive results is provided, confirming the usefulness of the attention mechanism for the explanation. Moreover, the authors of [3] point out that, while counter-examples can undermine the interpretability of the attention mechanism in general, for specific models and tasks an attention-based explanation can be useful in practice.

In this paper, we focus on the interpretability of a deep learning model for the classification of radiology reports of an Italian hospital (*Spedali Civili di Brescia*). We extract the attention weights and evaluate its behaviour with respect to the characteristics of the document. The identified explanations for the classification are compared with a set of manual annotations made by the domain experts in order to analyze the effectiveness of the attention mechanism in our context.

2 Classification of Radiology Reports

Our training set consists of 5,752 classified computed tomography reports, focusing on the chest and on the lung region. These reports were collected during the activity of the radiology departments of *Spedali Civili di Brescia* and anonymized. Given that a radiology exam can analyze several body parts and describe them in different sections, a custom algorithm extracts the introduction, the section regarding the chest and, eventually, the conclusions.

We analyse our reports according to two aspects: **Exam Type** (*First Exam* or *Follow-Up*) and **Result** (*Suspect* or *Negative*), which focuses on the possibility of the patient having a neoplastic lesion.

In the pre-processing phase we divide our reports in sentences and tokens (i.e. single words). Each word in our corpus is represented with a vector of length 200. These vectors are obtained by applying the Word2vec algorithm, using as training data our corpus of reports plus over 9,000 unclassified reports. More details for the pre-processing phase are showed in [6]. We train two models (one for Exam Type and one for Result) composed of: a **bidirectional LSTM layer** that processes the input sequence; an **attention mechanism** that weights the influence of the words for the classification task, and produces the document representation; an **output layer** that provides the classification and that is formed by a single neuron with sigmoid activation.

Our two classification models are evaluated using a test set consisting of almost 400 reports, which were labelled and manually annotated by the physicians [2]. For the Exam Type, the achieved performance is very high, with both accuracy and F-Score above 94%. Good performance are obtained also for the Suspect Level, with 75.3 in terms of accuracy and 73.9 in terms of F-Score. The main issues for this task are mainly due to differences of the distribution of the training data and the test data. For example, while the training data are made by over 65% of the *Negative* reports, in the test set *Negative* reports are only 19% of the total. Moreover, we conjecture that some cases can be evaluated differently depending on personal judgments of the physicians. A more detailed analysis of the performance of our models and a deeper discussion of the main issues related to these tasks are available in [6].

3 Analysis of the Attention Mechanism

For the classification models introduced in Sect. 2, we extract the attention weights α_i with $i \in [1, n]$ where n is the length of the document. Given that the sum of all the attention weights is equal to 1, and therefore the weights value is influenced by n, in order to facilitate the comparison of reports with different length, we normalize the attention weights between 0 and 1 through the min-max normalization.

The work by Vashishth et al. [8] introduced the concept of the attention mechanism as a gating unit. Analysing our attention models, we find a similar behaviour, where the text is mainly separated into (i) the important part, which has a normalized weight very close to 1, and (ii) the not important part, with a weight close to 0. Considering the average μ_w of the normalized weights for each sentence in our test set, we can group our sentences into three categories: **important sentences**, with $\mu_w > 0.75$; **intermediate sentences**, with $0.4 \leq \mu_w \leq 0.75$; and **not important sentences**, with $\mu_w < 0.4$. We found that our Suspect Model recognizes 56.5% of the sentences as important, 29.1% as non important, and the 14.4% as intermediate. A further analysis showed that most of the intermediate sentences are formed by a small subset of words (between 3 and 5, typically for describing a lesion with its characteristics) that have a high weight, while the remaining words that have small weights. The same can be said for the Exam Type Model that, however, considers non important a higher percentage of sentences (44%). Another considerable aspect is that the attention has the tendency to operate on entire sentences or complex expressions and not on single words.

This behaviour of the attention mechanism can be seen as a sort of selection process of the most relevant sentences in the document. Figure 1a shows how the report length influences the selection process in long reports (over 120 words, which is the median length of the reports in our dataset, in yellow) and short reports (under 120 words, in purple). In particuòar , the attention mechanism selects fewer sentences for the longer reports. The selection process can be very important in terms of interpretation of our system. In fact, when a report is long, it is very useful if only the most important parts are highlighted, because this saves time to the physicians who have to read it focusing their attention on what is most relevant. On the other hand, if the report is short, each sentence may contain relevant information and therefore it should be read in its entirety.

In the following, we evaluate how well the attention mechanism behaves in detecting the most important parts of a report. In particular we concentrate our analysis on the Suspect Model, which identifies potential neoplastic patients. To

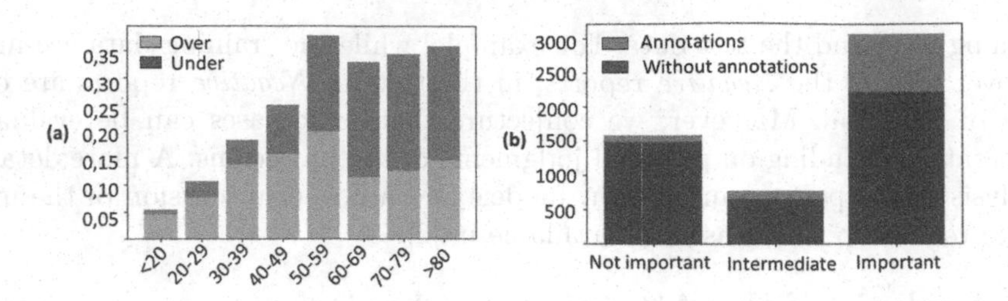

Fig. 1. (a) Fraction of reports over or under 120 words (on the y-axis) with respect to the percentage of important sentences (on the x-axis) according to the attention mechanism. (b) Not Important, Intermediate and Important sentences according to the attention mechanism. In orange, the sentences which contain at least one manual annotation, in the blue the ones without any annotation. (Color figure online)

do so, we compare the sentences highlighted by the attention mechanism to the manual annotations made by the physicians on our test set.

First of all, we point out that we could not use the manual annotations as they are, but had to first elaborate them. In fact, while the attention mechanism has the tendency to highlight entire sentences, a manual annotation consists of only 4 words on average. Therefore, a direct comparison is not possible. However, in our previous work [6], we found out that often manual annotations do not contain all the valuable information; for example, in the expression *nodule of 4 mm with irregular and spiculed margins*, the annotation includes only the words *nodule of 4 mm* without the description of the margins, which could be very important. Moreover, we observed that often in the same sentence there is more than one annotation. In order to cope with these issues, we consider a sentence relevant according to the manual annotators when it contains at least one manual annotation. We then compare such sentences with the important ones selected by the attention mechanism. As shown in Fig. 1b, 80.2% of the manual annotations are contained in the sentences considered important by the attention mechanism, which are 56% of all the sentences in our corpus; 10.8% are contained in the intermediate ones, which are 14%; and only 8.9% are contained in the not important ones, which are almost 30%. The larger fraction of the orange part in the last column with respect to the other two columns, suggests a strong correlation between what is highlighted by the attention and what is important according to the physicians (indicated by the manual annotations).

We conducted a further analysis to understand why some annotated sentences were not identified as important by the attention mechanism. While some of them are simply errors made by the model, the others can be grouped into the following three classes:

1. **Annotations in negative sentences**, consisting of expressions that exclude the presence of a concept, like for example *no secondary lesions.*
2. **Adenopathies and lymph nodes**, which, although important for evaluating the conditions of the patient, are not directly connected only to neoplastic

lesions, and are present in both *Suspect* and *Negative* reports. Therefore, it is understandable that the attention mechanism does not rely on these concepts.
3. **Rare expressions**. In some reports, very important expressions like *outcome of radiotherapy* or conclusive remarks on the patient's conditions are annotated. However, these cases are very rare in our training set, and they are not entirely captured by the attention mechanism.

As shown in Fig. 1b, the attention mechanism highlights many more sentences than the manual annotators. In fact, the blue part of the last column in the histogram corresponds to more than 2,000 sentences that are considered relevant only by the attention mechanism. This discrepancy can be due to the fact that, especially for the shorter reports, the attention mechanism has the tendency to include a large portion of the text. On the other hand, we remind that the manual annotations often do not include all the important aspects in the reports [6], due to the fact that the manual annotation process is a demanding task even for the domain experts.

4 Conclusions

We have investigated the behaviour of the attention mechanism in the context of a classification task for radiology reports written in Italian using deep learning techniques. Our analysis confirms that the attention mechanism works as a gating unit [8], and that it often highlights entire sentences instead of single words. We have then compared the explanation provided by the attention mechanism highlighting the most relevant parts of the text, and the most important sentences according to the manual annotators, showing that there is a significant correlation between them. This result confirms that the attention mechanism can provide a useful tool for the interpretability of a deep learning model.

References

1. Bahdanau, D., Cho, K., Bengio, Y.: Neural machine translation by jointly learning to align and translate. In: Proceedings of the 3rd International Conference on Learning Representations (2015)
2. Gerevini, A.E., et al.: Automatic classification of radiological reports for clinical care. Artif. Intell. Med. **91**, 72–81 (2018)
3. Jacovi, A., Goldberg, Y.: Towards faithfully interpretable NLP systems: how should we define and evaluate faithfulness? In: Proceedings of the 58th Annual Meeting of the ACL, pp. 4198–4205 (2020)
4. Jain, S., Wallace, B.C.: Attention is not explanation. In: Proceedings of the 2019 Conference of the North American Chapter of the Association for Computational Linguistics: Human Language Technologies, vol. 1, pp. 3543–3556 (2019)
5. Mullenbach, J., Wiegreffe, S., Duke, J., Sun, J., Eisenstein, J.: Explainable prediction of medical codes from clinical text. In: Proceedings of the 2018 Conference of the North American Chapter of the ACL, pp. 1101–1111 (2018)

6. Putelli, L., Gerevini, A.E., Lavelli, A., Olivato, M., Serina, I.: Deep learning for classification of radiology reports with a hierarchical schema. In: Proceedings of the 24th International Conference KES-2020. Procedia Computer Science, vol. 176. Elsevier (2020)
7. Serrano, S., Smith, N.A.: Is attention interpretable? In: Proceedings of the 57th Annual Meeting of the Association for Computational Linguistics, pp. 2931–2951 (2019)
8. Vashishth, S., Upadhyay, S., Tomar, G.S., Faruqui, M.: Attention interpretability across NLP tasks. CoRR arXiv:1909.11218 (2019)
9. Wiegreffe, S., Pinter, Y.: Attention is not not explanation. In: Proceedings of EMNLP-IJCNLP, pp. 11–20 (2019)

Evaluation of Encoder-Decoder Architectures for Automatic Skin Lesion Segmentation

José G. P. Lima(✉), Geraldo Braz Junior, João D. S. de Almeida, and Caio E. F. Matos

Computer Applied Group, Federal University of Maranhão, São Luis, Brazil
guilherme.jose@discente.ufma.br,
{geraldo,jdallyson,caioefalcao}@nca.ufma.br

Abstract. Melanoma is one of the most severe skin cancer types due to its high mortality rate, which can achieve 70%. An early diagnosis of the disease is crucial as it increases the ten-year survival rate up to 97%. The segmentation of skin lesions is one of the essential steps of the diagnosis process for accurate melanoma detection. However, even for specialist doctors, segmenting these lesions is costly and challenging due to the wide variety of stains, which can have irregular edges, different dimensions, and colors, and due to the high amounts of exams to analyze. This paper aims to compare encoder-decoder architectures based on popular convolutional neural networks to segmentation dermoscopic images in order to assist in the automatic diagnosis process.

Keywords: Melanoma · Fully convolutional network · U-net

1 Introduction

Skin Cancer is the most common type of cancer, accounting for one in every three cases worldwide [6]. Can be divided into two main groups: melanoma and non-melanoma. Although melanoma accounts for just 22% of cases [3], it is by far the most dangerous because it is more likely to grow and spread. The latest statistics available in the world show that cases have been rising each year at an alarming rate. In the United States, it is estimated that the number of new melanoma cases diagnosed in 2021 will increase by 5.8%, with 106,110 new cases of melanoma being diagnosed resulting in about 7,000 deaths [10].

One of the non-invasive ways of such diagnosis is through dermoscopy, which consists of a medical expert examining dermoscopic skin lesion images. However, even for specialist doctors, segmenting these lesions is costly and difficult due to the wide variety of stains that sometimes have irregular edges, different dimensions, and colors. Therefore, studies have been carried out to automatically target these injuries to assist medical professionals.

Federal University of Maranhão for the financial support.

A. Tucker et al. (Eds.): AIME 2021, LNAI 12721, pp. 373–377, 2021.
https://doi.org/10.1007/978-3-030-77211-6_43

In recent years, Convolutional Neural Networks (CNNs) have emerged as one of the most powerful tools in image processing showing promising results in multiple domains, including medical image analysis [8].

This work aims to evaluate multiple encoder-decoder deep neural network architectures for automatic segmentation of skin lesions. The main goals of this paper can be summarized as follows: (1) employ some of the most popular deep CNN architectures extensively used in the computer vision community for semantic image segmentation on dermoscopic images of skin lesions (2) evaluate the performance of different deep CNN models in pixel-wise image labeling. (3) Identify a high-performance CNN model from state of the art metrics and speed points of view so it can be effectively used in many real-life automatic image segmentation to assist in skin cancer diagnosis.

2 Proposed Methodology

This section describes the proposed methodology used for developing the work for the analysis of the segmentation models. Topics will also explain the encoder-decoder architectures implemented and the analysis metrics employed. Figure 1 shows the sequence of steps in the methodology of this work.

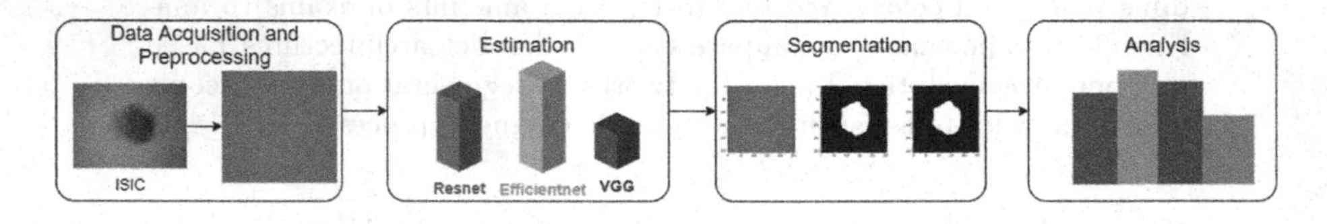

Fig. 1. Proposed methodology

The first step was resizing the images which initially range from 576 × 768 and 6748 × 4499 pixels to 256 × 256 pixels. CIELAB color space was used. Each input channel was then normalized to the range [0, 1].

Four different encoder-decoder models were trained and evaluated: Unet, FPN, PSPNet, and Linknet.

Unet is a fully convolutional network modified and extended so that it should work with very few training images and yield more precise segmentations [8]. It is composed of two parts. The first part is feature extraction, and the second part is upsampling, where upsampling operators replace pooling operators. These layers increase the resolution of the output. For better localization, features from the contracting path are combined with the upsampled output.

Figure 2 shows the evaluated U-net model's architecture. It contains the initial fully convolution layer with 32 filters, followed by mobile inverted bottleneck MBConv [9] convolutional blocks inherent to Efficientnetb1. Each decoder block consists of upsample, concatenate, conv, BN, ReLU. Conv is the convolutional layer, BN is the batch normalization layer, upsample the upsampling layer, and

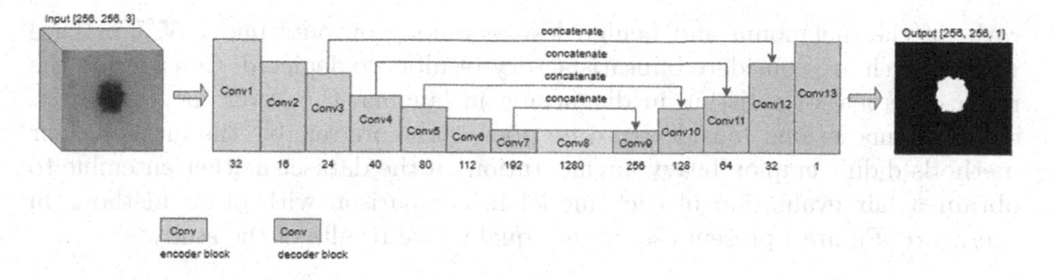

Fig. 2. Schematic architecture of the proposed U-Net for lesion segmentation.

ReLU the activation function. At the final layer, a final 1×1 convolution is used to map each 16-component feature vector to the desired number of classes and sigmoid activation. The yellow block Conv8 in Fig. 1 consists of MBConv, conv, BN, ReLU6 to connect the encoder and decoder paths.

Linknet [4], PSPNet [11] and FPN [7] are variations of fully convolutional networks, similar to U-NET, that are also evaluated in this work in order to compare with U-NET.

For all segmentation models, transfer learning was used by using models pre-trained on the ImageNet dataset. Taking advantage of data from the first set to extract information that may be useful when learning or even when directly making predictions in the second setting [2], the objective of using transfer learning was to decrease the training time and also result in lower generalization error.

Extensive experiments were performed using various pre-trained networks: ResNet, VGG, EfficientNet, DenseNet, Inception, MobileNet, SeNet, SE-ResNeXt, ResNeXt.

A combination of dice loss (DCL) and binary focal loss (BFL) was used for the loss function and defined as $L = DCL + BFL$.

The obtained segmentation results are evaluated with Mean thresholded Jaccard index (threshold = 0.65) and mean Dice Coefficient were the metrics used to evaluate each model's output to the ground truth image.

3 Results

In this work, we used the dataset provided by ISIC archive which obtained 4000 skin lesion images: 3000 benign lesions and 1000 malignant lesions. It was used 50% of those images for training and the other images to evaluate.

First we perform an estimation of backbones for feature extraction, resulting that EfficientNetb1 has the best result with 76.1% of Jaccard index, followed by Se-ResNeXt101 and EfficientNetb3.

Experiments conducted on the acquired ISIC database are depicted in Table 1 revealing that The Unet-based model achieved the highest metrics values. PSP-Net resulted in the worst performance out of all models tested, and that may be due to PSPNet lacking skip connections directly from the encoder to decoder, thus losing spatial information after successive convolutional layers. Also, the ratio of the melanoma and benign lesions among outputs under 65% of Dice value, which are considered unsatisfactory results are depicted. Considering the main objective of assisting in diagnosing melanoma, the lower the proportion of melanoma lesions considered to be poor, the more reliable the method. Our methods didn't employ heavy augmentation on the dataset neither ensemble to obtain a fair evaluation of each model in comparison with other methods in literature. Figure 3 presents some case qualitative results of the study.

Table 1. Obtained results and comparison of different methods on literature. Also, last two columns presents the amount of images under 65% Dice index threshold.

Author	Method	Jaccard (%)	Dice (%)	Melanoma (%)	Benign (%)
Author	PSPNet	74.26	82.92	22.5	77.5
	FPN	78.3	86.43	42.8	57.1
	Linknet	78.3	86.2	31.4	68.5
	Unet	**78.48**	**86.64**	33.3	66.6
Yuan et al. [5]	Unet-VGG16	77.2	–		
Amin et al. [1]	VGG-16	85.0	82.0		

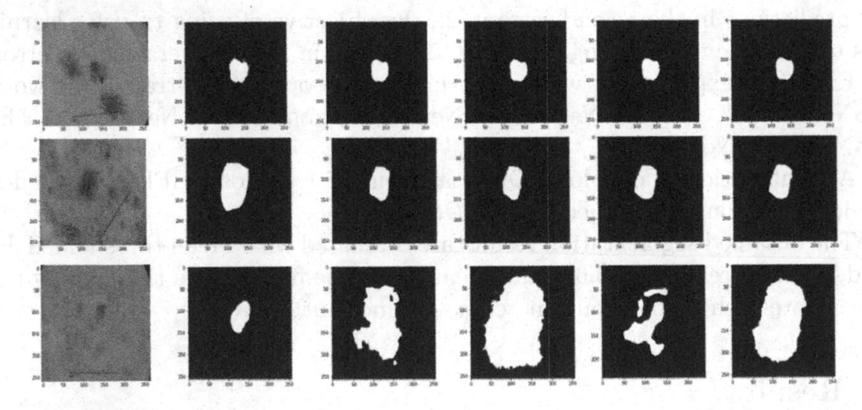

Fig. 3. Examples of each model's segmentation output. The first row represents a good output for all methods. The second row represents a medium outcome, and the third-row a bad output. The first column is the original RGB image. The second column shows the ground truth mask. The third column contains the FPN output, the fourth column the Linknet output, the fifth column the PSPNet output, and the sixth column the Unet output.

4 Conclusion

We trained and evaluated four CNN models for the segmentation of skin lesions in dermoscopic images: Unet, Linknet, FPN and PSPNet. Our results showed that CNN architectures with direct skip connections from the encoder to decoder path present better outcomes for the task of skin lesion segmentation. The best model evaluated, Unet with pre-trained Efficientnetb1 had encouraging results and low errors for the segmentation of melanoma proving to be reliable for medical usage. For future work, we would like to implement more complex CNNs presenting a broader evaluation with dispersion measures.

References

1. Amin, J., et al.: Integrated design of deep features fusion for localization and classification of skin cancer. Pattern Recogn. Lett. **131**, 63–70 (2020)
2. Bengio, Y., Goodfellow, I., Courville, A.: Deep Learning, vol. 1. MIT Press, Massachusetts (2017)
3. World Cancer Research Fund International: Diet, nutrition, physical activity and cancer: a global perspective (2018). www.dietandcancerreport.org. Accessed 17 Jan 2021
4. Chaurasia, A., Culurciello, E.: LinkNet: exploiting encoder representations for efficient semantic segmentation. In: 2017 IEEE Visual Communications and Image Processing (VCIP), pp. 1–4. IEEE (2017)
5. Codella, N., et al.: Skin lesion analysis toward melanoma detection 2018: a challenge hosted by the international skin imaging collaboration (ISIC). arXiv preprint arXiv:1902.03368 (2019)
6. Ge, Z., Demyanov, S., Chakravorty, R., Bowling, A., Garnavi, R.: Skin disease recognition using deep saliency features and multimodal learning of dermoscopy and clinical images. In: Descoteaux, M., Maier-Hein, L., Franz, A., Jannin, P., Collins, D.L., Duchesne, S. (eds.) MICCAI 2017. LNCS, vol. 10435, pp. 250–258. Springer, Cham (2017). https://doi.org/10.1007/978-3-319-66179-7_29
7. Lin, T., Dollár, P., Girshick, R., He, K., Hariharan, B., Belongie, S.: Feature pyramid networks for object detection. In: 2017 IEEE Conference on Computer Vision and Pattern Recognition (CVPR), pp. 936–944 (2017). https://doi.org/10.1109/CVPR.2017.106
8. Ronneberger, O., Fischer, P., Brox, T.: U-net: convolutional networks for biomedical image segmentation. In: Navab, N., Hornegger, J., Wells, W.M., Frangi, A.F. (eds.) MICCAI 2015. LNCS, vol. 9351, pp. 234–241. Springer, Cham (2015). https://doi.org/10.1007/978-3-319-24574-4_28
9. Sandler, M., Howard, A., Zhu, M., Zhmoginov, A., Chen, L.C.: Mobilenetv 2: Inverted residuals and linear bottlenecks. In: Proceedings of the IEEE conference on computer vision and pattern recognition. pp. 4510–4520 (2018)
10. Society, A.C.: Cancer facts and figures 2021 (2021). https://www.cancer.org/content/dam/cancer-org/research/cancer-facts-and-statistics/annual-cancer-facts-and-figures/2021/cancer-facts-and-figures-2021.pdf. Accessed 17 Jan 2021
11. Zhao, H., Shi, J., Qi, X., Wang, X., Jia, J.: Pyramid scene parsing network. In: CVPR (2017)

A Novel Deep Learning Model for COVID-19 Detection from Combined Heterogeneous X-ray and CT Chest Images

Amir Bouden[1], Ahmed Ghazi Blaiech[1,2](✉), Khaled Ben Khalifa[1,2], Asma Ben Abdallah[1,3], and Mohamed Hédi Bedoui[1]

[1] Laboratoire de Technologie et Imagerie Médicale, Faculté de Médecine de Monastir, Université de Monastir, 5019 Monastir, Tunisie

[2] Institut Supérieur des Sciences Appliquées et de Technologie de Sousse, Université de Sousse, 4003 Sousse, Tunisie

[3] Institut supérieur d'informatique et de Mathématiques, Université de Monastir, 5019 Monastir, Tunisie

Abstract. COVID-19 originally started in Wuhan city in China. The disease rapidly became a worldwide pandemic, causing a respiratory illness with symptoms such as coughing, fever, and in more severe cases difficulty in breathing. With the current testing processes, it is very difficult and sometimes impossible to manage and provide the necessary treatment to suspected patients since the number of the infected is rapidly increasing. Hence, the availability of an artificial intelligent driven system can be an assistive tool to provide accurate diagnosis using radiology imaging techniques. In this paper, we put forward a new deep learning architecture, which integrates the Nested Residual Connections (NRCs) in a DarkCovidNet model, called DarkCovidNet-NRC, in order to classify chest images and to detect COVID-19 cases. The proposed architecture is validated with the K-fold cross-validation technique on X-ray and CT chest datasets separately and then combined. The experimental results reveal that the suggested model performs very well in the medical classification task and it competes with the state of the art in multiple performance metrics by respectively achieving an accuracy and precision of 0.9609 and 0.978 on the combined dataset.

Keywords: COVID-19 · Deep learning · Classification · Combined heterogeneous chest images

1 Introduction

Since December 2019, one of the most life-threatening viruses has appeared. COVID-19 has caused a devastating effect on both daily life and public health. Thousands of lives are taken daily around the world. Unfortunately, there is no effective treatment to eliminate the virus, and so far, the vaccine has not given reassuring results. Several approaches and applications have been presented, implementing advanced artificial intelligence and medical resources, to diagnose the virus in its first stages [1, 2]. Many studies have been

A. Tucker et al. (Eds.): AIME 2021, LNAI 12721, pp. 378–383, 2021.
https://doi.org/10.1007/978-3-030-77211-6_44

proposed for the detection of COVID-19 by using machine learning algorithms like random forests, genetic algorithms, and Artificial Neural Networks (ANNs) [3–5]. ANNs have gained researchers' interest, in recent years, for its good performance to become the best solution for classification tasks. Particularly, Deep Learning (DL) approaches are widely used for successful classification. It has enabled the automatic extraction of complex data features at high levels of abstraction. The last approaches of DL, which are also characterized by their large number of hidden layers in their networks, provide the most efficient solutions to problems caused by massive calculations and allow machines to learn and predict object classes with more accuracy [6, 7]. Many research papers have contributed to solving the detection of the COVID-19 pandemic by using DL architectures as a solution [5, 8]. In [5], the authors put forward an advanced model for COVID-19 detection, named DarkCovidNet. This model was designed to provide accurate binary (COVID, non-COVID) and triple classification (COVID, non-COVID, Pneumonia) through X-ray images. The used dataset consisted of 500 healthy cases and 500 pneumonia ones. The DarkCovidNet architecture comprised 17 convolution layers. A Batch Normalization (BN) layer was adopted after every convolution layer in order to normalize each layer output. The implementation of DarkCovidNet achieved 98.08% on binary classification and 87.02% in triple classification after 100 epochs of training. In [8], an end-to-end system for COVID-19 and pneumonia infection detection was propounded. Both X-ray for COVID-19 detection and CT images for pneumonia detection, collected from different publicly available resources, were considered to evaluate the model. The Inception Recurrent Residual Neural Network (IRRCNN) was suggested for the detection of COVID-19, and the NABLA-N network model was used for the identification and segmentation of the infected regions. The IRRCNN architecture was composed of an input layer, five inception recurrent residual units, a global average pooling layer and a softmax output layer. The implemented NABLA-N network architecture was based on a U-Net template that contained two nested U-Nets inside it. The detection model showed around 84.67% testing accuracy from X-ray images and 98.78% accuracy in CT-images. All these studies used many DL approaches to COVID-19 detection, but they also utilized either the X-ray or CT datasets. In order to improve the performance of COVID-19 detection, we firstly combine the heterogeneous X-ray and CT chest datasets into a large one, and secondly, we investigate the most adequate DL architectures that combine the advantages of the best approaches and techniques in the literature in order to boost pre-processing and feature extraction suitable for heterogeneous images. In this context, we put forward a novel DL model, called DarkCovidNet-NRC, which integrates the Nested Residual Connections (NRCs) in the DarkCovidNet model. This paper is organized as follows: Sect. 2 describes the proposed DL model that will improve the performance of COVID-19 detection. Section 3 describes the heterogeneous dataset used for the experiment and the used implementation software environment, and it discusses the obtained implementation results. The last section concludes and gives some future perspectives of this work.

2 DarkCovidNet-NRC DL Architecture

In this paper, we investigate the most adequate DL architectures to come up with a better performing model. The proposed model, named DarkCovidNet-NRC, is a combination between the DarkCovidNet architecture and the NRC blocks, which are composed of nested residual blocks (residual blocks within a residual block). In fact, as described in [5], the DarkCovidNet model can achieve good accuracy as it provides efficient feature-map extraction through its convolution layers. The integration of residual blocks with skip-connections in the DarkCovidNet architecture can also make this model more robust and expandable and it can reduce the chances of overfitting, thus achieving a better performance of COVID-19 classification. The DarkCovidNet-NRC architecture is depicted in Fig. 1. It contains two single Dark Net (DN) blocks which each one contains one convolutional layer followed by BN and PReLU operations. Furthermore, the suggested architecture integrates four NRC blocks each one is a residual block composed of one DN block and two successive nested residual blocks. Moreover, this block provides a skip-connection from the first DN block to the output of the second nested residual block. Finally, a PReLU activation function is applied at the end of this block. The DarkCovidNet-NRC architecture ends by one DN block, a flattening layer and a softmax layer that produces the outputs. In the first six blocks, the channel size of convolutions increases twice. This size decreases to 2 in the last DN block, which is equal to number of targeted classes (COVID, non-COVID).

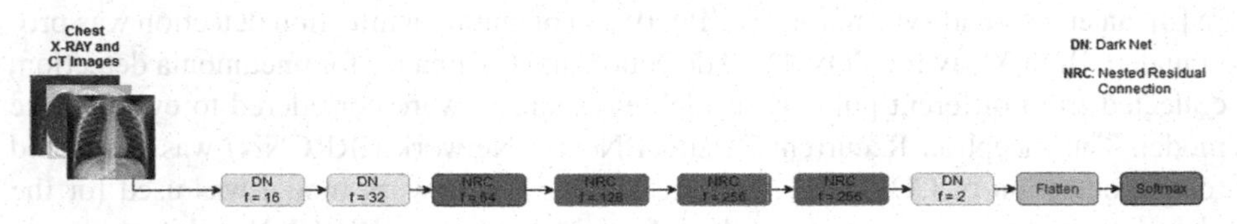

Fig. 1. DarkCovidNet-NRC architecture

3 Experiments and Results

In order to implement the suggested DL model for the COVI-19 detection, we first describe the used dataset, set up the need for software and hardware environments and present and discuss the implementation results.

3.1 Dataset

We use an open dataset, named "Extensive COVID-19 X-ray, and the CT Chest Images Dataset" published on 12/06/2020 by Walid El-Shafai and Fathi Abd El-Samie [9]. This open dataset has been collected from multiple sources and augmented with different techniques to end up containing over 17,599 annotated samples of COVID images and non-COVID ones. The X-ray dataset contains 5,500 images of non-COVID and 4,044 images of COVID, and the CT images are divided into 2,628 non-COVID images and 5,427 COVID ones.

3.2 Experiment Settings

The DL architectures are developed using Python language and Keras library. The experiments are achieved with the hardware implementation of an AMD R5 3600 @3.6 GHz CPU, a GTX 1060 6 GB GPU, and 16 GB of RAM. For training and validation, we preprocessed the images by dividing them by 255. The adaptive moment estimation (Adam) optimizer is used for weight parameter learning. We also use a learning rate of 0.001. The number of epochs to train is set to 50 while an early stopping is implemented if the model does not improve for 20 epochs. The batch size is set to 32. All these hyper-parameters are fixed to ensure the convergence of the network.

3.3 Results and Discussion

The implementation has been done to show the effectiveness of the proposed architectures for the detection of COVID-19 cases. The training method will be a categorical classification that implements the K-Fold cross validation technique with $K = 5$. In this section, we present and discuss the implementation results. At the end, a comparison with the state of art is done.

Implementation Results. The training process is done on three different 5-fold cross validation sessions. The first session is applied only on the X-ray dataset. The second session includes only the CT dataset, and in the final session we use the combination of X-ray and CT datasets. Table 1 presents the average of the 5-fold cross validation sessions of accuracy, precision, recall and F1-Score. The results from this table indicate that our model achieves better results on the mixed session, which proves the capability of our model in classifying combined heterogeneous X-ray and CT chest images. The accuracy and precision can respectively achieve 0.9609 and 0.9780 on the combined dataset using the proposed DL architecture.

Table 1. Performance measures of DarkCovidNet-NRC architecture of the three sessions

Datasets	Accuracy	Precision	Recall	F1-score
X-ray	0.9521	0.9680	0.9078	0.9359
CT	0.9562	0.9724	**0.9122**	**0.9403**
Mixed (X-ray, CT)	**0.9609**	**0.9780**	0.9024	0.9358

Discussion and Comparison with State of the Art. We implement other architectures recently and successfully used in the literature for the COVID-19 detection in order to compare and locate our model. Table 2 shows the average of the 5-fold cross validation of accuracy, precision, recall and F1-Score using DarkCovidNet, Mobilenet-v2 and VGG19 architectures validated on a mixed dataset. We note that the proposed model can compete and outperform the other models in accuracy and precision metrics. This shows that this model has a great capability of classifying COVID-19 images. The proposed architecture permits designing better DarkCovidNet architecture in integrating the NRC blocks, which enables feature reusability and facilitates the propagation of information for better classification performances.

Table 2. Summary of state-of-the-art results

DL architecture	Accuracy	Precision	Recall	F1-Score
DarkCovidNet [5]	0.9551	0.9549	**0.9553**	**0.9551**
Mobilenet-v2	0.9222	0.9068	0.8679	0.8869
VGG19	0.8722	0.8471	0.8762	0.8614
DarkCovidNet-NRC	**0.9609**	**0.9780**	0.9024	0.9386

4 Conclusion and Perspectives

In this paper, we have introduced a novel DL architecture, DarkCovidNet-NRC, which integrates the NRC in DarkCovidNet model. Indeed, we have used a large dataset for the implementation of the K-fold cross-validation technique on X-ray and CT chest datasets separately and then combined. The implementation results of DarkCovidNet-NRC have improved the performance of the detection of COVID-19 in many metrics using the combined heterogeneous datasets. We note that the new architecture competes with the state of the art and outperforms the literature in some metrics. In the future, we will improve this architecture to realize the classification of many classes related to lung diseases. Furthermore, we can use this model to classify other diseases.

References

1. Nemati, M., Ansary, J., Nemati, N.: Machine-learning approaches in COVID-19 survival analysis and discharge-time likelihood prediction using clinical data. Patterns **1**(5), 100074 (2020)
2. Doanvo, A., Qian, X., Ramjee, D., Piontkivska, H., Desai, A., Majumder, M.: Machine learning maps research needs in COVID-19 literature. Patterns **1**(9), 100123 (2020)
3. Iwendi, C., Bashir, A.K., Peshkar, A., Sujatha, R., Chatterjee, J.M., Pasupuleti, S., Mishra, R., Pillai, S., Jo, O.: COVID-19 patient health prediction using boosted random forest algorithm. Front. Public Health **8**, 357 (2020)
4. Albadr, M.A.A., Tiun, S., Ayob, M., Al-Dhief, F.T., Omar, K.O., Hamzah, F.A.: Optimised genetic algorithm-extreme learning machine approach for automatic COVID-19 detection. PLOS ONE **15**, e0242899 (2020)
5. Ozturk, T., Talo, M., Yildirim, E.A., Baloglu, U.B., Yildirim, O., Acharya, U.R.: Automated detection of COVID-19 cases using deep neural networks with X-ray images. Comput. Biol. Med. **121**, 103792 (2020)
6. Khessiba, S., Blaiech, A.G., Ben Khalifa, K., Ben Abdallah, A., Bedoui, M.H.: Innovative deep learning models for EEG-based vigilance detection. Neural Comput. Appl. (2020)

7. Boudegga, H., Elloumi, Y., Akil, M., Bedoui, M.H., Kachouri, R., Ben Abdallah, A.: Fast and efficient retinal blood vessel segmentation method based on deep learning network. Comput. Med. Imaging Graph. **90**, 101902 (2021)
8. Alom, M.Z., Rahman, M.S., Nasrin, M.S., Taha, T.M., Asari, V.K.: COVID_MTNet: COVID-19 detection with multi-task deep learning approaches. arXiv preprint (2020)
9. Extensive COVID-19 X-ray, and the CT chest images dataset. https://data.mendeley.com/datasets/8h65ywd2jr/3. Accessed 15 Apr 2021

An Experiment Environment for Definition, Training and Evaluation of Electrocardiogram-Based AI Models

Nils Gumpfer[1], Joshua Prim[1], Dimitri Grün[2], Jennifer Hannig[1], Till Keller[2], and Michael Guckert[1](✉)

[1] Cognitive Information Systems, Kompetenzzentrum für Informationstechnologie, Technische Hochschule Mittelhessen, 61169 Friedberg, Germany
michael.guckert@kite.thm.de

[2] Department of Internal Medicine I, Cardiology, Justus-Liebig-University Gießen, 35390 Gießen, Germany

Abstract. The use of artificial intelligence (AI) for analysis of electrocardiogram (ECG) data has recently gained much interest in the AI and medical communities. The discussed models have shown to be able to deliver high diagnostic sensitivity and specificity for detection of various cardiac diseases including rhythm disorders and ischemic events. However, the experiments leading to these results are often difficult to reproduce outside of the original experimental setup and researchers who want to externally validate such results or use them as starting points for new experiments are forced to develop their own models from scratch. We therefore propose a software environment that enables to build, train and evaluate AI models for ECG classification in a reproducible manner and offers sharing of experiment configurations among researchers. The environment further provides simple connection of publicly available data sources of validated ECG recordings. It offers various validation techniques such as bootstrapping and cross-validation. A proof of concept is given for a deep learning model consisting of a convolutional neural network for the classification of acute myocardial infarction based on ECG data.

Keywords: Experiment · Environment · Reproducibility · Electrocardiogram · Keras

1 Introduction

The electrocardiogram (ECG) is a widely used diagnostic tool in cardiology providing comprehensive information about heart activity in a non-invasive way. Artificial intelligence (AI) based models have shown to leverage these information and to deliver results comparable to that of human interpreters, as we have shown in a recent meta-analysis [3] and previous research [4]. Practical applications of such models in wearable devices have demonstrated advances for early detection

A. Tucker et al. (Eds.): AIME 2021, LNAI 12721, pp. 384–388, 2021.
https://doi.org/10.1007/978-3-030-77211-6_45

of atrial fibrillation [1]. Development of AI models for analysis of ECG recordings has therefore gained much interest among researchers.

However, published results are often hard to reproduce outside of the original experimental setup, as source code is often not made available. This urges other researchers to start from scratch, leading to slightly different implementations, often yielding differing results from that originally published. Although some researchers share code and provide software environments for training of neural networks, these environments often do not support full reproducibility. We argue, that for this, the complete experiment setup (including data splits, data preprocessing steps, model architecture, training parameters, evaluation, and metric calculation) needs to be publicly available in form of configurations containing fully controllable parameters that, if left unchanged, will always deliver identical results. We therefore propose a software environment that enables to build, train and evaluate AI models for ECG classification in a reproducible manner and offers sharing of experiment configurations among researchers.

This paper is structured as follows: In Sect. 2, we describe the essential parts and processes of the environment that are necessary to define, train and evaluate your models. In Sect. 3, we demonstrate the usage of the environment at a use case using publicly available data from PTB-XL [5] and building an ECG based convolutional neural network for classification of myocardial infarction.

2 Components

2.1 Configuration

The parameter configuration of each conducted experiment should be fully transparent and controllable. Therefore, our environment works with central configuration files in INI format to configure the process chain from preprocessing to evaluation. This central configuration contains all parameters that are necessary to load, preprocess and split data, as well as to build, compile and train the model, and finally to evaluate its results. Experiment configurations can easily be copied to variate parameters in the course of experiment series. Furthermore, such configurations can be shared among researchers and results can easily be reproduced. Experiment configurations are structured into sections, grouping parameters by tasks, e.g. parameters for data preprocessing and those needed to build the model. A detailed description of all parameters is provided in the github repository (see *Code Availability* at the end of this paper).

2.2 Data Sources and Snapshots

The environment integrates access to PTB-XL [5] and other PhysioNet [2] databases through the waveform database (WFDB) API. Reproducibility of experiments requires version control. PhysioNet databases are versioned by default, but other data sources may not. Our environment therefore provides snapshot creation to save local, versioned working copies of the database contents (see Fig. 1). These snapshots are used as unified representations for further data processing. Snapshots contain records consisting of ECG data that may be enriched with additional clinical parameters that further describe its medical context (e.g. age and sex of patient, diagnoses). Users can also use local files as data source. Currently, CSV files and HL7-formatted XML files are supported. Additional clinical parameters can be provided in JSON format.

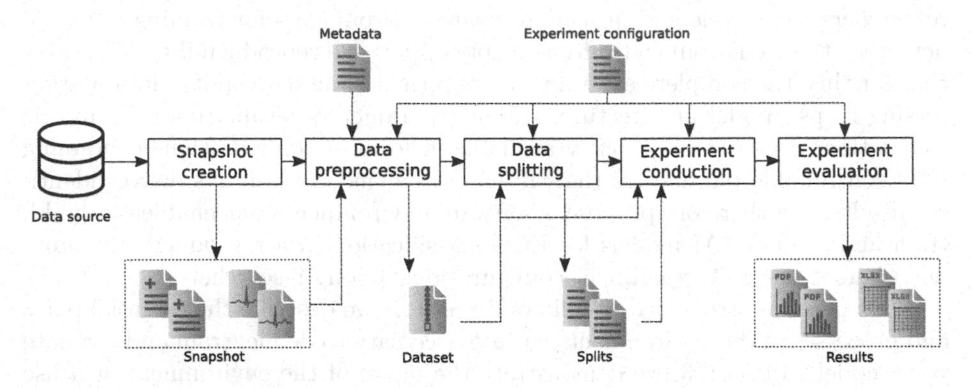

Fig. 1. Process overview. Snapshots are created from data sources and contain ECGs and clinical parameters. Data preprocessing is performed on snapshots, producing a dataset. The dataset is split into training, validation and test records. The model is compiled and records are loaded from the dataset for training and evaluation.

2.3 Data Preprocessing and Splitting

Preprocessing steps are performed on the snapshot. ECGs require preprocessing, including optional lead selection and value scaling. Clinical parameters need to be validated, cleaned, normalized, and transformed into one-hot-encodings before they can serve as input or label. These steps are controlled by rules defined in a metadata file in JSON format. This file is data source-specific and describes valid values or ranges for each parameter, as well as imputation and replacement rules, and specifies how one-hot encodings are performed. Based on these rules, the preprocessing becomes fully controllable and settings can easily be shared. Resulting from the preprocessing, all records are combined to a ready-to-use dataset in PICKLED format. Depending on the validation method chosen for the experiment, different data splits are necessary, e.g. k-fold cross-validation requires k different splits into training and validation records. To be able to reuse and share how the records were split, the environment persists these splits as JSON files. The following validation and split methods are currently supported by the environment: ratio-based split, k-fold cross-validation, n-bootstrapping, and repeated versions of each type. In case of repetitions, the training is repeated with a different weight initialization of the neural network, but with the same splits. Besides splitting into training and validation records, an additional test set can be held back in advance. The record IDs of the test set are then persisted to an additional split file. Stratification based on one target variable is possible.

2.4 Experiment Conduction

The conduction of an experiment involves data loading, model building, compilation, and training. At first, the prepared dataset is loaded and optional subsampling is applied. Subsampling increases the number of training samples and normalises their length. Model construction is based on model shells which inflate to their final shape through parameters specified in the experiment configuration. Model shells define a general model structure without specific layer numbers or options. Users can create their own model shells based on the prototype provided with the environment.

2.5 Experiment Evaluation

Performance metrics are logged during training for each epoch. When subsampling is used, the calculation of these metrics is performed on subsample and sample level. In the first case, the confusion matrix is generated based on all subsamples, each treated like a single record. In the second case, the predictions for all subsamples of one original sample are first averaged before the confusion matrix is generated. Based on the confusion matrix, relevant metrics can be calculated, e.g. sensitivity, specificity, area under receiver-operator characteristic curve (AUC), diagnostic odds ratio, F1-score, Youden's J-statistic, and accuracy. The metrics are calculated target class-specific, as defined in the experiment configuration. The best-performing models with respect to the target metric specified in the experiment configuration are selected. For k-fold cross-validation, this yields k models, and n models for n-bootstrapping. Sensitivity and specificity thresholds can be defined to filter results. Filtered results, mean metrics, histograms and boxplots for each metric, classification table, and receiver-operating characteristic curves can be exported to PDF and XLSX formats.

3 Proof of Concept

As proof of concept we use a simple convolutional architecture to classify ECGs with myocardial infarction and healthy controls from PTB-XL [5] and evaluate the performance based on bootstrapping ($n = 100$). The script `download_ptbxl.sh` creates a snapshot from PTB-XL. A configuration file `experiments/ptbxl_poc.ini` contains necessary general hyperparameters (e.g. optimizer, learning rate, loss function) and model-specific parameters (e.g. number of layers, number of neurons). To link all process steps, snapshot and dataset name, metadata file, and splits are defined as well. Further, required metrics and thresholds for evaluation are listed. We only use a subset (I-AVF) of 12 leads. In preparation for the experiment, the data is preprocessed via `python3 preprocessing_runner.py -e ptbxl_poc`. Based on the resulting dataset, 100 different bootstrapping splits are generated via `python3 split_runner.py -e ptbxl_poc`. The training is then started via `python3 experiment_runner.py -e ptbxl_poc`. After training, the evaluation is performed automatically.

The achieved performance metrics for AUC, sensitivity and specificity are 0.96 ± 0.01, $79.49 \pm 3.35\%$, $98.54 \pm 0.42\%$, respectively. All files related to the proof of concept can be found in the github repository (see *Code Availability*).

4 Conclusion and Future Work

The proposed environment enables researchers to share experiment configurations as well as intermediate results such as datasets. This enables research groups to exchange their data and configurations more easily and to mutually reproduce results. With this environment, we aim to support faster and more densely connected research. The presented concept is currently limited to AI models for ECG classification, but can be extended to other time-series based settings, and with some effort also to image-based concepts. Beside general improvements, we plan to extend the environment by methods for explainability and a graphical user-interface.

Code Availability
The code for the environment presented in this work is available from github at https://github.com/nilsgumpfer/experiment-environment-ecg-ai.

References

1. Bumgarner, J.M., et al.: Smartwatch algorithm for automated detection of atrial fibrillation. J. Am. Coll. Cardiol. **71**(21), 2381–2388 (2018). https://doi.org/10.1016/j.jacc.2018.03.003
2. Goldberger, A.L., et al.: PhysioBank, PhysioToolkit, and PhysioNet. Circulation **101**(23), e215–e220 (2000). https://doi.org/10.1161/01.CIR.101.23.e215
3. Grün, D., et al.: Identifying heart failure in ECG data with artificial intelligence-a meta-analysis. Front. Digital Health **2**, 67 (2020). https://doi.org/10.3389/fdgth.2020.584555
4. Gumpfer, N., Grün, D., Hannig, J., Keller, T., Guckert, M.: Detecting myocardial scar using electrocardiogram data and deep neural networks. Biol. Chem. (2020). https://doi.org/10.1515/hsz-2020-0169
5. Wagner, P., et al.: PTB-XL, a large publicly available electrocardiography dataset. Sci. Data **7**(1), 154 (2020). https://doi.org/10.1038/s41597-020-0495-6

Enhancing the Value of Counterfactual Explanations for Deep Learning

Yan Jia(✉), John McDermid, and Ibrahim Habli

University of York, York, UK
{yan.jia,john.mcdermid,ibrahim.habli}@york.ac.uk

Abstract. Counterfactual examples can be used to explain a specific clinical prediction from a deep learning model by identifying what kind of feature changes would produce a different result, i.e. flipping the prediction's classification. On-going research seeks to refine the metrics for discovering counterfactual examples, given a specific input to a deep learning model. Our work enhances this by using feature importance to reveal how much individual feature changes in the counterfactual example contribute to flipping the prediction's classification, compared with the original. Our approach does not depend on the specific metrics used for generating the counterfactual examples, so it is general. It can be used either to gain further insight when the counterfactual examples have already been generated or to influence the generation of the counterfactual examples. We illustrate this novel approach with a healthcare example.

Keywords: Explainability · Deep learning · Counterfactual examples

1 Introduction

Clinical predictions based on Machine Learning (ML) are having an increasingly profound impact on the safety and quality of healthcare services [10], e.g. by recommending treatments. Our focus in this paper is on using explainability for ML-based systems to assist a clinician in achieving a desired healthcare goal.

Much work on explainability for ML-based models focuses on feature importance explanations, which score or rank the input features, conveying the relative importance of each input feature to the model output (or prediction) [7]. However, this does not help model users to understand what they should do in order to achieve a desired goal. More recently, Wachter et al. [12] introduced counterfactual explainability which produces counterfactual examples that identify what changes in inputs to the ML model would be needed to reverse (or "flip") the ML model prediction. In this paper, we are interested in identifying changes in ML model inputs or patient conditions, that would enable a clinician to achieve a desired goal for a given patient.

Supported by the Assuring Autonomy International Programme.

A. Tucker et al. (Eds.): AIME 2021, LNAI 12721, pp. 389–394, 2021.
https://doi.org/10.1007/978-3-030-77211-6_46

The counterfactual examples should be close to the initial inputs to the model as smaller changes from the initial inputs are more likely to be achievable. Thus the approach should measure how far the predicted outcome of the counterfactual is from the desired outcome and the *distance* from the counterfactual to the initial input. When there are many features, searching for counterfactuals which combine changes in multiple inputs is computationally expensive, so it is necessary to find efficient solutions and to make simplifying assumptions, e.g. that the effects of small changes in inputs are additive in order to flip the prediction.

In some situations, it is desirable to provide a set of *diverse* counterfactuals, e.g. alternative changes in treatment, so that a user can choose which one to implement [5]. Our work builds on this idea and seeks to provide more insight into the different counterfactual examples. Specifically, our work enhances the value of counterfactual explanations for deep learning classifiers by revealing how much each input feature change in the counterfactual example contributes to flipping the decision. This novel combination of diverse counterfactual explanations and feature importance gives insight that enables users to choose which alternative to implement – thus making the ML models more actionable.

2 Background

Counterfactual explanations have been studied in philosophy and psychology and the work of Kahneman and Tversky in the 1970s and 1980s [4] presages many aspects of counterfactuals now addressed in ML. The introduction of counterfactual explanations for ML is more recent [12] but there is already some evidence that users prefer counterfactuals over feature importance methods [1].

Counterfactual explanations were formalised by Wachter et al. [12]. Generally, given an input x, an ML classifier f, and a distance metric d, a counterfactual explanation x' which produces the desired output y can be generated by solving the optimisation problem:

$$x' = argmin\{yloss(f(x'), y) + d(x, x')\} \tag{1}$$

where $yloss$ "pushes" the counterfactual x' towards a different classification than the initial input x, and the second term keeps the counterfactual x' close to the initial input x. There are four desirable properties for identifying good counterfactuals [7]. First, they should achieve the desired outcome as closely as possible, which is related to the first term in Eq. 1. Second, the counterfactuals should be as close as possible to the original instance, which is related to the second term in Eq. 1, i.e. the distance measure. Third, the counterfactuals should be *sparse*, i.e. an ideal counterfactual needs to change only a small number of features from the original instance. Fourth, it is desirable to have *diverse* counterfactuals. On-going research seeks to incorporate these properties in the loss function and optimisation methods. An overview of existing counterfactual explanation methods for ML is provided by Verma et al. [11].

3 Method

Our method combines feature importance with counterfactuals. Specifically, it uses DeepLIFT (Deep Learning Important FeaTures) [9] to assign a contribution score to each feature that changed in a counterfactual example. This can help users to understand how much individual feature changes in the counterfactual example contribute to flipping of the prediction's classification compared with the original instance. Where diverse counterfactual examples are available, the feature importance can help to choose between them.

DeepLIFT is an additive feature attribution method, developed specifically for use with deep neural networks (NNs). DeepLIFT compares the activation of each neuron for the input features of interest to its "reference activation" and attributes to each input a contribution score according to the difference. The "reference activation" is a user-defined reference input representing a background value. In order to enhance the value of counterfactual explanations, we assign a contribution score to each feature that changed in the counterfactual examples using DeepLIFT where the initial or original input features provides the "reference activation"[1].

We chose DeepLIFT because it compares the counterfactual examples to the initial instance and assigns the contribution scores according to the difference in the predictions. In addition, it considers both positive and negative contributions of features, hence identifying the sign of dependencies between the input features and the output. Further, the contribution score is generated by a single backwards pass through the NN so the scores can be generated efficiently.

If there are many features in the counterfactual example that have a very low contribution score, e.g. less than 1%, then that example might be discarded. This facilitates the identification of sparse counterfactual examples which is particularly important when choosing between diverse counterfactuals (see Sect. 2).

4 Clinical Example

In Intensive Care Units (ICUs), mechanical ventilation is a common intervention that consumes a significant proportion of ICU resources [13]. It is of critical importance to determine the right time to wean the patient from mechanical support. However, assessing a patient's readiness for weaning is a complex clinical task and it is potentially beneficial to use ML to assist clinicians [6]. Our example uses Convolutional NN (CNN) based on the MIMIC-III data set [3] to predict readiness for weaning in the next hour. 25 patient features are included in the model as shown in Table 1. The predicted outcome is the probability of weaning readiness in the next hour with 0.5 as the threshold (0 means wean; 1 means continue). The CNN architecture and details of this example can be found in [2].

We illustrate our method with a patient's record at a particular time as the original instance to generate the counterfactual examples using DiCE (Diverse

[1] The contribution score for the features that didn't change in the counterfactual examples is zero, due to the way DeepLIFT works.

Table 1. Counterfactual examples for a given original instance with contribution scores (shown in blue and in parentheses)

Features	Original instance	Counterfactual examples			
		1	2	3	4
Admit type	Emergency	—	—	—	—
Ethnicity	White	—	—	—	—
Gender	Female	—	—	—	—
Age	78.2	—	—	—	—
Admission weight	86.5	—	—	—	—
Heart rate	119	—	—	—	—
Respiratory rate	24	21.9 ($\leq$ 0.01)	—	24.1 ($\leq$ 0.01)	21.7 ($\leq$ 0.01)
SpO2	98	—	—	96 ($\leq$ 0.01)	—
Inspired O2 fraction	100	—	—	—	—
PEEP set	10	1.1 (−0.23)	9.2 ($\leq$ 0.01)	2 (−0.2)	5.1 (−0.12)
Mean airway pressure	14	—	15.2 ($\leq$ 0.01)	—	14.8 ($\leq$ 0.01)
Tidal volume (observed)	541	—	540.1 ($\leq$ 0.01)	541.9 ($\leq$ 0.01)	541.9 ($\leq$ 0.01)
PH (arterial)	7.46	—	7.49 ($\leq$ 0.01)	—	—
Respiratory rate(Spont)	0	—	13.1 (−0.06)	—	—
Richmond-RAS scale	−1	—	0 (−0.32)	—	2 (−0.37)
Peak Insp. pressure	21	—	—	—	—
O2 flow	5	—	7.3 (−0.01)	—	2.4 (0.02)
Plateau pressure	19	—	—	—	—
Arterial O2 pressure	124	123.6 ($\leq$ 0.01)	123.6 ($\leq$ 0.01)	123.6 ($\leq$ 0.01)	124.3 ($\leq$ 0.01)
Arterial CO2 Pressure	33	—	—	—	—
Blood pressure (systolic)	101	—	—	—	—
Blood pressure (diastolic)	65	—	—	—	—
Blood pressure (mean)	76	—	—	—	—
Spontaneous breathing trials	0	1 (−0.06)	1 (−0.06)	1 (−0.07)	—
Ventilator mode	18	9 (−0.38)	1 (−0.44)	1 (−0.52)	—
Predicted outcome	0.93	0.27	0.04	0.14	0.46

Counterfactual Explanation) [8]. DiCE can generate multiple diverse counterfactuals and works for any differentiable model. Thus it is widely applicable given the characteristics of commonly used deep learning methods. Four counterfactuals are shown in Table 1 along with the original instance, where "—" means the feature in the counterfactuals is not changed from the original instance. In order to enhance the value of the counterfactual examples, each changed feature in the counterfactuals is assigned with a contribution score (shown in parentheses and in blue) to gain insight into how much it contributes to flipping the prediction. For example, in Example 1, the sum of contribution score from changing PEEP set, Spontaneous breathing trials, and Ventilator Mode is 0.67, which is the difference between the original prediction and the new prediction. The changes of Respiratory Rate and Arterial O2 pressure in the counterfactual Example 1 contribute less than 1% each, which is negligible.

The benefit of adding the contribution score in the counterfactual examples is twofold. First, it can help the user to choose which example to implement. In our four counterfactual examples, Example 1 is attractive as it avoids a lot of unnecessary changes which make little contribution by comparison with the others, especially Example 2. Also, it helps the users to prioritise the changes with high contribution scores. Second, it can also help to generate sparse counterfactuals through post filtering. For example, we can add constraints that if the contribution score in the counterfactuals is less than 1%, then the feature is left unchanged. In counterfactual Example 4, when the features Respiratory Rate and Arterial O2 pressure are kept the same as the original input, the new prediction score is the same as the counterfactual Example 4 to two decimal places. Thus, this will improve the sparsity of the counterfactual.

5 Conclusion

We have introduced a novel method to enhance the value of counterfactual explanations by revealing how much individual feature changes in the counterfactual example(s) contribute to flipping the prediction's classification. Our method uses DeepLIFT to generate contribution scores for the features in the counterfactual examples. We illustrated the method to show how it can help in choosing between diverse counterfactuals generated by DiCE, potentially enabling identification of sparse counterfactuals to implement, i.e. making the counterfactual more readily actionable. Although we have used a specific healthcare example and DiCE for producing the counterfactual examples to illustrate the method, we believe it is general as it does not depend on the specific metrics used for generating the counterfactual examples. Future work will include exploration of further examples and more extensive assessment of the method in a clinical setting.

References

1. Fernandez, C., Provost, F., Han, X.: Explaining data-driven decisions made by AI systems: the counterfactual approach. arXiv preprint arXiv:2001.07417 (2020)
2. Jia, Y., Kaul, C., Lawton, T., Murray-Smith, R., Habli, I.: Prediction of weaning from mechanical ventilation using convolutional neural networks. Artif. Intell. Med. (2020 (Submitted))
3. Johnson, A.E., et al.: MIMIC-III, a freely accessible critical care database. Sci. Data **3**(1), 1–9 (2016)
4. Kahneman, D., Tversky, A.: The simulation heuristic. Stanford University CA Department of Psychology, Technical Report (1981)
5. Kunaver, M., Požrl, T.: Diversity in recommender systems-a survey. Knowl.-Based Syst. **123**, 154–162 (2017)
6. Kuo, H.J., Chiu, H.W., Lee, C.N., Chen, T.T., Chang, C.C., Bien, M.Y.: Improvement in the prediction of ventilator weaning outcomes by an artificial neural network in a medical ICU. Respir. Care **60**(11), 1560–1569 (2015)
7. Molnar, C.: Interpretable Machine Learning. Lulu. com, Morrisville (2020)
8. Mothilal, R.K., Sharma, A., Tan, C.: Explaining machine learning classifiers through diverse counterfactual explanations. In: Proceedings of the 2020 Conference on Fairness, Accountability, and Transparency, pp. 607–617 (2020)
9. Shrikumar, A., Greenside, P., Kundaje, A.: Learning important features through propagating activation differences. In: International Conference on Machine Learning, pp. 3145–3153. PMLR (2017)

10. Topol, E.J.: High-performance medicine: the convergence of human and artificial intelligence. Nat. Med. **25**(1), 44–56 (2019)
11. Verma, S., Dickerson, J., Hines, K.: Counterfactual explanations for machine learning: a review. arXiv preprint arXiv:2010.10596 (2020)
12. Wachter, S., Mittelstadt, B., Russell, C.: Counterfactual explanations without opening the black box: automated decisions and the GDPR. Harv. J. Law Technol. **31**, 841 (2017)
13. Wunsch, H., Wagner, J., Herlim, M., Chong, D., Kramer, A., Halpern, S.D.: ICU occupancy and mechanical ventilator use in the United States. Crit. Care Med. **41**(12), 2712–2719 (2013)

Natural Language Processing

A Multi-instance Multi-label Weakly Supervised Approach for Dealing with Emerging MeSH Descriptors

Nikolaos Mylonas(✉), Stamatis Karlos, and Grigorios Tsoumakas

Department of Informatics, Aristotle University of Thessaloniki, 54124 Thessaloniki, Greece
{myloniko,stkarlos,greg}@csd.auth.gr

Abstract. The constant evolution of Medical Subject Headings (MeSH) vocabulary and specifically the changes in its descriptors brings forth a number of issues that need automation. The main one being that changed descriptors often lack proper ground truth articles. Therefore, the learning models which demand strong supervision are not directly applicable, settling the predictions on such changes not a straightforward task. The importance of this problem is also enforced by its multi-label nature and the fine-grained character of the examined class-descriptors, factors that demand a lot of human resources. In this work, we alleviate these issues through retrieving insights from a source of information about those descriptors present in MeSH in order to create a weakly-labeled train set. Furthermore, we exploit short-text information per article, implementing an averaging transformation on the corresponding sentence embeddings, applying a similarity mechanism for assigning weak-labels to our formatted data set, thus we named our approach **WeakMeSH**. The benefits of applying the proposed end-to-end approach are examined on a large-scale subset of the BioASQ 2018 data set consisting of 900 thousand instances, investigating two separate groups of MeSH changes: brand new and complex changes. Our performance tested on BioASQ 2020 data set against several other approaches that can either distill weak information on their own or apply alternative transformations against the proposed one was proven highly competitive.

Keywords: Weakly supervised learning · MeSH indexing · Multiple-instance learning · Sentence and word embeddings · Similarity threshold tuning

1 Introduction

MEDLINE contains more than 26 million citations to journal articles related mainly to biomedicine and more generally to life sciences. A key property of MEDLINE is that articles are indexed with an average of 13 out of the more than 28,000 descriptors of the Medical Subject Headings (MeSH)[1] thesaurus.

[1] https://www.nlm.nih.gov/mesh/meshhome.html.

A. Tucker et al. (Eds.): AIME 2021, LNAI 12721, pp. 397–407, 2021.
https://doi.org/10.1007/978-3-030-77211-6_47

This enables semantic search and retrieval of articles. However, it comes at a significant cost in time and money, as indexing is mainly a manual process conducted by human experts. Therefore a lot of research has been devoted towards methods and tools for supporting indexers in accomplishing their task faster and better [3,9,12,17], with the BioASQ challenge being an important driving force of this progress [1].

MeSH is not set in stone. On the contrary, it changes all the time, in accordance with the evolution of our biomedical knowledge. Yearly MeSH updates include the introduction of new MeSH descriptors, the withdrawal of existing ones, updates in the hierarchical structure of existing descriptors and even more subtle changes involving the concepts and terms that are associated with the descriptors [17]. This paper focuses on the new MeSH descriptors that arise each year and the challenge they introduce to supervised machine learning models, due to the lack of training examples.

To address this challenge we propose a novel approach for obtaining weak supervision, called *WeakMeSH*, that: i) takes advantage of label provenance knowledge that is available within the meta-data of MeSH in order to focus on the most relevant MeSH articles for each new descriptor, ii) employs a multi-instance representation for these articles by considering state-of-the-art embeddings for each sentence of their abstract, and iii) weakly labels these articles based on the maximum similarity of each descriptor across all sentences of their abstract, thresholded by an unsupervised component.

In addition, we contribute a real-world multi-instance multi-label benchmark data set with new labels suitable for experimentation with weakly supervised (WSL), as well as zero-shot (ZSL) learning methods. The majority of such existing methods use data sets, where new labels are constructed by removing the ground truth of existing labels and/or concern easily separable classes compared to the existing ones [18]. Such *artificial* data sets are often aligned with core assumptions of the corresponding proposed methods. This is far from the actual real-world case of MEDLINE that we contribute here, where new descriptors of a naturally evolving thesaurus are not easily separable from existing ones.

To deal with this benchmark we pair our weak labeling approach with simple multi-instance (average of the sentences embeddings of the article) and multi-label (binary relevance) transformations. Experimental results against state-of-the-art weakly supervised methods for text classification are promising. Our approach, data set and experiments are available online[2].

The rest of this paper is organized as follows. Section 2 provides a notation of the tackled problem and summarizes some recently demonstrated state-of-the-art WSL works. Next, we discuss the separate stages of our proposed method, while Sect. 4 reveals information about the exploited data set. The experimental procedure and the produced results are placed in Sect. 5, before we conclude and propose some future directions in Sect. 6.

[2] https://github.com/intelligence-csd-auth-gr/WeakMeSH.

2 Related Work

To the best of our knowledge not much research has been conducted about how the yearly changes in MeSH affect existing article annotations. One such work is [2], where 14 different versions of MeSH thesaurus were used to annotate 5000 random articles and then those annotations were used to calculate the difference between the indexing of articles for any 2 successive years. Their findings suggest that changes in MeSH versions have a big impact in article indexing even if the changes in MeSH are minor ones, setting the problem of biomedical article annotation as a very challenging one.

Three distinct categories of WSL are discussed in [19]: Incomplete, inexact and inaccurate supervision. The first one concerns situations where there are many unlabeled data, but not enough labeled data to train a good model. The second one concerns situations where the supervised information is inexact, as in the case of having a label for a bag of instances, instead of a single instance. The last one concerns situations where noisy information is present in the feature and/or target space. Our examined problem can be categorized as a hybrid one between the last two categories. This is due to the fact that we treat each MEDLINE article as a bag of separate sentences to obtain our weakly-labels based on a stochastic process that may inject noise on the target space.

Focusing on the WSL literature concerning textual data, we distinguish two different manners of tackling this kind of learning: i) extrapolate the semantic meaning of classes into the Label space (Y) for creating new instances, ii) learn a predictive function between the input space (X) and Y, based on noisy training examples, that can still generalize well on unseen data. We point out the most recent related approaches.

Exploitation of seed words is the most popular strategy regarding the first of the aforementioned categories. This external knowledge source, which may be provided by users without necessarily much expertise, is usually directly available and can trigger weak supervision over unlabeled documents. The main ambition here is the augmentation of the instances that are related to each label due to the scarcity of available training instances. A dataless approach based on self-training, seed words occurrence and bayesian models was proposed in [6], while a more recent work that employs self-training based on DNNs is found in [7]. That approach, called *WeST Class* (Weakly Supervised Text Classification), exploits label descriptors, class-related keywords or beforehand labeled documents for fitting a class-distribution model. These models facilitate the generation of pseudo documents over which the DNNs – either convolutional or recurrent variants – are trained before the assignment of probabilistic labels to the unknown test set takes placed on a transductive fashion. Both of these works have been applied to data sets with limited labels (2, 4, 10 or 20) which are mainly relevant to news articles (contain labels like politics, sports etc.). Seed-guided solutions seem effective for easily discriminated entities, but this does not always hold on coarse-grained label spaces such as MeSH.

The second category consists of methods that try to learn a mapping function under the existence of noisy instances. Snuba [16] assumes that a small set of labeled data is provided, together with a large set of unlabeled data. It iteratively labels the unlabeled data probabilistically using multiple simple classifiers trained from the labeled data. These classifiers consider different small subsets of the features (typically less than 4) and are based on standard algorithms (decision stump, logistic regression, k nearest neighbors). A multi-instance approach for predicting aspect ratings in reviews was presented in [13]. Each review was represented as a bag of sentences. The key idea in this approach is to represent each bag as a weighted average of the representations of the sentences. A regularized regression model is used to jointly learn these weights together with the parameters for predicting the ratings of a review aspect from the weighted average of the sentences. Cost-sensitive classifiers were employed in [4], in the sense of learning accurate models under the existence of weakly annotated instances. The latter process is based on the decisions of a committee of learning functions or algorithms from which the relative costs are generated by an unsupervised method.

3 Obtaining Weak Supervision with WeakMeSH

With each yearly update of MeSH, a number of new descriptors get introduced. In some cases, such as when a supplementary concept record (SCR) of MeSH is being promoted to a MeSH descriptor, existing MEDLINE articles get automatically re-indexed with these descriptors. In other cases however, new descriptors come without ground truth annotations and we cannot use supervised machine learning algorithms.

To address this issue, we introduce **WeakMeSH**, an approach to obtain weak labels for such new descriptors. Our approach takes as input a data set of biomedical abstracts from MEDLINE and a set of new MeSH descriptors, for which there is no ground truth annotation in the data set. **WeakMeSH** weakly labels biomedical articles in two stages: i) candidate labels generation based on descriptor provenance knowledge, ii) label filtering based on multi-instance semantic similarity.

3.1 Candidate Labels Generation

For each biomedical article, each of the new descriptors is theoretically a candidate for weak labeling. Typically, a measure of semantic similarity between the article and the descriptors is employed for assigning the weak labels [3]. We also do this in the second stage of **WeakMeSH**. However, given the complex hierarchically organized biomedical knowledge of MeSH, we employ a novel knowledge-based first stage that considers a subset of the new descriptors, based on provenance information found in the meta-data of MeSH [11]. This information points to existing descriptors that were associated with the meaning of a new descriptor in the past. In particular, we consider the following two fields in the records of new MeSH descriptors:

- Previous Indexing (PI), refers to one or more older descriptors used for indexing articles that *could* be relevant to the new descriptor in previous years. Being indexed with a PI is a necessary, but not sufficient, condition for an article to be considered relevant to the new descriptor. Note also that this field is not present in every new descriptor.
- Public Mesh Note (PMN), refers to an old descriptor that is related in some way to the newly introduced one. This can be through a parent-child relation in the MeSH tree hierarchy, the novel descriptor previously being a SCR for the old one or by having similar meanings. The presence of this field in a new descriptor, signifies that it was already present inside the MeSH vocabulary, but not as a descriptor.

As an example, existing descriptors *Sexuality* and *Reproductive Health* could be hosting the meaning of the new descriptor *Sexual Health* that was introduced in 2018, with the former being a PI and the latter a PMN of *Sexual Health*[3].

For each biomedical article, we consider as candidate weak labels those new descriptors, whose PI(s) or PMN appear in the ground truth annotations of the article.

3.2 Multi-instance Semantic Similarity

Since each article is not always related to its PI(s) and PMN, assigning every candidate weak label to that article would introduce a lot of label noise. To deal with this issue, the second stage of **WeakMeSH** considers the semantic similarity of each article, with each candidate weak label.

In particular, we employ BioBERT [5], a variant of the BERT language model fine-tuned on biomedical data with state-of-the-art results in several downstream tasks. BioBERT produces embedding vectors in $\mathcal{R}^{768}$ for both words and sentences. We obtain a word or sentence embedding for each new descriptor, depending on the number of words it contains. For the articles, we follow a multi-instance paradigm, treating the abstract of each article as a bag of sentences and obtaining one embedding per sentence. Multi-instance representations are particularly useful for multi-label data [20], as in our case, since each sentence may be associated with a different descriptor.

Given the multi-instance representation of the abstract of an article as a set of sentences S, along with a set of candidate weak labels C, **WeakMeSH** computes the cosine similarity between the embeddings of each sentence $s \in S$ and the embedding of each candidate label $c \in C$. For each candidate label $c \in C$ we take the maximum of the computed similarities across all sentences in S. A candidate label is then considered as weak label if this maximum similarity is above a threshold, t. Equation 1 shows formally the final set of weak labels.

$$\{c \in C : \max_{s \in S} cosine\left(BioBERT(c), BioBERT(s)\right) > t\} \tag{1}$$

[3] https://meshb.nlm.nih.gov/record/ui?ui=D000074384.

The requirement of a similarity threshold is considered as a weak point for end-to-end AI tools. Arbitrarily set thresholds are usually provided by human users or estimated through applying cross-validation procedures. None of these approaches are acceptable in our case, due to the typically large number of new descriptors and the shortage of ground truth instances.

To avoid this, we use a novel approach based on Gaussian Mixture Models (GMMs) [14], in order to automatically calculate a separate threshold t for each new descriptor. We first compute the maximum similarity of the embedding of each new descriptor with the embeddings of the sentences of each one of the articles that were indexed with at least one of its PIs or PMN. We assume that these maximum similarities are coming from two populations, one for relevant and one for irrelevant articles with respect to the descriptor. We therefore fit a GMM with two components on the distribution of these maximum similarities. Finally we take as threshold the average of the two means of the corresponding sub-populations (see bottom right part of Fig. 1).

3.3 Multi-instance Multi-label Learning from Weak Supervision

After obtaining weak labels for the articles, we proceed with a simple transformation of the multi-instance article representation to a single-instance one. In particular, we represent each of the articles with a vector in $\mathcal{R}^{768}$ computed through the average of the BioBERT embeddings of its sentences. Using this representation strategy we can then employ any standard multi-label learning algorithm to learn a model that will be able to predict the new descriptors in new articles. Figure 1 depicts the overall architecture of such an approach building upon **WeakMeSH** to obtain weak labels.

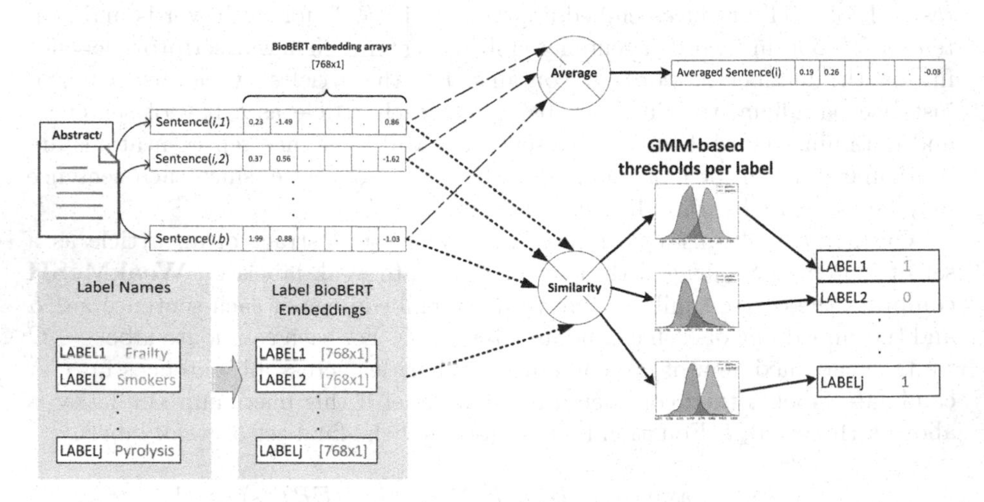

Fig. 1. Distillation of input and label space by WeakMeSH for creating weakly labeled instances

4 A Real-World Benchmark for Weakly Supervised Learning

The data set that we contribute and use in our experiments comes from the BioASQ challenge[4], more specifically the BioASQ 2018 and BioASQ 2020 data sets, with the former being used for training and a part of the latter for testing. These data sets contain articles published up to their corresponding year. Furthermore they use the MeSH vocabulary of the same year. The reason for choosing the 2018 and 2020 data sets instead of the 2018 and 2019 ones, is that many of the new descriptors introduced in 2019 are not present in the BioASQ 2019 data set and thus we would not be able to fully assess our method's accuracy.

Since our method focused on novel descriptors that are not automatically indexed in existing articles, we had to single out those specific ones from the list of all new descriptors between the aforementioned years. To do so, we searched for new descriptors that appear as labels on articles present in BioASQ 2020 that are absent in BioASQ 2018. In total, 450 novel descriptors were found. Out of them, 399 are completely new ones, while the rest 51 are produced by some type of complex change. This means that the participant labels of the former subset appear for the first time in the MeSH 2019 or MeSH 2020 vocabulary, whereas the corresponding labels of the latter subset were previously a part of the vocabulary, but not as descriptors. For computational simplicity, we decided to focus on the top 100 most frequent new descriptors on the test set, since their appearances sum up to 44,938 out of the 57,582 of all appearances (78%), leaving us with 88 that appeared for the first time into the last variant of MeSH (*brand new*), and 12 who became descriptors by a more complicated procedure (*complex change*).

After an appropriate discarding stage, where we only keep the descriptors that have at least one PI or PMN, we were left with 62 final descriptors used for our experiments. All the removed labels belong to the *brand new* group, since the PMN field is always available for the labels in the *complex change* group. Using the PI(s) and PMN for each one of the 62 new descriptors we singled out articles labeled with at least one of them (previous host data set). The final data set consisted of 900,000 labeled articles from BioASQ 2018. For the test set, we found 32,908 articles from the BioASQ 2020 version labeled with at least one of the 62 new descriptors. Furthermore, our test set is imbalanced – since the individual frequency of several labels is quite scarce – putting additional obstacles towards accurate predictions.

5 Experimental Setup and Results

This Section discusses the experiments we performed to evaluate our proposed method along with the produced results. We compare these results with those

[4] http://bioasq.org/.

of two state-of-the-art approaches for WSL, namely WeST [7] and WMIR [13], as well as a related ZSL method introduced in our previous work [10]. Note that one domain-independent zero-shot learning method called EZSL [15] was also implemented, as well as a variation of our method that completely disregards the information about PI and PMN, but the results they yielded were very low and as such won't be shown here. Similar behavior was recorded by Snuba [16], whose demand of accurate train data prevent it from achieving competitive performance. We also used two more strategies for representing our training data that can be directly compared to our own representation.

1. **Prime:** Each article is represented as the BioBERT embedding of its abstract's sentence(s) with the maximum similarity to each examined descriptor(s). This way the most salient part of a text segment is selected to represent the total entity. A similar competitor was used in [13]. More than one instances may occur during this stage from distilling each abstract, or even none depending on the number of sentences that pass the t mentioned during the weakly-labeling stage. When drawing a decision for an unknown instance with this approach, the final prediction is the combination of the predictions for each one of its sentences that surpass an arbitrarily defined confidence threshold (*Conf*).
2. **Extended Prime:** Each article is represented into the $\mathcal{R}^{1536}$ space, concatenating two BioBERT Embeddings. The first is found as in **Prime**, while the second corresponds to the averaged embedding transformation of the remaining sentences per abstract. The training and predicting procedure takes place as previously.

Any probabilistic classifier can be combined with the proposed method, *Prime* and *Extended Prime*. The information about those probabilities is then used to measure the confidence of the predictions and extract the final decisions. It is worth noting we only show the best results produced by the Logistic Regression (*LogReg*) among some linear and bayesian classifiers that were examined concerning all the three of them. For the two last approaches, the best results were achieved for $Conf = 0.70$.

For dealing with the multi-label nature of our data set, we decided to use the well known problem transformation method called Binary Relevance. The reason for this choice is twofold. First the simplicity of the method along with its ease of use made it an adequate choice for our approach. Seeing that our main goal is to propose a method for finding possible relevant train examples for new MeSH descriptors in older articles, along with ways of representing the large abstracts without convoluting the information present in them and not delve too deep on how these examples will be used for training. Second we wanted to showcase that even a simple multi-label classification approach trained on our produced weakly-labeled train set, can still compete with other more complex state-of-the-art approaches.

We also discuss here some implementation details of the rest compared algorithms. For *WeST Class*, the authors had 2 sets of parameters in their original paper, with 3 different modes for obtaining weakly labeled examples. For the above reason we only show the produced results for the best included combination (parameters setting: 'agnews', input mode: 'label', model: 'CNN'). In case of *WMIR*, we tried to adjust the included hyperparameters for avoiding over/under-fitting phenomena, training one model per label providing the weakly-labeled data set that was created from the proposed approach. We also balanced the training data set per label boosting its performance.

The macro averaged F1 score of all methods participating in our experiments can be found in Table 1. As we can see our method clearly outperforms the other approaches as far as predictive ability goes. This can be attributed to the fact that by using provenance knowledge about the new descriptors we reduce the inherent noise of the collected train data set. This fact, in combination with the averaged embedding approach that aggregates equivalently each abstract's sentence, let us to represent all the relevant information of the article efficiently. In contrast, *WeST* class creates weakly labeled examples based solely on the test set, thus ignoring information like PI and PMN which is present inside the previous host data set. This leads to a smaller number of train examples which proves to be inefficient in cases with a large amount of different labels such as our own. In case of *WMIR*, they chose to represent each article as the weighted sum of its sentences with the weights being assigned based on a regression problem solved for each bag of sentences and the bag's weakly assigned label. This approach is inherently single-label thus when applied to a multi-label problem it can lead to a large amount of 'noisy' weights thus reducing overall performance. Moreover, one main point of this work was to be interpretable, which seems to sacrifice some of its predictive ability when faced with complex Label spaces. The performance of *Extended Prime* recorded a slight improvement over *Prime*, though needing more computational resources. In total, the strategy of **WeakMeSH** to average each bag of sentences for handling the weakly-annotated input instances seems to bridge the gap between train and test sets.

Table 1. Comparison results based on F1-Score (Macro) performance metric

Approach	Macro-averaged F1 score		
	All	Brand new	Complex change
WeakMeSH	**0.532**	**0.501**	**0.14**
Extended prime	0.452	0.439	0.115
Prime	0.444	0.433	0.12
WeST class [7]	0.322	0.307	0.091
ZSLbioSentMax [10]	0.303	0.294	0.093
WMIR [13]	0.26	0.258	0.078

We should mention that since out of the 62 descriptors only 11 of them were "complex change" ones, the number of instances in our test set with them as labels was pretty small compared to the "brand new" subset. As a result we cannot draw definite conclusions on each method's performance concerning that subset.

6 Conclusions and Next Steps

We investigated the use of weakly-supervised learning for MeSH indexing, where finding ground truth data for emerging descriptors is not always feasible. To that end we proposed the use of explicit provenance information to aid in detecting possible relevant data for each new descriptor from past MEDLINE articles. We also presented a semantic similarity-based approach for measuring the relatedness of the detected data with their relevant novel descriptors and assign weak labels. This approach treats each MeSH article as a bag of sentences and measures the similarity for each of them separately, before averaging these sentences in order to represent each article inside our weakly-labeled training set. For facilitating our experiments, we sampled a large-scale data set that satisfies the conditions which accompany a real-world multi-instance multi-label problem with an evolutionary behavior. This is included in our repository, providing thus a benchmark data set to AI and ML communities. The produced results show that our approach outperforms similar weakly-supervised learning methods that do not make use of provenance information as well as approaches that use the same weakly-labeled training set we created but represent the data differently.

Of course, this work does not come without limitations. The underlying relationship between labels are not exploited for reduction of noisy annotations, as well as other hierarchical information that categorize each label on an initial fine-level. Therefore, the computation of anchors/prototypes per separate label indicator for reducing the effect of the noise of the weakly annotated instances should be examined as future work [8].

Acknowledgements. The research work was supported by the Hellenic Foundation for Research and Innovation (H.F.R.I.) under the "First Call for H.F.R.I. Research Projects to support Faculty members and Researchers and the procurement of high-cost research equipment grant" (Project Number: 514)

References

1. Balikas, G., Krithara, A., Partalas, I., Paliouras, G.: BioASQ: a challenge on large-scale biomedical semantic indexing and question answering. In: Müller, H., Jimenez del Toro, O.A., Hanbury, A., Langs, G., Foncubierta Rodríguez, A. (eds.) Multimodal Retrieval in the Medical Domain. LNCS, vol. 9059, pp. 26–39. Springer, Cham (2015). https://doi.org/10.1007/978-3-319-24471-6_3
2. Cardoso, S.D., et al.: Leveraging the impact of ontology evolution on semantic annotations. In: Blomqvist, E., Ciancarini, P., Poggi, F., Vitali, F. (eds.) EKAW 2016. LNCS (LNAI), vol. 10024, pp. 68–82. Springer, Cham (2016). https://doi.org/10.1007/978-3-319-49004-5_5
3. Dai, S., You, R., Lu, Z., Huang, X., Mamitsuka, H., Zhu, S.: FullMeSH: improving large-scale MeSH indexing with full text. Bioinform **36**(5), 1533–1541 (2020)

4. Jain, S., R., K., Kuo, T., Bhargava, S., Lin, G., Hsu, C.: Weakly supervised learning of biomedical information extraction from curated data. BMC Bioinform. **17**(S-1), 1–12 (2016)
5. Lee, J., et al.: BioBERT: a pre-trained biomedical language representation model for biomedical text mining. Bioinformatics (2019)
6. Li, X., Yang, B.: A pseudo label based dataless Naive Bayes algorithm for text classification with seed words. In: Proceedings of the 27th International Conference on Computational Linguistics. pp. 1908–1917. ACM, Santa Fe, New Mexico, USA, August 2018
7. Meng, Y., Shen, J., Zhang, C., Han, J.: Weakly-supervised neural text classification. In: Cuzzocrea, A., et al. (eds.) CIKM, pp. 983–992. ACM (2018)
8. Mikalsen, K.Ø., et al.: Using anchors from free text in electronic health records to diagnose postoperative delirium. Comput. Meth. Programs Biomed. **152**, 105–114 (2017)
9. Mork, J., Aronson, A., Demner-Fushman, D.: 12 years on - is the NLM medical text indexer still useful and relevant? J. Biomed. Semant. (2017). https://doi.org/10.1186/s13326-017-0113-5
10. Mylonas, N., Karlos, S., Tsoumakas, G.: Zero-shot classification of biomedical articles with emerging mesh descriptors. In: 11th Hellenic Conference on Artificial Intelligence, pp. 175–184. SETN 2020. Association for Computing Machinery, New York, NY, USA (2020)
11. Nentidis, A., Krithara, A., Tsoumakas, G., Paliouras, G.: What is all this new mesh about? exploring the semantic provenance of new descriptors in the mesh thesaurus (2021)
12. Papanikolaou, Y., Tsoumakas, G., Laliotis, M., Markantonatos, N., Vlahavas, I.: Large-scale online semantic indexing of biomedical articles via an ensemble of multi-label classification models. J. Biomed. Semant. **8**(1), 1–13 (2017). https://doi.org/10.1186/s13326-017-0150-0
13. Pappas, N., Popescu-Belis, A.: Explicit document modeling through weighted multiple-instance learning. J. Artif. Intell. Res. **58**, 591–626 (2017). https://doi.org/10.1613/jair.5240
14. Reynolds, D.: Gaussian Mixture Models. Encycl. Biometrics, **741**, 659–663 (2009)
15. Romera-Paredes, B., Torr, P.H.S.: An embarrassingly simple approach to zero-shot learning. In: Bach, F.R., Blei, D.M. (eds.) ICML, Lille, France. JMLR Workshop and Conference Proceedings, vol. 37, pp. 2152–2161. JMLR.org (2015)
16. Varma, P., Ré, C.: Snuba: automating weak supervision to label training data. Proc. VLDB Endow. **12**(3), 223–236 (2018)
17. Xun, G., Jha, K., Zhang, A.: MeSHProbeNet-P: improving large-scale MeSH indexing with personalizable MeSH probes. ACM Trans. Knowl. Discov. Data **15**(1), 14 (2020)
18. Yin, W., Hay, J., Roth, D.: Benchmarking zero-shot text classification: datasets, evaluation and entailment approach. In: Inui, K., Jiang, J., Ng, V., Wan, X. (eds.) EMNLP-IJCNLP, pp. 3912–3921. ACM (2019)
19. Zhou, Z.H.: A brief introduction to weakly supervised learning. Nat. Sci. Rev. (2018). https://doi.org/10.1093/nsr/nwx106
20. Zhou, Z.H., Zhang, M.L., Huang, S.J., Li, Y.F.: Multi-instance multi-label learning. Artif. Intell. **176**(1), 2291–2320 (2012). https://doi.org/10.1016/j.artint.2011.10.002

Demographic Aware Probabilistic Medical Knowledge Graph Embeddings of Electronic Medical Records

Aynur Guluzade(✉), Endri Kacupaj, and Maria Maleshkova

University of Bonn, Bonn, Germany
s6ayguluu@uni-bonn.de, {kacupaj,maleshkova}@cs.uni-bonn.de

Abstract. Medical knowledge graphs (KGs) constructed from Electronic Medical Records (EMR) contain abundant information about patients and medical entities. The utilization of KG embedding models on these data has proven to be efficient for different medical tasks. However, existing models do not properly incorporate patient demographics and most of them ignore the probabilistic features of the medical KG. In this paper, we propose DARLING (**D**emographic **A**ware p**R**obabi**L**istic med**I**cal k**N**owledge embeddin**G**), a demographic-aware medical KG embedding framework that explicitly incorporates demographics in the medical entities space by associating patient demographics with a corresponding hyperplane. Our framework leverages the probabilistic features within the medical entities for learning their representations through demographic guidance. We evaluate DARLING through link prediction for treatments and medicines, on a medical KG constructed from EMR data, and illustrate its superior performance compared to existing KG embedding models.

Keywords: Demographics · Probabilistic medical knowledge graph · Knowledge graph embedding · Electronic medical records

1 Introduction

In recent years, knowledge graphs (KGs) have been established for medical assistance as the underlying core component of clinical decision support systems (CDSSs) [16,21] and self-diagnostic symptom checkers [14]. Those KGs are often extracted from sources such as Electronic Medical Records (EMR) and store clinical information into a set of triples for representing medical entities as nodes and relations as the edges between them. Medical KG-based applications have been reported in different scenarios, such as treatment recommendations [3], medicine recommendations [7], and drug-to-drug similarity measurements [2]. Those applications usually are performed through a link prediction process and can be divided into two steps: 1) learn embeddings of medical entities and relations, 2) make predictions/recommendations according to these embeddings.

A. Tucker et al. (Eds.): AIME 2021, LNAI 12721, pp. 408–417, 2021.
https://doi.org/10.1007/978-3-030-77211-6_48

Several approaches rely on medical KG embeddings for recommendation tasks, work from Gong et al. [7] proposed a medicine recommendation framework that embeds medical entities such as diseases, medicines, patients, and their corresponding relations into a shared lower dimensional space. The authors use the embeddings to decompose the task into a link prediction process while considering the patient's diagnosis and adverse drug reactions. Chen et al. [3] proposed a framework operating on medical KG for thyroid treatment recommendation with cold start based on TransD [10] and network embeddings with a hierarchical structure.

While recent research [2,3,7,12] employs traditional KG embedding methods [1,10,13,18] as a first step for representing patients and medical entities, it lacks the consideration of demographic meta-data. However, such information is very advantageous and even necessary for medical tasks. Several works [9,17,20] have investigated and shown the importance and development of demographics in different medical tasks and challenges. Considering this, we argue that incorporating demographics as part of medical KGs allows us to retain patients' generic information and construct more accurate representations for medical entities.

Hence, in this paper, we propose DARLING (**D**emographic **A**ware p**R**obabi**L**istic med**I**cal k**N**owledge embeddin**G**) – the first demographic-aware medical KG embedding framework that explicitly incorporates demographics in the medical entities space by associating patient demographics with a corresponding hyperplane. Our framework leverages probabilistic features within medical entities for learning their representations through demographic guidance. We evaluate DARLING on link prediction for treatments and medicines, where it achieves improved results compared to multiple existing KG embedding models on standard metrics. Furthermore, for evaluation purposes, we construct a medical KG from the MIMIC-III [11] data, which comprises medical elements relating to patient admissions, such as demographics, disease diagnosis, treatment procedures, etc. The medical KG contains diseases, treatments and medicines, where our construction method automatically links all extracted entities with existing biomedical knowledge graphs, including ICD-9 [15] ontology and DrugBank [19]. With our work we make the following key contributions to the state of the art:

- We propose DARLING, the first demographic-aware framework for learning probabilistic medical KG embeddings.
- We provide a method to construct a medical KG with demographics metadata that links all extracted entities with existing biomedical knowledge graphs.
- We demonstrate DARLING's effectiveness through extensive experiments and show its superior performance via link prediction on treatments and medicines. We also illustrate the sensitivity of different demographic categories in our framework.

To facilitate reproducibility and reuse our framework implementation, alongside the medical knowledge graph construction method, the results are also publicly available[1].

[1] https://github.com/AynurGuluzade/DARLING.

The rest of the paper is structured as follows: Section 2 summarises the related work and Sect. 3 presents the proposed DARLING framework. Section 4 describes the experiments, including the experimental setup and the evaluation. Finally, we conclude in Sect. 5.

2 Related Work

Our work lies at the intersection of medical KG embeddings and graph-based embedding approaches that employ patient demographics. In this section, we describe previous efforts and refer to different approaches.

Current efforts on knowledge graphs have concentrated on automatic knowledge base completion and population. Multiple KGs [5,6] have been constructed from vast volumes of medical databases over the last years. Medical KGs contain medical facts of medicines and diseases and provide a pathway for medical discovery and applications, such as effective treatment and medicine recommendation. Unfortunately, such medical KGs suffer from severe data incompleteness problems, which impedes their application in clinical medicine. Celebi et al. [2] proposed a KG embedding approach for drug-to-drug interaction prediction in a realistic scenario. Hettige et al. [8] proposed an EMR embedding framework that introduces a graph-based data structure to capture visit-code associations in an attributed bipartite graph and the temporal sequencing visits through a point process. Moreover, Choi et al. [4] proposed another approach that learns the representations for both medical codes and visits from large datasets. Both works [4,8] consider patient demographics on medical codes; however, they only focus on visit-code embeddings and do not directly harness medical KG embeddings or any medical task. Hence, specific medical information is insufficiently tailored. To the best of our knowledge, none of the existing approaches employs demographics as an essential factor for medical KG embeddings.

3 DARLING

This section presents a detailed description of DARLING (Fig. 1). In this work, we use a boldface lower-case letter $\boldsymbol{x}$ to denote a vector, $\|\boldsymbol{x}\|_p$ to represent its l_p norm and d for the embedding dimension. Please refer to the appendix[2] for the background of our framework, detailed experiment results, etc.

3.1 Probabilistic Medical Knowledge Graph with Demographics

A KG can be denoted as a set of triples $\mathcal{K} \subseteq \mathcal{E} \times \mathcal{R} \times \mathcal{E}$ where $\mathcal{E}$ and $\mathcal{R}$ are the set of entities and relations respectively. More precisely, the KG comprises triples $(h, r, t) \in \mathcal{K}$ in which $h, t \in \mathcal{E}$ represent a triples' respective head and tail entities and $r \in \mathcal{R}$ represents its relation. The direction of a relationship indicates the roles of the entities, i.e., head or tail entity. In our scenario, we

[2] https://github.com/AynurGuluzade/DARLING/blob/main/paper/Appendix.pdf.

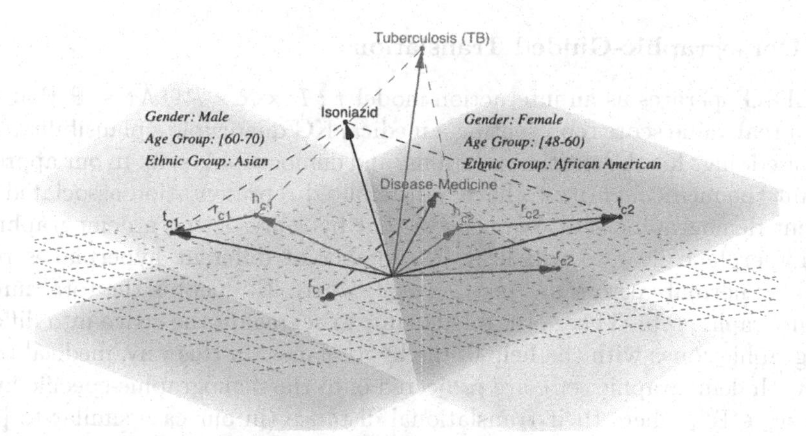

Fig. 1. The medical triple (Tuberculosis (TB), Disease-Medicine, Isoniazid) is valid for both demographic sets c_1 and c_2. $\boldsymbol{h}_{(c1)}, \boldsymbol{r}_{(c1)}$ and $\boldsymbol{t}_{(c1)}$ are the projections of the triple onto the demographic hyperplane c_1. As we can observe, the hyperplanes encode demographic information such as gender, age group, and ethnic group. DARLING learns demographic-aware representations of medical entities and relations by minimizing the translational distance $\|\boldsymbol{h}_c + \boldsymbol{r}_c - \boldsymbol{t}_c\|_2$ based on the triple probability score.

consider medical entities, extracted from EMR data, such as disease diagnosis, treatment procedures, and medicines; therefore $\mathcal{E} \subseteq \mathcal{D} \cup \mathcal{P} \cup \mathcal{M}$, where $\mathcal{D}$ is the set of diagnosis, $\mathcal{P}$ is the set of procedures and $\mathcal{M}$ is the set of medicines. In particular, we construct triples with $h \in \mathcal{D}$ and $t \in \mathcal{P} \cup \mathcal{M}$. Hence, our medical KG is denoted as $\mathcal{K} \subseteq \mathcal{D} \times \mathcal{R} \times \mathcal{P} \cup \mathcal{M}$.

We incorporate patient demographic meta-data such as gender, age group and ethnic group, by adding a new demographic dimension to the KG triples. We define the demographic set as $\mathcal{C} \subseteq \mathcal{G} \times \mathcal{A} \times \mathcal{T}$, where $\mathcal{G}$ is the set of genders, $\mathcal{A}$ the set of age groups and $\mathcal{T}$ the set of ethnic groups. Our medical KG contains quadruples $(h, r, t, c) \in \mathcal{K}$ where $c \in \mathcal{C}$ is a demographic set for which the corresponding medical triple (h, r, t) holds. We aim to incorporate the demographic meta-fact c directly into our learning algorithm, to learn demographic-aware embeddings of the KG elements.

Moreover, inspired by [12], we strengthen each quadruple (h, r, t, c) existence by introducing a statistical probability that indicates how likely is the particular triple (h, r, t) to appear with the demographic set c. Specifically, we associate each quadruple with the probability $p(h, r, t, c)$ which is calculated as, $p(h, r, t, c) = \mathcal{N}_{(h,r,t,c)}/\mathcal{N}_h$, where $\mathcal{N}_{(h,r,t,c)}$ is the number of EMR admissions that the quadruple (h, r, t, c) was extracted, while $\mathcal{N}_h$ is the number of admissions that contain the medical entity h. In contrast to [12], our probability value is calculated by considering also the demographic set c and not only the triple (h, r, t).

3.2 Demographic-Guided Translation

DARLING operates as an interaction model $f : \mathcal{D} \times \mathcal{R} \times \mathcal{P} \cup \mathcal{M} \rightarrow \mathbb{R}$ that computes a real-value score representing a medical KG quadruple's plausibility, given the embeddings for the entities, relations, and demographic sets. In our approach, we want the medical entity to have a distributed representation associated with different demographic sets. We achieve that by representing a demographic set as a hyperplane i.e., we will have $|\mathcal{C}|$ number of different hyperplanes represented by normal vectors $\boldsymbol{w}_1, \boldsymbol{w}_2, \ldots, \boldsymbol{w}_{|\mathcal{C}|}$, where $|\mathcal{C}|$ denotes the total number of demographic sets. Therefore, we attempt to segregate the space into different demographic zones with the help of the hyperplanes. In this way, medical triples valid with demographic set c are projected onto the demographic-specific hyperplane $\boldsymbol{w}_c \in \mathbb{R}^d$, where their translational distance (in our case similar to [1]) is minimized. Figure 1 illustrates an example where the medical triple (h, r, t) is valid for both demographic sets c_1 and c_2. Hence it is projected onto the hyperplanes corresponding to those demographic sets. Using the triple embeddings $\boldsymbol{h}, \boldsymbol{r}, \boldsymbol{t} \in \mathbb{R}^d$, the projected representations on $\boldsymbol{w}_c$ are computed as:

$$\boldsymbol{h}_c = \boldsymbol{h} - \boldsymbol{w}_c^\top \boldsymbol{h} \boldsymbol{w}_c, \qquad \boldsymbol{r}_c = \boldsymbol{r} - \boldsymbol{w}_c^\top \boldsymbol{r} \boldsymbol{w}_c, \qquad \boldsymbol{t}_c = \boldsymbol{t} - \boldsymbol{w}_c^\top \boldsymbol{t} \boldsymbol{w}_c, \tag{1}$$

where $\boldsymbol{h}_c, \boldsymbol{r}_c, \boldsymbol{t}_c \in \mathbb{R}^d$. Consequently, we expect that a positive triple, valid with demographic set c, will have the mapping as $\boldsymbol{h}_c + \boldsymbol{r}_c \approx \boldsymbol{t}_c$. Accordingly, our scoring function is defined as:

$$f_c(h, r, t) = \|\boldsymbol{h}_c + \boldsymbol{r}_c - \boldsymbol{t}_c\|_p, \tag{2}$$

where $p \in \{1, 2\}$ is a hyper-parameter. Alongside the entity and relation embeddings, we also learn $\{\boldsymbol{w}_c\}_{c=1}^{|C|}$ for each demographic set c. Furthermore, by projecting the triple onto its demographic hyperplane, we incorporate demographic knowledge into the entity and relation embeddings, i.e., the same distributed representation will have a different role in different demographic sets.

3.3 Optimization Through Probability Score

DARLING employs a margin-based pairwise ranking loss to differentiate between correct/positive and incorrect/negative triples. The negative triples are obtained by corrupting the positive one; thus, the pairs often share common head or tail entities and relations. Formally, we aim to minimize the following loss function:

$$\mathcal{L} = \sum_{c \in [C]} \sum_{x \in \mathcal{D}_c^+} \sum_{y \in \mathcal{D}_c^-} \max(0, g_c(x) - g_c(y) + \gamma), \tag{3}$$

with respect to the entity, relation and demographic set vectors. $\mathcal{D}_c^+$ is the set of valid triples with demographic set c, the negative triples are drawn from the set $\mathcal{D}_c^-$, and γ is a margin separating correct and incorrect triples.

Unlike existing approaches that employ pairwise ranking loss, DARLING does not directly use the score function $f_c(h, r, t)$. Instead, we utilize probability

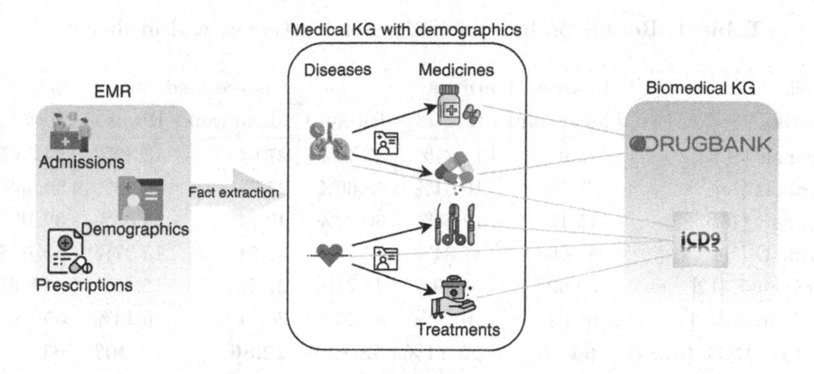

Fig. 2. Overview of our Medical KG construction process.

values p for estimating the scores for optimization. First, we introduce a mapping function or probability score function f_p that allows us to map each quadruple probability value $p(h, r, t, c)$ into a score. This function is defined as:

$$f_p(h, r, t, c) = \lambda \ln p(h, r, t, c)^{-1}, \tag{4}$$

where λ is a scaling factor. To avoid a 0 denominator for negative quadruples, we introduce a constant probability value $\varepsilon_n > 0$. Furthermore, we set a minimum probability value for positive quadruples as ε_p, where $\varepsilon_p > \varepsilon_n$. Second, we define the function $g_c()$ as the absolute difference of the probability score f_p and the score function f_c. Formally, this is described as:

$$g_c(h, r, t) = |f_p(h, r, t, c) - f_c(h, r, t)|. \tag{5}$$

In this way, the DARLING optimization process allows us to learn representations that would satisfy each quadruple probability value. Specifically, the quadruple entities with a high probability value will have representations closer in space compared to those with a low probability.

4 Experiments

4.1 Datasets and Medical Knowledge Graph Construction

We perform experiments on a real EMR dataset – MIMIC-III [11], and two biomedical knowledge graphs, DrugBank [19] and the ICD-9 [15] ontology. MIMIC III (Medical Information Mart for Intensive Care) comprises information related to patients admitted to critical care units at a large tertiary care hospital. The dataset contains distinct information about 46,520 patients, 58,976 admissions, and 1,517,702 prescription records associated with 6,985 diagnosis, 2,032 procedures and 4,525 medicines. For our work, we extract patient demographics, disease diagnosis, treatment procedures, and medicines. We link the extracted medicines to DrugBank, which is a bioinformatics resource that consists of medicine-related entities. The DrugBank KG contains 8,054 medicines,

Table 1. Results on link prediction for treatments and medicines.

Task	Disease-Treatment			Disease-Medicine		
Methods	Mean rank	Hits@3	Hits@10	Mean rank	Hits@3	Hits@10
TransE [1]	73.94	15.71%	47.40%	27.04	15.89%	54.33%
TransH [18]	75.56	16.51%	48.60%	27.71	16.23%	55.46%
TransR [13]	115.12	12.64%	30.34%	45.74	14.33%	39.16%
TransD [10]	84.66	17.34%	47.64%	33.51	15.97%	55.76%
PrTransE [12]	69.69	16.29%	47.21%	27.51	15.45%	54.80%
PrTransH [12]	69.01	16.89%	47.25%	26.71	16.14%	55.73%
DARLING (ours)	**64.65**	**22.71%**	**52.19%**	**22.86**	**26.90%**	**61.73%**

4,038 other related entities (e.g., protein or drug targets) and 21 relationships. Moreover, we link extracted diseases and treatments with ICD-9 ontology (International Classification of Diseases, Ninth Revision) which contains 13,000 international standard codes of diagnosis and procedures. We connect MIMIC-III, DrugBank, and the ICD-9 ontology by constructing the medical KG (c.f. Fig. 2).

4.2 Models for Comparison

For evaluating the performance of our framework, we compare against the following methods: TransE [1] is a simple but effective translation-based model. A major advantage of TransE is its computational efficiency, which enables its use for large-scale KGs. TransH [18] is an extension of TransE where each relation is represented by a hyperplane. Our proposed framework, DARLING also adjusts TransE in a similar way by treating the demographic sets as hyperplanes. TransR [13] explicitly considers entities and relations as different objects and therefore represents them in different vector spaces. TransD [10] is similar to TransR, however, instead of performing the same relation-specific projection for all entity embeddings, entity-relation-specific projection matrices are constructed. PrTransE & PrTransH [12] are extensions of TransE and TransH, which introduce triple probabilities. Our optimization via probability score was inspired by this work, however, our approach differs considerably in multiple aspects.

4.3 Results

We apply the following frequently used metrics to summarize the overall performance: 1) Mean rank (MR): which represents the average rank of the test triples, where smaller values indicate better performance. 2) Hits@K: which denotes the ratio of the test triples that have been ranked among the top-k triples, where larger values indicate better performance. We report results for $k = \{3, 10\}$.

Table 1 illustrates the results for treatment and medicine prediction from our constructed medical KG. As indicated, DARLING outperforms all other

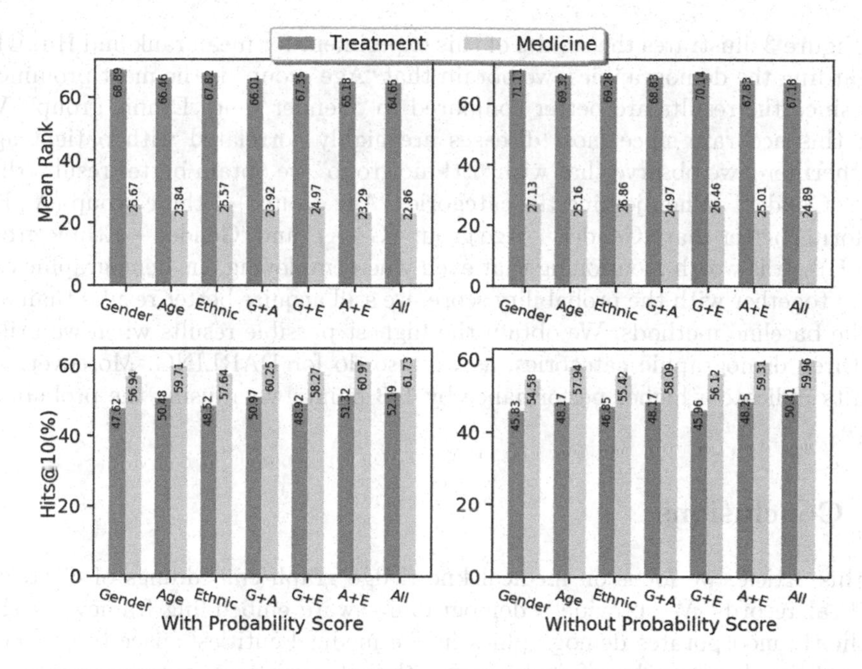

Fig. 3. DARLING results based on different demographic combinations with and without probability score.

traditional KG embedding approaches on both mean rank and Hits@K. More precisely, for Hits@3 and Hits@10, our framework has an absolute difference of at least 5 points from the baselines, while for Hits@3 on medicine prediction, this difference is increased to 10 absolute points. DARLING also has lower mean rank compared to all other baselines. The improved performance arises from DARLING's ability to generate more accurate embeddings of the medical entities. The demographic-based hyperplanes that our framework produces allow it to categorize diseases, treatments and medicines into "subspaces" where it learns their embeddings. Furthermore, the quadruple probability score provides some weighting for the distance between the entities. In contrast, all the baselines project the medical entities in one or two spaces (since the number of relations is two). As demonstrated by our results, this is not sufficient to represent all entities accurately.

4.4 Demographic and Probability Score Sensitivity

We perform experiments to identify the demographic and probability score sensitivity of our framework. In particular, we aim to recognize which demographic category (gender, age group, ethnic group) is more effective and whether the probability score impacts our results. To do so, we adjust DARLING to construct hyperplanes with all demographics individually and with all possible combinations. At the same time, we run the framework by including and excluding the probability scores.

Figure 3 illustrates the results of this experiment for mean rank and Hits@10. Regarding the demographics, we obtain that "Age group" is the most prominent one since the results are better compared to "Gender" and "Ethnic group". We find this accurate since most diseases are highly correlated with patient age. Furthermore, we observe that with "Ethnic group" we obtain better results than with "Gender". When joining the categories "Age group + Ethnic group (A+E)" performs better than "Gender + Age group (G+A)" and "Gender + Ethnic group (G+E)". It is worth mentioning that even when employing one demographic category together with the probability score, we still acquire better results than any of the baseline methods. We obtain the highest possible results when we utilise all three demographic categories, as we also do for DARLING. Moreover, the results indicate a higher performance by 1–3 points when using the probability score.

5 Conclusions

In this article, we focus on medical knowledge graph embeddings of electronic medical records. We provide a demographic-aware embedding framework that explicitly incorporates demographics in the medical entities' space by associating patient demographics (gender, age, ethnicity) with a corresponding hyperplane. Our framework leverages probabilistic features of entities for learning their embeddings through demographic guidance. Furthermore, for evaluating our approach, we present a method to construct a medical KG from EMR data and automatically link all extracted entities with existing biomedical knowledge graphs (ICD-9 and DrugBank). We empirically show that our model achieves the best results in link prediction for treatments and medicines, compared to other traditional KG embedding approaches. Moreover, we perform a demographic sensitivity experiment and discover that age is the demographic category that significantly affects our high results. We also show the importance of probability score in our framework. For future work, we intend to employ our framework as an embedding method for existing medical recommendation systems.

References

1. Bordes, A., Usunier, N., Garcia-Duran, A., Weston, J., Yakhnenko, O.: Translating embeddings for modeling multi-relational data. In: NIPS. Curran Associates, Inc. (2013)
2. Celebi, R., Uyar, H., Yasar, E., Gumus, O., Dikenelli, O., Dumontier, M.: Evaluation of knowledge graph embedding approaches for drug-drug interaction prediction in realistic settings. BMC Bioinform. **20**, 1–14 (2019)
3. Chen, D., Ma, C., Wu, Y.: Clinical knowledge graph embeddings with hierarchical structure for thyroid treatment recommendation. In: IEEE DASC/PiCom/CBDCom/CyberSciTech (2019)
4. Choi, E., et al.: Multi-layer representation learning for medical concepts. In: 22nd ACM SIGKDD. Association for Computing Machinery (2016)

5. Dumontier, M., et al.: Bio2RDF release 3: a larger connected network of linked data for the life sciences. In: ISWC-PD 2014 (2014)
6. Ernst, P., Siu, A., Weikum, G.: KnowLife: a versatile approach for constructing a large knowledge graph for biomedical sciences. BMC Bioinform. **16**, 157 (2015). https://doi.org/10.1186/s12859-015-0549-5
7. Gong, F., Wang, M., Wang, H., Wang, S., Liu, M.: Smr: Medical knowledge graph embedding for safe medicine recommendation. Big Data Res. (2021)
8. Hettige, B., Wang, W., Li, Y.F., Le, S., Buntine, W.: MedGraph: structural and temporal representation learning of electronic medical records. In: ECAI Digital - 2020. IOS Press (2020)
9. Ihra, G.C., et al.: Development of demographics and outcome of very old critically ill patients admitted to intensive care units. Intensive Care Med. **38**, 620–626 (2012). https://doi.org/10.1007/s00134-012-2474-7
10. Ji, G., He, S., Xu, L., Liu, K., Zhao, J.: Knowledge graph embedding via dynamic mapping matrix. In: 53rd ACL-IJCNLP. Association for Computational Linguistics (2015)
11. Johnson, A.E., et al.: MIMIC-III, a freely accessible critical care database. Sci. Data **3**, 1–9 (2016)
12. Li, L., et al.: A method to learn embedding of a probabilistic medical knowledge graph: algorithm development. JMIR Med. Inform. **8**(5), e17645 (2020)
13. Lin, Y., Liu, Z., Sun, M., Liu, Y., Zhu, X.: Learning entity and relation embeddings for knowledge graph completion. In: 28th AAAI. AAAI Press (2015)
14. Rotmensch, M., Halpern, Y., Tlimat, A., Horng, S., Sontag, D.: Learning a health knowledge graph from electronic medical records. Sci. Rep. **7**, 1–11 (2017)
15. Schriml, L.M., et al.: Disease ontology: a backbone for disease semantic integration. Nucleic Acids Res. **40**, D940–D946 (2012)
16. Sherimon, P.C., Krishnan, R.: OntoDiabetic: an ontology-based clinical decision support system for diabetic patients. Arabian Journal for Science and Engineering **41**(3), 1145–1160 (2015). https://doi.org/10.1007/s13369-015-1959-4
17. Standish, L.J., et al.: Alternative medicine use in HIV-positive men and women: demographics, utilization patterns and health status. AIDS Care **13**, 197–208 (2001)
18. Wang, Z., Zhang, J., Feng, J., Chen, Z.: Knowledge graph embedding by translating on hyperplanes. In: 28th AAAI. AAAI Press (2014)
19. Wishart, D.S., et al.: DrugBank 5.0: a major update to the DrugBank database for 2018. Nucleic Acids Res. **46**, D1074–D1082 (2017)
20. Huang, X., Ribeiro, J.D., Musacchio, K.M., Franklin, J.C.: Demographics as predictors of suicidal thoughts and behaviors: a meta-analysis. PLoS One **12**, e0180793 (2017)
21. Xiang, X., Wang, Z., Jia, Y., Fang, B.: Knowledge graph-based clinical decision support system reasoning: a survey. In: IEEE 4th DSC (2019)

Modeling and Representation by Graphs of the Reasoning of an Emergency Doctor: Symptom Checker MedVir

Loïc Etienne[1], Francis Faux[2](✉), and Olivier Roecker[1]

[1] Société Medical Intelligence Service, 75013 Paris, France
{loic.etienne,o.roecker}@mis-medvir.fr
[2] IRIT, Université Paul Sabatier, 31062 Cedex 9 Toulouse, France
francis.faux@irit.fr

Abstract. This article deals with the symptom checker MedVir which is modeled on the reasoning of an emergency physician. His reasoning is very particular because he often has no knowledge of the patient and he doesn't have much time to evaluate the situation. He needs to make decisions rapidly based on diagnostic hypotheses and an estimation of the severity of the patient's condition. We present a ten step model of the reasoning of an emergency physician by a four layer network composed with what we call a "neuronal entity" and a question prioritization algorithm which checks the most important questions. This "neuronal entity" generalizes the neuron concept but differs from those usually used in machine learning. Visualization by graphs displays all the characteristics of each neuron and each synapse thickness corresponds to the argumentative strength of a question. Hence, these graphs could be very useful in the training of physicians and health professionals.

Keywords: Medical reasoning · Symptom checker · Decision under uncertainty · Neural network · Differential diagnosis

1 Introduction

Symptom checkers are online applications whose goals are both to help doctors to obtain diagnoses hypothesis and to give patients tools to check their own symptoms and to self-triage remotely. A symptom checker first assesses the patient by asking a limited series of questions in order to avoid patients filling in a long symptom questionnaire. Finally the diagnosis process outputs a list of potential diseases that the patient may have. The design goal of a symptom checker is then to achieve high diagnosis accuracy when only a limited number of symptoms inquiries can be made. Many online symptom checkers have been developed such as Babylonhealth, WebMD, Mayo Clinic, ADA, infermedika, IBM watson, Isabel. Recently, a large number of web-based COVID-19 symptom checkers and chatbots have been developed but with highly varying conclusions [13].

A. Tucker et al. (Eds.): AIME 2021, LNAI 12721, pp. 418–427, 2021.
https://doi.org/10.1007/978-3-030-77211-6_49

In previous works expert systems have been widely applied in the medical domain of diagnosis due to their ability to construct explanations for their lines of reasoning [2,3,9,16]. However, expert systems have many weaknesses that include lack of common sense knowledge, narrow focus and restricted knowledge, inability to respond creatively to unusual situations and difficulty in adapting to changing environment [8,15]. In order to improve accuracy of symptom checking while also making a limited number of inquiries, medical expert systems based on fuzzy logic were recently proposed in order to cope with uncertainty and imprecision of patient testimony and medical data [1,4,6]. In the domain of mental health Mohmmadi et al. have designed a web-based system for diagnosing depression [12]. Das et al. developed a web based medical system for diagnosing disease using fuzzy logic and intuitionistic fuzzy logic [5]. Ochab proposed an expert system to support an early prediction of the Bronchopulmonary Dysplasia in extremely premature infants [14]. Methods based on machine learning such as Bayesian inference or decision trees were also used to perform diagnoses but their weakness lies in the use of approximated scheme and often result in compromised accuracy [11]. Recently new approaches based on machine learning methods allowed to improve diagnosis accuracy. We can mention the interesting work of Kao et al. [10] which employ a hierarchical reinforcement learning scheme including contextual information.

The aim of this paper is to present the physician reasoning model used by the symptom checker MedVir based on fuzzy logic in the context of emergency medicine. The originality of MedVir (https://medvir.fr) is based on its design. The doctors who designed it are all emergency physicians (Drs Etienne, Chaumont, Welhoff, Jeannerod). In 1987, a minitel service was available to the French public (3615 Ecran santé), then www.docteurclic.com on the Internet, which have since carried out 450,000 teleconsultations. The collection of patient language and the study of doctors questionnaires made it possible to model the thought of the emergency doctors. The construction stages were as follows: creation of a glossary of symptoms (13,000 words and expressions obtained from 1 million requests) creation of an ontology of symptoms, selection of diagnoses covering 80% of common pathologies from Evidence Based Medicine (EBM), addressing all specialties, all age groups, and covering all patient symptoms (headache, fever, vomiting, etc.). All the documentation collected manually over several years by the design physicians was carried out on recognized French and international knowledge bases (medline, Cochrane, learned societies,...), and allowed the constitution of each present diagnosis.

This paper deals with all the reasoning stages as well as a visualization of this reasoning model via a four-layer graph which describes clinical information that helps physicians in their diagnostic tasks. Each layer is composed of what we call a "neuronal entity", a concept which generalizes the concept of neuron but which differs from that usually considered in machine learning.

The rest of the paper is organized as follows. In Sect. 2 we present the reasoning model of a physician in the context of emergency medicine. In this section we introduce the neuronal entity concept used in the previous model, we detail the question prioritization algorithm and the scenario management. Graphs obtained by sql queries that allow to list the most relevant questions are presented in Sect. 3. Finally, we present our concluding remarks and perspectives.

2 Reasoning Model of an Emergency Physician

An emergency doctor is only faced with a limited number of diagnoses. Its exercise (in regulation over the telephone or in intervention) is particular: no knowledge of the patient or his medical data, uncertain answers from the patient, need for rapid decision-making based on diagnostic hypotheses and an estimate of the severity of the patient's condition. (Ex: burns when urinating generally have a medium severity, but if they are accompanied by chills, the severity becomes important because it raises fear of pyelonephritis and not simple cystitis).

Preliminary remarks concerning the patient's complaint and the doctor's reasoning. Anyone is able to determine which symptom bothers them the most or which worries them the most. In front of a set of symptoms (urinary disorders, fever, pain in the back for example), the doctor always asks what is the symptom which led the patient to consult (for example pain in the back). This is the main symptom. If the patient is unconscious or cannot speak this in itself is a symptom. The other symptoms expressed by the patient in addition to the main symptom are accompanying signs, and it often happens that the doctor reasons from a main symptom that is not the one chosen by the patient (urinary disorders for example). Although he has changed his main symptom as the starting point for his reasoning, the doctor keeps in mind the other accompanying signs. It is thanks to this prioritization that the doctor, in a minimum of questions (and therefore in a minimum of time which is an urgent need) can consider the most probable diagnostic hypotheses. It is this particularity of emergency medicine that the symptôm checker MedVir has taken into account.

The steps in physician reasoning are summarized in the sequence below.

1. Collection of the patient's complaint,
2. Determination of symptoms
3. Selection of the main symptom
4. Identification of possible symptoms related to the main symptom,
5. Determining for each symptom of the elements that characterize it (group of questions)
6. Rapid exploration of gravity and hypotheses diagnostics from vital questions allowing to rule out a life-threatening emergency,
7. Refining the severity of questions and diagnostic hypotheses with major and minor questions
8. Differential diagnosis
9. Diagnostic hypotheses and final severity
10. Decision

Each of these stages includes elements of uncertainty which are: the meaning of the words used, the difficulty sometimes in selecting the main symptom, the interpretation of the complaint by the doctor, the questioning which is necessarily operator-dependent, the absence of clinical examination and additional examinations (medical regulation), with however the need to make a differential diagnosis, simply with questions. A symptom checker is subject to exactly these constraints. The doctor's reasoning is therefore logical and standardized, with decision making under uncertainty.

The Fig. 1 presents the modeling of the reasoning of an emergency physician in 10 steps.

Steps 1 to 6 included determine in a minimum of questions determined by the question prioritization algorithm (QPA), detailed in Sect. 2.2, the most serious diagnoses (Ex: pyelonephritis) as well as the most serious frequent (eg cystitis). In steps 7 to 9 the QPA algorithm checks that diagnoses have not been forgotten by means of so-called refinement questions. Decision-making is carried out in step 10.

The organization of this set constitutes an informational network with 4 layers (Fig. 1): 835 diagnoses [D], 206 symptoms [S], 66 question groups [G], and 4583 questions/sub questions [Q]).

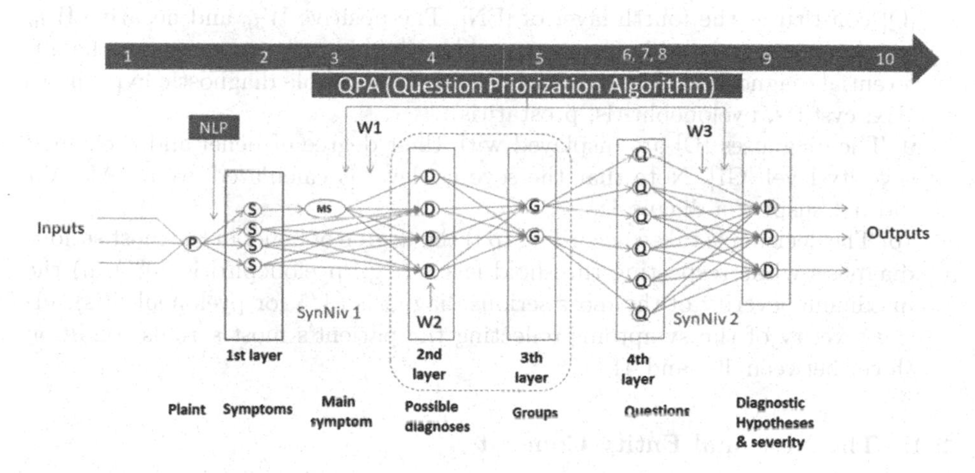

Fig. 1. Reasoning model in emergency medicine

The detailed description of the model is given below:

- NLP (Natural Language Processing) analyzes the patient's verbatim (the complaint) [P] which, along with age, sex, prior to onset of signs and geolocation, constitutes the input data. He compares this verbatim (Ex: "it burns me while urinating, I feel a bit of a bug today, and I shiver") to the basis of synonymy of MedVir, which makes it possible to isolate the symptoms [S] of the patient (urinary disorder, fever and chills).
- The patient is invited to select among these symptoms [S] the main symptom [MS] (the one that bothers him or that worries him the most, i.e., chills).

- The [MS] lists all possible diagnostics [D] where it is present. For each diagnosis i considered (Ex: sepsis, flu, pyelonephritis, etc.) and at any time of the examination, the weight of the Synapse [SynNiv1] is calculated $W_{1i} = W_{1ip} + W_{1in}$ for each symptom selected by the patient ([MS] (chills) + all [S](fever, urinary disorder)). This W_{1i} weight is normalized to a W_{1iN} weight such as $W_{1iN} = \frac{W_{1i}}{\Sigma_{i=1}^{n} W_{1i}}$. (Ex: If for the pyelonephritis diagnosis the total sum of all its constitutive symptoms is 100, and the sum of the W_{1i} of all the symptoms described by the patient is 15, the degree of belief in pyelonephritis at this time of l questioning is 15%).
- Each diagnosis i includes risk factors (history, recent or old events, etc.) which each have a weight $W_{2i} = W_{2ip} + W_{2in}$. The elements selected by the patient by checking the boxes which correspond to him are added to the weight W_{1i} (Ex: if the patient has had a history of pyelonephritis, which adds a weight of 4, the degree of belief is then increased to 19 % (15 + 4 = 19/100).
- A single link connects the second layer to the groups of questions [G] which make up the third layer of [EN].
- The three steps 6, 7 and 8 are handled by the prioritization algorithm, which decides the order of the questions asked. Each question and sub-question [Q] constitutes the fourth layer of [EN]. The positive W_{3p} and negative W_{3n} weights are calculated in the synapse [SynNiv2] in order to carry out the differential diagnosis and to limit the number of probable diagnostic hypotheses (Ex: cystitis, pyelonephritis, prostatitis).
- 9. The diagnoses [D] are displayed with their degree of belief and their final severity level [Gf]. Note that the severity level is calculated even if MedVir did not suspect a diagnosis.
- 10. The decision is taken according to i) the degree of belief in the most serious diagnosis while respecting threshold levels (e.g.: pyelonephritis 30%); ii) the maximum severity of the most serious diagnosis (4/5 for pyelonephritis); iii) the severity of the symptoms reflecting the patient's most serious condition (fever between 40° and 41 °C).

2.1 The Neuronal Entity Concept

In order to model (Fig. 1) the emergency doctor reasoning we introduce the notion of "neuronal entity" which is an extension of the concept of neuron and which includes both numerical data (synaptic weights), but also criteria for belonging to classes defined in the database (symptom, characteristics of the symptom, diagnosis), a semantic meaning (type of word, synonymy, medical specialty), a link with the reality of the human body (body area, organs, functions, human activities, etc.), and the populations by age group to which this entity belongs (Ex: the word burn while urinating is a complaint related to the urinary disorder symptom located in the lower abdomen, belonging to urinary function, to the organs bladder, urethra, ureter, kidney, etc., affecting all populations except infants, felt as pain, belonging to the urology specialty, etc.) Each of the elements D, S, G, Q of the reasoning model therefore constitutes an "neuronal entity" [EN] which has a mixed numerical function for the calculation of the synaptic weight and semantic for the definition of each entity. Each [EN] is linked to the others [EN] by logical links from Evidence Based Medicine.

Stimulation of [EN] is performed by the patient who selects the symptom or who answered the question. This stimulation of the [EN] brings a positive weight and a negative weight in Questions Prioritization Algorithm [QPA]. These respective weight was directly correlated with scientific data for each diagnosis (incidence and prevalence for each age group) recovered in EBM (Cochrane, Medline, learned societies, etc.). The positive weight increases the diagnostic belief (positive diagnosis) (Ex: If the EBM in 85% of adults who experienced pyelonephritis had backache, the weight Wp of the symptom backache in the pyelonephritis diagnosis and for this population is 8.5). The negative weight reduces this belief or invalidate this diagnosis (differential diagnosis). (Ex: If in EBM 100% of adults who experienced pyelonephritis had fever, absence of fever totally invalid diagnosis of pyelonephritis in this population. Similarly, if 40% of adults do not have. Presented chills, the weight Wn of the symptom shivers down pyelonephritis diagnosis is −4).

There are currently 6,687 neural entities linked by a total of 379,602 different digital synapses.

2.2 Questions Priorization Algorithm [QPA]

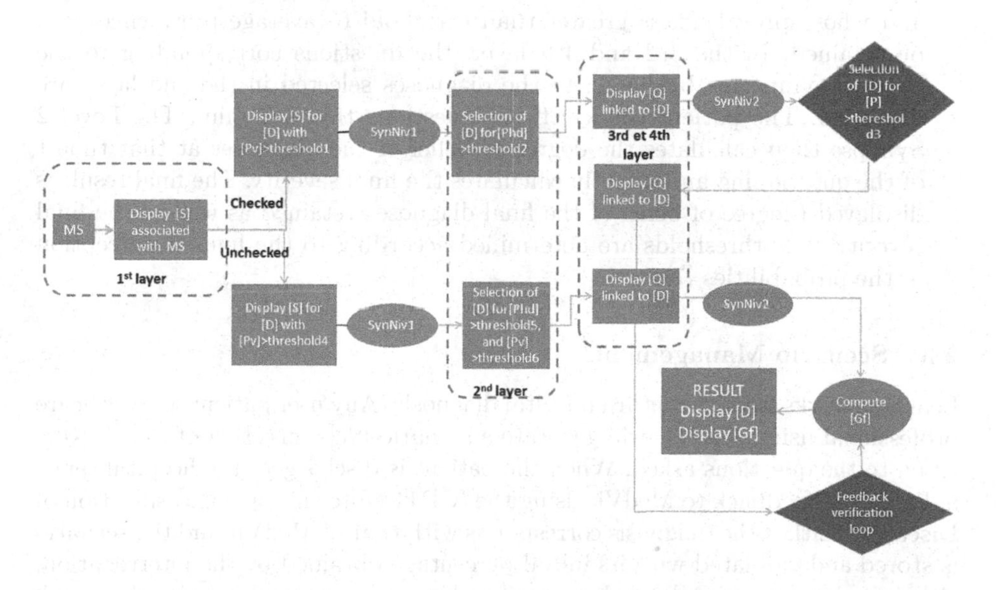

Fig. 2. Questions priorization algorithm

In the first layer, the main symptom is determined by the patient. The patient checks off the symptoms he wants. Two routes are possible:

1. Pathway for checked symptoms: For any checked symptom, the QPA suggests symptoms belonging to diagnoses whose prevalence is greater than threshold 1 (high prevalence diagnosis). For each of these diagnoses, the Level 1 Synapse calculates, at this time t of the questioning, the degree of belief of each. In the 2nd layer, only diagnoses whose degree of belief is greater than threshold 2 (diagnosis of which the degree of belief is high) will be retained. In the 3rd and 4th layer, the questions corresponding to the checked symptoms belonging to the diagnoses selected in the 2nd layer are displayed. The patient checks off the questions that suit him. The level 2 Synapse then calculates the degree of belief of the diagnoses at this time t of the questioning and selects the diagnoses whose degree of belief is greater than threshold 3 (diagnoses of average degree of belief). Additional questions are then asked to which the patient answers. The system then checks through a feedback loop all the other diagnoses whose degree of belief is less than threshold 3. The severity is calculated at this instant t of the questioning and the final result is displayed (degree of belief of the final diagnoses selected as well as the final gravity).
2. Pathway for unchecked symptoms: For any checked symptom, the QPA suggests symptoms belonging to diagnoses whose prevalence is greater than threshold 4 (diagnosis of average prevalence). For each of these diagnoses, the Level 1 Synapse calculates, at this point in the questioning, each person's degree of belief. In the 2nd layer, only diagnoses whose degree of belief is greater than threshold 5 (diagnosis of which the degree of belief is high) and whose prevalence is greater than threshold 6 (average prevalence) will be retained. In the 3rd and 4th layer, the questions corresponding to the checked symptoms belonging to the diagnoses selected in the 2nd layer are displayed. The patient checks off the questions that suit him. The Level 2 Synapse then calculates the degree of belief of the diagnoses at that time t of the questioning and directly calculates the final severity. The final result is displayed (degree of belief of the final diagnoses retained as well as the final severity. The thresholds are determined according to the human perception of the probabilities [7].

2.3 Scenario Management

Learning works from the return hospital diagnosis. Any user patient or healthcare professional using MedVir will generate a scenario [Sc] consisting of the answers given to the questions asked. When the patient is discharged, the hospital sends a diagnostic feedback to MedVir using the ICD11 (International Classification of Diseases). Either the diagnosis corresponds with that of MedVir and the scenario is stored and validated with its initial percentage obtained by the interrogation; either it does not match, and supervised learning can improve questions, ask more, or remove some. All scenarios (selected neural entities and value of each synapse) are stored in the database. Statistical monitoring by a Khi2 test will make it possible to know after how many scenarios for a so-called closed diagnosis (using only its original data) can be considered to have sufficient elements to be considered reliable. This closed diagnostic is then opened, and all the correct diagnostic returns are stored there in an automatic, unsupervised manner. The new data then replaces the old ones and the system therefore learns automatically in an unsupervised way. Each new scenario then increments all the synapses that were called upon during the interrogation.

3 Visualisation by Graphs

The graphs are obtained by simply querying the database using sql queries. The graph in Fig. 3 represents all the diagnoses [D] considered by MedVir in urology. In the center are the [EN] of the diagnostics (1st layer). Around (2nd layer) are the [EN] of the symptoms. These two types of [EN] are linked together by Level 1 Synapses. Around (3rd layer) is the layer of question groups [D], each of which is linked to the layer of [S] by a link which is not numerical but semantic (the meaning of the question asked). Finally the 4th layer is that of questions/sub-questions [Q]. We have the ability to filter the diagnoses (one or more) that we are looking for. For example pyelonephritis in the Fig. 4.

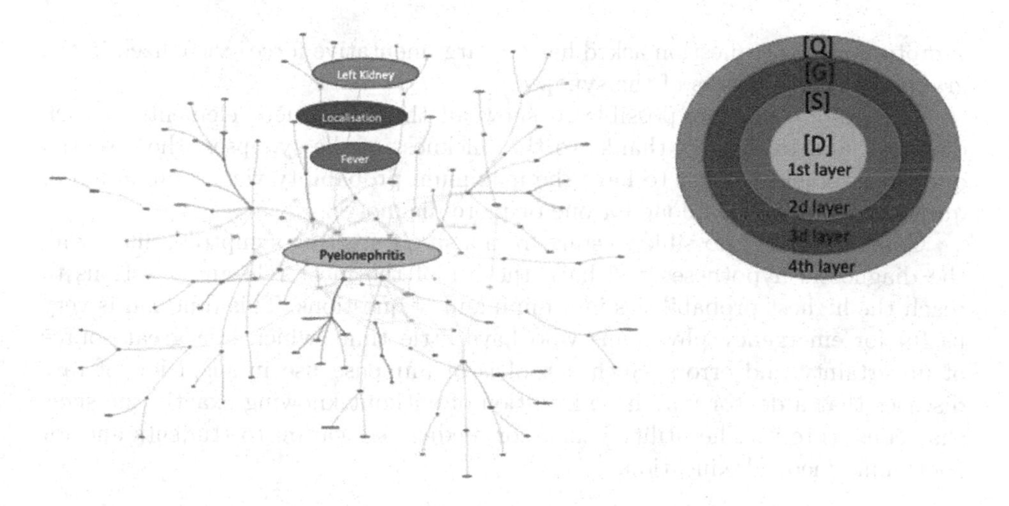

Fig. 3. Urology graph

3.1 Pyelonephritis Graph

Zooming in on this graph displays all the characteristics of each neuron and each synapse. The thickness of the synapses corresponds to the argumentative strength of a question (see below). Zooming in on the urology graph displays all the characteristics of each neuronal entity [D], [S] [G] and [Q]. The SynNiv1 synapse is located between the diagnosis [D] (here pyelonephritis) and the symptoms [S] which characterize it. The synapse between each symptom [S] and the question group [G] is not numerical but semantic in nature. Finally, the SynNiv2 synapse is located between the group [G] and the questions [Q] attached to each

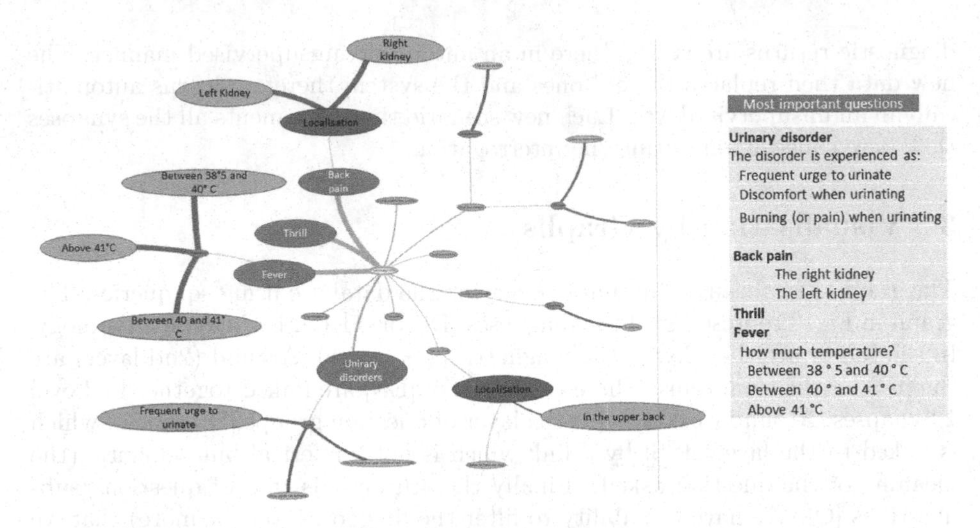

Fig. 4. Pyelonephritis graph

symptom [S]. Any question asked has an "argumentative force" visualized in the graphs by the thickness of the synapse.

The graph makes it possible to show all the constituent elements of each diagnosis and to display, thanks to the thickness of the synapses, the essential questions to ask in order to have the maximum probability with a minimum of questions. This can be done for one or more diagnoses.

Conversely, it is possible to start from a single starting symptom, display all the diagnostic hypotheses and have the list of the most relevant questions to reach the highest probabilities in a minimum of questions. This function is very useful for emergency physicians who have little time, which is a great source of uncertainty and errors. Such a tool is of immense use in suspicion of rare diseases that a doctor may have intuition of without knowing exactly the signs that constitute it. The utility is also for medical education to students and for continuing medical education.

4 Conclusion and Perspectives

In this paper a ten step model of an emergency physician reasoning has been presented. This model is attained by a four layer network composed of neuronal entities and a question prioritization algorithm. An experiment carried out in 2012 at the Lariboisière hospital in Paris in prof. Plaisance service showed on 397 patients that simple questioning by MedVir was 100% reliable with regard to the assessment of severity, and allowed in 87% of cases to suspect the diagnosis made subsequently by the hospital after clinical examination and additional examinations. A larger-scale study will be developed with the innovation unit of APHP (Assistance Publique Hôpitaux de Paris), the results of which will be available in the last quarter of 2021. In addition, this 4-layer model will be supplemented by a 6-layer system intended to provide a pathophysiological explanation of each symptom in the context of each diagnosis at all times during the interrogation, of great utility for the training of medical students.

References

1. Abu-Nasser, B.: Medical expert systems survey. Int. J. Eng. Inf. Syst. (IJEAIS) **1**(7), 218–224 (2017)
2. Adlassnig, K.: Fuzzy set theory in medical diagnosis. IEEE Trans. Syst. Man Cybern. **16**(2), 260–265 (1986)
3. Adlassnig, K.: Fuzzy systems in medicine. In: Proceedings of the 2nd International Conference in Fuzzy Logic and Technology, Leicester, United Kingdom, pp. 2–5 (2001)
4. Ahmadi, H., Gholamzadeh, M., Shahmoradi, L., Nilashi, M., Rashvand, P.: Diseases diagnosis using fuzzy logic methods: a systematic and meta-analysis review. Comput. Methods Programs Biomed. **161**, 145–172 (2018)
5. Das, S., Guha, D., Dutta, B.: Medical diagnosis with the aid of using fuzzy logic and intuitionistic fuzzy logic. Appl. Intell. **45**(3), 850–867 (2016)
6. Drweesh, Z., Al-Bakry, A.: A web/mobile decision support system to improve medical diagnosis using a combination of k-mean clustering and fuzzy logic. TELKOMNIKA (Telecommun. Comput. Electron. Control) **17**, 3145 (2019)
7. Gallistel, C., Krishan, M., Liu, Y., Miller, R., Latham, P.: The perception of probability. Psychol. Rev. **121**, 96–123 (2014)
8. Giarratano, J., Riley, G.: Expert Systems: Principles and Programming. Fourth ed. PWS-Kent (2004)
9. Holman, J.G., Cookson, M.J.: Expert systems for medical applications. J. Med. Eng. Technol. **11**(4), 151–159 (1987)
10. Kao, H.C., Tang, K., Chang, E.: Context-aware symptom checking for disease diagnosis using hierarchical reinforcement learning. In: AAAI (2018)
11. Kononenko, I.: Machine learning for medical diagnosis: history, state of the art and perspective. Artif. Intell. Med. **23**, 89–109 (2001)
12. Mohammadi Motlagh, H., Minaei, B., Fard, A.: Design and implementation of a web-based fuzzy expert system for diagnosing depressive disorder. Appl. Intell. **48**(5), 1302–1313 (2017)
13. Munsch, N., et al.: Diagnostic accuracy of web-based covid-19 symptom checkers: comparison study. J. Med. Internet Res. **22**(10), e21299 (2020)
14. Ochab, M., Wajs, W.: Expert system supporting an early prediction of the bronchopulmonary dysplasia. Comput. Biol. Med. **69**, 236–244 (2016)
15. Shang, Y.: Expert systems. The Electrical Engineering Handbook, pp. 367–377 (2005)
16. Singla, J., Grover, D., Bhandari, A.: Medical expert systems for diagnosis of various diseases. Int. J. Comput. Appl. **93**, 36–43 (2014)

Effect of Depth Order on Iterative Nested Named Entity Recognition Models

Perceval Wajsbürt[1(✉)], Yoann Taillé[1,2], and Xavier Tannier[1]

[1] Sorbonne Université, Inserm, LIMICS, Paris, France
perceval.wajsburt@sorbonne-universite.fr
[2] SCAI, Sorbonne Université, Paris, France

Abstract. This paper studies the effect of the order of depth of mention on nested named entity recognition (NER) models. NER is an essential task in the extraction of biomedical information, and nested entities are common since medical concepts can assemble to form larger entities. Conventional NER systems only predict disjointed entities. Thus, iterative models for nested NER use multiple predictions to enumerate all entities, imposing a predefined order from largest to smallest or smallest to largest. We design an order-agnostic iterative model and a procedure to choose a custom order during training and prediction. We propose a modification of the Transformer architecture to take into account the entities predicted in the previous steps. We provide a set of experiments to study the model's capabilities and the effects of the order on performance. Finally, we show that the smallest to largest order gives the best results.

Keywords: Named entity recognition · Biomedical · Nested entities

1 Introduction

Biomedical concept recognition is a classical and essential task of natural language processing for biomedical applications [11], aiming to extract information such as symptoms, treatments, proteins, genes, dates, and durations from free text. Classic methods assume that entities are disjoint and formulate the problem as a sequence segmentation task, using word tagging schemes. However, in a real-world scenario, entities can compose or overlap, thus breaking the assumption that they are disjoint. For example, a temporal event "after anesthesia" contains the nested treatment entity "anesthesia."

moment
treatment
Checking empty vials after anesthesia

A class of methods deals with this problem of nested named entity recognition with a cascade of flat (non-nested) named entity recognition layers for different depths, i.e., predict the entities at a given depth iteratively predicting large entities or short entities first. The predictions of a given depth are used as additional input for the next prediction. We can argue whether the depth order matters during the training of such a model: is it easier for the model to predict large entities first and look inside its previous predictions

A. Tucker et al. (Eds.): AIME 2021, LNAI 12721, pp. 428–432, 2021.
https://doi.org/10.1007/978-3-030-77211-6_50

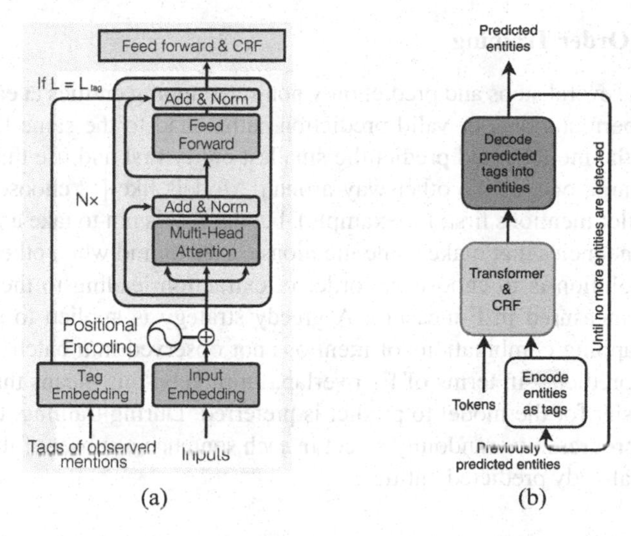

Fig. 1. (a) modified Transformer conditioned on the tags of the previously observed entities after layer L_{tag} - adapted from [15]; (b) Global model prediction diagram

for smaller ones, or predict small entities and compose them to build larger ones? To answer this question, we design an order-agnostic auto-regressive model based on the Transformer encoder architecture and a procedure to let it choose a custom order.

This work was originally designed to address Task 3 of the DEFT 2020 evaluation campaign (in French), and we further evaluated it on the classical GENIA dataset (in English). This DEFT tasks deals with the detection of named entities in texts describing clinical cases [3]. More details about the challenge are presented in [1], in which we describe our 3 official submissions. To our knowledge, our work is the first to evaluate a nested biomedical NER system in French, which constitutes an additional contribution, as resources for languages other than English are very scarce [10]. More details about our work can be found at https://arxiv.org/abs/2104.01037.

2 Method

2.1 Model

Our model is an auto-regressive encoder-only Transformer [15] taking as input a sequence of words and a list of entity mentions already extracted (empty list at the first iteration) and predicts a list of new mentions. The entities predicted at each iteration do not overlap, but all the entities predicted at the end of the prediction may overlap.

We handle entities in the form of tags assigned to each token with the conventional BIO or BIOUL formats by embedding each tag into a vector space and summing the embeddings of different tags at the same position. The sentences are tokenized and represented with a Transformer model. The output of the Transformer are fed into a linear CRF predicting flat entities (Fig. 1a). At each iteration, the model receives the tags predicted at the previous iterations (Fig. 1b).

2.2 Greedy Order Training

We proceed in several steps and predict only non-overlapping entities at each run. However, several permutations, or valid prediction paths, lead to the same list of entities. For example, the model could predict the smallest entity first and use that information to detect the large one, or the other way around. Models like [4] choose a strategy in advance (smaller mentions first, for example), but the risk is not to take advantage of all the inter-dependencies that make some mentions easier to find when others are known.

Another solution is to choose the order of extraction leading to the model's best performance, measured in F-measure. A greedy strategy is applied to select, among the non-overlapping combinations of mentions not observed in a batch, the closest to the mentions predicted in terms of F1 overlap. Intuitively, this means that a combination that is easier for the model to predict is preferred. During training, to simulate an extraction in progress, we randomly select in each sentence a subset of the entities and label them as already predicted entities.

2.3 Model Parameters

We initialize the Transformer with CamemBERT [9] weights for DEFT and BioBERT [7] for GENIA. We use a dropout of 0.25, we optimize the parameters by backpropagation with Adam [6], over 40 epochs for DEFT and 10 epochs for GENIA. We use a linear decay learning rate schedule with a 10% warmup and two initial learning rates: $4 \cdot 10^{-5}$ for the Transformer and $9 \cdot 10^{-3}$ for other parameters. We insert the tag embeddings at layer $L_{tag} = 6$ for BERT base and 19 for BERT large.

3 Experiments and Discussions

3.1 Datasets

We conduct experiments on the DEFT [1] and GENIA [5] datasets, which present texts with different languages and different types and depths of entities. We perform splits following [2] for GENIA and keep 10% of the training data as a validation set for DEFT.

3.2 Baselines

We compare our results against a simple flat NER model composed of a Transformer and a CRF that can only predict non-nested mentions. Since a choice is required during training as to which mentions should be predicted, we evaluate three modes: we only recover the shortest mention in a nested group, or the largest, or let the model decide greedily. We also compare our model against the state-of-the-art models on GENIA and the other participants' models on DEFT.

Table 1. Model performance on DEFT and GENIA datasets

	GENIA			DEFT 3.1			DEFT 3.2			DEFT
	P	R	F1	P	F	F1	P	R	F1	F1
Ju et al. [4]	0.785	0.713	0.747							
Wang and Lu [16]	0.770	0.733	0.751							
Sohrab and Miwa [13]	0.932	0.642	0.771							
Lin et al. [8]	0.758	0.739	0.748							
Shibuya and Hovy [12]	0.763	0.747	0.755							
[14], BERT+Flair			0.783							
[17], BERT+Flair	0.803	0.783	0.793							
HESGE, BERT large				0.702	0.624	0.660	0.788	0.725	0.755	
Median DEFT						0.456			0.615	
Flat short entities, BERT base	0.793	0.698	0.742	0.609	0.228	0.332	0.757	0.669	0.710	0.623
Flat large entities, BERT base	0.815	0.707	0.757	0.609	0.608	0.609	0.722	0.315	0.439	0.509
Flat greedy, BERT base	0.814	0.710	0.758	0.608	0.348	0.443	0.785	0.605	0.683	0.620
(our) greedy, BERT base	0.812	0.721	0.764	0.626	0.609	0.618	0.762	0.742	0.751	0.713
(our) large→short, BERT base	0.802	0.718	0.758	0.626	0.606	0.616	0.741	0.747	0.744	0.708
(our) short→large, BERT base	0.803	0.734	0.767	0.611	0.619	0.615	0.756	0.745	0.751	0.712
(our) short→large, BERT large	0.793	0.745	0.768	0.661	0.660	0.660	0.781	0.776	0.778	0.745

3.3 Results

On the DEFT task 3.1, our model obtains the best F1 result of 0.66. On the DEFT task 3.2, the same model obtains a F1 of 0.778. Flat NER models lose between 10 and 20 points in F1, due to the large number of nested mentions.

On the GENIA dataset, our best model reaches 0.768 F1 with BioBERT large. We hypothesize that our method ranks lower on the latter dataset because it only uses BERT instead of BERT and other word features, and that the insertion of tags directly in BERT architecture may lead to loose some of the pretrained model abilities. We can also observe that flat NER is competitive with iterative models, which can be explained by the low ratio of nested mentions in the dataset.

We study the effect that forced prediction order during training has on model performance. We compared three prediction modes: top to bottom, bottom to top, and greedy decoding. In the top to bottom mode, given a previously predicted entity at depth D, we force the model to predict a named entity located at depth D+1. In the bottom to top mode, we use the inverse depth as training order. Finally, in greedy decoding mode, we let the model choose the mentions by selecting those closest to its prediction.

We can observe that the short-to-large training order obtains the highest performance on both GENIA and DEFT validation splits. The large-to-short depth training order obtains the lowest accuracy. We hypothesize that learning to detect the smallest, and often easier, entities first leads the model to learn how to compose new entities from small entities. On the other hand, learning to predict large, and often more difficult, mentions first, must lead the model to overfit on these large mentions and fail to recover smaller nested mentions when the largest ones are wrongly predicted. The greedy training reaches an intermediate performance, so we conclude that a learned prediction order is suboptimal.

4 Conclusion

This paper proposes an architecture to perform named entity recognition based on iterative predictions and dynamic mention matching during training. We also provided insights into the model behavior and showed that training depth mention order impacts performance on auto-regressive layered named entity recognition models, and short-to-large order obtains the best results.

References

1. Cardon, R., Grabar, N., Grouin, C., Hamon, T.: Presentation of the DEFT 2020 Challenge: open domain textual similarity and precise information extraction from clinical cases. In: Actes de TALN (2020)
2. Finkel, J.R., Manning, C.D.: Nested named entity recognition. In: Proceedings of the 2009 Conference on Empirical Methods in Natural Language Processing (2009)
3. Grabar, N., Claveau, V., Dalloux, C.: CAS: French Corpus with Clinical Cases. In: Proceedings of the Ninth International Workshop on Health Text Mining and Information Analysis
4. Ju, M., Miwa, M., Ananiadou, S.: A neural layered model for nested named entity recognition. In: Proceedings of the 2018 NAACL Conference (2018)
5. Kim, J.D., Ohta, T., Tateisi, Y., Tsujii, J.: GENIA corpus - A semantically annotated corpus for bio-textmining. Bioinformatics **19**(SUPPL. 1), i180–i182 (2003)
6. Kingma, D.P., Ba, J.L.: Adam: a method for stochastic optimization. In: 3rd International Conference on Learning Representations, ICLR (2015)
7. Lee, J., et al.: BioBERT: a pre-trained biomedical language representation model for biomedical text mining. Bioinformatics **36**(4), 1234–1240 (2020)
8. Lin, H., Lu, Y., Han, X., Sun, L.: Sequence-to-nuggets: nested entity mention detection via anchor-region networks. In: Proceedings of the 57th Annual Meeting of the Association for Computational Linguistics (2019)
9. Martin, L., et al.: CamemBERT: a tasty french language model. In: Proceedings of the 58th Meeting of the Association for Computational Linguistics (2019)
10. Névéol, A., et al.: Clinical information extraction at the CLEF eHealth evaluation lab 2016. CEUR Workshop (2016)
11. Raghavan, P., Chen, J.L., Fosler-Lussier, E., Lai, A.M.: How essential are unstructured clinical narratives and information fusion to clinical trial recruitment? AMIA Joint Summits on Trans. Sci. Proc. (2014)
12. Shibuya, T., Hovy, E.: Nested named entity recognition via second-best sequence learning and decoding. Transactions of the ACL 8 (2019)
13. Sohrab, M.G., Miwa, M.: Deep exhaustive model for nested named entity recognition. In: Proceedings of the 2018 Conference EMNLP (2018)
14. Straková, J., Straka, M., Hajic, J.: Neural architectures for nested NER through linearization. In: Proceedings of the 57th ACL Meeting (2019)
15. Vaswani, A., et al.: Attention is all you need. Adv. Neural Inf. Process. Syst. (2017)
16. Wang, B., Lu, W.: Neural segmental hypergraphs for overlapping mention recognition. In: Proceedings of the 2018 EMNLP Conference (2018)
17. Wang, J., Shou, L., Chen, K., Chen, G.: Pyramid: A layered model for nested named entity recognition. In: Proceedings of the 58th ACL Meeting (2020)

The Effectiveness of Phrase Skip-Gram in Primary Care NLP for the Prediction of Lung Cancer

Torec T. Luik[1], Miguel Rios[1], Ameen Abu-Hanna[1], Henk C. P. M. van Weert[2], and Martijn C. Schut[1(✉)]

[1] Department of Medical Informatics, Amsterdam Public Health Research Institute, Amsterdam UMC, University of Amsterdam, Amsterdam, Netherlands
{t.t.luik,m.c.schut}@amsterdamumc.nl
[2] Department of General Practice/Family Medicine, Amsterdam Public Health Research Institute, Amsterdam UMC, University of Amsterdam, Amsterdam, Netherlands

Abstract. Neural models that use context-dependency in the learned text are computationally expensive. We compare the effectiveness (predictive performance) and efficiency (computational effort) of a context-independent Phrase Skip-Gram (PSG) model and a contextualized Hierarchical Attention Network (HAN) model for early prediction of lung cancer using free-text patient files from Dutch primary care physicians. The performance of PSG (AUROC 0.74 (0.69–0.79)) was comparable to HAN (AUROC 0.73 (0.68–0.78)); it achieved better calibration; had much less parameters (301 versus > 300k) and much faster (36 versus 460 s). This demonstrates an important case in which the complex contextualized neural models were not required.

Keywords: Prediction models · Deep learning · Word embeddings · N-Grams · Phrase skip-gram · Cancer · Primary care

1 Introduction

As not much progress has been made during the last decades in early detection of cancer we need a new approach. Improvement in early detection of cancer might come from new, so far unknown, cues that might be present in the patient's consultation notes. This information is not readily accessible for processing, but recent developments in natural language processing (NLP) provide ways to learn representations of free-text to capture relevant semantics allowing their use as predictors in prediction models.

The aim of this study is to compare the effectiveness (predictive performance) and efficiency (computational effort) of the context-independent model Phrase Skip-Gram [1] and the contextualized model Hierarchical Attention Network [2] for early prediction of lung cancer with Dutch free-text from primary care patient files.

A. Tucker et al. (Eds.): AIME 2021, LNAI 12721, pp. 433–437, 2021.
https://doi.org/10.1007/978-3-030-77211-6_51

2 Materials and Methods

Population. The Amsterdam UMC primary care HAG-net database contained at the time data of six primary care practices with over 100,000 historical patients. Per consultation visit the primary care physician (PCP) writes four types (S, O, A, and P) of free-text notes in Dutch for Subjective [reason for encounter and symptoms], Objective [signs or findings], Analysis [diagnosis] and Plan [actions]. Patients over 30 years old that had historical data were included. Lung cancer was identified by the ICPC code R84 (Bronchus/lung malignancy). For each patient, we included two years of data. For patients diagnosed with lung cancer, this period was from two years prior up to one month before the diagnosis (in order to reduce suspicion bias). For patients without an R84 diagnosis, we use the same period (two years data with one month offset), but up to their last visit. Patients that did not visit the PCP in the selected two years were excluded, as they have no input data.

Data Extraction. We de-identified the notes with a modified version of DEDUCE [3], then notes were tokenized into words, and white spaces were stripped. A token was added to indicate the type of note (S, O, A, or P). We applied skip-gram pre-training [4] and phrase detection to concatenate multiple tokens. Data was randomly split into a stratified 60-20-20% training/validation/test sets. Performance was assessed by the Brier score, AUROC, AUPRC, and calibration curves. Confidence intervals and statistical differences were obtained with percentile bootstrapping with 2,000 repetitions. Computational effort was measured by the number of trainable parameters, and CPU time on inference. Model selection was based on the AUROC on the validation set.

Learning Algorithms. We implemented PSG by extending the publicly available code from gensim[1] to work with *n*-grams where $n > 2$. We computed PSG from bi-grams up to 5-grams to find the best size of phrases given performance. Patient's notes were represented as a bag-of-*n*-grams (i.e., words and phrases) with frozen pre-trained embeddings, and the average of the embeddings was used to compute a final patient representation for a linear prediction layer (see Fig. 1a). For PSG we performed manual tuning. For our implementation of HAN (see Fig. 1b), the first set of layers represents words in a patient note, and the second layer represents a sequence of such notes for the final patient representation. The encoders use bidirectional Gated Recurrent Unit (BiGRU) [5]. The HAN hyper-parameters were optimized with Botorch [6] (parameter values are reported in supplementary Table 1). We used a class weighting strategy, inversely proportional to the class's occurrence rate in the training set, to deal with class imbalance, and recalibrated the predicted probabilities accordingly. We used a desktop computer with an Intel® Xeon® W-2175 CPU @ 2.50 GHz (14 cores), 64 GB RAM and an NVIDIA Quadro P4000 GPU with 8 GB memory.

Supplementary materials are available on https://osf.io/7uks8/.

[1] https://radimrehurek.com/gensim/models/phrases.html

3 Results

We included 58,169 patients older than 30 years of which 450 (0.77%) were diagnosed with lung cancer. The majority of the population (68%) was between 30 and 60 years old. Lung cancer patients had more visit notes (84) than patients without lung cancer (59). Supplementary Table 2 shows descriptive statistics of the study population. Table 1a shows the predictive performances of 3-grams PSG and HAN on the test set. (For n = 3, PSG performance on the validation set was highest; see Supplementary Table 3). No statistically significant differences were measured between PSG and HAN's performance.

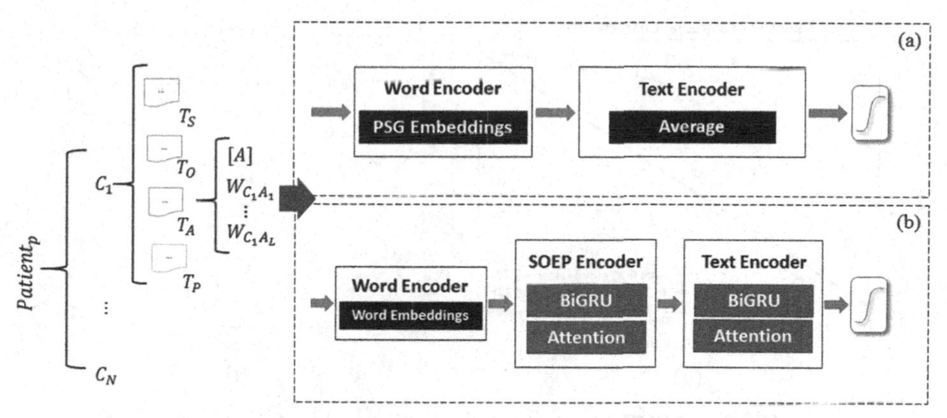

Fig. 1. (a) Architecture of the Phrase Skip-Gram model. (b) Architecture of the Hierarchical Attention Network model. A patient has consultation notes [C], [T] is a SOAP section, and [W] denotes words or phrases.

Table 1. Predictive performance (a) and computational effort statistics (b) of 3-grams Phrase Skip-Gram and Hierarchical Attention Network on the test set.

Model	PSG	HAN
(a) Performance		
AUROC (95% CI)	0.742 (0.690 – 0.788)	0.730 (0.679 – 0.775)
AUPRC (95% CI)	0.020 (0.014 – 0.028)	0.021 (0.014 – 0.033)
Brier (95% CI)	0.008 (0.006 – 0.009)	0.008 (0.006 – 0.009)
(b) Effort		
Trainable parameters	301	316,954
Frozen parameters	88,644,900	53,606,700
Test patient encoding time (s)	36.01	N/A
Test inference time (s)	0.382	460
Test inference total time (s)	36.39	460

Table 1b shows the computational effort for 3-grams PSG and HAN in terms of number of parameters and required inference time. Inference time was split into patient encoding (only done by PSG and only required once) and prediction. HAN needs to estimate 1,000 times as many parameters as PSG, and PSG (36 s) was over ten times faster than HAN (460 s).

Figure 2a shows the ROC curves of the 3-grams PSG and HAN on the test set. Figure 2b shows the calibration curves of PSG and HAN models. PSG is better calibrated than HAN. The predictions from HAN are overconfident. Brier scores for the PSG were 0.0077 (95% CI: 0.0062–0.0092) and for HAN is 0.0077 (95% CI: 0.0062–0.0093), which are comparable to the Brier score of a model predicting the mean probability.

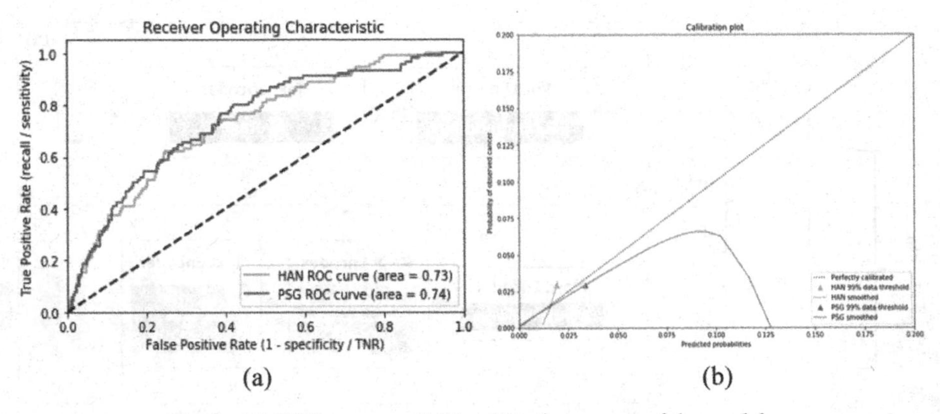

Fig. 2. (a) ROC curves and (b) calibration curves of the models.

4 Conclusions

Our study shows that in our context of oncological prediction models that use primary care free text, simpler context-independent neural models are a viable alternative to the prevailing more complex contextualized neural models. We observed that the PSG achieves similar predictive performance in terms of overall performance, discrimination and calibration to the HAN but at a fraction of the computational effort. The training effort difference between PSG and HAN is even greater than the inference effort, but not reported as they ran on different hardware (CPU vs GPU). Particularities of the data that may explain these results are: the notes are typically quite short; there are not many notes per patient, the lung cancer patients tend to have many notes; and the language style of the PCP resembles shorthand and contains many typographical errors and abbreviations.

In related work, Gao et al. [7] use an HAN with word embeddings for the sequential representation of cancer pathology reports in the context of predicting the primary site of a tumor. As for neural models for text representation, Agibetove et al. [8] compared a simple word representation model based on bag-of-n-grams to a complex sequential model (recurrent neural networks) for classifying sentences on biomedical papers. Like in our work, they observed that the simple representation model achieves a comparable performance to the sequential models with faster training times.

Our work has limitations. Both coding and timing of registration of the diagnoses in primary care is variable, which potentially leads to considering the wrong moments of diagnosis in our study and as we did not use validated diagnosis some patients might be erroneously classified as having lung cancer. We did not include temporal information (dates, order) of the consultations, nor did we include objective measurements like laboratory investigations.. We used bootstrapping to obtain the distribution of AUC differences, but validation was restricted to one random test set. The PSG is effective and efficient for clinical prediction but produces context-independent representations, which conflates the different meanings of a word into one vector, and it ignores the sequential nature of events in free-text.

As for future work, our prediction model can be improved by taking temporal aspects of the visits into account. Data from multiple centers can increase model performance and enable external validation. Choices for different offsets and patient history horizons may impact model performance.

But, irrespective of the mentioned limitations, an important implication of the study is that free-text prediction analyses can be run by institutions on local computational infrastructure as they may not have access to high performance computing infrastructure which have high operating costs, may not be used for confidential data and have limited accessibility for non-technical users.

References

1. Mikolov, T., et al.: Distributed representations of words and phrases and their compositionality. arXiv preprint arXiv:1310.4546 (2013)
2. Yang, Z., et al.: Hierarchical attention networks for document classification. In: Proceedings of the 2016 Conference of the North American Chapter of the Association for Computational Linguistics: Human Language Technologies (2016)
3. Menger, V., et al.: DEDUCE: a pattern matching method for automatic de-identification of Dutch medical text. Telem. Inform. **35**(4), 727–736 (2018)
4. Mikolov, T., et al.: Efficient estimation of word representations in vector space. arXiv preprint arXiv:1301.3781 (2013)
5. Bahdanau, D., Cho, K., Bengio, Y.: Neural machine translation by jointly learning to align and translate. arXiv preprint arXiv:1409.0473 (2014)
6. Balandat, M., et al.: BoTorch: a framework for efficient Monte-Carlo Bayesian optimization. Adv. Neural Inf. Process. Syst. **33** (2020)
7. Gao, S., et al.: Hierarchical attention networks for information extraction from cancer pathology reports. J. Am. Med. Inform. Assoc. **25**(3), 321–330 (2018)
8. Agibetov, A., et al.: Fast and scalable neural embedding models for biomedical sentence classification. BMC Bioinform. **19**(1), 1–9 (2018)

Customized Neural Predictive Medical Text: A Use-Case on Caregivers

John Pavlopoulos(✉) and Panagiotis Papapetrou

Department of Computer and System Sciences, Stockholm University, Stockholm, Sweden
{ioannis,panagiotis}@dsv.su.se

Abstract. Predictive text can speed up authoring of everyday tasks, such as writing an SMS or a URL. When deployed in a clinical setting, it can enable practitioners to compile diagnostic text reports in a speedier manner, hence allowing them to be more time-efficient when examining patients. The language used by medical practitioners when authoring clinical reports is, however, far from common, not only between practitioners but also between medical units. In this paper, we demonstrate this clinical language variation, by showing that a model trained on texts written by some physicians may not work for predicting the text of others. We use a dataset created out of the clinical notes of 17 caregivers to show that language models trained on the notes of each caregiver outperform the ones trained with texts from several ones.

Keywords: Language modeling · Predictive medical text

1 Introduction

The benefits of predictive text have been highlighted in many studies and apply to a wide range of everyday tasks and problems, one of which is producing predictive medical text in a clinical setting [8]. The main objective of predictive text is to generate the next block of text in an online and interactive manner, with *block* typically referring to a text chunk of various granularity, such as characters (or keystrokes), words, or sentences [3]. For the case of medical text, this problem is also referred to as *predictive medical text* [8].

In this paper, we highlight the benefits of predictive medical text, especially under extenuating circumstances of time pressure when hospitals are flooded with incoming patients, such as during a pandemic. In such situations the probability of errors when writing a clinical report (e.g., a discharge or any diagnostic report) increases due to lack of time or tiredness. It has been recently demonstrated that using an RNN-based language model built on clinical text [8] can achieve promising predictive performance in terms of accuracy. Nonetheless, the language used by medical practitioners when authoring clinical reports is far from common, not only between practitioners but also between medical units. Our paper pinpoints this clinical language variation by demonstrating that a

A. Tucker et al. (Eds.): AIME 2021, LNAI 12721, pp. 438–443, 2021.
https://doi.org/10.1007/978-3-030-77211-6_52

```
tolerating feeds over extended feeding times
w/o gi problems except as noted above/benign.
```

Fig. 1. Excerpt from the notes of a caregiver, where highlighted words have been suggested by a neural clinical predictive keyboard.

model trained on clinical text written by a particular group of physicians (e.g., of a specific medical unit) may not work for predicting the text of another group (Fig. 1).

Statistical language modeling, applied to clinical notes, leads to substantial keystroke reductions, and hence saving typing time for the clinician [2,11]. Neural language models, which outperform their statistical predecessors [1], lead to even greater time improvements [8]. The advantages of applying neural language modeling to medicine are still under investigation, and span from simple spelling correction in clinical notes [7,10] to predicting viral mutations related to SARS-CoV-2 [4]. In this work we focus on next word prediction in a medical setting, following the work of [8,9]. Neural predictive text is improved when structured information from electronic health data is used, such as the gender and the age of the patient [9]. Moreover, neural predictive text outperforms statistical solutions when applied to medical text, while there are benefits for clinicians even if only the most frequent words are used for prediction by the neural model [8].

Contributions. We focus on two weaknesses in the literature on predictive medical text. First, the potential of clinical predictive text is disregarded, introducing medical errors. The impact of a language mistake in a clinical report depends on the nature of the term. If the term is medical it could lead to wrong medication or treatment. Second, no work to date has explored the effect of customising the language model to the authoring physician. Hence, we build on these two directions, by (1) benchmarking the use of Long-short Term Memory (LSTM) [5] networks for next word prediction during the generation of clinical text, (2) assessing the proposed model in terms of word and medical accuracy, (3) showing that the exploitation of author-specific clinical text for building customized models per caregiver can lead to significant improvements in terms of predictive performance compared to using models trained on the whole corpus.

2 Empirical Evaluation

We present our empirical evaluation on caregiver specific datasets in terms of word accuracy and medical F1. We first provide a description of the used dataset followed by our experimental setup and results.

Datasets. We used the notes of the caregivers (CGs) from the (MIMIC-III) [6] database of 38,597 adult patients admitted between 2001–2008 to critical care units at Beth Israel Deaconess Medical Center in Boston, Massachusetts. The database comprises 1,912 CGs plus one with missing ID, with less than 150

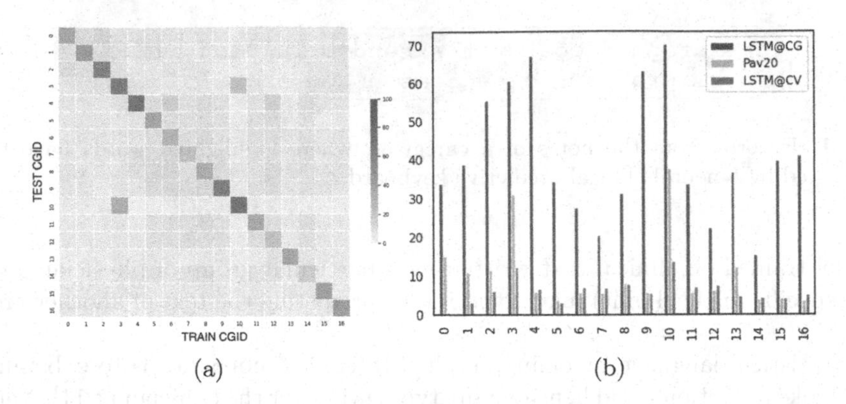

Fig. 2. Accuracy of each LSTM, trained per caregiver (CG; x axis). Blue bars (b) and the diagonal (a) refer to prediction of the next word in texts of the same CG. The accuracy of an LSTM trained on texts of all CGs, as in [8], is shown in orange (b). In green (b), it is trained on texts of all but the CG in question (cross validation; CV). (Color figure online)

of them having more than 1,000 notes each. We ranked the CGs based on the number of notes and kept the top 100. We filtered out notes with less than ten sentences and CGs who had less than 2,000 or more than 10,000 notes. We also filtered out CGs who had less than 2,000 word types overall. This resulted in 17 CGs each with 26 sentences per note on average (max: 136).[1] We sampled 10 sentences per note using 1,000 notes for training and 10 notes for testing. The train/test sentences per CG were concatenated, leading to 17 datasets, one per CG. We will refer to this dataset as MIMIC-CG.

Evaluation Measures. We employed word accuracy [8] and $F1$ combined with a confusion set of 2,061 medical terms (HARVARD lexicon).[2]

Models. We trained an LSTM [5] on texts of all caregivers. This is similar to the work of [8], where models disregarded author information, hence, this is our baseline, referred to as PAV20. In order to investigate whether caregivers could benefit from Customized language models, we trained 17 LSTMs, one per caregiver (LSTM@CG). Each model was trained on the same number of sentences as PAV20 with the same parameters. Following the work of [8,9], we used 50 dimensions for all hidden representations. We used a vocabulary of the 5,000 most frequent words; a context window of 5 preceding words; uniformly initialized word embeddings of 200 dimensions; a single-layer feed-forward neural network of 100 dimensions and a RELU activation before the softmax; Adam optimization and categorical cross entropy; batch size 128; 10% validation split; early stopping of 100 epochs with patience of 3 epochs and validation accuracy monitoring. In a second line of experiments, we employed leave-one-out cross validation. We created one more model per caregiver by training an LSTM on

[1] Sentence splitting was performed with NLTK's Punkt sentence tokeniser.
[2] https://www.health.harvard.edu/a-through-c.

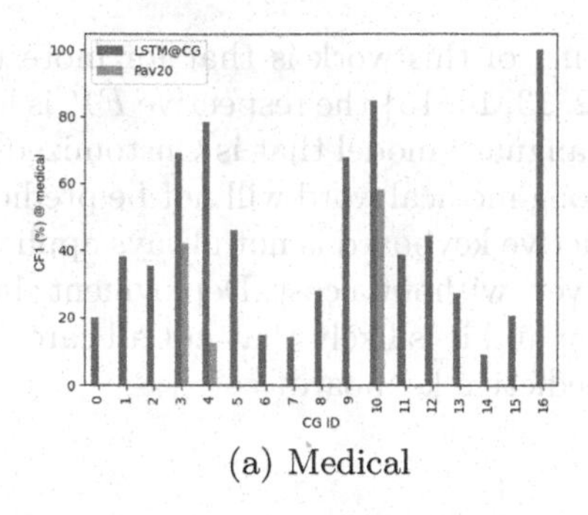

(a) Medical

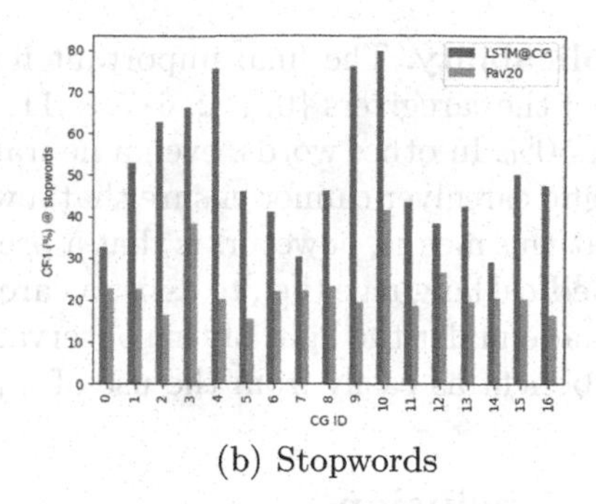

(b) Stopwords

Fig. 3. CF1 using a medical (a) and a stopwords (b) confusion set. Blue bars correspond to predictive medical text produced by LSTM@CG trained on the notes of the tested caregiver. Orange bars indicate the ones trained on notes from all caregivers. Dummy caregiver IDs are shown on the x axis. (Color figure online)

texts of all caregivers but the one in question resulting in 17 more models and 35 overall (referred to as LSTM@CV), which are compared to our baseline (PAV20).

Results. Each of the LSTM@CG models was assessed on each of the MIMIC-CG evaluation datasets (see Fig. 2a). Excluding a pair of caregivers (with dummy IDs 3 and 10), the rest perform well only to their own notes. This is reasonable given that CGs employ a different medical language defined by the patients they treat and their conditions, among other aspects. Our models consistently outperform PAV20 (trained disregarding author information). This is observed in Fig. 2b, where the blue bars (Customized models) are always higher than the orange bars (PAV20). Note that PAV20 is trained on notes of all CGs, including the one the model is being tested. When these are removed (shown in green bars), PAV20 performs much worse. In Fig. 3a we present the $F1^c$ score by using a medical confusion set (HARVARD lexicon; see Sect. 2) for all Customized models (one per CG) and the baseline model of [8]. In Fig. 3b we present the $F1^c$ score by using a confusion set of stopwords.[3] That is, we evaluate all the language models as for their predictions on stopwords instead of medical terms.

Customization. As regards medical language, i.e., the one used by caregivers, Customized models trained only on texts of the caregiver in question are advantageous. Not only next word prediction is improved in general, as shown in Fig. 2b and Fig. 3b, but Customized models also lead to safer medical use (Fig. 3a), compared to baseline models that are trained on standard corpora (e.g., disregarding the ID of the author). However, we also note that the Customized models only perform well when they are applied on notes of the same caregiver (see Fig. 2a). Despite the fact that each Customized model generalises well when texts of the same caregiver are examined (the diagonal of the heatmap), typically all models fail when they are assessed on texts of other caregivers.

[3] We used Punkt from NLTK (https://www.nltk.org/).

Applicability. The final important finding of this work is that for more than half of the caregivers $\{0, 1, 2, 6, 7, 8, 11, 12, 13, 14, 15\}$ the respective $F1^c$ is lower than 50%. In other words, even a neural language model that is Customized to a specific caregiver cannot ensure that a wrong medical word will not be predicted. What this means, however, is that a predictive keyboard is not always applicable to medical language (i.e., to assist a caregiver) without a cost. Deployment should be made under the light of this observation and it is likely that not all caregivers can benefit as easily from the use of a predictive keyboard.

3 Conclusion

In this paper, we highlight the need for customized predictive medical keyboard and provide a caregiver-specific solution using an LSTM model trained on clinical text produced by each caregiver. Our proposed model achieves substantial improvement compared to a recently proposed baseline solution in terms of word accuracy and medical F1 score. Directions for future work include the integration of biomedical, pre-trained word embeddings; assessing Transformer-based solutions; better tokenisation and sentence splitting during the text pre-processing step; and applying a manual evaluation of the solution by involving medical practitioners and qualitatively analysing the kinds of errors that are generated.

References

1. Dauphin, Y.N., Fan, A., Auli, M., Grangier, D.: Language modeling with gated convolutional networks. In: Proceedings of the 34th International Conference on Machine Learning-Volume 70, pp. 933–941. JMLR.org (2017)
2. Eng, J., Eisner, J.M.: Radiology report entry with automatic phrase completion driven by language modeling. Radiographics **24**(5), 1493–1501 (2004)
3. Gelšvartas, J., Simutis, R., Maskeliūnas, R.: User adaptive text predictor for mentally disabled Huntington's patients. Intell. Neurosci. **2016** (2016)
4. Hie, B., Zhong, E.D., Berger, B., Bryson, B.: Learning the language of viral evolution and escape. Science **371**(6526), 284–288 (2021)
5. Hochreiter, S., Schmidhuber, J.: Long short-term memory. Neural Comput. **9**(8), 1735–1780 (1997)
6. Johnson, A., et al.: MIMIC-III, a freely accessible critical care database. Scientific Data **3**, 160035 (2016)
7. Patrick, J., Sabbagh, M., Jain, S., Zheng, H.: Spelling correction in clinical notes with emphasis on first suggestion accuracy. In: 2nd Workshop on Building and Evaluating Resources for Biomedical Text Mining, pp. 1–8 (2010)
8. Pavlopoulos, J., Papapetrou, P.: Clinical predictive keyboard using statistical and neural language modeling. In: 2020 IEEE 33rd International Symposium on Computer-Based Medical Systems (CBMS), pp. 293–296. IEEE (2020)
9. Spithourakis, G.P., Petersen, S.E., Riedel, S.: Clinical text prediction with numerically grounded conditional language models. arXiv preprint arXiv:1610.06370 (2016)

10. Yazdani, A., Ghazisaeedi, M., Ahmadinejad, N., Giti, M., Amjadi, H., Nahvijou, A.: Automated misspelling detection and correction in Persian clinical text. J. Digital Imag. 1–8 (2019)
11. Yazdani, A., Safdari, R., Golkar, A., Kalhori, S.R.N.: Words prediction based on n-gram model for free-text entry in electronic health records. Health Inf. Sci. Syst. **7**(1), 6 (2019)

RETRACTED CHAPTER: Outlier Detection for GP Referrals in Otorhinolaryngology

Chee Keong Wee[1(✉)] and Nathan Wee[2]

[1] Digital Application Services, eHealth Queensland, Brisbane, QLD, Australia
ck.wee@health.qld.gov.au
[2] Faculty of Science, University of Queensland, Brisbane, QLD, Australia
nathan.wee@uq.net.au

The authors have retracted this paper because its preparation involved unauthorised use of confidential Queensland Health patients' data. The scientific content has been removed for legal reasons and to protect patient privacy. All authors agree to this retraction.

A. Tucker et al. (Eds.): AIME 2021, LNAI 12721, pp. 444–451, 2021.
https://doi.org/10.1007/978-3-030-77211-6_53

The Champollion Project: Automatic Structuration of Clinical Features from Medical Records

[illegible]

Abstract. Cancer is one of the leading causes of mortality worldwide and as population ages, the burden is growing. [illegible] ... we can [illegible].

Keywords: [illegible]

1 Introduction

The global cancer incidence [illegible] ... worldwide in 2020 [1]. In the current digital medical era, [illegible] ... various types of [illegible] images, structured clinical data, written health records. Extracting and [illegible] insights from all these data and producing [illegible] ... with [illegible].

Medical practices change over time. [illegible]. Our experience shows that [illegible] data have to be manually [illegible] and curated before [illegible]

[illegible] AIME 2021, LNAI [illegible], pp. [illegible], 2021.
https://doi.org/10.1007/978-3-030-77211-6_[illegible]

The Champollion Project: Automatic Structuration of Clinical Features from Medical Records

Oliver Hijano Cubelos(✉), Thomas Balezeau, and Julien Guerin

Institut Curie, 26, rue d'Ulm, 75005 Paris, France
oliver.hijano-cubelos@curie.fr

Abstract. Cancer is one of the leading causes of mortality worldwide and as populations age, the burden is growing. Treating increasing numbers of patients enables us to gather detailed medical records. Databases with exhaustive, high quality structured data are thus an essential resource for cancer researchers and provide invaluable information to clinicians whenever they need to treat their patients. In addition, these databases fuel our data strategy as the cornerstone of our digital healthcare ecosystem and they provide crucial support for the development of Artificial Intelligence-related projects. Feeding such databases and registries requires manual curation to ensure their quality over time. Finding alternatives to manual structuration is essential because around 80% of the relevant clinical information is contained in open text and it is costly to maintain teams of curators given the growing volumes of data generated every year. In this article we describe an Artificial Intelligence system developed at Institut Curie, capable of structuring clinical features from unstructured Electronic Health Records. Our system allows us to structure clinical data with reduced manual labor and with accuracy comparable to that of expert clinicians, empowering our data ecosystem and improving the support we can give to clinicians and researchers.

Keywords: Data mining · NLP · Machine learning · Healthcare · Structuration

1 Introduction

The global cancer incidence is estimated at 19.3 million cases and 10 million deaths worldwide in 2020 [1]. In the current digital medicine era, petabytes of data are generated every year. Looking after our patients generates data in multiple formats; genomic data, various types of medical images, structured clinical data, written health records, etc. Extracting and analyzing valuable insights from all these data and producing clean, retrospective databases with exhaustive data is a major challenge.

Medical practices change over time. Health data is produced for patient care and is usually not suitable for a straightforward usage in a research context. From a longitudinal research point of view, this is a critical issue, because the data quality level is insufficient. Our experience shows that most data have to be manually extracted and curated before it

A. Tucker et al. (Eds.): AIME 2021, LNAI 12721, pp. 452–456, 2021.
https://doi.org/10.1007/978-3-030-77211-6_54

can be used and analyzed. With the growing amounts of health data collected in various systems, computational approaches can come to the rescue.

In this contribution, we discuss the Champollion project: a pipeline to automatically structure data from our Electronic Health Records (EHR) using Machine Learning (ML) methods. As data processing technologies evolve, new powerful opportunities are available for extracting and structuring data. The great potential of this tool is enhanced given that in our experience around 80% of the key clinical information is contained in the text of health records. This data structuration has a major contribution in several fields such as the assistance to the screening of clinical trials, the early detection of errors in the medical records or the creation of databases for research purposes.

2 Methods

At the turn of the century, organizations such as our hospital initiated the migration towards digital health records. Today, our healthcare information system includes over 16 million documents related to individual medical records. Our aim is to extract data from original text automatically, mainly clinical features from EHRs. The rationale is to be able to recover new insights from EHRs whenever we need new data for a specific project. ML approaches are of particular interest because there is no need to take care of the specific lexicon used in the medical field, the diverse text formatting or the vast data heterogeneity. The purpose of our computational models is to imitate clinicians' expertise reading and interpreting medical records.

Building a generic artificial intelligence algorithm (AI) capable of extracting any number of clinical features from any kind of text is out of scope. The optimal strategy is to build and train multiple classifier models separately. Each model having the unique objective of mining one specific set of data. This divide-and-conquer strategy allows us to train one model for each clinical feature. Extracting data and inferring essential medical information are some of the challenges faced by Natural Language Processing (NLP), with quality in the process of the uttermost importance for research and clinical studies.

Both rule-based and ML algorithms have been implemented and refined over time. Building these algorithms is a particularly difficult task when dealing with medical records. These documents do not follow common syntactic structures and common typographies. Word variation is common, not only synonyms but also misspellings. Report structures are highly variable, particularly those derived from scanned documents. Medical vocabulary is rich with numerous spelling variations; many ambiguous acronyms are often used. Chronological order is critical, most sentences are context-dependent and the context needs to be handled for each identified concept (negative form, hypothesis, medical history). Physicians' reports are not often easily related to standard international terminologies. In extreme cases, some sentences might mean different things depending on the doctor's specialization or scale classification used.

Instead of working on the whole corpus of text, our models work with single sentences matching one or more keywords. Our implementation can be seen as a sequential chain of processes. We start defining a single structuration project for a given clinical feature. We then create a database of manually annotated sentences related to that feature. This database is then used to train an ML Classifier which learns from our clinicians' expertise. The classifier is finally used to extract the clinical feature from the whole corpus of text. The complete automatic structuration pipeline is shown in Fig. 1. In order to annotate sentences, we have used the *Doccano* open source text annotation tool [2]. Slight modifications of the source code have allowed us to create an appealing graphical user interface that can be used by doctors and clinicians to annotate sentences. The *Doccano* interface is connected to our EHR databases so that the annotation process and the subsequent models training are completely automatic.

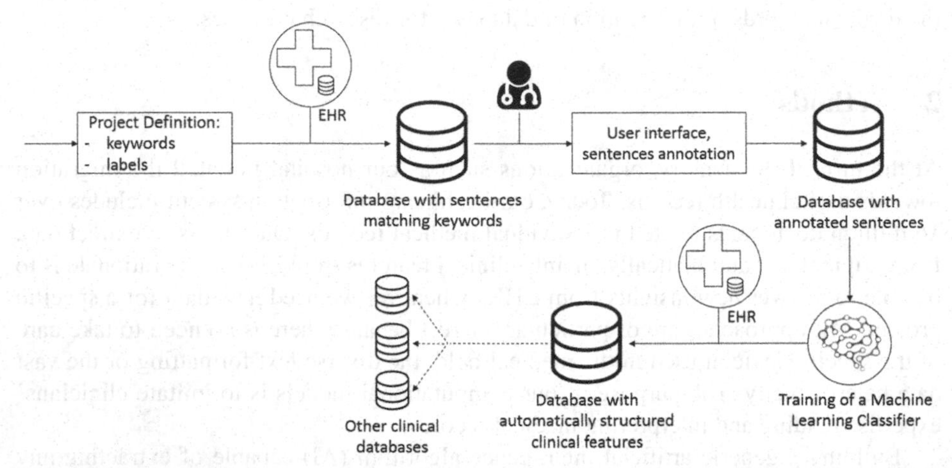

Fig. 1. General overview of the Champollion pipeline. A project is defined by a list of keywords and a list of labels. A random sample of sentences is extracted from the whole corpus of Electronic Health Records (EHR) and stored in a database. A web user interface enables clinicians to annotate sentences manually using a set of labels previously defined. Annotated sentences are stored in another database and used as a Training Dataset to optimize classifier models. Trained models are then used on the whole corpus of text to classify new sentences and store clinical features in a clinical database. Finally, extracted features can be compared and validated with other databases when available.

2.1 Project Definition

The first step of our automatic structuration process is to define a specific project for a given clinical feature. This is arguably the most critical step; in our experience structuration quality drops when projects arc ill-defined. Examples of projects are: classification of inflammatory/non-inflammatory cancer, finding hormone receptors' status (positive, negative) or assessing the pathologic staging of breast cancer (scored from 0 to IV). A project is defined by a set of labels and a list of keywords.

Keywords. A collection of words and/or regular expressions that are used to extract sentences from the whole corpus of text. They have to be exhaustive in order to classify all the existing sentences which are somewhat related to the clinical feature in question.

Labels. Every sentence is classified by a unique label. Thus, the list of labels has to be complete and exhaustive, including a label for uninformative sentences.

2.2 Sentences Annotation – Training Datasets

Once a project is defined, the next step is to build a dataset of annotated sentences that will be used to train a classifier model. Our strategy is to use the already defined list of keywords to extract 5,000 random sentences from the whole corpus of medical records and storing them in a database. Clinicians then use an ergonomic user interface to annotate some of these sentences.

2.3 Training Machine Learning Classifiers

A classifier model is composed of two elements; a vectorizer and an ML algorithm. The vectorizer transforms string sentences into mathematical vectors that are subsequently digested by the algorithm. Depending on the project, some vectorizers will be more appropriate than others. ML algorithms output a single label given an input vector. We have collected a list of the best performing Machine Learning classifiers known in the literature (Linear Regressors, SVMs, Decision Trees, Neural Nets, AdaBoost, etc.) and we have created several algorithms exploring each of their internal hyperparameters. The different combinations between vectorizers and algorithms yield an exhaustive list of classifier models. After Training The Models, We Evaluate Their Performance Using 10-Fold Cross Validation, Based On The Annotated Sentences Of The Project. The best performing model will be retained and used to classify the whole EHR corpus and structure the corresponding clinical features.

2.4 Automatic Structuration of Data

For any given project, we use the previously defined list of keywords to extract exhaustively all the matching sentences from the whole corpus of health records. The corresponding classifier model (trained in previous steps) classifies each sentence separately and each sentence is tagged with a single label. Since each sentence belongs to a particular health record with a specific date, we can then compile all the results for any given patient in a timeline and infer the true value of the clinical feature. Automatically extracted clinical features are structured in a secure MySQL database and are made available to Institut Curie's clinicians and researchers for analysis.

2.5 External Validation

For some projects we have databases with already structured data for all the patients or a sub-cohort of them. For example, estrogen/progesterone status has been manually structured in our databases since the year 2017. We have set up some automatic validation reports assessing the quality of the automatic extraction, which helps us to manually fine-tune some projects (adding/removing keywords or labels). Manually curated validation databases are a very valuable resource to fine-tune our projects and models.

3 Results and Conclusions

The quality and performance of the classifiers depend on several factors, such as the complexity of the project, the number of labels or the amount of annotated sentences. In Table 1 we show the score of four different projects we have in production. The score is measured as the percentage of correct classifications. Overall, the performance is quite high. Other projects in development show similar promising results.

Table 1. Evaluation of performance of Machine Learning Classifiers.

Project	Labels	# annotated sentences	Score
Inflammatory cancer			
The objective is to assess whether a particular tumor is inflammatory or not	Yes, No, n/a	1846	95%
Genetic mutations			
Mutation status of a given a set of genes whose mutations (or absence thereof) are known to impact the evolution of the tumor	Mutated, Not mutated, n/a	1031	93%
PD-L1 immunohistochemistry			
Percentage of colored tumor cells	A percentage in the text, n/a	508	97%
Performance status			
Described using the Zubrod scale [3]	Score 0–4, n/a	480	98%

In summary, written health records are gold mines in terms of data. We have built a tool capable of imitating clinicians' classification procedures in order to extract data around any given clinical feature. When clinicians and algorithms are comparable in accuracy, computers have the advantage of speed. This tool allows us to structure clean data that can then be queried and evaluated by clinicians and researchers alike.

References

1. GLOBOCAN 2020: New Global Cancer Data. https://www.uicc.org/news/globocan-2020-new-global-cancer-data
2. doccano, an open source text annotation tool for humans. https://github.com/doccano/doccano
3. Oken, M.M., Creech, R.H., Tormey, D.C., Horton, J., Davis, T.E., McFadden, E., Carbone, P.P.: Toxicity and response criteria of the Eastern Cooperative Oncology Group. Am. J. Clin. Oncol. **5**(6), 649–656 (1982)

Knowledge Representation and Rule Mining

Modelling and Assessment of One-Drug Dose Titration

David Riaño[1](✉) and Aida Kamišalić[2]

[1] Universitat Rovira i Virgili, Av Paisos Catalans 26, 43007 Tarragona, Spain
david.riano@urv.cat
[2] Faculty of Electrical Engineering and Computer Science, University of Maribor, Koroška cesta 46, 2000 Maribor, Slovenia
aida.kamisalic@um.si

Abstract. In health-care, medical errors are quantified. Among them, wrong dose prescriptions occur. Drug dose titration (DT) is the process by which dosage is progressively adjusted to the patient till a steady dose is reached. Depending on the clinical disease, drug, and patient, dose titration can follow different procedures. Once modeled, these procedures can serve for clinical homogenization, standardization, decision support and retrospective analysis. Here, we propose a language to model dose titration procedures. The language was used to formalize single-drug titration of chronic and acute cases, and perform retrospective analysis of the drug titration processes on 1,000 cases treated with Bisoprolol and 2,430 cases treated with Ramipril, in order to identify different types of drug titration deviations from standard DT methods.

Keywords: Drug dose titration · Medication errors · Knowledge representation · Evidence-based medicine.

1 Introduction

Only in the US, 7,000 to 9,000 people die due to medication errors and hundreds of thousands of additional medication errors occur with milder consequences [19]. Furthermore, drug-related problems such as adverse drug events, adverse drug reactions, and medication errors represent a major issue leading to hospitalization, especially in adult and elderly patients [1].

This accounts for $40 billion per year to correct the effect on the health of these patients. Among these errors, improper doses refer to overdoses (33%), underdoses (16%), or extra doses and dose omissions (16%) [5]. The WHO identifies health care professionals as one of the factors of wrong dose prescriptions and suggests the use of automated information systems such as computerized provider order entry (CPOE) with decision support as a potential solution to the problem [14]. This appreciation is observed in several recent studies such as [5] or [6]. However, these studies are made on punctual prescriptions and not on a gradual process of dose adjustment.

A. Tucker et al. (Eds.): AIME 2021, LNAI 12721, pp. 459–468, 2021.
https://doi.org/10.1007/978-3-030-77211-6_55

Dose titration (DT) is the process of adjusting the dose of a medication for the maximum benefit of the patient causing no adverse effects [13]. This optimal dose is called steady dose. A cautious approach to DT consists on starting with a low dose and progressively increasing the dose till the steady dose is reached, an adverse effect is observed, or the maximum recommended dose is reached.

Even though other dosing regimens are available, such as response-guided titration, still DT is being predominantly used [18]. Several studies covering different medical domains are investigating DT regimens [4,7,9,10] as well as proposing patient specific drug regimen models [11,12,20]. For example, Landry et al. [9] studied the efficacy and tolerance of DT vs other dosing methods, such as the age-based method and the fixed dose method in psychiatric treatments. In order to reduce the risk of over- or under-dosing, the method predominantly used and recommended is DT [9].

Modelling DT and developing tools which are based on these models could contribute to homogenize the DT processes, to generate DT standards, to identify and quantify prescription errors for clinical quality analysis, and for DT benchmark. On the other hand, having clinical data on concrete DT processes would allow us to determine the clinical evidence associated with the DT models. Additionally, once validated, DT models could be integrated with other more general clinical practice models [15] for a better medical care.

Literature review detected a lack of computerised methods for modelling DT. Following our previous work [8], here (in Sect. 2) we propose a way of constructively modelling DT procedures and confronting retrospective data about real prescriptions with the models obtained, in order to detect anomalous DT actions. We focus on the titration of one drug and formalize three approaches to DT: *single-drug* DT as the basic iteration of dose increments till the steady dose is reached [13], *chronic single-drug* DT or continued DT for long-term care involving periodic reassessment of the dose, and *acute single-drug* DT or finite DT for short-term care involving a given number of encounters for dose reconsideration.

DT models are useful tools to identify some common medical errors [2,19] concerning incorrect dose prescriptions, incorrect duration of DT processes, and wrong reassessment times. In order to confirm this utility, in Sect. 3, we include the results obtained after the application of the *single-drug* DT model to detect medical errors in the titration procedures described in two synthesized data sets. Synthesizing clinical data is a common practice when real-world data is not available or difficult to obtain [3], and it brings several benefits to real-world data processing [16,21]: accessibility, cost-efficiency, test efficiency, patient privacy protection, completeness, benchmarking, and validation. We adapted the software in [16] to synthesize 1,000 and 2,430 patient DT processes on the bisoprolol and ramipril drugs for a technical validation of the DT models and also to demonstrate the benefits of DT models in the detection of abnormal DT actions.

2 A 3-Step Modelling of Drug Dose Titration Procedures

Our analysis of DT publications (e.g., [4,7,9–13,20]) concluded with the identification of five basic formal components that can be combined to model multiple titration procedures, which on their turn, can be combined to form more complex meta-models, which describe real clinical behaviors such as single DT of chronic and acute patients, among others. We used a formalism that is similar to the ones that can be found in multiple languages to represent clinical practice guidelines to allow the easy adaptation of our methods to these languages.

2.1 Basic Components

We have identified seven basic functional components for DT formalization (see Fig. 1). Two of them, *begin* and *end*, are used to determine the points where the DT procedure starts and finishes, respectively. Five additional components are basic formal DT components used to describe the DT flow: *inquire* which is used to determine consultation points where some clinical information should be obtained (e.g., asking whether the patient is pregnant or not), *fork* which is used to define alternative (or different) DT procedures depending on the clinical characteristics of the patient (e.g., determine among alternative treatments or deciding whether a treatment has to come to an end or not). *Action* is used to state drug-dose prescription orders. Four types of such prescriptions are possible: INIT to assign initial dose, INC to increase the current dose, MAX to assign maximum dose, and NULL to cancel a drug treatment. *Delay* is another basic component to define min-max time delays. Delays are used, for example, to allow a drug to take effect after administration or to define when the dose should be reconsidered. *Join* is used to make two or more DT procedures to converge in a single common DT flow, for example, go back to a regular dose when a patient leaves a critical situation that required higher doses.

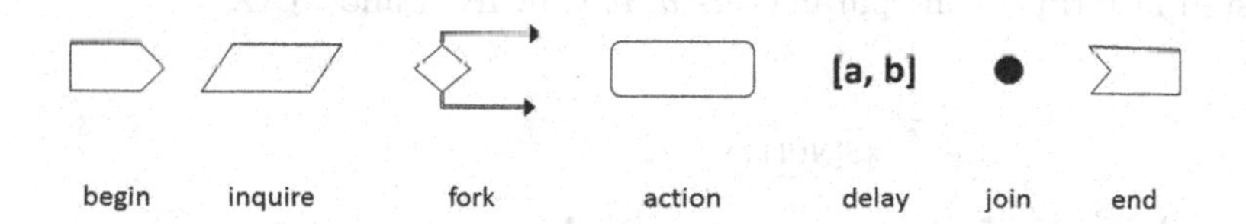

Fig. 1. Basic elements in the definition of DT procedures.

2.2 Basic Procedures

Components in the previous section are combined to describe basic DT procedures. These procedures are the basic blocks used to formalize standard DT procedures such as single-drug DT, chronic single-drug DT, or acute single-drug DT, which will be introduced in Sect. 2.3.

In this work, we leverage four basic DT procedures: initiate treatment, basic DT, cancel treatment, and delay. These are represented in Figs. 2 and 3.

Initiate Treatment. Figure 2a describes the basic steps to initiate a drug treatment. In terms of DT, it consists of assigning an initial dosage (INIT) of a given drug d, if a clinical condition c is observed, and then wait between a and b time. If the clinical condition c is not observed, *do-nothing* (i.e., keep the current dose steady) is the treatment initiated. This procedure can be adjusted in terms of the parameters a, b, c, d, and INIT. For instance, $(a, b, c, d, \text{INIT}) =$ (2w, 4w, HT, bisoprolol, 5 mg) would describe the initiation of a treatment of hypertension (HT) with bisoprolol 5 mg, whose effect is expected to be observed after 2 to 4 weeks.

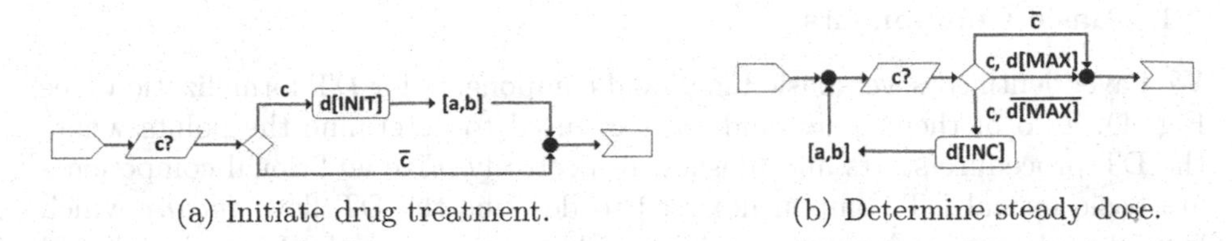

(a) Initiate drug treatment. (b) Determine steady dose.

Fig. 2. Initiate treatment and basic DT procedures. Parameters: a, b, c, d, INIT, INC, and MAX stand for min-max time delay after DT adjustment, condition for this drug treatment, drug name, and initial, increment and maximum dose for this drug, respectively.

Basic DT. Figure 2b formalizes the DT loop involved in determining a patient's steady dose. That is to say, while clinical condition c is observed (e.g., unstable high blood pressure), and the maximum dose (MAX) of the selected drug d (e.g., bisoprolol) is not reached, the dose is increased (INC) and a time in the range $[a, b]$ is left before checking whether the steady dose has been reached or not. The loop is finished either if the clinical condition c is not observed (i.e., the dose reached the expected effect) or the maximum dose MAX is reached (i.e., the drug was unable to solve the health problem). This basic procedure can be adjusted in terms of the parameters a, b, c, d, INC, and MAX.

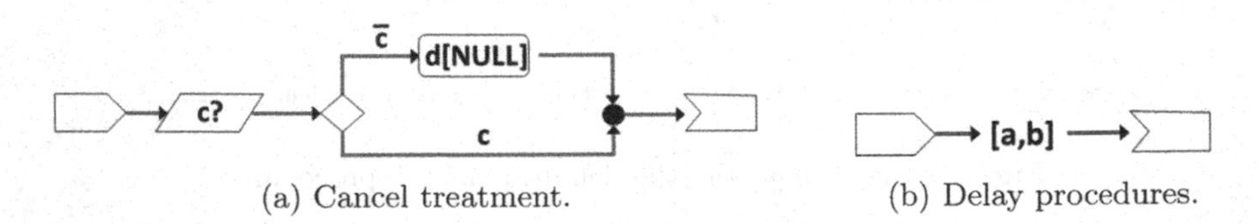

(a) Cancel treatment. (b) Delay procedures.

Fig. 3. Cancel treatment and delay DT procedures. Parameters: a, b, c, d, and NULL stand for a min-max time delay, the condition for drug treatment, drug's name, and zero-dose drug treatment, respectively.

Cancel Treatment. Under specific circumstances, the administration of one drug must come to an end. Figure 3a represents this process. Therefore, if the clinical condition c that justifies the treatment with drug d is not observed, the drug administration is cancelled (i.e., the dose is made null). For safety reasons, if condition c persists, cancel treatment is forced to have no effect. However, if it was required, other implementations of this procedure, such as forcing dose cancellation either if c is satisfied or not, or introducing a time delay after cancellation could be done. Here, the cancel treatment procedure is adjusted in terms of the parameters c and d. NULL dose is equivalent to zero dose.

Delay. Although time delays are part of some of the previous basic DT procedures, an independent delay procedure is required, for instance, to define the time for future follow-up actions to reconsider patient's treatment. Figure 3b represents this process. The procedure depends only on the parameters a and b, as minimum and maximum delay times.

2.3 Meta-models

In [8], we introduced the eTTD language, a way to formalize procedural knowledge in medicine by means of extended time-transition diagrams. Here, we use eTTD as a meta-model language to describe drug DT tasks by combining the basic procedures introduced in Sect. 2.2. Although many other DT tasks can be defined with this formalism, this paper is focused on the description of three essential titration tasks: single-drug DT, chronic single-drug DT, and acute single-drug DT. See the corresponding eTTDs in Fig. 4.

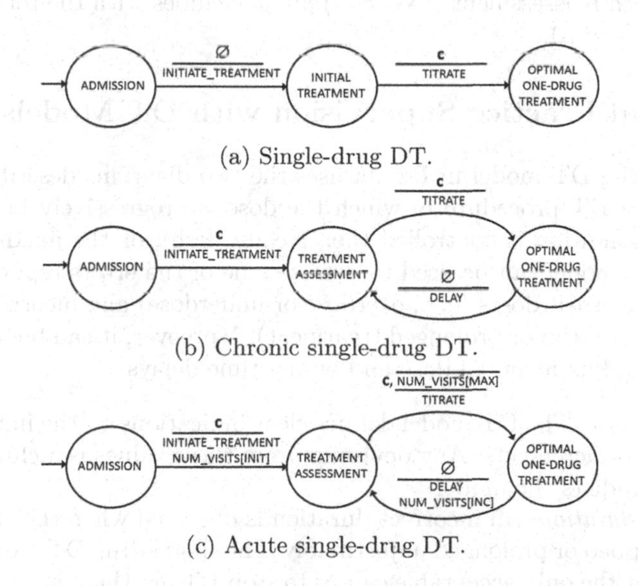

Fig. 4. Meta-models for single-drug, chronic single-drug, and acute single-drug DTs.

Single-Drug DT. Dose titration of one-drug treatment consists of two sequential steps in which the first one starts the treatment with an initial dose, and the second one iterates dose increments until the steady dose is found. In Fig. 4a, this behavior has three possible states: *admission* in which the patient is previous to drug assignment, *initial treatment* in which the patient is receiving the initial dose of the selected drug (e.g., 5 mg of bisoprolol once a day), and *optimal one-drug treatment* when the patient has reached the steady dose. Note that actions in the edges correspond to basic procedures in Sect. 2.2, once the corresponding parameters are instantiated.

Chronic Single-Drug DT. The treatment of chronic conditions such as hypertension (HT) or chronic heart failure (HF) implies a long-term DT in which steady dose has to be reconsidered from time to time. Figure 4b formalizes this procedure with an eTTD in which the *optimal one-drug treatment* state of the patient has to be considered after a time delay and the treatment (i.e., the correct dose) reassessed.

Acute Single-Drug DT. Acute cases require the application of a drug for a short-term period (usually days or weeks). In these cases, reassessment of the dose is done a specific number of times. Figure 4c accomplishes this with NUM_VISITS, a counter that starts with the first visit (INIT = 1), is incremented in each reassessment (INC = 1) and concludes with the total number of n visits (MAX = n).

3 Clinical Practice Supervision with DT Models

The single-drug DT model in Fig. 4a uses the two diagrams described in Fig. 2 to formalize a DT procedure in which the dose is progressively increased until the patient condition is controlled (i.e., steady dose), or the maximum dose is reached. This model can be used to detect some of the errors reported in [2,19] concerning incorrect doses (e.g., overdose or underdose) and incorrect durations (e.g., premature stop or prolonged treatment). Moreover, it can be used to check for incorrect adjustments of dose and wrong time delays.

- *Incorrect dose*: The DT model defines clear indications on the initial dose and the allowed increments. Any deviation from these values is a clinical decision that contradicts the model.
- *Incorrect duration*: An incorrect duration is observed when the DT process is either stopped or prolonged unjustifiably. The single-drug DT model in Fig. 4a states that the only acceptable causes to stop DT are that the patient reaches the steady dose (i.e., the clinical condition is under control) or the maximum dose (i.e., it was impossible to control the patient with the use of this drug). Any other discharge represents an unjustified cessation of titration. Similarly, DT is incorrectly continued if it does not stop when the patient reaches either the steady or the maximum doses.

- *Incorrect dose adjustment*: In this study, DT actions correspond to dose increments. An incorrect dose increment corresponds to episodes where the patient condition is stabilized and the clinical order is to keep increasing the dose.
- *Wrong time delay*: The DT model determines two delays, one that is produced when the patient receives the initial dose, and another one when doses are incremented. Delays are formalized as time intervals. Any clinical action that does not respect these intervals should be considered as a wrong time delay order.

The ability to detect such DT errors by the introduced models is analyzed in two case studies: (1) Bisoprolol dose titration[1] and (2) Ramipril dose titration[2].

Bisoprolol is a beta blocker commonly used for heart diseases such as hypertension (HT) or heart failure (HF). Bisoprolol DT procedure is formalized in Fig. 5. The recommended doses are as follows. Initial dose: 5 mg orally once a day (od). Dose titration: If desired response is not achieved, may increase the dose to 10 mg, then 20 mg if necessary. Maximum dose: 20 mg per day.

Ramipril is an ACE inhibitor medication used to treat HT, HF, and diabetic kidney disease (DKD). Recommended doses are: Initial 2.5 mg od, increment: 5 mg od, and max: 10 mg od.

DT revision of both drugs is recommended every 2 to 4 weeks.

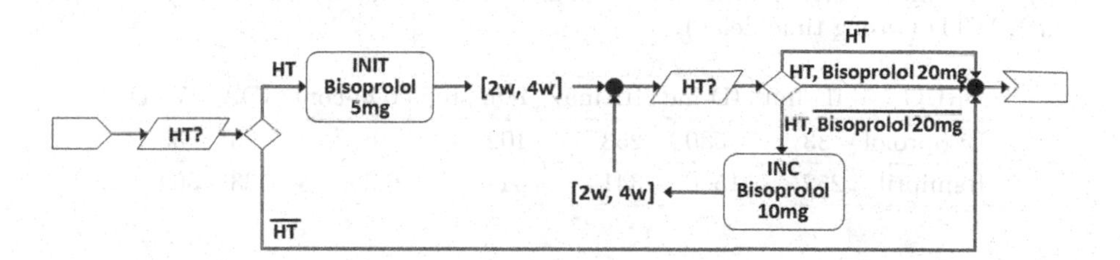

Fig. 5. Bisoprolol DT model.

The software in [16] was adapted to synthesize data about DT clinical processes. For Bisoprolol, DT processes corresponding to 1,000 patients with HT were generated. For Ramipril, 2,430 HF cases were synthesized. These data sets are made available at http://banzai-deim.urv.cat/repositories/Data/titration/. Table 1 shows some descriptive information on these data sets.

In both data sets, the average number of DT actions per case is almost two, with some patients having long DT processes involving up to 10 DT actions (Max. column) which could drive to overdose. The NCD column shows the number of patients whose DT finished before the clinical cause for medication was under control (i.e. the patient still had uncontrolled HT or HF, respectively). These represent 11.9% of HT cases and 12.8% of HF cases. The table also shows

[1] https://www.drugs.com/dosage/bisoprolol.html.
[2] https://www.drugs.com/dosage/ramipril.html.

Table 1. Description of data sets in terms of the number of patients (N), number of episodes of care (N-eoc), average and maximum number of DT encounters per patient (Avg., Max.), number of patients discharged ill (NCD or not controlled discharges), and number of patients discharged with maximum dose (MDD or maximum dosage discharged).

DRUG	N	N-eoc	Avg.	Max.	NCD	MDD
Bisoprolol	1,000	1926	1.9260	10	119	120
Ramipril	2,430	4613	1.8984	10	767	311

(MDD column) the number of discharged cases with overdose: 12% of the cases finished the DT process with Bisoprolol doses above 20 mg, the maximum dose, and 12.8% of the cases ended with Ramipril overdoses.

When the *single-drug* DT eTTD for bisoprolol in Fig. 5 (and the corresponding eTTD for ramipril) are confronted to the respective data sets, they are able to detect the DT deviations that are quantified in Table 2.

Table 2. Quantification of DT errors: ID-init (incorrect dose: INIT), ID-inc (incorrect dose: INC), ID-max (incorrect dose: MAX), IDu-stop (incorrect duration: premature stop), IDu-cont (incorrect duration: prolonged treatment), IDA (incorrect dose adjustment), WTD (wrong time delay).

DRUG	ID-init	ID-inc	ID-max	IDu-stop	IDu-cont	IDA	WTD
Bisoprolol	33	530	253	102	98	96	55
Ramipril	252	1506	341	644	637	533	364

Thirty three (or 3.3%) of the HT cases started with an incorrect initial dose, as well as 9.9% of the HF cases (ID-init column). We also detected 530 and 1,506 cases with wrong dose increments or increments which are different from the recommendations. And registered 253 DT moments in which some HT patient was overdosed, regardless whether this situation was corrected afterwards or not, and 341 maximum dose overshoots in the HF cases. In column IDu-stop, the number of incorrect premature DT stops are given (i.e., number of cases whose DT process is finished with the clinical condition uncontrolled and still room to increase the dose). The incorrectly prolonged treatments is quantified in column IDu-cont. These are the cases for which the DT process continues even if the medical problem is under control or there is no margin for DT. Column IDA contains the number of times that a patient clinical condition is solved but the physician prescribes a dose increment, with respective percentages 9.6% and 21.93% of the total cases. And column WTD counts the number of times a patient DT is reassessed after a time that is not in the range determined by the official DT model, which in our case is between 2 and 4 weeks.

4 Conclusions

Numerous studies indicate the extent of drug-related problems and its consequences for the society, especially influencing older and comorbid patients [1]. There is a need for tools to validate DT process to prevent medication errors.

We propose a versatile DT process representation language that can be used to model multiple DT strategies, whereby here three among them were presented: single-drug DT, chronic single-drug DT, and acute single-drug DT. These models can be used to detect wrong DT actions in clinical databases. Presented case studies show the capacity of these models to detect incorrect dose prescriptions, incorrect DT process durations, incorrect dose readjustments, as well as wrong DT time delays. Therefore proposed methodology could be of much interest in the detection of medication errors.

Our future work provisions building other DT models such as models for alternative drug treatments and multi-drug treatments. Furthermore, side-effects and intolerances will be incorporated in the DT modelling [8]. The system will be retrospectively and prospectively tested on real-world data from the Community Healthcare Center Maribor, Slovenia. Since the major risk factors associated with drug-related problems are old age, polypharmacy and comorbidities [1], we will give a special emphasis on complex patient treatments due to the comorbidities [17].

Acknowledgements. The authors acknowledge financial support from the Slovenian Research Agency (Research Core Funding No. P2-0057) and the Spanish Ministry of Science and Innovation (Funding Code PID2019-105789RB-I00).

References

1. Al Hamid, A., Ghaleb, M., Aljadhey, H., Aslanpour, Z.: A systematic review of hospitalization resulting from medicine-related problems in adult patients. Br. J. Clin. Pharmacol. **78**(2), 202–17 (2014). https://doi.org/10.1111/bcp.12293
2. Aronson, J.K.: Medication errors: definitions and classification. Br. J. Clin. Pharmacol. **67**(6), 599–604 (2009). https://doi.org/10.1111/j.1365-2125.2009.03415.x
3. Buczak, A.L., Babin, S., Moniz, L.: Data-driven approach for creating synthetic electronic medical records. BMC Med. Inform. Decis. Mak. **10**, 59 (2010). https://doi.org/10.1186/1472-6947-10-59
4. Carroll, R., Mudge, A., Suna, J., Denaro, C., Atherton, J.: Prescribing and up-titration in recently hospitalized heart failure patients attending a disease management program. Int. J. Cardiol. **216**, 121–27 (2016). https://doi.org/10.1016/j.ijcard.2016.04.084
5. Corny, J., et al.: A machine learning-based clinical decision support system to identify prescriptions with a high risk of medication error. J. Am. Med. Inform. Assoc. **27**(11), 1688–1694 (2020)
6. Gates, P.J., Meyerson, S.A., Baysari, M.T., Westbrook, J.I.: The prevalence of dose errors among paediatric patients in hospital wards with and without health information technology: a systematic review and meta-analysis. Drug Safety **42**(1), 13–25 (2018). https://doi.org/10.1007/s40264-018-0715-6

7. Hickey, A., et al.: Improving medication titration in heart failure by embedding a structured medication titration plan. Int. J. Cardiol. **224**, 99–106 (2016). https://doi.org/10.1016/j.ijcard.2016.09.001
8. Kamišalić, A., Riaño, D., Kert, S., Welzer, T., Nemec Zlatolas, L.: Multi-level medical knowledge formalization to support medical practice for chronic diseases. Data Knowl. Eng. **119**, 36–57 (2019). https://doi.org/10.1055/s-0040-1702016
9. Landry, M., Lafrenière, S., Patry, S., Potvin, S., Lemasson, M.: The clinical relevance of dose titration in electroconvulsive therapy: a systematic review of the literature. Psychiat. Res. **294** (2020). https://doi.org/10.1016/j.psychres.2020.113497
10. Michel, M.C., Staskin, D.: Understanding dose titration: overactive bladder treatment with fesoterodine as an example. Eur. Urol. Suppl. **10**, 8–13 (2011). https://doi.org/10.1016/j.eursup.2011.01.004
11. Miftahurrohmah, B., Iriawan, N., Wulandari, C., Dharmawan, Y.S.: Individual control optimization of drug dosage using individual Bayesian pharmacokinetics model approach. Proc. Comput. Sci. **161**, 593–600 (2019). https://doi.org/10.1016/j.procs.2019.11.161
12. Mirinejad, H., Gaweda, A.E., Brier, M.E., Zurada, J.M., Inanc, T.: Individualized drug dosing using RBF-Galerkin method: case of anemia management in chronic kidney disease. Comput. Methods Progr. Biomed. **148**, 45–53 (2017). https://doi.org/10.1016/j.cmpb.2017.06.008
13. Maxwell, S.: Chapter 2: therapeutics and good prescribing: choosing a dosing regime. In: Walker, B.R., Colledge, N.R., Ralston, S.H., Penman, I.D. (eds.) Davidson's Principles and Practice of Medicine, p. 34. Elsevier Health Sciences (2013). ISBN 978-0-7020-5103-6
14. Medication Errors: Technical Series on Safer Primary Care. Geneva: World Health Organization (2016). Licence: CC BY-NC-SA 3.0 IGO
15. Riaño, D., Bohada, J.A., Collado, A., Lopez-Vallverdu, J.A.: MPM: a knowledge-based functional model of medical practice. J. Biomed. Inform. **46**(3), 379–87 (2013). https://doi.org/10.1016/j.jbi.2013.01.007
16. Riaño, D., Fernández-Pérez, A.: Simulation-based episodes of care data synthetization for chronic disease patients. In: Riaño, D., Lenz, R., Reichert, M. (eds.) KR4HC/ProHealth -2016. LNCS (LNAI), vol. 10096, pp. 36–50. Springer, Cham (2017). https://doi.org/10.1007/978-3-319-55014-5_3
17. Riaño, D., Ortega, W.: Computer technologies to integrate medical treatments to manage multimorbidity. J. Biomed. Inform. **75**, 1–13 (2017). https://doi.org/10.1016/j.jbi.2017.09.009
18. Schuck, R.N., Pacanowski, M., Kim, S., Madabushi, R., Zineh, I.: Use of titration as a therapeutic individualization strategy: an analysis of food and drug administration-approved drugs. Clin. Transl. Sci. **12**(3), 236–39 (2019). https://doi.org/10.1111/cts.12626
19. Tariq, R. A., Vashisht, R., Sinha, A., et al.: Medication dispensing errors and prevention. In: StatPearls [Internet]. Treasure Island (FL): StatPearls. https://www.ncbi.nlm.nih.gov/books/NBK519065/. Accessed Jan 2021
20. Truda, G., Marais, P.: Evaluating warfarin dosing models on multiple datasets with a novel software framework and evolutionary optimisation. J. Biomed. Inform. (2019). https://doi.org/10.1016/j.jbi.2020.103634
21. Wang, Z., Myles, P., Tucker, A.: Generating and evaluating cross-sectional synthetic electronic healthcare data: preserving data utility and patient privacy. Comput. Intell. 1–33 (2021). https://doi.org/10.1111/coin.12427

TransICD: Transformer Based Code-Wise Attention Model for Explainable ICD Coding

Biplob Biswas[1], Thai-Hoang Pham[1,2], and Ping Zhang[1,2](✉)

[1] Department of Computer Science and Engineering, The Ohio State University, Columbus, OH 43210, USA
{biswas.102,pham.375,zhang.10631}@osu.edu
[2] Department of Biomedical Informatics, The Ohio State University, Columbus, OH 43210, USA

Abstract. International Classification of Disease (ICD) coding procedure which refers to tagging medical notes with diagnosis codes has been shown to be effective and crucial to the billing system in medical sector. Currently, ICD codes are assigned to a clinical note manually which is likely to cause many errors. Moreover, training skilled coders also requires time and human resources. Therefore, automating the ICD code determination process is an important task. With the advancement of artificial intelligence theory and computational hardware, machine learning approach has emerged as a suitable solution to automate this process. In this project, we apply a transformer-based architecture to capture the interdependence among the tokens of a document and then use a code-wise attention mechanism to learn code-specific representations of the entire document. Finally, they are fed to separate dense layers for corresponding code prediction. Furthermore, to handle the imbalance in the code frequency of clinical datasets, we employ a label distribution aware margin (LDAM) loss function. The experimental results on the MIMIC-III dataset show that our proposed model outperforms other baselines by a significant margin. In particular, our best setting achieves a micro-AUC score of 0.923 compared to 0.868 of bidirectional recurrent neural networks. We also show that by using the code-wise attention mechanism, the model can provide more insights about its prediction, and thus it can support clinicians to make reliable decisions. Our code is available online (https://github.com/biplob1ly/TransICD).

Keywords: ICD · Multi-label classification · Transformer-based model

1 Introduction

The International Classification of Diseases (ICD) is a health care classification system maintained by the World Health Organization (WHO) [23], that provides a unique code for each disease, symptom, sign and so on. Over 100 countries around the world use ICD codes and in the United States alone, the healthcare coding market is a billion-dollar industry [7]. In manual ICD coding, professional coders use patients' clinical records representing diagnoses and

A. Tucker et al. (Eds.): AIME 2021, LNAI 12721, pp. 469–478, 2021.
https://doi.org/10.1007/978-3-030-77211-6_56

procedures performed during patients' visits to assign codes. While it serves purposes including billing, reimbursement and epidemiological studies, the task is expensive, time-consuming and error-prone. Fortunately, the advent of machine learning approaches has paved the way for automatic ICD coding. Figure 1 illustrates an example of such ICD coding process where the coding model takes clinical text as input and outputs predicted ICD codes. It also shows that the model puts attention to subtext (highlighted in red) that is relevant to a disease, e.g. 'gastrointestinal bleeding' is related to the disease *Acute posthemorrhagic anemia' (ICD-9 code: 586).*

However, the task poses a couple of challenges. First, with more than 15,000 codes in ICD-9, it is a multi-label classification problem of high dimensional label space. Second, the majority of the codes are associated with rare diseases and hence, used infrequently, resulting in an imbalance in the dataset. Third, clinical records are noisy, lengthy and contain a large amount of medical vocabulary.

Previous well-known models [16,19] employed methods such as CNNs, LSTMs to automate ICD coding. However, CNNs and LSTMs have a weakness to encode the long sequence of discharge summaries (average token count before preprocessing ≈1500). On the other hand, a self-attention based transformer [21] model processes a sequence as a whole and thus can avoid long term dependency issue of LSTMs. Unfortunately, most pre-trained transformer models such as off-the-shelf BERTs [1,6,11] have a limitation of a smaller sequence length and the usual ones [6] experience a lot of out-of-vocabulary (OOV) words in representing clinical text. Training a transformer encoder with a pre-trained CBOW (Continuous Bag Of Words) [15] embedding of clinical tokens can mitigate both the problem of limited sequence length and OOV words. With this intuition, in this work, we present an end-to-end deep-learning model for ICD coding. Here are our contributions:

- We propose an ICD coding model that utilizes transformer encoder to obtain contextual representation of tokens in a clinical note. Aggregating those representations, we employ the structured self-attention mechanism [13] to extract label-specific hidden representations of an entire note.
- To address the long-tailed distribution of ICD codes, we apply a label distribution aware margin (LDAM) [4] loss function. For evaluation, we make a comparative analysis of our model with the well-known models on the benchmark MIMIC-III dataset [8].
- Finally, we present a case study to demonstrate visualizable attention to label-specific subtext indicating interpretability of our coding process.

2 Related Works

The study of automatic ICD coding can be traced back to the late 1990s [10,12]. Last two decades have seen quite a good number of ICD coding models with various approaches from both feature-based classical machine learning and deep learning technique. Most of these studies addressed the task as a multi-label classification problem.

Larkey and Croft [10] adopted an ensemble of K-nearest neighbors, relevance feedback and Bayesian independence to identify ICD code of a discharge

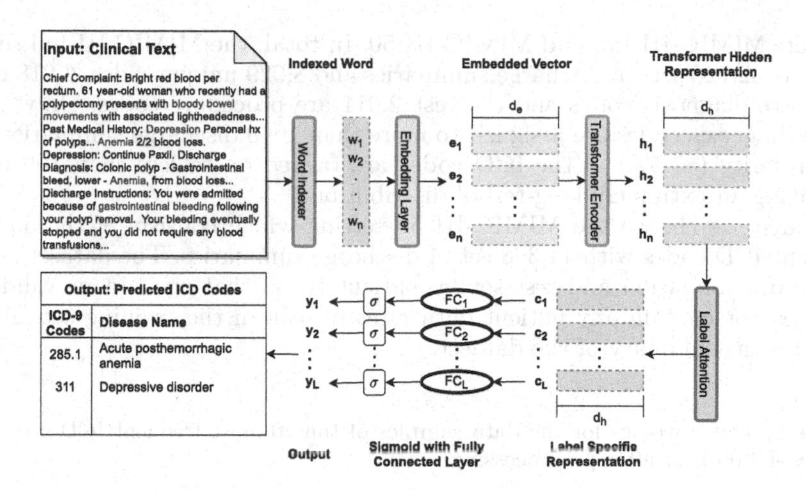

Fig. 1. The framework of the proposed ICD coding model. The model takes clinical text as input and passes it through embedding layer, transformer encoding layer, label attention layer and finally through dense layer to predict corresponding codes. (Color figure online)

summary. Both de Lima et al. [12] and Perotte et al. [17] proposed hierarchical models to capture the hierarchical relationship of ICD codes. However, the former study uses cosine similarity between the discharge summaries while the latter one employs SVM for prediction.

In the last few years, different variations of neural networks have been applied to this task. Ayyar et al. [2] and Shi et al. [19] utilized word and character level LSTM (C-LSTM-Att) respectively to capture the long-distance relationships within a clinical text. Mullenbach [16] employed the baseline models such as Logistic Regression (LR), CNN [9], Bi-GRU [5] on the MIMIC datasets for ICD coding and presented a convolutional attention network (CAML) that achieved a state of the art results. In another work [3], the authors introduced a hierarchical attention as part of a GRU-architecture that provides interpretability. Wang et al. put forward a label embedding attentive model (LEAM) [22] that encodes labels (i.e. codes) and words in the same representational space and uses cosine similarity between them for label prediction. However, being motivated by the recent success of transformer-based models [1,6,11], in our ICD coding task, we train one such encoder from scratch to circumvent sequence length limitation and learn better token representation.

3 Dataset

MIMIC-III [8] is one of the benchmark datasets that provides ICU medical records and is widely used in ICD coding prediction. Each record of it includes a discharge summary describing diagnoses and procedures that took place during a patient's stay and is labeled with a set of ICD-9 codes by professional coders. Following previous works [16], we prepare two common settings of the dataset: MIMIC-III full and MIMIC-III 50. In total, the MIMIC-III full setting contains 52,726 sets of discharge summaries and 8,929 unique codes. 6,918 of the codes are diagnosis codes and the rest 2,011 are procedure codes. Only 1.84% of the diagnosis codes are assigned to more than 1000 discharge summaries, and the majority (87.5%) of the ICD codes are tagged on to less than 100 notes, indicating an extremely long-tail of distribution.

Hence, we choose the MIMIC-III 50 setting which consists of the 50 most frequent ICD codes with 11,368 set of discharge summaries. The dataset is split into train, validation and test set by patient ID so that the test or validation set does not contain any patient data already seen in the training set. Table 1 provides the summary of the dataset.

Table 1. The statistics for the data samples of the 50 most frequent ICD-9 codes in MIMIC-III dataset after preprocessing.

Split	# Samples	# Unique Codes	# Mean Tokens	# Mean Codes	# Stdev of Code freq
Train	8,066	50	922	5.69	577.89
Validation	1,573	50	1,115	5.88	121.01
Test	1,729	50	1,133	6.03	136.93

Preprocessing. For each discharge summary sample, we lowercase and tokenize the text, remove punctuations, numbers, English stopwords, and any token with less than three characters. After that, we stem them with Snowball stemmer and replace any remaining digits with character 'n' which converts tokens such as '350 mg' to 'nnnmg'. From the resulting distribution of token count per record, we observe that more than 98% of the discharge summaries are bound within 2500 tokens. So we use 2500 as the maximum length of token sequence for training. We exploit word2vec CBOW method [15] to obtain word embeddings of size, $d_e = 128$ by training the entire discharge summary set. Finally, we extract a vocabulary of 123916 tokens from training set and augment it with 'PAD' and 'UNK' token for padding and out of vocabulary words respectively.

4 Methods

4.1 Problem Formulation

Since each discharge summary sample can have multiple ICD codes associated with it, we approach the code prediction task as a multi-label classification problem. Given a clinical record with token sequence, $\mathbf{W} = [w_1, w_2, ..., w_n]$, our objective is to determine $y_{l \in L} \in \{0, 1\}$ where L is the set of labels i.e. ICD-9 codes.

4.2 Transformer Based Label Attention Model

We leverage the concept of multi-headed self-attention, popularly known as transformer, to encode the tokens of the clinical notes. Figure 1 illustrates the overall architecture of our model. The following subsections describe the model framework in detail.

Embedding Layer. Considering an input clinical note, $\mathbf{W} = [w_1, w_2, \ldots, w_n]^T$, where w_i is the vocabulary index of the i-th word and n is the maximum possible length, we map them to the pre-trained embeddings (Sect. 3). This provides us with a matrix representation of the document, $\mathbf{E} = [\mathbf{e}_1, \mathbf{e}_2, \ldots, \mathbf{e}_n]^T$ where $\mathbf{e}_i \in \mathbb{R}^{d_e}$ is the word embedding vector for the i-th word.

Transformer Encoder Layer. The word embeddings, $\mathbf{E} \in \mathbb{R}^{n \times d_e}$ of a clinical note is fed into a transformer encoder which employs multi-headed self attention mechanism [21] to the sequence as a whole and provides us with contextual word representations, $\mathbf{H} \in \mathbb{R}^{n \times d_h}$. Mathematically:

$$\mathbf{H} = \text{TransformerEncoder}(\mathbf{E}) \tag{1}$$

where $\mathbf{H} = [\mathbf{h}_1, \mathbf{h}_2, \ldots, \mathbf{h}_n]^T$.

Code-Specific Attention Model. Being a multi-label classification task, it demands further processing of the encoded representation, $\mathbf{H} \in \mathbb{R}^{n \times d_h}$ to produce a code-wise representation. To this end, we apply a structured self-attention mechanism on $\mathbf{H}$. First, the attention weights, $a_l \in \mathbb{R}^n$ corresponding to tokens of a note for label l is computed by:

$$\mathbf{a}_l = \text{Softmax}\,(\tanh(\mathbf{HU})\mathbf{v}_l) \tag{2}$$

$$\mathbf{c}_l = \mathbf{H}^T \mathbf{a}_l \tag{3}$$

where $\mathbf{U} \in \mathbb{R}^{d_h \times d_a}$ and $\mathbf{v}_l \in \mathbb{R}^{d_a}$ are trainable parameters and d_a is a hyper parameter. Next, we multiply the contextual representation $\mathbf{H}$ and the attention scores $\mathbf{a}_l$ to produce a fixed length code-specific document representation $\mathbf{c}_l$ for each label $l \in L$ (Eq. 3). Intuitively, $\mathbf{c}_l \in \mathbb{R}^{d_h}$ encodes information sensitive to label l. Finally, we concatenate this attended document representation $\mathbf{c}_l$ for all labels to obtain $\mathbf{C} = [\mathbf{c}_1, \mathbf{c}_2, \ldots, \mathbf{c}_L]^T \in \mathbb{R}^{L \times d_h}$

Multi-label Classification. To compute the probability for label l, we feed the corresponding label-wise document representation $\mathbf{c}_l$ to a single layer fully connected network with a one node in the output layer followed by a sigmoid activation function (Eq. 4). Having the probability score, We use a threshold of 0.5 to predict the binary output $\in \{0, 1\}$. For training, we adopt multi-label binary cross-entropy as loss function (Eq. 5).

$$\hat{y}_l = \sigma(\mathbf{Z}\mathbf{c}_l + \mathbf{b}) \tag{4}$$

$$L_{BCE}(y, \hat{y}) = -\sum_{l=1}^{L}[y_l \log(\hat{y}_l) + (1 - y_l)\log(1 - \hat{y}_l)] \tag{5}$$

To address the long-tailed distribution of ICD codes in the dataset, following previous work of Song et al. [20], we employ label-distribution-aware margin (LDAM) [4], where the probability score is computed by Eq. 6.

$$\hat{y}_l^m = \sigma(\mathbf{Z}\mathbf{c}_l + \mathbf{b} - \mathbf{1}(y_l = 1)\Delta_l) \tag{6}$$

$$L_{LDAM} = L_{BCE}(y, \hat{y}^m) \tag{7}$$

where function $\mathbf{1}(.)$ is 1 if $y_l = 1$. $\Delta_l = \frac{C}{n_l^{1/4}}$ and C is a constant and n_l is the total count of training notes having l as true label. Finally, we obtain LDAM loss using Eq. 7.

5 Training Details

A search for optimal hyper-parameter leads us to the following setting of values: {Encoder layer: 2, Attention head: 8, Epochs: 30, Learning rate: 0.001, Dropout rate: 0.1}. We also set $d_a = 2 * d_e$ and $C = 3$. We train the models on an NVIDIA Tesla P100 (Pascal). In our best setting, each epoch takes around 168 s.

6 Evaluation

To evaluate our model, we utilize commonly used metrics such as micro-averaged and macro-averaged area under the ROC curve (AUC) and F1 score. As specified by Manning et al. [14], macro-averaged values are computed by averaging metrics calculated per label. On the other hand, micro-averaged values are computed considering each pair (document, code) as a separate prediction. The macro-averaged values are usually low in this task as they put more emphasis on infrequent label prediction. We also include precision at k (P@k) which computes the fraction of the true labels that are present in our top-k predictions. As the average number of codes per note is around 5.8, we choose $k = 5$ for evaluation.

6.1 Results

Table 2 provides a comparison of our proposed ICD coding model to the previous methods on the top-50 frequent ICD codes of the MIMIC-III dataset. The scores are in percentage and are measured on the held-out test set with the aforementioned hyperparameter setting (Sect. 5). We ran our model five times and use different random seeds in each run to initialize the model parameters. We present the means and standard deviations of these five runs as our final result of the proposed TransICD model. The low standard deviations indicate that our model consistently performs well, and thus it is stable.

Our proposed TransICD model produced the highest scores on micro-F1, macro-AUC, and micro-AUC, whereas the result in macro-F1 and precision@5 are comparable to the corresponding best score. Table 2 also shows that we achieved a substantial improvement from all the baselines including the recurrent networks (Bi-GRU, C-LSTM-Att) and convolutional models (CNN, CAML).

Table 2. Test set results (in %) of the proposed models on the MIMIC-III 50 dataset. Models marked with * are ours and values with boldface are the best in the corresponding column.

Models	AUC		F1		P@5
	Macro	Micro	Macro	Micro	
Logistic Regression (LR)	82.9	86.4	47.7	53.3	54.6
Bi-GRU	82.8	86.8	48.4	54.9	59.1
C-MemNN [18]	83.3	-	-	-	42.0
C-LSTM-Att [19]	-	90.0	-	53.2	-
CNN [9]	87.6	90.7	**57.6**	62.5	**62.0**
CAML [16]	87.5	90.9	53.2	61.4	60.9
LEAM [22]	88.1	91.2	54.0	61.9	61.2
*Transformer	85.2	88.9	47.8	56.3	56.5
*Transformer + Attention	88.2	91.1	49.4	59.3	59.6
*TransICD(Transformer + Attention + L_{LDAM})	**89.4** $\pm$ 0.1	**92.3** $\pm$ 0.1	56.2 $\pm$ 0.4	**64.4** $\pm$ 0.3	61.7 $\pm$ 0.3

In fact, our basic transformer model (without attention) that simply uses mean pooling over the encoded token vectors for document representation outperforms logistic regression (LR) by at least 2.3% in macro-AUC, 2.5% in micro-AUC, 0.1% in macro-F1, 3.0% in micro-F1, and 1.9% in precision@5. We believe this is due to the transformer encoder's superior ability to capture the long-term dependency of the tokens in contrast to that of recurrent units or hand-crafted feature extraction.

With code-wise attention and LDAM loss, our best setting TransICD exceeds the strong baseline LEAM [22] in macro-AUC by 1.3%, in micro-AUC by 1.1%, in macro-F1 by 2.2%, in micro-F1 by 2.5% and in precision@5 by 0.5%. In macro-F1, our model takes a back seat only to CNN [9]. The overall low scores in this metric also signify that the models struggle in predicting rare codes. The precision@5 of our model indicates that out of 5 predictions with the top probabilities, on average 61.9% i.e. 3.085 are correct. The score is relatively higher than most of the other baselines except CNN.

Comparing previous models, we observe that logistic regression (LR), being a conventional machine learning model, performs worse than all other neural networks. Further inspection reveals that the attention-based models result in a significant improvement over the normal ones of the same kind. For instance, CAML outperforms the regular CNN.

Ablation Study. The contribution of different components of our model can be recognized from the bottom three rows of Table 2. First, we notice a substantial drop in every metric when label-distribution aware margin (LDAM) loss is not adopted. In another way, LDAM improves the performance in AUC by (macro-1.2%, micro-1.2%), F1 by (macro-6.8%, micro-5.1%), and precision@5 by 2.1%. This clearly demonstrates that LDAM loss played a powerful role to counter the imbalanced frequency of the labels. Moreover, instead of the label attention, if we simply use mean pooling of the token representations from our transformer encoder to encode the entire document, we end up having the same hidden vector for all the labels. This further hurts the performance of the model. Putting differently, extending basic transformer model with code-wise attention increases AUC score by (macro-3.0%, micro-2.2%), F1 score by (macro-1.9%, micro-3%), and precision@5 by 3.1%. This corroborates that extraction of code-specific representation of a document does improve the corresponding label prediction.

6.2 Distribution of Scores

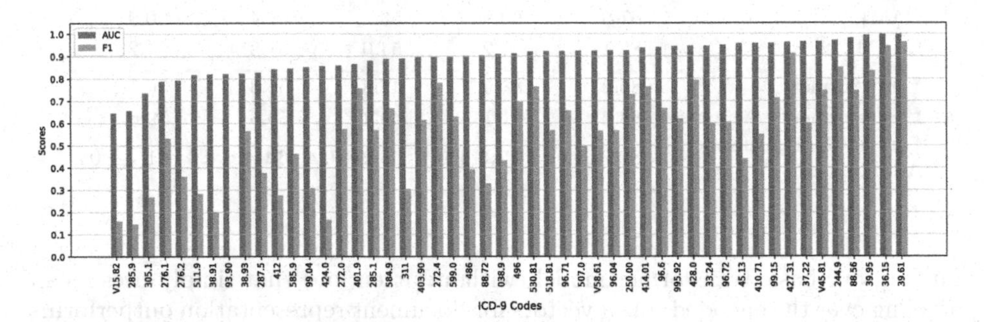

Fig. 2. AUC and F1 scores across the top-50 frequent ICD-9 codes of MIMIC-III dataset

Our model achieves higher AUC scores for many ICD codes. Specifically, for 90% of the codes, our model attains an AUC higher than 0.8 and for 56% of them, we have an AUC higher than 0.9. On the other side, an AUC score lower than 0.7 is seen for only 4% of the codes. We notice that some of the low scoring ICD codes such as V15.82, 305.1, 276.1 are also the least frequent ones in the training set. Another observation shows misclassification among closely related codes. For instance, *Tobacco use disorder (ICD: 305.1)* and *Arterial catheterization (ICD: 38.91)* are seen to be very frequently mislabeled as *History of tobacco use (ICD: V15.82)* and *Venous catheterization, not elsewhere classified (ICD: 38.93)* respectively. Above all, most frequent wrongly classified codes such as 401.9, 96.04 are also found to be the dominant ones in the training set indicating a bias towards them. A naive random oversampling of the dataset can be a way to get rid of such bias. Analyzing F1 scores, we find a relatively smaller number of the codes (10%) having F1 score greater than 0.8. We present the individual AUC and F1 score of the most frequent 50 codes in Fig. 2.

6.3 Visualization

For high-stakes prediction applications such as healthcare, there has been an increasing demand to explain the prediction of a model in a way that humans can understand. Although an automated model is set to reduce human labor, being able to observe which parts of a text are contributing to the final prediction

(a) first namen diagnosis ascending descending aortic aneurysm coronary artery disease hypertension mitral valve prolapse history rheumatic fever elevated hemoglobin anc drug rash secondary ciprofloxacin preoperative urinary tract infection condition good instructions driving one month

(b) first namen diagnosis ascending descending aortic aneurysm coronary artery disease hypertension mitral valve prolapse history rheumatic fever elevated hemoglobin anc drug rash secondary ciprofloxacin preoperative urinary tract infection condition good instructions driving one month

Fig. 3. Visualization of the model attending on an excerpt from a discharge summary for label- (a) *Urinary tract infection (ICD: 599.0)* and (b) *Single internal mammary-coronary artery bypass (ICD: 36.15)*. Darker color indicates higher attention.

provides reliability and transparency. In Fig. 3, we provide such visualization of our code-wise attention model where an excerpt of a note is highlighted with attention scores corresponding to two different labels.

Figure Subsect. 6.3 shows that for disease *Urinary tract infection (ICD: 599.0)*, our model successfully puts high attention to the closely related words- 'urinary tract infection'. However, the model ignores the same words while predicting for label *Single internal mammary-coronary artery bypass (ICD: 36.15)* as illustrated in Figure Subsect. 6.3 because they are not relevant for the latter label. On the other hand, being associated with the latter label, 'coronary artery' is seen to gain more attention in Figure Subsect. 6.3, although the same bi-gram is not attended for the former label in Figure Subsect. 6.3.

All these suggest that the reasoning of our model is highly correlated to the features that a human would have looked for while tagging a note with ICD codes. Consequently, we believe, this model would help clinicians in the ICD coding process with higher reliability and transparency.

7 Conclusion

The study proposes a transformer-based deep learning method to predict ICD codes from discharge summaries representing diagnoses and procedures conducted during patients' stay in hospital. We adopt LDAM loss to counter the imbalanced dataset and employ a code-wise attention mechanism for more accurate multi-label predictions. Our visualization report illustrates that the model attends to the relevant features and hence provides evidence for reliability. For future work, we will focus on a larger dataset containing more or even all the ICD codes.

References

1. Alsentzer, E., et al.: Publicly available clinical BERT embeddings. In: Proceedings of Clinical NLP, pp. 72–78, June 2019
2. Ayyar, S., Don, O., Iv, W.: Tagging patient notes with ICD-9 codes. In: Proceedings of NeurIPS, pp. 1–8 (2016)

3. Baumel, T., Nassour-Kassis, J., Elhadad, M., Elhadad, N.: Multi-label classification of patient notes a case study on ICD code assignment. ArXiv abs/1709.09587 (2018)
4. Cao, K., Wei, C., Gaidon, A., Aréchiga, N., Ma, T.: Learning imbalanced datasets with label-distribution-aware margin loss. In: Wallach, H.M., Larochelle, H., Beygelzimer, A., d'Alché-Buc, F., Fox, E.B., Garnett, R. (eds.) Proceedings of NeurIPS, pp. 1565–1576 (2019)
5. Cho, K., et al.: Learning phrase representations using RNN encoder-decoder for statistical machine translation. In: Proceedings of EMNLP, pp. 1724–1734, October 2014
6. Devlin, J., Chang, M.W., Lee, K., Toutanova, K.: BERT: pre-training of deep bidirectional transformers for language understanding. arXiv preprint arXiv:1810.04805 (2018)
7. Grand View Research: U.S. Medical Coding Market Size, Share and Trends Analysis Report By Classification System (ICD, HCPCS, CPT), By Component (In-house, Outsourced), And Segment Forecasts, 2019–2025 (2019). https://www.grandviewresearch.com/industry-analysis/us-medical-coding-market
8. Johnson, A.E., et al.: MIMIC-III, a freely accessible critical care database. Scientific Data **3**(1), 160035 (2016)
9. Kim, Y.: Convolutional neural networks for sentence classification. In: Proceedings of EMNLP, pp. 1746–1751, October 2014
10. Larkey, L.S., Croft, W.B.: Combining classifiers in text categorization. In: Proceedings of SIGIR, SIGIR 1996, pp. 289–297 (1996)
11. Lee, J., et al.: BioBERT: a pre-trained biomedical language representation model for biomedical text mining. Bioinformatics **36**(4), 1234–1240 (2019)
12. de Lima, L.R.S., Laender, A.H.F., Ribeiro-Neto, B.A.: A hierarchical approach to the automatic categorization of medical documents. In: Proceedings of CIKM, CIKM 1998, pp. 132–139 (1998)
13. Lin, Z., et al.: A structured self-attentive sentence embedding. CoRR abs/1703.03130 (2017)
14. Manning, C.D., Raghavan, P., Schütze, H.: Introduction to Information Retrieval. Cambridge University Press, Cambridge (2008)
15. Mikolov, T., Chen, K., Corrado, G., Dean, J.: Efficient estimation of word representations in vector space. arXiv preprint arXiv:1301.3781 (2013)
16. Mullenbach, J., Wiegreffe, S., Duke, J., Sun, J., Eisenstein, J.: Explainable prediction of medical codes from clinical text. In: Proceedings of NAACL-HLT, pp. 1101–1111, June 2018
17. Perotte, A., Pivovarov, R., Natarajan, K., Weiskopf, N., Wood, F., Elhadad, N.: Diagnosis code assignment: models and evaluation metrics. JAMIA **21**, 231–237 (2013)
18. Prakash, A., et al.: Condensed memory networks for clinical diagnostic inferencing. In: Proceedings of AAAI, AAAI 2017, pp. 3274–3280. AAAI Press (2017)
19. Shi, H., Xie, P., Hu, Z., Zhang, M., Xing, E.P.: Towards automated ICD coding using deep learning. CoRR abs/1711.04075 (2017)
20. Song, C., Zhang, S., Sadoughi, N., Xie, P., Xing, E.: Generalized zero-shot text classification for ICD coding. In: Bessiere, C. (ed.) Proceedings of IJCAI, pp. 4018–4024, July 2020
21. Vaswani, A., et al.: Attention is all you need. In: Guyon, I., et al. (eds.) Advances in NeurIPS, vol. 30, pp. 5998–6008. Curran Associates, Inc. (2017)
22. Wang, G., et al.: Joint embedding of words and labels for text classification. In: Proceedings of ACL, pp. 2321–2331, July 2018
23. WHO: International classification of diseases (ICD) information sheet: World health organization (2014). https://www.who.int/classifications/icd/factsheet/en/

Improving Prediction of Low-Prior Clinical Events with Simultaneous General Patient-State Representation Learning

Matthew Barren(✉) and Milos Hauskrecht(✉)

University of Pittsburgh, Pittsburgh, PA 15260, USA
{mpb43,milos}@pitt.edu

Abstract. Low-prior targets are common among many important clinical events, which introduces the challenge of having enough data to support learning of their predictive models. Many prior works have addressed this problem by first building a general patient-state representation model, and then adapting it to a new low-prior prediction target. In this schema, there is potential for the predictive performance to be hindered by the misalignment between the general patient-state model and the target task. To overcome this challenge, we propose a new method that simultaneously optimizes a shared model through multi-task learning of both the low-prior supervised target and general purpose patient-state representation (GPSR). More specifically, our method improves prediction performance of a low-prior task by jointly optimizing a shared model that combines the loss of the target event and a broad range of generic clinical events. We study the approach in the context of Recurrent Neural Networks (RNNs). Through extensive experiments on multiple clinical event targets using MIMIC-III [8] data, we show that the inclusion of general patient-state representation tasks during model training improves the prediction of individual low-prior targets.

Keywords: Simultaneous learning · Low-prior events · General patient-state representation · Weighted loss · LSTM · RNN

1 Introduction

Across machine learning domains, many important events are difficult to predict because of their low-prior probability. This situation is frequent in clinical event prediction, where severe events and interventions are both uncommon and imperative to foresee. To some degree, low-priors are a constraint of the task definition. For example, the prediction of first sepsis onset will have at most one positive instance per a patient hospitalization, and therefore it is constrained by design. Additionally, in temporal modeling this prior is further reduced by the frequency (e.g. predict every two hours) and time horizon of prediction.

A. Tucker et al. (Eds.): AIME 2021, LNAI 12721, pp. 479–490, 2021.
https://doi.org/10.1007/978-3-030-77211-6_57

Previous machine learning works have utilized general patient-state representations (GPSRs) [4,12,14,17,18] or transfer learning [5] as methods to deal with low-prior events. However, in both cases, there is a potential for predictive performance to be hindered by the misalignment of the general purpose model and the low-prior target. For example, in GPSR learning, it is possible that the extracted representation features obfuscate the signals from the raw inputs that are highly important for accurately predicting a septic patient.

In order to improve the prediction performance of low-prior clinical events, we propose a new method that simultaneously trains a shared model that can support both low-prior target prediction and general patient-state representation tasks. Accordingly, the parameters of the model are optimized through a two-component loss function. To better tune the model to the desired low-prior clinical task, a weight parameter is used to adjust the influence between the low-prior target and GPSR. Thus, a GPSR is learned jointly to aid a specific prediction target instead of being used as an upstream step to accommodate it. We explore our method in the context of recurrent neural networks (RNNs) with long short-term memory cells (LSTM). LSTMs have been used to define both a GPSR models for clinical sequences [11,14], as well as, a model for predicting single events from past clinical sequences [22].

We explore the benefits of our simultaneous learning method experimentally using clinical data derived from the MIMIC-III database [8] predicting three low-prior events: 72 h mortality, 6 h sepsis onset, and 2 h norepinephrine administration. These targets have priors that range from 0.0013 to 0.0109. The GPSR component of the LSTM model is defined as a broad range of clinical lab and vital sign events that are one-hot encoded to normal and abnormal values. Through extensive experiments we show that simultaneously optimizing the GPSR and the low-prior prediction task leads to models with improved prediction performance as measured by the area under the precision-recall curve (AUPRC). In addition, two ablation studies reducing the event priors and samples in the training data demonstrate the robustness of our approach.

2 Related Work

General Patient-State Representation Learning. General patient-state representations are often desirable for their ability to compress complex data into a lower dimensional representation with the goal of accurately representing the signals that are inherent to the patient-state. General patient-state representation models include a wide range of standard matrix factorization approaches and modern neural architectures models. Examples of GPSR models include Singular Value Decomposition (SVD) [15], autoencoder architectures [17,18], recurrent neural network models [4,11,12], attention mechanisms [2], and composites of the previously mentioned paradigms [14]. For example, the authors of DeepPatient used a denoising autoencoder (AE) to learn a patient representation over time windows of clinical observations and applied it predict patient diseases. The concept of autoencoding has further been applied in the space of sequence models, such as LSTMs [4,12,14].

Task Specific Model Learning. Task specific model learning optimizes parameters based only on a supervised target(s). Models are trained using a supervised loss. Especially suitable for this purpose, are autoregressive models that provide an end-to-end framework for defining the model, inputs, outputs, and target task loss. LSTMs applied to clinical tasks have been found to provide strong predictive performance, such as predicting chronic kidney injury [22].

Simultaneous GPSR and Tasks-Specific Learning. Simultaneous learning of two different task paradigms has been explored in the space of topic models. Supervised Latent Dirichlet Allocation (sLDA) combines the objective function of the expectation of a token belonging to a topic with a supervised task to guide topic learning [1]. In prediction focused sLDA (pf-sLDA) [19], the authors used weighting to examine the balance between topic and supervised tasks. In our work, we take a similar approach to pf-sLDA by jointly modeling a supervised task with a general patient representation. In contrast to the work of Ren et al., we solely focus on the performance of the supervised task, and use the general patient representation as support to improve prediction performance.

3 Methodology

3.1 Model Definition

Our objective is to learn a model $f : X \rightarrow Y$ that can predict a future target event Y from past observations X. Since past observations grow in time, X is often replaced with a fixed length summary vector S. A summary vector can be developed from a number of different strategies, such as using feature templates that featurize time-series of all clinical variables defining X [6,7] or by compressing observations to a low-dimensional space using SVD [15]. RNNs have had success with learning clinical targets by segmenting past observations X to the current time, t, into sequences of observations $X_1, X_2, \ldots, X_t$. A summary state is then defined as S_t representing the hidden state of the RNN. In this

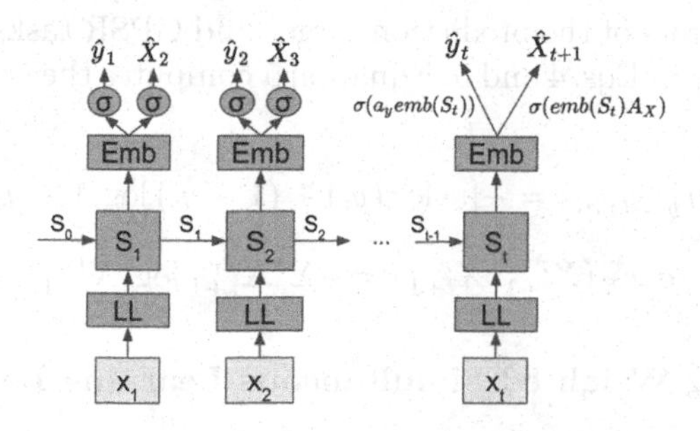

Fig. 1. The shared low-prior clinical prediction task and GPSR task architecture. LL in the model architecture stands for linear layer.

work, we consider an RNN model with Long Short-Term Memory units [3] to represent the predictive model f.

Since learning a model f for a low-prior target is sensitive to the training data size, we propose to aid its learning with a GPSR model. A GPSR model is not tailored to cover one specific prediction task, instead it tries to represent the overall data sequences and their characteristics. In this work, we consider an LSTM-based model of GPSRs. This GPSR is defined to predict future clinical observations based on the past events, $g : X_1, X_2, \ldots, X_t \rightarrow X_{t+1}$. Each one-hot encoded generic clinical task in X_{t+1} can be represented as a set of multi-class targets $R = (r_1, r_2, ..., r_d)$, where each clinical event, r_i, is a set of discretized class values for that clinical event, $r_i = (c_1, c_2, ..., c_m)$. The reason for choosing a GPSR LSTM-based model is that it can be aligned with the LSTM event prediction model f. Briefly, the GPSR LSTM model can be summarized in terms of a state, S_t, similarly to the LSTM model of the model f, but instead of just one target event, it predicts a broad range of clinical events.

The key idea of our approach is to have f and g share a set of parameters to learn both tasks simultaneously. More specifically, f and g can be defined on the same summary state, S_t, and thus can be redefined to $f' : S_t \rightarrow Y_t$ and $g' : S_t \rightarrow X_{t+1}$. The summary state can further be defined as a function of the current input and past state, $m : X_t, S_{t-1} \rightarrow S_t$. In our model a shared embedding layer, $emb(S_t)$, is applied to the summary state S_t before prediction outputs are computed. The $emb(S_t)$ layer is a linear layer with a rectified linear unit (ReLU) activation. Thus the current state can be computed from Eq. 1, which is a function of the shared architecture between the low-prior target and the GPSR. The sigmoid of the dot product between the target weight vector, a_y, and S_t are computed to get the low-prior prediction, Eq. 2. Similarly, the next-step patient representation can be computed from the sigmoid of the state vector S_t being multiplied to the GPSR weight matrix, A_X, Eq. 3. Figure 1 graphically depicts the described model definition.

$$S_t = m(X_t, S_{t-1}) \tag{1}$$

$$\hat{y}_t = f'(S_t) = \sigma(a_y emb(S_t)) \tag{2}$$

$$\hat{X}_{t+1} = g'(S_t) = \sigma(emb(S_t)A_X) \tag{3}$$

Further, the errors of the prediction target and GPSR tasks can be computed using cross entropy in Eqs. 4 and 5. Equation 5 computes the error for each GPSR task, r, in X.

$$err_y(\hat{y}_t, y_t) = -[y_t \log(\hat{y}_t) + (1 - y_t)\log(1 - \hat{y}_t)] \tag{4}$$

$$err_X^r(\hat{X}_{t+1}^r, X_{t+1}^r) = -\Sigma_c^r X_{t+1}^{r,c} \log(\hat{X}_{t+1}^{r,c}) \tag{5}$$

3.2 Optimizing Weighted Simultaneous Learning Loss

Similar to Ren et al. [19], weighting is applied to control the parameter learning, weights and LSTM gates, influence between the two objectives, f' and g'.

The loss function, Eq. 6, uses a hyperparameter, p, to weight the errors between the low-prior target and the GPSR tasks. Setting $p = 1$ results in a low-prior event only driven model, and conversely, $p = 0$ will yield parameter updates based only on the GPSR tasks. For optimizing model weights, Adaptive moment with decoupled weight decay was used (AdamW) [13].

$$l(X_{t+1}) = p(err_y(\hat{y}_t, y_t)) + (1-p)\frac{\Sigma_r^R err_X^r(\hat{X}_{t+1}^r, X_{t+1}^r)}{|R|} \tag{6}$$

4 Experiments

4.1 Simultaneous Model Architectures

In this paper, two simultaneous model architectures are proposed. A third was trained/evaluated, but excluded from the results due to it's similar performance to **Evt+GPSR** model. They each use the same base architecture shown in Fig. 1, but the latter model extends the network with additional layers that are task specific. The proposed models are the following:

- Low-Prior Event Target and GPSR (**Evt+GPSR**) (Fig. 1)
- Low-Prior Event with Linear Layer and GPSR with Multi-task Linear Layer (**EvtLL+GPSR-MTLL**) (Fig. 2d)

4.2 Baseline Model Architectures

Three baseline models are used to compare with the proposed simultaneous models. The same general model structure given in Fig. 1 is used for each with some modifications to their respective prediction objective.

- **Supervised model (RNN Spv.)** low-prior task-specific model (Fig. 2a)
- **RNN Embedding** is a GPSR model that is trained to forecast the generic clinical events. The supervised target is then learned using a single linear layer based on the features learned from the embedding model. For the experiments, this model is trained to each prediction time horizon to align with the low-prior prediction targets (Fig. 2b).
- **RNN Residual** uses the learned RNN Embedding model with additional residual layers. The learned embedding layer continues to be optimized during supervised training to allow for additional tuning to the target. The residual layers attempt to learn the low-prior target signal from raw inputs that were not captured in the GPSR embedding model (Fig. 2c).

Thus, **RNN Spv.** is equivalent to **Evt+GPSR** if the loss weight value is set equal to 1.0, $p = 1$, where only the supervised low-prior task influences parameter learning. Likewise, the other baselines utilize GPSR learning in a sequential fashion where the target is not simultaneously considered with the GPSR tasks.

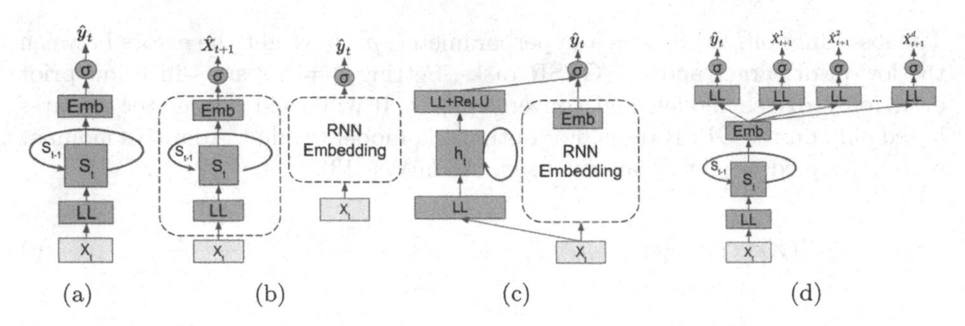

Fig. 2. Figures 2a, 2b, **and** 2c are the baseline prediction models, **RNN Spv.**, **RNN Embedding**, and **RNN Residual**. Likewise, Fig. 2d is the proposed model with an extended structure, **RNN EvtLL+GPSR-MTLL**

4.3 Low-Prior Targets

Experiments were conducted using MIMIC-III's [8] electronic healthcare record data set. The data set included ICU patients of 18 years of age and older with an inpatient time that exceeds both 24 h and the prediction horizon. Target statistics can be found in Table 1. The prediction time horizons for mortality, sepsis, and intravenous (IV) norepinephrine were 72, 6, and 2 h, respectively. The separation time between each instance in a sequence is the same as the prediction horizon except for the mortality task, which uses 24 h sequence intervals. Sepsis prediction targets were generated based on Physionet's competition [20]. The IV Norepinephrine task is a prediction of a new medication administration. A patient may have multiple administrations in a single hospital stay, and to determine a new delivery, the half-life of the medication was compared to statistics of subsequent drug time intervals. Given the short half-life of 2.5 min [21] and the distribution of subsequent administration intervals, a holdout period of 2 h after drug delivery is applied before predictions may resume (i.e. during drug administration, prediction is suspended).

Table 1. Data set statistics for each low-prior event.

	Data Sets											
	Mortality				**Norepinephrine**				**Sepsis**			
	Adms	**# Pos**	**# Neg**	**Prior**	**Adms**	**# Pos**	**# Neg**	**Prior**	**Adms**	**# Pos**	**# Neg**	**Prior**
Train	8803	700	82,797	0.0084	11, 694	948	1,156,503	0.0008	11,694	506	384,097	0.0013
Valid	2363	214	18,544	0.0114	3, 141	370	248,929	0.0013	3141	140	90,687	0.0015
Test	4850	425	38,626	0.0109	6, 535	750	589,720	0.0013	6,535	287	189,092	0.0015

4.4 Inputs and GPSR Tasks

The inputs for each task are 191 lab and vital sign observations that are discretized to a one-hot encoding of normal/abnormal or normal/abnormal low/abnormal high. This discretization is based on a knowledge base of normal ranges that were compiled from [9,10,16]. A last value carry forward (LVCF) method is applied for each observation relative to the prediction point, and in the event no prior observation exists, the encoding positions for that observation remain zero.

The generic clinical tasks for the GPSR were 189 of the 191 laboratory and vital sign observations. Two observations were excluded as a task because their presence by definition is abnormal. A LVCF was also used for the target classes, and in the event that no value exists, a normal class is imputed. For this paper, the time horizons of the generic clinical tasks aligned with the respective prediction target time. For example the 6 h sepsis target had GPSR tasks of 6 h time horizons too. This was done for all models that utilized generic clinical tasks for their GPSR.

4.5 Model Training and Selection

Since AUPRC is the primary metric for evaluating low-prior event performance, AUROC was used to determine early stopping to avoid biasing model selection on a single evaluation metric. All models were trained over a number of epochs with early stopping based on validation AUROC. Dropout was applied for the linear layers. For each model architecture the same set of layer sizes were explored along with regularization parameters on the supervised output parameters over multiple iterations. For example, both **RNN Spv.** and **Evt+GPSR** explored the same set of layer configurations. The best performing average validation AUROC determined the model hyperparameters. The GPSR embedding model was trained with early stopping based on the tolerance of the validation loss.

4.6 Weighted Loss Selection

Initially, each proposed model structure hyperparameters (e.g. layer sizes) were selected based on AUROC validation performance with a weighted loss of 0.9, $p = 0.9$. After selecting structure hyperparameters, p was iterated over to find the best loss weight according to the validation AUROC. Figure 3 demonstrates this search, and Table 2 show the selected p for each model and target.

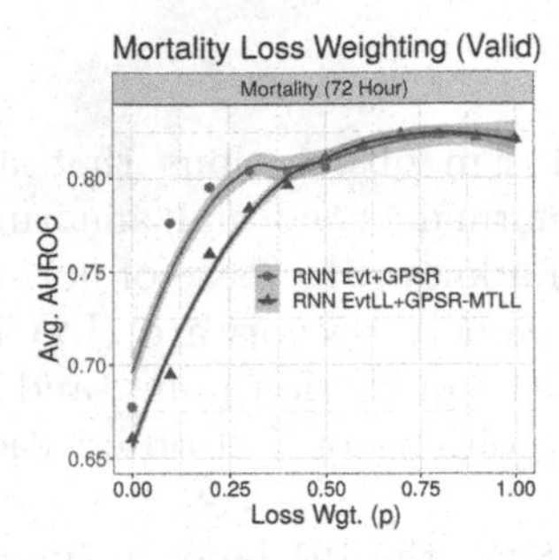

Fig. 3. Loss weighting search results for mortality validation set AUROC.

Table 2. Best loss weighting results for all targets based on the validation set AUROC.

Model Name	Loss Weighting Mortality	Norepi.	Sepsis
RNN Evt+GPSR	0.8	0.8	0.8
RNN EvtLL+GPSR	0.9	0.9	0.9
RNN EvtLL+GPSR-MTLL	0.8	0.9	0.9

5 Results and Discussion

Since we are interested in increasing the performance of low-prior event prediction, the area under the precision-recall curve (AUPRC) is the primary metric used for evaluation. AUROC is also important to ensure that the overall predictive performance is not being heavily sacrificed for better low-prior event prediction, and thus there are additional plots to demonstrate the AUROC performance. Each model is compared under three different conditions (i) average performance, (ii) average performance with a reduced prior likelihood of the positive class, and (iii) average performance over a reduced sample size.

5.1 Predictive Performance

Figure 4a, shows an improvement in AUPRC performance over the candidate models when compared to the baselines and the prior likelihood. This performance increase is particularly notable in the prediction of IV Norepinephrine. Figure 5a demonstrates that the proposed simultaneous learning models are maintaining a competitive if not stronger AUROC performance compared to the baselines. Based on the results for these three tasks, learning a GPSR simultaneously with a low-prior event provides a competitive to an improved prediction performance. Further, this suggests that low-prior clinical events benefit from the additional signal learning of generic clinical tasks.

5.2 Reduction of Prior Likelihood

By reducing the prior likelihood of each event, the models can be examined as the low-prior prediction target becomes increasingly more challenging to discern. The prior likelihoods of both the training and validation sets were reduced, while the test set remained at the same likelihood. This was performed for 7 iterations of randomly selected positive sequences for each model, and the selections were held constant across models to give a fair comparison. Additionally, the proposed

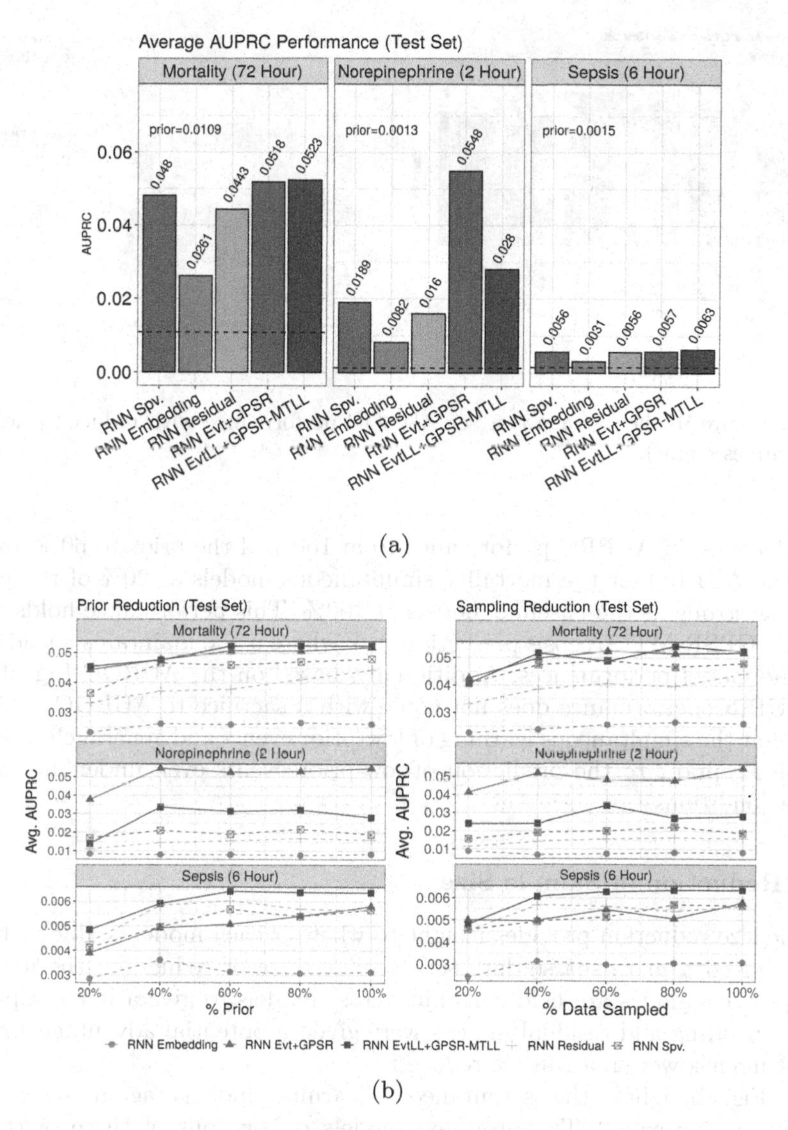

Fig. 4. Figure 4a the average test set AUPRC performance of each proposed and baseline model with a dashed line indicating the test set prior. **Figure** 4b (left)the average test set AUPRC when reducing the train/valid prior, and (right) the average test set AUPRC when reducing train/valid samples.

models' loss weight, p, was held constant, but hypothetically a more optimal p could have been rendered from this prior reduction.

In Fig. 4b(left), there is a strong AUPRC performance for the majority of prediction events and prior reductions over the two proposed simultaneous learning models. Particularly for norepinephrine and mortality prediction, there is

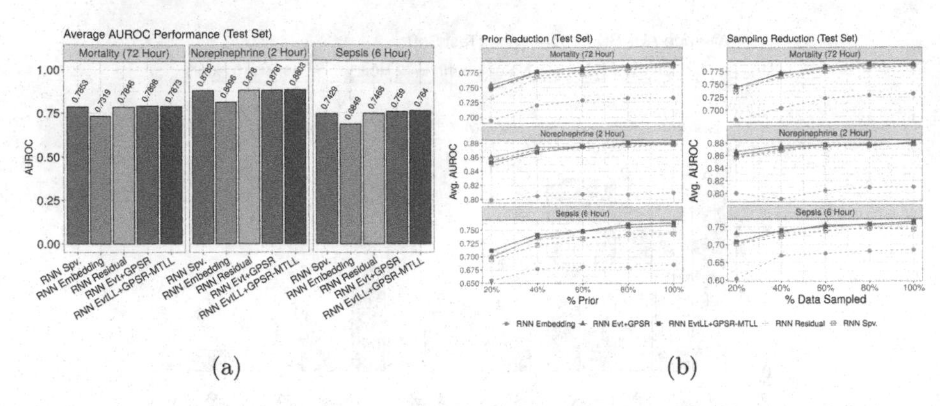

Fig. 5. Figure 5b the average test set AUROC performance with reduced prior (left) and samples (right).

little decrease in AUPRC performance from 100% of the prior to 60%. In addition, the AUPRC for the mortality simultaneous models at 20% of the prior is about as strong as the baseline models at 100%. This performance holds for the EvtLL+GPSR-MTLL sepsis prediction task where it too maintains a lead on the baseline model performances. Additionally, based on the AUROC Fig. 5b(left), the AUPRC performance does not come with a sacrifice to AUROC. This suggests that the simultaneous learning of low-prior events and general clinical tasks provides support to the prediction of low-prior events even under increasingly sparse conditions.

5.3 Reduction of Sample Size

Sample size reduction provides insight to whether each model is able to be predictive given a more sparse data set. Similar to prior reduction, the iterations of sequence samples are held constant across models to give a fair comparison. The embedding and residual models were given a potential advantage since the GPSR models were not sample reduced.

In Fig. 4b(right), the simultaneous learning models again show strong AUPRC performance. The proposed models on two out of three of the low-prior events have near consistent AUPRC performance up to 40% of the sample size. In addition, the AUROC performance shown in Fig. 5b(right) for the two proposed models maintains a competitive edge over the baselines. Therefore, simultaneous learning of a GPSR to improve low-prior event prediction maintains a competitive edge in a reduced sample data set.

6 Conclusion

Based on the results for these three clinical targets, weighted simultaneous learning of a low-prior event and GPSR improves the prediction of the low-prior task. This prediction improvement is sustained throughout the ablation studies, reduced prior and sample size. This suggests that the predictive signal from forecasting generic clinical tasks provides additional support to the low-prior event, and this predictive benefit can be further capitalized when the low-prior target is simultaneously optimized with the patient representation.

Acknowledgment. The work presented was supported by NIH grant R01GM088224. The content of this paper is solely the responsibility of the authors and does not necessarily represent the official views of NIH.

References

1. Blei, D.M., et al.: Supervised topic models. arXiv preprint arXiv:1003.0783 (2010)
2. Choi, E., et al.: Retain: An interpretable predictive model for healthcare using reverse time attention mechanism. arXiv preprint arXiv:1608.05745 (2016)
3. Gers, F.A., et al.: Learning to forget: Continual prediction with LSTM (1999)
4. Gupta, P., et al.: Using features from pre-trained TimeNet for clinical predictions. In: KHD@ IJCAI (2018)
5. Gupta, P., et al.: Transfer learning for clinical time series analysis using deep neural networks. J. Healthcare Inf. Res. **4**(2), 112–137 (2020)
6. Hauskrecht, M., et al.: Outlier detection for patient monitoring and alerting. JBI **46**(1), 47–55 (2013)
7. Hauskrecht, M., et al.: Outlier-based detection of unusual patient-management actions: an ICU study. JBI **64**, 211–221 (2016)
8. Johnson, A.E., et al.: Mimic-iii, a freely accessible critical care database. Sci. Data **3**, 1–9 (2016)
9. Kratz, A., et al.: Laboratory reference values. NEJM **351**, 1548–1564 (2004)
10. Laposata, M.: Laposata's Laboratory Medicine Diagnosis of Disease in Clinical Laboratory Third Edition. McGraw-Hill Education (2019)
11. Lee, J.M., et al.: Modeling multivariate clinical event time-series with recurrent temporal mechanisms. AIME, p. 102021 (2021)
12. Lei, L., et al.: An effective patient representation learning for time-series prediction tasks based on EHRs. In: 2018 IEEE BIBM, pp. 885–892. IEEE (2018)
13. Loshchilov, I., et al.: Decoupled weight decay regularization. arXiv preprint arXiv:1711.05101 (2017)
14. Lyu, X., et al.: Improving clinical predictions through unsupervised time series representation learning. arXiv preprint arXiv:1812.00490 (2018)
15. Malakouti, S., Hauskrecht, M.: Predicting patient's diagnoses and diagnostic categories from clinical-events in EHR data. In: Riaño, D., Wilk, S., ten Teije, A. (eds.) AIME 2019. LNCS (LNAI), vol. 11526, pp. 125–130. Springer, Cham (2019). https://doi.org/10.1007/978-3-030-21642-9_17
16. McDonald, C.J., et al.: LOINC, a universal standard for identifying laboratory observations: a 5-year update. Clin. Chem. **49**(4), 624–633 (2003)
17. Miotto, R., et al.: Deep patient: an unsupervised representation to predict the future of patients from the electronic health records. Sci. Rep. **6**(1), 1–10 (2016)
18. Rajkomar, et al.: Scalable and accurate deep learning with electronic health records. NPJ Digit. Med. **1**(1), 1–10 (2018)
19. Ren, J., et al.: Prediction focused topic models for electronic health records. arXiv preprint arXiv:1911.08551 (2019)

20. Reyna, M.A., et al.: Early prediction of sepsis from clinical data: the physionet/computing in cardiology challenge 2019. In: 2019 Computing in Cardiology (CinC), p. 1. IEEE (2019)
21. Smith, M.D., et al.: Norepinephrine. StatPearls [Internet] (2019)
22. Tomašev, N., et al.: A clinically applicable approach to continuous prediction of future acute kidney injury. Nature **572**(7767), 116–119 (2019)

Identifying Symptom Clusters Through Association Rule Mining

Mikayla Biggs[1](✉), Carla Floricel[3], Lisanne Van Dijk[2], Abdallah S. R. Mohamed[2], C. David Fuller[2], G. Elisabeta Marai[3], Xinhua Zhang[3], and Guadalupe Canahuate[1](✉)

[1] University of Iowa, Iowa City, IA, USA
{mikayla-biggs,guadalupe-canahuate}@uiowa.edu
[2] University of Texas MD Anderson Cancer Center, Houston, TX, USA
[3] University of Illinois in Chicago, Chicago, IL, USA

Abstract. Cancer patients experience many symptoms throughout their cancer treatment and sometimes suffer from lasting effects post-treatment. Patient-Reported Outcome (PRO) surveys provide a means for monitoring the patient's symptoms during and after treatment. Symptom cluster (SC) research seeks to understand these symptoms and their relationships to define new treatment and disease management methods to improve patient's quality of life. This paper introduces association rule mining (ARM) as a novel alternative for identifying symptom clusters. We compare the results to prior research and find that while some of the SCs are similar, ARM uncovers more nuanced relationships between symptoms such as anchor symptoms that serve as connections between interference and cancer-specific symptoms.

Keywords: Association rule mining · Symptom clusters · PRO

1 Introduction

Cancer patients experience a range of symptoms during and after treatment [1–3]. Research on these symptoms, their prevalence, relationships, and progression can improve disease prognosis and inform the appropriate treatment [4,5]. Symptom cluster (SC) research aims to identify co-occurring symptoms (e.g., pain, fatigue, dry mouth) and to understand the underlying mechanisms that drive these clusters [6]. This research is facilitated by increasingly available Patient-Reported Outcome (PRO) data, collected via questionnaires, that allows patients to rate the occurrence and severity of their symptoms.

The M.D. Anderson Symptom Inventory (MDASI) [7], and its head-and-neck (HN) cancer module [8], are short, validated questionnaires that patients record each visit. Three key groups comprise the 28 MDASI-HN survey questions: 13 core items for common symptoms to all cancers, nine items specific to HN, and six items regarding symptom interference with daily activity. Patients rate their symptoms using a 0–10 scale, from "not present" to "as bad as you can imagine"

A. Tucker et al. (Eds.): AIME 2021, LNAI 12721, pp. 491–496, 2021.
https://doi.org/10.1007/978-3-030-77211-6_58

(core and HN), respectively from "did not interfere" to "interfered completely" (interference). Preliminary SCs in the MDASI-HN data have been identified using factor and cluster analysis [9,10].

This paper introduces association rule mining (ARM) [11] as an alternative for identifying symptom clusters. To the best of our knowledge, this is the first ARM application in the SC domain. This work's main contribution is to offer an alternative methodology for defining new and interesting relationships for SC research using PRO data. We model each PRO response as a patient transaction and process PROs during and after treatment to identify acute and late symptom clusters, respectively. We furthermore model the severity of the symptoms. We present a graph-based visualization for the most significant association rules to identify symptom clusters for both acute and late stages. Finally, we evaluate this methodology on a real HN cancer patient dataset.

2 Modeling Symptom Clusters with ARM

The ARM approach can use any PRO; in this work, we focus on the MDASI-HN questionnaire. The M.D. Anderson Symptom Inventory (MDASI) is a multi-symptom patient-reported outcome measure to assess both the severity of cancer symptoms and symptom interference with daily life. Table 1 shows a sample of the symptoms described in the MDASI-HN survey and the short symptom labels used to refer to the MDASI-HN symptoms to improve readability.

ARM has two steps: the first one is to identify frequent item-sets (FIS) from the data, and the second is to generate the association rules from the FIS. The Apriori algorithm identifies the frequent items in the data set using a set of core metrics. Support is a measure of absolute frequency, i.e., the fraction of sets that contain items A and B. Confidence ($A \rightarrow B$) is a measure of correlative frequency. It tells us how often the items A and B occur together, given the number times A occurs. Lift indicates the strength of a rule over the random occurrence of A and B. The higher the lift, the more significant the association. A lift greater than 1.0 implies that the relationship between the antecedent and the consequent is more significant than expected if the two were independent. With a lift of 1.0, we can say that the relationships appear as expected and are not significantly associated. For example, with the rule $\{fatigue\} \rightarrow \{drowsy\}$ with 50% support, and 80% confidence we could say that these two symptoms

Table 1. The 28 MDASI-HN symptoms organized into 3 symptom categories

Category	Symptom labels
Common cancer	Pain, fatigue, nausea, sleep, distress, SOB, memory, appetite, drowsy, drymouth, sad, vomit
Head & Neck	Numb, mucus, swallow, choke, voice, skin, constipation, taste, mucositis, teeth
Interference	General_activity, mood, work, enjoy, relations, walking, enjoy

are experienced together by 50% of the patients, and "if a patient experiences fatigue, they are 80% likely to experience drowsiness'.'

Since symptom severity is non-binary data, we generate two categories for each symptom and use the labels low and severe to distinguish them. For one questionnaire, symptoms with a rating greater than 0 are considered occurring symptoms. A symptom is *low* if the patient rated its severity less than five and *severe* otherwise. The data models the transactions with one unique PRO for each patient, and the two items being "bought" together, indicating low or severe, are concurrent symptoms. We consider symptom clusters at two different time points. Acute symptoms refer to symptoms experienced during treatment (about six weeks from the start of treatment). For late symptoms, patients survey the PROs up to 18-months post-treatment. Symptoms with missing scores (NaN) were replaced with 0 s. Patients with no PRO recorded during the acute or late phases were not included in the time frame analysis.

3 Experimental Results

The dataset used for these experiments consists of MDASI-HN responses for a cohort of 823 patients. The patient surveys were broken into acute and late time points with two items per symptom (low and severe) used to capture the severity of the symptoms. A total of 643 patients had at least one acute PRO, and 745 patients had at least one late PRO. Figure 1 shows the symptom's overall support for low and severe symptoms during the acute and late time frames. As shown, in the acute stage, many patients experienced both low and severe symptoms during treatment. In contrast, symptoms experienced in the late stage have a lower severity than during the acute phase. We used minimum support of 20% for both the acute and late as it is the minimum cutoff between both stages for consistency in our analysis of each.

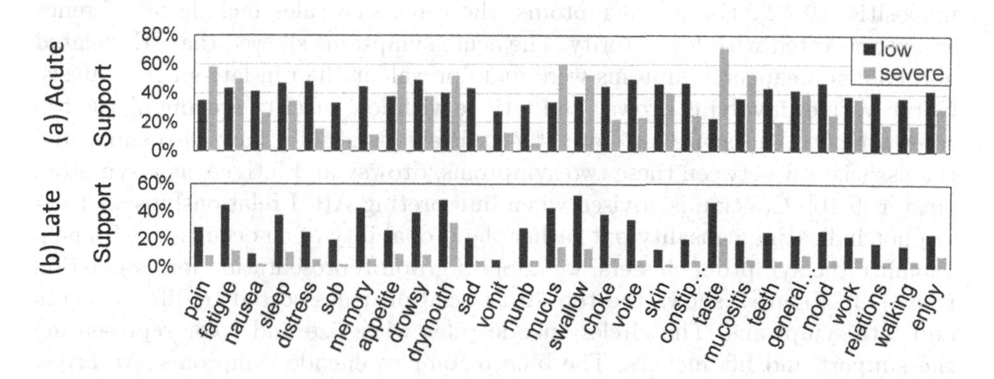

Fig. 1. Symptom Severity in the (a) acute and (b) late stages. Acute: > half of patients experience low severity symptoms, while a sizable 20% experience severe symptoms. Late: patients experience mostly low rated symptoms with highest prevalence in fatigue, drymouth, swallow, and taste.

Table 2. Top association rules for acute and late symptoms. Top five rules for each stage with the highest lift. The symptom's subscripts l and s stand for low and severe ratings, respectively.

Acute Stage				Late Stage			
antecendent	consequent	confidence	lift	antecendent	consequent	confidence	lift
$\{\text{pain}_s, \textit{taste}_s\}$	$\{\text{mucositis}_s\}$	.85	2.82	$\{\text{general\_activity}_l\}$	$\{\text{work}_l\}$	.79	2.96
$\{\text{mucus}_s, \textit{taste}_s\}$	$\{\text{swallow}_s\}$	.77	2.71	$\{\text{enjoy}_l\}$	$\{\text{mood}_l\}$	.75	2.84
$\{\text{swallow}_s, \textit{taste}_s\}$	$\{\text{mucus}_s\}$	.89	2.70	$\{\text{fatigue}_l, \textit{swallow}_l\}$	$\{\text{pain}_l\}$	.77	2.35
$\{\text{mucus}_s, \textit{taste}_s\}$	$\{\text{drymouth}_s\}$	.75	2.64	$\{\text{pain}_l, \textit{fatigue}_l\}$	$\{\text{swallow}_l\}$	.80	2.28
$\{\text{drowsy}_l\}$	$\{\text{fatigue}_l\}$	.76	2.19	$\{\text{drowsy}_l\}$	$\{\text{fatigue}_l\}$	.83	2.19

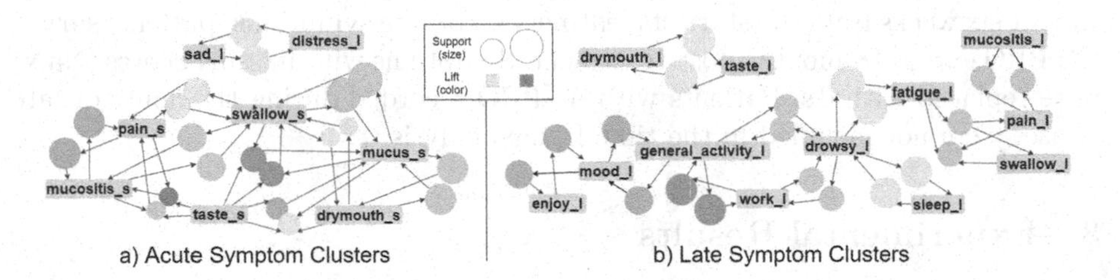

Fig. 2. Symptoms Association Rule Graph. The graph encoding shows the top 20 association rules for (a) acute and (b) late symptoms. In the acute state there is a large cluster of severe symptoms. In the late stage, drowsy and fatigue appear to be anchor symptoms connecting a cluster of interference symptoms with a cluster of cancer symptoms.

Table 2 shows the top 5 association rules with the highest lift for the acute and late stages. The top rule for the acute stage involves pain, taste, and mucositis. While this association is clinically valid, since mucositis presents as small painful oral ulcers in patients, it notably could interfere with oral functions like taste. Other studies have shown pain to cluster more closely to fatigue than mucositis [10,12]. For late symptoms, the top three rules include interference symptoms rated with low severity. The acute symptoms showed that HN-related and common cancer symptoms were more prevalent than in late-stage analysis. Notably, rules involving drowsy and fatigue with low severity are among the top rules for both the acute and late stages. Previous studies have also supported the association between these two symptoms, drowsy and fatigue, as a symptom cluster [9,10]. Caution is advised when interpreting ARM relationships, as rules are not indicating causality but rather the probability of co-occurrence. To help visualize the symptom clusters, we adopt a graph representation for association rules [13]. Figure 2 shows the top 20 association rules sorted by lift for acute and late symptoms. The circles encode rules with size and color representing the support and lift metrics. The blue rectangles encode symptoms. An arrow pointing towards a circle means that the associated symptom is an antecedent for the association rule. If the arrow points towards a symptom, that symptom is the consequent for the association rule.

For acute symptoms, two clusters are consistent with previously reported clusters for HN cancer [10]. For late symptoms, there are four identifiable clusters. Interestingly, drowsy and fatigue seem to be anchor symptoms between interference and HN-related symptoms, a relationship that more traditional approaches for symptom cluster research cannot capture. Furthermore, we found that pain is associated with both mucositis and fatigue. These findings highlight that symptoms could appear in different clusters with the ARM algorithm, providing a more accurate model for the complex relationships between symptoms. In contrast, highly occurring symptoms would cluster together earlier when symptoms are *partitioned* into clusters, as in hierarchical clustering.

4 Conclusion

We introduce association rule mining as a powerful approach to identify patient symptom clusters and uncover interesting relationships between symptoms. Our approach models PRO data as transactions, visualizes the most significant association rules in symptom clusters, and captures the severity of symptoms in both acute and late stages. When applied to PRO data from head and neck cancer patients, this approach correctly identified higher symptom prevalence and severity during treatment and a gradual decrease after treatment. The new acute symptom clusters found include severely rated HN-related and common cancer symptoms. The late symptom clusters found include more interference symptoms and low severity symptoms. Our analysis identifies new anchor symptom clusters that connect interference and HN-related symptoms, offering new opportunities for targeted interventions that could positively affect cancer patients' quality of life while supporting previously identified SCs. In the future, we plan to include clinical variables such as staging, dose, and organs at risk [14,15] into the ARM analysis to determine whether patient characteristics are related to individual symptoms or symptoms clusters.

References

1. Christopherson, K.M., et al.: Chronic radiation-associated dysphagia in oropharyngeal cancer survivors. Clinic. Transl. Rad. Oncology **18**, 16–22 (2019)
2. Wentzel, A., et al.: Precision toxicity correlates of tumor spatial proximity to organs at risk in cancer patients. Radiother. Oncol. **148**, 245–251 (2020)
3. Wentzel, A., et al.: Cohort-based T-SSIM visual computing for radiation therapy prediction and exploration. IEEE Trans. Vis. Comp. Graph. **26**(1), 949–959 (2019)
4. Marai, G.E., et al.: Precision risk analysis of cancer therapy with interactive nomograms and survival plots. IEEE Trans. Vis. Comp. Graph. **25**(4), 1732–1745 (2018)
5. Sheu, T., et al.: Conditional survival analysis of patients with locally advanced laryngeal cancer. Sci. Rep. **7**, 43928 (2017)
6. Miaskowski, C., et al.: Advancing symptom science through symptom cluster research. J. Nat. Cancer Instit. **109**(4) (2017)
7. Cleeland, C., et al.: Assessing symptom distress in cancer patients: the M.D. Anderson Symptom Inventory. Cancer **89**, 1634–46 (2000)

8. Rosenthal, D.I., et al.: Measuring head and neck cancer symptom burden. Head Neck J. Sci. Specialt. **29**(10), 923–931 (2007)
9. Skerman, H.M., et al.: Multivariate methods to identify cancer-related symptom clusters. Res. Nurs. Health **32**(3), 345–360 (2009)
10. Rosenthal, D.I., et al.: Patterns of symptom burden during radiotherapy or concurrent chemoradiotherapy for H&N cancer. Cancer **120**(13), 1975–1984 (2014)
11. Agrawal, R., Srikant, R., et al.: Fast algorithms for mining association rules. In: Proceedings 20th International Conference on Very Large Data Bases, VLDB, vol. 1215, pp. 487–499 (1994)
12. Kirkova, J., Aktas, A., Walsh, D., Davis, M.P.: Cancer symptom clusters: clinical and research methodology. J. Palliat. Med. **14**(10), 1149–1166 (2011)
13. Hahsler, M.: arulesviz: interactive visualization of association rules with r. R J. **9**(2), 163 (2017)
14. Tosado, J., et al.: Clustering of largely right-censored oropharyngeal HNC patients to improve outcome prediction. Sci. Rep. **10**(1), 1–14 (2020)
15. Luciani, T., et al.: A spatial neighborhood methodology for computing & analyzing lymph node carcinoma similarity in precision medicine. J. Biomed. Info. **5** (2020)

A Probabilistic Approach to Extract Qualitative Knowledge for Early Prediction of Gestational Diabetes

Athresh Karanam[1(✉)], Alexander L. Hayes[2], Harsha Kokel[1], David M. Haas[2], Predrag Radivojac[3], and Sriraam Natarajan[1]

[1] The University of Texas at Dallas, Richardson, USA
bxk180004@utdallas.edu
[2] Indiana University Bloomington, Bloomington, USA
[3] Northeastern University, Boston, USA

Abstract. Qualitative influence statements are often provided a priori to guide learning; we answer a challenging reverse task and automatically extract them from a learned probabilistic model. We apply our Qualitative Knowledge Extraction method toward early prediction of gestational diabetes on clinical study data. Our empirical results demonstrate that the extracted rules are both interpretable and valid.

1 Introduction

The nuMoM2b (Nulliparous Pregnancy Outcomes Study: Monitoring Mothers-to-Be) study [3] aims to identify early warning signs of adverse pregnancy outcomes, design interventions, and assist with decision-making. Since 2010, eight research sites in the United States followed up with women throughout their pregnancies—collecting routine clinical information, exercise data, and food they ate. Using this data, we consider learning to explain the relationship between gestational diabetes mellitus (GDM) and some common risk factors.

A common way to employ knowledge in machine learning and AI is via the use of qualitative relationships that express how changes in a (subset of) feature(s)/risk factor(s) affect the target. These rules were mainly used as "inductive bias" apriori to learning since they are both intuitive and natural in many domains. We address the challenging "reverse task". Can we extract these rules from data? To this effect, in the context of nuMoM2b, we propose a two step

A. Karanam and A. L. Hayes—Equal contribution.

Electronic supplementary material The online version of this chapter (https://doi.org/10.1007/978-3-030-77211-6_59) contains supplementary material, which is available to authorized users.

A. Tucker et al. (Eds.): AIME 2021, LNAI 12721, pp. 497–502, 2021.
https://doi.org/10.1007/978-3-030-77211-6_59

process. First we learn a joint probability distribution over all the variables including the target (GDM). In the second step, the constraints are extracted by reasoning over this joint probability distribution. We demonstrate in our experiments that such an approach yields rules that are both intuitive and valid (as validated by our clinical expert Dr. David Haas). We first explain these constraints before outlining our approach and presenting our learned rules.

2 Extracting Qualitative Influences

A qualitative influence (QI) statement outlines how a change in one or more factor(s) would influence another factor [8]. We focus on two types of QI: *Monotonicity and Synergy* [1,5,9]. *Monotonicity* represents a direct relationship between two variables: "As BMI increases, neck circumference increases" indicates that the probability of greater neck circumference increases with increase in BMI. Specifically, a *monotonic influence* (MI) of variable X on variable Y, denoted by $X^{M+}_{\prec}Y$ (or its inverse $X^{M-}_{\prec}Y$), indicates that higher values of X stochastically result in higher (or lower) values of Y. *Synergy* represents interactions among influences. Two variables synergistically influence a third if their joint influence is greater than their separate, statistically independent influences. Synergy can capture influences like "Increase in BMI increases the risk of high blood pressure in patients with family history of hypertension more than patients without family history." Formally, a *synergistic influence* (SI) of two variables A and B on variable Y, denoted by $A, B^{S+}_{\prec}Y$, indicates that increasing the value of A has greater effect on Y for higher value of B than the lower value of B. Both A and B should necessarily have same monotonic relationship with Y.[1] Similarly, a *sub-synergistic influence* (sub-SI), denoted by $A, B^{S-}_{\prec}Y$, indicates that while A and B have increasing monotonic influence on Y, the joint influence is lesser than their separate, statistically independent influence.

2.1 Proposed Approach

Given: A data set $\mathcal{D}$ consisting of examples in the form of risk factors $\mathbf{X}$ and binary target Y (in this case: GDM).

To Do: Learn a set of QIs that explain the effect of $\mathbf{X}$ on Y.

We use X_a to denote the ath variable in the feature set $\mathbf{X}$. x^i_a denotes a particular value of variable X_a and $|X_a|$ denotes the number of discrete values X_a takes. We assume that the joint distribution (P) over the set of random variables $\mathbf{X}$ is known (we learn this joint distribution in our empirical evaluation

[1] Without loss of generality, assume the variables in synergistic relation have monotonically increasing impact.

using a causal learning algorithm). For brevity, we restrict the description of our method to extracting positive MIs and SIs, $\prec^{M+}$ and $\prec^{S+}$. The *degree of monotonic influence*, δ_a, of $X_a \in \boldsymbol{X}$ on Y is defined as

$$\delta_a = I_{(C_a>0)} \cdot \sum_j \sum_{j'>j} \sum_k \frac{P(Y \leq k|X_a = x_a^j) - P(Y \leq k|X_a = x_a^{j'})}{|X_a|} \tag{1}$$

where,

$$C_a = \prod_j \prod_{j'>j} \prod_k \max(P(Y \leq k|X_a = x_a^j) - P(Y \leq k|X_a = x_a^{j'}) + \epsilon_m, 0) \tag{2}$$

For monotonicity to hold, we require $P(Y \leq k|X_a = x_a^j) + \epsilon_m \geq P(Y \leq k|X_a = x_a^{j'})$ for all pairs of configurations of $X_a, (j, j')$ with $j' > j$ at any given threshold value k. Here the monotonic slack ϵ_m allows violating a constraint within a chosen margin. The degree of MI, δ_a, in Eq. 1 measures the cumulative difference in the probability that the target variable Y is less than a threshold k given X_a at two different values x_a^j and $x_a^{j'}$.

We extend the concept of degree of MI to SI by conditioning on a pair of variables instead of a single variable. First, consider the difference in the effect of changing X_a from x_a^i to $x_a^{i'}$ on Y under the context of two different values of X_b (x_b^j and $x_b^{j'}$). We define this as

$$\begin{aligned}\phi_{a,b}^{i,i',j,j'} = \sum_k & P(Y \leq k|X_a = x_a^i, X_b = x_b^j) - P(Y \leq k|X_a = x_a^{i'}, X_b = x_b^j) - \\ & P(Y \leq k|X_a = x_a^i, X_b = x_b^{j'}) + P(Y \leq k|X_a = x_a^{i'}, X_b = x_b^{j'})\end{aligned}$$

For synergy to hold, we require $\phi_{a,b}^{i,i',j,j'} + \epsilon_s$ to be non-negative for all $i' > i$ and $j' > j$. Where ϵ_s is the synergistic slack. We define the *degree of synergistic influence*, $\delta_{a,b}$, of variables $X_a \in \boldsymbol{X}$ and $X_b \in \boldsymbol{X}$ on $Y \in \boldsymbol{X}$ as the cumulative difference in degrees of context specific influence of X_a on Y in the context of X_b. It is given by

$$\delta_{a,b} = I_{(C_{a,b}>0)} \cdot \sum_i \sum_{i'>i} \sum_j \sum_{j'>j} \frac{\phi_{a,b}^{i,i',j,j'}}{|X_a| \cdot |X_b|} \tag{3}$$

where,

$$C_{a,b} = \prod_i \prod_{i'>i} \prod_j \prod_{j'>j} \max(\phi_{a,b}^{i,i',j,j'} + \epsilon_s, 0) \tag{4}$$

We employ both definitions to learn QIs in Algorithm 1, Qualitative Knowledge Extraction (QuaKE). The algorithm assumes the existence of a joint distribution [6] over ordinal features, which we learn using a causal probabilistic learning algorithm (PC) [2,7]. We chose PC algorithm to verify our hypothesis that the use of a causal model will yield causally interpretable qualitative relationships. We calculate the degree of MI of every variable $X_a \in \boldsymbol{X}$ on Y and SI of every pair of variables $X_a, X_b \in \boldsymbol{X}$ on Y. The MI rules $X_{a \prec}^{M+} Y$ are extracted if their corresponding degree of MI δ_a are above a pre-defined threshold T_m. Similarly, the synergistic rules $X_a, X_b{}_{\prec}^{S+} Y$ are extracted if their corresponding degree of SI $\delta_{a,b}$ are above a pre-defined threshold T_s.

Algorithm 1: QuaKE

```
input  : P, Y, X, ε_m, ε_s, T_m, T_s
output: Rules R
initialize: R ← ∅
for a ← 0 to (|X| − 1) do
    compute δ_a using Eq. 1
    if δ_a ≥ T_m then
        R ← (X_a ≺^{M+} Y) ∪ R
    for b ← a + 1 to (|X| − 1) do
        compute δ_{a,b} using Eq. 3
        if δ_{a,b} ≥ T_s then
            R ← (X_a, X_b ≺^{S+} Y) ∪ R
// Decreasing cases
return R
```

3 Learning Qualitative Influences for GDM Modeling

The nuMoM2b study tracked pregnancies of 10,037 women near 8 sites in the United States. We excluded 817 cases where women were already diagnosed with diabetes; and we evaluate our proposed method for extracting QIs using 8 features[2] of the remaining 9,220 women. 7 features had inherent ordering of categories whereas *Race* had no obvious ordering. We use an ordering based on previous studies [4] on the effect of *Race* on *GDM*.

We pose and answer the following questions: **(Q1)** Does QuaKE extract high-quality rules that align with background knowledge in this domain? **(Q2)** Does QuaKE help uncover QI statements in cases where prior knowledge is uncertain?

We compare learned rules with those from our clinical expert, *Dr. Haas.* W.r.t GDM, these could either be increasing, decreasing, no effect, or unknown. Since Algorithm 1 assumes a complete joint distribution P is available, we consider two factorizations of P. The first learns a causal model [2] and the other (baseline) estimates the probabilities directly from data. Alternative baselines might have included rules extracted from decision trees, rule mining, or Bayesian rule learning—but each induce conjunctive rules of the form $(x_1 \wedge x_2 \wedge ... \wedge x_n) \implies y$, making their exact connection to the QI statements tenuous.

All rules are presented in Table 1. The "Prior" knowledge refers to the rules provided by our expert. We compare these to the rules extracted by QuaKE and baseline (Data Alone). QuaKE's precision compared to expert advice is 0.923 ± 0; whereas the precision of our unstructured baseline is 0.636 ± 0. Precision of each

[2] Refer to the supplementary material for details on the data and features: https://starling.utdallas.edu/papers/QuaKE/.

Table 1. Comparision of QI from prior knowledge (PK), QuaKE and Data Alone. ✓/✗ represents that this relationship does/not exist respectively while ? represents unknown influence. The three groups of rows show: (1) MI, (2) SI, and (3) sub-SI. Colors highlight rules recovered by QuaKE and show (a.) coherent with the PK and baseline (b.) contradicting the baseline (c.) coherent with baseline but contradicts the PK.

Rule	Prior Knowledge	QuaKE	Data Alone
$BMI \preceq^{M+} GDM$	✓	✓	✓
$Age \preceq^{M+} GDM$	✓	✓	✓
$Race \preceq^{M+} GDM$	✓	✓	✗
$Education \preceq^{M+} GDM$	✓	✓	✗
$Gravidity \preceq^{M+} GDM$	✓	✓	✗
$Smoked3months \preceq^{M+} GDM$	✓	✗	✗
$SmokedEver \preceq^{M+} GDM$	✓	✗	✗
$Age, BMI \preceq^{S+} GDM$	✓	✓	✓
$Age, Smoked3months \preceq^{S+} GDM$	✓	✓	✓
$BMI, SmokedEver \preceq^{S+} GDM$	✓	✓	✓
$Education, Smoked3months \preceq^{S+} GDM$	?	✓	✓
$BMI, Gravidity \preceq^{S+} GDM$	✓	✓	✗
$BMI, Smoked3months \preceq^{S+} GDM$	✓	✗	✓
$Age, SmokedEver \preceq^{S+} GDM$	✓	✗	✗
$BMI, Education \preceq^{S+} GDM$	✗	✓	✓
$Education, SmokedEver \preceq^{S+} GDM$	?	✗	✗
$Age, Education \preceq^{S-} GDM$	✓	✓	✓
$BMI, Smoked3months \preceq^{S-} GDM$	✗	✓	✗
$Age, SmokedEver \preceq^{S-} GDM$	✗	✗	✓
$BMI, Gravidity \preceq^{S-} GDM$	✗	✗	✓
$Gravidity, SmokedEver \preceq^{S-} GDM$	✗	✗	✓
$Education, SmokedEver \preceq^{S-} GDM$	?	✗	✓
$Age, Gravidity \preceq^{S-} GDM$	✓	✗	✗

method was consistent across five stratified cross validation folds. This affirms **Q1**: QuaKE can extract high-quality rules aligning with prior knowledge.

Since we have formalized degree of the QIs in Eqs. 1 and 3, we can analyze rules that were highly uncertain according to the prior knowledge. Two of the synergistic relations involving smoking and education had an unknown effect with relation to GDM. $Education, Smoked3months \preceq^{S+} GDM$ was a high-confidence rule extracted by QuaKE and the baseline. We speculate that this could be either due to the high correlation between *Education* and *Age*, or related to an unobserved relationship between education and socioeconomic status. Note that both these results are especially interesting since we found only a weak monotonic relationship between smoking and GDM more generally. We use this to answer **Q2**: our approach can identify potentially interesting cases where prior knowledge is uncertain.

Discussion and Conclusion: We considered the problem of learning interpretable and explainable qualitative rules for modeling GDM. To this effect, we learned a causal (probabilistic) model and recovered the knowledge by applying the rules. Our results indicate that most of our rules are in line with the prior knowledge of our expert and some interesting influence relationships appear that are worth investigating. Incorporating richer domain knowledge, automatically refining the rules, identifying broader relationships and scaling to larger feature sets are interesting future research directions.

Acknowledgements. We gratefully acknowledge the support of 1R01HD101246 from NICHD and Precision Health Initiative of Indiana University. Thanks to Rashika Ramola, Rafael Guerrero for data processing and discussions.

References

1. Altendorf, E.E., Restificar, A.C., Dietterich, T.G.: Learning from sparse data by exploiting monotonicity constraints. In: UAI, pp. 18–26 (2005)
2. Colombo, D., Maathuis, M.H.: Order-independent constraint-based causal structure learning. J. Mach. Learn. Res. **15**(1), 3741–3782 (2014)
3. Haas, D.M., Parker, C.B., et al.: A description of the methods of the nulliparous pregnancy outcomes study: monitoring mothers-to-be (nuMoM2b). Am. J. Obstet. Gynecol. **212**(4), 539.e1–539.e24 (2015)
4. Hedderson, M.M., Darbinian, J.A., Ferrara, A.: Disparities in the risk of gestational diabetes by race-ethnicity and country of birth. Paediatr. Perinat. Epidemiol. **24**(5), 441–448 (2010)
5. Kokel, H., Odom, P., Yang, S., Natarajan, S.: A unified framework for knowledge intensive gradient boosting: leveraging human experts for noisy sparse domains. In: AAAI, vol. 34, pp. 4460–4468 (2020)
6. Pearl, J.: Probabilistic Reasoning in Intelligent Systems: Networks of Plausible Inference. Morgan Kaufmann, Burlington (1988)
7. Spirtes, P., Glymour, C.: An algorithm for fast recovery of sparse causal graphs. Soc. Sci. Comput. Rev. **9**(1), 62–72 (1991)
8. Wellman, M.P.: Fundamental concepts of qualitative probabilistic networks. Artif. Intell. **44**(3), 257–303 (1990)
9. Yang, S., Natarajan, S.: Knowledge intensive learning: combining qualitative constraints with causal independence for parameter learning in probabilistic models. In: Blockeel, H., Kersting, K., Nijssen, S., Železný, F. (eds.) ECML PKDD 2013. LNCS (LNAI), vol. 8189, pp. 580–595. Springer, Heidelberg (2013). https://doi.org/10.1007/978-3-642-40991-2_37

Author Index

Printed in the United States
by Baker & Taylor Publisher Services